Performance of Cement-Based Materials in Aggressive Aqueous Environments

RILEM STATE-OF-THE-ART REPORTS

Volume 10

RILEM, The International Union of Laboratories and Experts in Construction Materials, Systems and Structures, founded in 1947, is a non-governmental scientific association whose goal is to contribute to progress in the construction sciences, techniques and industries, essentially by means of the communication it fosters between research and practice. RILEM's focus is on construction materials and their use in building and civil engineering structures, covering all phases of the building process from manufacture to use and recycling of materials. More information on RILEM and its previous publications can be found on www.RILEM.net.

The RILEM State-of-the-Art Reports (STAR) are produced by the Technical Committees. They represent one of the most important outputs that RILEM generates – high level scientific and engineering reports that provide cutting edge knowledge in a given field. The work of the TCs is one of RILEM's key functions.

Members of a TC are experts in their field and give their time freely to share their expertise. As a result, the broader scientific community benefits greatly from RILEM's activities.

RILEM's stated objective is to disseminate this information as widely as possible to the scientific community. RILEM therefore considers the STAR reports of its TCs as of highest importance, and encourages their publication whenever possible.

The information in this and similar reports is mostly pre-normative in the sense that it provides the underlying scientific fundamentals on which standards and codes of practice are based. Without such a solid scientific basis, construction practice will be less than efficient or economical.

It is RILEM's hope that this information will be of wide use to the scientific community.

rilem

For further volumes:
http://www.springer.com/series/8780

Mark Alexander • Alexandra Bertron
Nele De Belie
Editors

Performance of Cement-Based Materials in Aggressive Aqueous Environments

State-of-the-Art Report, RILEM
TC 211-PAE

Springer

Editors
Mark Alexander
Department of Civil Engineering
University of Cape Town
Rondebosch, South Africa

Alexandra Bertron
LMDC (Laboratoire Matériaux et Durabilité des Constructions)
UPS-INSA Génie Civil
Université de Toulouse, France

Nele De Belie
Magnel Laboratory for Concrete Research
Faculty of Engineering and Architecture
Ghent University
Ghent, Belgium

ISSN 2213-204X
ISSN 2213-2031 (electronic)
ISBN 978-94-007-5412-6
ISBN 978-94-007-5413-3 (eBook)
DOI 10.1007/978-94-007-5413-3
Springer Science+Business Media Dordrecht

Library of Congress Control Number: 2012952586

Printed on acid-free paper

Springer is part of Springer Science+Business Media (www.springer.com)

Preface

Concrete and cement-based materials are the most widely used building materials globally, finding application in such diverse situations as buildings of all kinds, civil engineering structures, water supply and drainage infrastructure, agro-industrial infrastructure, transport infrastructure, and so on. Concrete is economical and watertight, provides good mechanical strength, ensures good thermal inertia, and conforms to the health standards imposed by industry, making it a material of choice for many applications. However, many of the environments to which concrete is exposed are highly aggressive due to various chemical and/or biological components, and the demands made on concrete to perform in such environments are likely to grow substantially in years to come as the built infrastructure expands and environmental pollution increases. These environments are often aqueous in nature, comprising natural environments such as soft waters, sea waters or ground waters, or man-made such as industrial environments, and treated water or wastewater and polluted environments.

In such environments, concrete is subjected to processes of degradation involving ion addition and exchange reactions, leading to breakdown of the matrix microstructure and weakening of the material. Sometimes this can be extremely rapid and serious such as in acidic environments, while in other cases degradation of the concrete matrix occurs over long periods, for example in mild or soft waters. However, in almost all cases, the scale of the problem means that the consequences are severe. Concrete failure can adversely affect the well-being of human communities and disturb ecological balances. There are also large direct costs from the need to maintain and sometimes replace deteriorated infrastructure, and indirect costs linked with suspension of production or reduction in transport mobility during maintenance work. This all places a great burden on society.

With these problems in mind, RILEM Technical Committee TC 211-PAE[1] was formed in 2005. Its aim was to address performance of cement-based materials in aggressive aqueous environments, where the main form of degradation relates to

[1] Performance of cement-based materials in aggressive aqueous environments.

hydrate alteration and transformation of mineralogy by ion exchange between the material and the environment, involving mass exchanges. One of the TC's objectives was to consider the scientific nature of the problem. Another objective was to attempt to provide engineering approaches for addressing these problems in practice. The TC had the ambitious aim of seeking to furnish an understanding of concrete performance over a broad range of environments in terms of fundamental mechanisms and processes. It sought to provide a generic framework for these processes, as well as guidelines for material selection and precautionary measures for adequate performance in aggressive environments. Construction and design standards were recognized as important. The TC took a critical approach to existing standards, with a view to providing a framework within which improved standards and guidelines could be written. The work of the technical committee has addressed many of these aims.

Focus of TC

The focus of the TC's work was on three main groups of aggressive waters:

- Moderate to highly saline waters (other than marine waters), e.g. sulfate-bearing waters, magnesium-bearing waters
- Soft or acidic waters, e.g. pure natural waters, organic waters, drainage effluents such as sewers, acidic waters of various origins (inorganic and organic acids)
- Marine waters (but with chloride ingress not being a major focus)

There was a lesser focus on highly aggressive industrial effluents. The approach was to recognise that all aqueous environments are aggressive to cement-based materials to a certain degree; therefore aqueous solutions should be considered as part of a continuum from pure (ion-free) waters to highly saline solutions, with intervening solutions of various salinities and ionic constituents. Indeed, some less saline solutions may be more aggressive than more highly saline solutions. In addition, the mobility of waters in regard to the attacked structures is important. While the scope of aggressive environments considered was fairly broad, certain deterioration aspects were specifically excluded from the work, such as rebar corrosion (extensively covered in other RILEM Technical Committees), alkali-aggregate reaction, and physical and cavitational damage.

RILEM TC 211-PAE met as a Committee for the first time in September 2005. Since then the committee has held several meetings and organised two public events: an International Workshop in Ghent in September 2007,[2] and a Conference in Toulouse in June 2009.[3] Both of these events yielded valuable Proceedings which have contributed to this STAR report. Numerous informal communications have, of course, been held over the years between members of the TC.

[2] Proceedings of the RILEM TC 211-PAE Workshop, Ghent, September 2007. (RILEM PRO 057, 120 pp., e-ISBN: 978-2-35158-059-2). Ed. N De Belie.

[3] Proceedings of the RILEM TC 211-PAE Final Conference, Toulouse, June 2009. (2 volumes, 598 pp. ISBN 978-2-35158-071-4). Eds. M.G. Alexander & A. Bertron.

This State-of-the-Art Report (STAR) is the compilation of the work of multiple authors who worked in groups to produce the needed knowledge and information. Four Working Groups were organised, each tackling a different aspect:

I. Nature of degradation and deterioration mechanisms, including kinetics, of cement-based materials and different cementitious systems in aggressive aqueous environments, considering the compositions of real environments
II. Modelling of deterioration and Service Life Modelling in such environments
III. Test methods to assess performance of cement-based materials in such environments, which can be used to characterise and rate relative performance as well as to inform long term predictions
IV. Engineering implications and consequences of deterioration in aggressive aqueous environments, and engineering approaches to the problem

Details of TC members are given at the end of this Preface. Members of the Working Groups are indicated in the following table. Each group comprised Contributing and Associate members, allowing coordination and cross-referencing between the various working groups.

Working group	Contributing members	Associate members
I Nature of the problem: Characterisation and degradation aspects	Convenors: Alexandra Bertron, Fredrik P. Glasser Josée Duchesne Gilles Escadeillas Esperanza Menéndez Manu Santhanam Thomas Matschei	Karen L. Scrivener, Nele De Belie, R. Doug Hooton, Barbara Lothenbach
II Modeling	Convenor: Eric Samson Patrick Le Bescop Barbara Lothenbach Kenneth A. Snyder Emmanuel Guillon	Fredrik P. Glasser
III Test methods	Convenor: Nele De Belie Alexandra Bertron Marta Castellote Karen L. Scrivener R. Doug Hooton Mark G. Alexander Manu Santhanam Kim Van Tittelboom Q. Yuan	Patrick Le Bescop, Clyde Fourie
IV Engineering	Convenor: Anjan Chatterjee Alaster Goyns	Mark G. Alexander, Clyde Fourie

Each Working Group contributed a part to the STAR report, which thus has four parts. Parts I and III comprise multiple chapters, while Parts II and IV have one chapter each, giving a total of 16 chapters. The main contributing authors are acknowledged at the start of each chapter. Chapter drafts underwent a review process by other assigned members of the TC. These reviewers are also acknowledged in each chapter. The part and chapter titles are as follows:

Part I. Mechanisms of degradation of cementitious materials in aggressive aqueous environments

1. General introduction to Part I
2. Sulfate attack of concrete
3. Magnesium attack of cementitious materials in marine environments
4. Leaching of cementitious materials by pure water and strong acids (HCl and HNO_3)
5. Ammonium nitrate attack on cementitious materials
6. Attack of cementitious materials by organic acids in agricultural and agrofood effluents

Part II. Modeling degradation of cementitious materials in aggressive aqueous environments

7. Modeling degradation of cementitious materials in aggressive aqueous environments

Part III. Methods for testing concrete degradation in aggressive aqueous environments

8. General considerations
9. Tests for leaching and degradation in soft or carbonated waters
10. Test methods for resistance of concrete to sulfate attack – a critical review
11. Testing for degradation by inorganic acids
12. Bacteriogenic sulfuric acid attack of cementitious materials in sewage systems
13. Test methods for chloride transport in concrete
14. Test methods for magnesium attack
15. Methods for testing cementitious materials exposed to organic acids

Part IV. Cementitious materials performance in aggressive aqueous environments – engineering perspectives

16. Cementitious materials performance in aggressive aqueous environments – engineering perspectives

I wish to acknowledge with sincere gratitude the Convenors of the Working Groups for facilitating the work of their groups. I also wish to acknowledge the contributions of the members of the working groups. All these individuals have contributed large amounts of time and effort in drafting what is hoped will be a valuable addition to the literature in this area.

I wish specifically to acknowledge the tremendous assistance and efforts of my co-editors: Dr Alexandra Bertron and Prof Nele De Belie. They have gone well beyond the call of duty in ensuring an excellent outcome to the STAR, besides acting as convenors of their own working groups. I am much obliged to them!

Prof Mark G. Alexander (University of Cape Town, South Africa)
Chair of RILEM Technical Committee TC-211 PAE

Mark G. Alexander
Department of Civil Engineering,
University of Cape Town, Cape Town, South Africa

Alexandra Bertron
Université de Toulouse; UPS-INSA; LMDC (Laboratoire
Matériaux et Durabilité des Constructions);
135, avenue de Rangueil; F-31 077 Toulouse Cedex 04, France

Nele De Belie
Magnel Laboratory for Concrete Research,
Department of Structural Engineering, Ghent University,
Technologiepark – Zwijnaarde 904, B-9052 Ghent, Belgium

RILEM TC 211-PAE Members

Prof Mark Alexander (*Chair*). Department of Civil Engineering, University of Cape Town, PO Rondebosch, 7701, South Africa. Email: mark.alexander@.uct.ac.za

Dr Eric Samson (*Secretary*). SIMCO Technologies Inc., 1400 Boul. du Parc Technologique, Suite 203, Québec (Qc), Canada, G1P 4R7. Email: esamson@simcotechnologies.com

Dr Alexandra Bertron. Université de Toulouse, UPS-INSA, LMDC (Laboratoire Matériaux et Durabilité des Constructions), 135, Avenue de Rangueil, 31077 Toulouse Cedex 4 – France. Email: alexandra.bertron@insa-toulouse.fr

Dr Marta Castellote. Instituto Eduardo Torroja, C/Serrano Galvache, 4. 28033 Madrid, Spain. Email: martaca@ietcc.csic.es

Dr Anjan Chatterjee. Conmat Technologies Private Limited, CF 318, Sector 1, Salt Lake, Kolkata 700064, India. Email: anjan.k.chatterjee@gmail.com; conmat_technologies@india.com

Prof Nele De Belie. Magnel Laboratory for Concrete Research, Ghent University, Technologiepark Zwijnaarde 904, B-9052 Ghent, Belgium. Email: nele.debelie@ugent.be

Prof Josée Duchesne. Department of Geology and Geological Engineering, CRIB, Université Laval, Québec, Canada G1K 7P4. Email : josee.duchesne@ggl.ulaval.ca

Prof Gilles Escadeillas, Université de Toulouse, UPS-INSA, LMDC (Laboratoire Matériaux et Durabilité des Constructions), 135, Avenue de Rangueil, 31077 Toulouse Cedex 4 – France. Email: gilles.escadeillas@insa-toulouse.fr

Prof Fred Glasser. University of Aberdeen, King's College, Engineering & Physical Sciences School, Chemistry, Aberdeen AB24 3UE, UK . Email: f.p.glasser@abdn.ac.uk

Mr Alaster Goyns. PIPES Consultancy, PO Box 12519, Clubview, 0014, South Africa. Email: mcegg@pix.co.za

Dr Emmanuel Guillon. Lafarge Centre de Recherches, 95 rue Montmurier, BP 15, F-38291 Saint-Quentin Fallavier, France. Email: emmanuel.guillon@pole-technologique.lafarge.com

Prof R. Doug Hooton. Department of Civil Engineering, University of Toronto, 35 St George St, Toronto, Ont., Canada M5S 1A4. Email: hooton@civ.utoronto.ca

Dr Patrick Le Bescop. CEA, DEN, DPC, SECR, Laboratoire d'Etude du Comportement des Bétons et des Argiles, F-91191 Gif-sur-Yvette, France. Email: patrick.le-bescop@cea.fr

Dr Barbara Lothenbach. EMPA, Ueberlandstrasse 129, CH-8600 Duebendorf, Switzerland. Email: barbara.lothenbach@empa.ch

Dr Esperanza Menéndez. Instituto Eduardo Torroja, C/Serrano Galvache, 4. 28033 Madrid, Spain. Email: emm@ietcc.csic.es

Dr Manu Santhanam. Department of Civil Engineering, IIT Madras, India. Email: manus@iitm.ac.in

Prof Karen Scrivener. Laboratory of Construction Materials, EPFL, CH-1015 Lausanne, Switzerland. Email: Karen.scrivener@epfl.ch

Dr Kenneth A. Snyder. NIST, 100 Bureau Drive, Bldg 226, Gaithersburg, MD 20899–8615, USA. Email: kenneth.snyder@nist.gov

Student Member

Dr Kim Van Tittelboom. Magnel Laboratory for Concrete Research, Ghent University, Technologiepark Zwijnaarde 904, B-9052 Ghent, Belgium. Email: kim.vantittelboom@ugent.be

Corresponding member

Mr Clyde Fourie. 1 Yarrow Rd., Milnerton, 7441 Cape Town, South Africa. Email: clydefourie@yahoo.com

Contributors

Dr Thomas Matschei. Research and Development (Holcim Group Support Ltd.), Im Schachen CH-5113, Holderbank, Switzerland
Email: thomas.matschei@holcim.com
(T. Matschei was an author in Part I but not formally a member of the TC.)

Dr Q Yuan. School of Civil Engineering, Central South University, 22 Shaoshan Nan Lu, 410075, Changsha, China. Email: dhxyyuan@yahoo.com.cn

Acknowledgements

The authors gratefully acknowledge the provision of material in various figures and tables in the Report. Sources are given below:

Academic Press
Chapter 15: Figures 15.6, 15.8

Agricultural Engineering Research Institute
Chapter 16: Table 16.5

American Ceramic Society
Chapter 7: Figure 7.1

American Concrete Institute
Chapter 10: Figure 10.3
Chapter 16: Table 16.12

American Society of Civil Engineers
Chapter 7: Figure 7.4
Chapter 12: Figure 12.4

Associated Cement Companies
Chapter 16: Table 16.4

Bureau of India Standard
Chapter 16: Tables 16.6, 16.7, 16.8

Canadian Standards Association
Chapter 16: Table 16.13

Elsevier
Chapter 4: Figures 4.1, 4.2, 4.3, 4.4, 4.5, 4.6, 4.8, 4.9, 4.10, 4.11
Chapter 5: Figures 5.1, 5.2, 5.3, 5.4, 5.7, 5.8, 5.9, 5.10, 5.11, 5.12, 5.14
Chapter 6: Figures 6.3, 6.8, 6.9, 6.10, 6.11, 6.12.

Chapter 7: Tables 7.1, 7.2, 7.3, 7.4; Figures 7.1, 7.2, 7.3, 7.4.
Chapter 10: Figure 10.2
Chapter 15: Figures 15.4, 15.10, 15.12

Geological Society of America
Chapter 2: Table 2.1

Portland Cement Association
Chapter 10: Figure 10.1

RILEM
Chapter 4: Figure 4.7
Chapter 5: Figures 5.6, 5.13
Chapter 6: Figures 6.4, 6.5, 6.6, 6.14, 6.15
Chapter 7: Table 7.2
Chapter 8: Figure 8.2
Chapter 11: Figures 11.2, 11.7, 11.10
Chapter 15: Figures 15.5, 15.10
Chapter 16: Tables 16.19, 16.20

Spon Press, Taylor & Francis Group
Chapter 2: Figures 2.6, 2.22
Chapter 16: Figure 16.6

Springer
Chapter 7: Table 7.3
Chapter 16: Table 16.15

Standards Australia
Chapter 16: Table 16.14

The Concrete Centre
Chapter 16: Figures 16.2; Tables 16.9, 16.10, 16.11

Contents

Part I
Mechanisms of Degradation of Cementitious Materials in Aggressive Aqueous Environments

Authors of chapters in Part I

Chapter 1:
A. Bertron

Chapter 2:
E. Menéndez
T. Matschei
F.P. Glasser

Chapter 3:
M. Santhanam

Chapter 4:
J. Duchesne
A. Bertron

Chapter 5:
G. Escadeillas

Chapter 6:
A. Bertron
J. Duchesne

Reviewers for Part I

Mark G. Alexander, Alexandra Bertron, Gilles Escadeillas, Fredrik P. Glasser, R. Doug Hooton, Barbara Lothenbach

Chapter 1
General Introduction to Part I

Alexandra Bertron

Keywords Concrete structures • Aggressive environments • Aqueous media • Chemical species • Diffusion • Ordinary cements • Degradation mechanisms

Concrete structures are exposed to aggressive aqueous environments in very varied situations, and the range of aggressive species is wide. Among numerous examples, foundations and/or parts of structures in contact with groundwaters, soils or wastewaters may be subject to sulfate attack, and harbour and other maritime facilities are exposed to the action of seawater involving the combined action of magnesium, sulfates and chlorides. Dams, waterpipes and other structures exposed to the action of soft or carbonated waters suffer leaching and/or carbonation. Many agricultural and agrofood facilities (storage silos, animal houses etc.) suffer severe degradation notably linked with organic acids in waste water. The understanding of the mechanisms of degradation of cementitious matrices by these aggressive media is an essential step toward the development of concrete that performs well in these environments and toward the increase of the service life or the safety of the structures and facilities.

Concrete is a multiphase, porous, strongly basic material. The pH of the interstitial solution is around 13.[1] Thus, because of its mineralogical and chemical nature, concrete is naturally in dis-equilibrium with its environment. The characteristics of the pore network, dimensions and connectivity of the capillary porosity in particular, determine the transfer of aggressive species inside the matrix. Diffusion and permeation are the two essential mechanisms of these transfers, the former being dominant in the case of aqueous environments. Nevertheless, the chemical nature

[1] Except for low-pH concretes.

A. Bertron (✉)
LMDC (Laboratoire Matériaux et Durabilité des Constructions), Université de Toulouse; UPS, INSA, 135, avenue de Rangueil, F-31 077 Toulouse Cedex 04, France
e-mail: alexandra.bertron@insa-toulouse.fr

M. Alexander et al. (eds.), *Performance of Cement-Based Materials in Aggressive Aqueous Environments*, RILEM State-of-the-Art Reports 10,
DOI 10.1007/978-94-007-5413-3_1, © RILEM 2013

of the matrix is also a key parameter of the resistance of concrete to aggressive aqueous environments. The chemical composition of the hydrated phases (C-S-H, $Ca(OH)_2$, AFm and AFt for hydrated Portland cement) and their proportions inside the matrix – which depend mainly on the composition of the binder (cement and supplementary cementitious materials) – mostly determine the chemical stability of the matrix.

Water plays a key role in the interactions between the cementitious matrix and the aggressive aqueous media: it is a driving force and an essential participant in the chemical reactions between the external aggressive species and the cementitious phases. Chemical reactions are of two types: dissolution of the hydrated and anhydrous phases, and/or precipitation of newly formed products. These reactions often proceed from the outside of the concrete, progressing with time towards the interior, creating an altered zone in the process. The properties of the altered zone are generally poorer (physical and mechanical characteristics in particular) than those of the sound zone. Besides, the formation of new products when formed, may in some cases induce supplementary deleterious effects in the matrix. Crystallization pressure may occur, which in turn induces cracks in the matrix. These cracks modify not only the mechanical properties of the matrix, but also the transport properties, which induces higher kinetics of penetration of the aggressive species. Ultimately, the external aggressive species progress and reach the reinforcement, leading to the destruction of the structures. These are only general mechanisms of the interaction between chemical external species and the cementitious matrix, the phenomenology of the attack being highly dependent on the nature of the aggressive compound.

The aim of Part I is to consider in detail the chemical and mineralogical mechanisms of the attack of concrete by aggressive aqueous solutions. The environments under consideration are: (i) sulfates, (ii) magnesium attack in the context of marine environments, (iii) pure water and strong acids leading to leaching, (iv) ammonium nitrate, and (v) organic acids. The attack by dissolved CO_2 (soft water) and chloride do not have dedicated sections although the action of chlorides on the matrix is covered in the section on magnesium attack. For each section, where relevant, authors have attempted to give information about the composition of real media and concentration of aggressive species under consideration. Degradation mechanisms are then reviewed on the basis of laboratory studies mainly, but also of in-situ data and modeling. Mechanisms and consequences of the degradations in terms of microstructural changes, formation of new compounds, and physical and mechanical properties of the cementitious matrix are detailed. Common binders are mainly considered. Finally, aspects of kinetics are in some cases provided. It must be kept in mind that all the results presented are highly dependent on the experimental conditions or on the assumptions and hypothesis of the models implemented, these aspects being developed in Parts II and III of this Report. In all the chapters of this Part, authors have endeavoured to highlight the limits of knowledge and, if need be, the possible lack of consensus.

The first aggressive medium to be considered in this Part is that of sulfates (Chap. 2). External sulfate attack mainly occurs in concrete in contact with sulfate-bearing soils and groundwaters, seawater and wastewaters. The attack by sulfate is highly dependent on the associated cation, and so the mechanisms are reviewed as a function of the relevant cation. Attack by calcium, sodium, potassium, ammonium and magnesium sulfates are thus considered separately. In the case of sodium, the authors also discuss, on the basis of computer calculation, the influence of sodium sulfate concentration on the alteration mechanisms of Portland cement matrices, with important implications for the design of tests for sulfate attack.

The attack by sulfuric acid is also described as well as the issue of thaumasite formation. Although external sulfate attack is mainly developed, internal sulfate attack is also covered – delayed ettringite formation on the one hand, and the presence of sulfides (pyrite or pyrrhotite) in aggregates on the other. Finally, the impact of the crystallization of sulfate salts related to wet/dry cycles to which the material can be subjected is considered.

The attack by magnesium in the context of sea water is examined in Chap. 3. The presence of sodium chloride and of magnesium sulfate in marine waters leads to a combined attack of magnesium and sulfate ions influenced by the presence of chloride. The alteration mechanisms in such conditions are detailed as well as the influence of the concentration of aggressive species of the attacking solution on these mechanisms. The effect of the composition of the cementitious matrix on the resistance of the material to the attack is also examined.

Leaching by pure water and strong acids that produce soluble calcium salts (HNO_3 and HCl) is reviewed in Chap. 4. Concrete is exposed to acidic media and pure water in real situations: acids occur naturally in urban and industrial situations and structures such as water pipes, dams or wastewater pipes may be subjected to the action of low-mineralised, flowing or static waters. In laboratory conditions, leaching of concrete by pure water is used to investigate the long-term behaviour of nuclear waste repository structures. In Chap. 4, the mechanisms of leaching and the effects on the matrix are detailed notably as a function of the pH of the aggressive solution. Data on the kinetics of progression of the degradation front is provided and some factors influencing the kinetics are reviewed.

Chapter 5 is concerned with attack by ammonium nitrate. Although this species is encountered in-situ in industrial and agricultural contexts (in fertilizers for example), this type of damage on concrete is quite rare. Concrete is mainly exposed to ammonium nitrate in laboratory conditions in the context of accelerated degradation tests (to simulate pure water leaching), ammonium nitrate being among the most aggressive chemical agents for concrete and the consequences of the attack being similar to that by pure water. Mechanisms and kinetics of degradation are reviewed and the acceleration of the kinetics with respect to leaching by pure water is discussed. Architecture of most common laboratory tests involving ammonium nitrate is also described.

The attack by organic acids in agricultural and agrofood industries is considered in Chap. 6. Organic acids are either present naturally in the agricultural products themselves (i.e. plants or fruit) and/or are intermediary or final products of the metabolism of microorganisms growing in the effluents. The composition of the real media in terms of nature, pH and concentration of the acids is first identified. Chemical data on the wide range of organic acids is provided. Then, the mechanisms of degradation by organic acids are examined by distinguishing acids with soluble salts from acids with slightly soluble to insoluble salts. The aggressiveness of the acids is correlated with the various properties of the acids and their salts. The few studies in the literature considering mixtures of acids are finally reviewed.

Chapter 2
Sulfate Attack of Concrete

Esperanza Menéndez, Thomas Matschei, and Fredrik P. Glasser

Contents

E. Menéndez (✉)
Institute Eduardo Torroja of Construction Science (CSIC), C/Serrano Galvache, 4, Madrid, Spain
e-mail: emm@ietcc.csic.es

T. Matschei
Research and Development, (Holcim Group Support Ltd.), Im Schachen CH-5113, Holderbank, Switzerland
e-mail: thomas.matschei@holcim.com

F.P. Glasser
King's College, Engineering & Physical Sciences, University of Aberdeen, Meston Building, Old Aberdeen AB24 3UE, Scotland
e-mail: f.p.glasser@abdn.ac.uk

M. Alexander et al. (eds.), *Performance of Cement-Based Materials in Aggressive Aqueous Environments*, RILEM State-of-the-Art Reports 10,
DOI 10.1007/978-94-007-5413-3_2, © RILEM 2013

Keywords External/internal sulfate attack • Alkaline sulfate • Sulfuric acid • Sea water • Sulfides • Ettringite • Thaumasite • Delayed ettringite formation

2.1 Introduction

Cements have long been known to undergo deterioration in sulfate-rich service environments. Yet the mechanism of attack has not been universally agreed and, arguably of even greater importance, test procedures to measure sulfate attack and specifications for sulfate-resistant formulations are not agreed, or are disputed. We also lack reliable mitigation procedures.

The first relevant fact is that Portland cement is not stable in virtually all service environments, but reacts. Thus metastability and reaction have to be seen as normal, not exceptional. In some cases, the products of attack are benign: mechanical properties and dimensional stability are largely unaffected and we are therefore not much concerned about the process. An example is plain (unreinforced) concrete which, upon carbonation, continues to perform well. But other processes are more damaging. Thus sulfate attack is often accompanied by a volume expansion, as a consequence of which the concrete cracks and the process of degradation becomes self-accelerating as cracks propagate and spalling occurs.

Many investigators have sought to distinguish between the *chemical* and *physical* aspects of sulfate attack. This is not practicable and the difficulty of making clear-cut distinctions hinders our understanding of the processes involved: while we may not be in a position mathematically to couple physical and chemical properties and processes, this must remain an ultimate goal. Hence we try to maintain a unified approach.

It is also true that we are influenced by the approaches taken in the past. Investigators have broadly made three types of approach: (i) to obtain a broad understanding of sulfate attack through study of actual concretes removed from service, and to compare the performance in service conditions of different types and formulations of concrete, (ii) to devise tests, usually accelerated tests conducted in the laboratory, where conditions can be controlled, which can be used to predict the behaviour of concrete and (iii) modelling of durability by means of calculation. As will be shown, all three approaches have value. Indeed the RILEM TC 211-PAE format has provided the opportunity to set into context the contributions made from all three approaches.

In the quest for unity and simplicity, we cannot overlook the complex interactions of equilibrium and kinetics. Although internal sulfate attack is possible, this review concentrates on externally-induced attack; this requires diffusion of chemical into (and out of) concrete. Also, diffusion processes are in the realm of kinetics. Nevertheless, equilibrium describes the direction of reactions. In our opening statement, we noted the metastability of cement in its service environment. This metastability is not a feature unique to cement: many construction materials, *e.g*. steel, are similarly metastable in their service environments. Their *serviceability* depends, in large measure, on reaction kinetics. However cement differs from steel in two important respects: it is porous, and its corrosion does not generally involve electrochemical processes. Thus differences in approach are needed to develop and quantify cement reactions.

It is also necessary to take a holistic view of the chemistry involved: 'sulfate attack' is a misnomer inasmuch as electroneutrality must be maintained, locally as well as globally. As a consequence sulfate must always be associated with a cation or cations and the nature of the cation(s) has a strong effect on the nature of the reaction, as well as upon reaction kinetics. Thus a universal model cannot adequately embrace attack by sulfates as diverse as sulfuric acid, sodium sulfate and magnesium sulfate. The differences resulting from the chemical nature of the cation are very real and are not just pedantic devices introduced by academics intent on obfuscation.

This chapter does not pretend that we know all the answers. Indeed, it highlights areas of knowledge deficiencies. But it does attempt to provide guidance to those seeking to interpret field evidence and diagnose sulfate attack, to explore some of the limitations of conventional testing procedures, and briefly to illustrate progress in modelling the attack process and its consequences.

2.2 General Aspects of External Sulfate Attack

Sulfate attack in concrete originates mainly from its exterior, but it can also occur internally by a mechanism termed delayed ettringite formation (DEF) (see Sect. 2.5.1). External attack often arises from an increased sulfate concentration in the service environment. The sulfate may be normal or as a consequence of disturbance, as by aeration with oxidation of sulfides in soil, in aggregate, etc.

The damage to concrete structures resulting from external sulfate attack is related to the following factors:

- Transport of ions to the interior of the concrete through the pore system.
- Chemical reactions between sulfate and the solid hydration products of cement.
- Generation of stress in the interior of the concrete as a result of the formation of expansive products.
- Mechanical response of the bulk material resulting from these stresses, such as cracking.
- Characteristics of the concrete components and their proportions.

Many concretes are in contact with large quantities of sulfates but without significant problems, when made with suitable cements and with good qualities, such as compactness and low permeability.

To analyze the behaviour of concrete in contact with high sulfate contents, it is necessary to take into account different questions such as the absorption of sulfate from solution, the diffusion of sulfates into the pore structure, and the nature and consequences of chemical reaction between the sulfates and the hydration products.

As a consequence of the characteristics of sulfate attack, tests to predict the behaviour of concrete often focus on a particular mechanism under defined conditions, e.g. internal attack or external attack under either constant exposure or partial and cyclic exposure (Ferraris et al. 2006; ASTM C 1012-04 2004; Skalny et al. 2002; Clifton et al. 1999).

The main factor that affects the timeline of the concrete degradation resulting from sulfate attack is the penetration of the ions, either by absorption or by diffusion or some combination, depending partly on the saturation level of the concrete.

The main consequence of sulfate attack is the reaction of sulfate ions with cement hydration products to form gypsum and ettringite, or, in general, modification of the C-S-H structure. This latter reaction leads to the destruction of the hydration product that constitutes the backbone of the cement paste and forms the binder of concrete. The exact mechanisms of attack are influenced by the cations associated with sulfate, e.g., Mg, Ca, or Na. The role of the cation cannot be ignored.

The degree and the kinetics of the deterioration also depend on environmental factors (Skalny et al. 2002; Brown 1981) such as:

- The sulfate content in the soil or water in contact with concrete.
- Whether the concrete is subject to a wet-dry cycle or to constant immersion.
- The restrictions, if any, on the availability of sulfate ions which might affect the deterioration of the specimen, e.g. flowing water, soil, seawater which furnishes an infinite reserve of sulfate or stagnant water, etc., with a limited reservoir of sulfate.
- The pH of the solution in contact with the structure, which influences the mechanisms of deterioration.

The main factors in analyzing the problem are to know or estimate (Ferraris et al. 2006):

- The diffusion coefficient of the matrix.
- Its sorptivity coefficient for sulfate.
- Chemical reactions between cement paste and sulfate ions and the mass balances, allowing for dissolution.
- Impact of cations associated with sulfate.

The first two bullet points relate to properties of the material and depend on the mix design of the concrete and on the specific cement used, while the latter properties depend on the composition of the cement and the geochemistry of the service environment. Depending on the mechanism and kinetics, it can be conceptually possible that a "non-sulfate resistant" cement concrete with low sorptivity (or low diffusion coefficient) may have a longer service life than the same cement used in a high sorptivity (or high diffusion coefficient) concrete.

Ferraris et al. (2006) propose a number of factors to be taken into account in predicting the behaviour of the concrete in contact with sulfates: diffusion, absorption and wet-dry cycles. These will be discussed in turn.

2.2.1 *Diffusion*

In the case of external sulfate attack, the degradation mechanism depends partly on the rate of diffusive transport. Therefore, to characterize sulfate attack accurately, the diffusive transport of relevant ionic species must also be known.

The two approaches to modelling diffusive transport for the ith species are (i) empirical or (ii) physical. Empirical approaches (Tyrrell and Harris 1984) are typically based on Fick's law (Crank 1975): for flux j, effective diffusion coefficient D, and concentration c,

$$j_i = -D\frac{\partial c_i}{\partial x} = -D\ \nabla c_i \tag{2.1}$$

This approach has limited applicability in systems where the relative concentration of the available species can vary dramatically or reaction products can change the pore structure and hence the value of D.

In the physical approach, ionic transport and chemical reactions are treated separately: the transport equation is based on modelling transport within the pore fluid. On this scale, the interactions between the diffusing species are relatively well known. One drawback to this approach is that the physical properties of the pore solution are required, such as density and viscosity. However, the advantage of the physical approach is that since all the interactions are known, any number of diffusing species can be modelled using only two material parameters: capillary porosity and the formation factor.

The transport equation is based on the electro-diffusion equation (Helfferich 1962; Rubinstein 1990). This equation has been the basis of a number of studies of concrete materials (Samson and Marchand 1999; Samson et al. 1999; Tang 1999a, b; Truc et al. 2000a, b).

Ferraris et al. (2006) explain the application of the flux equation to bulk materials, by a combination of the flux equation for diffusive transport of the i-th ionic species in an electrolytic solution, with the chemical potential. And, taking into account the formation factor (ratio of the pore solution electrical conductivity to the bulk concrete electrical conductivity) (Helfferich 1962; Rubinstein 1990; Collins 1961).

$$j_i = -\frac{D_i^o}{Y}\left(1 + \frac{\partial ln\gamma_i}{\partial lnc_i}\right)\nabla c_i - z_i u_i c_i \nabla \varphi_D \tag{2.2}$$

Where, j_i is the flux, D_i^o is the self-diffusion coefficient (diffusion coefficient of species i, when the chemical potential gradient is equal to zero), Y is the formation factor, γ_i is the molar activity coefficient, c_i is the concentration, z_i is species valence, u_i is mobility and φ_D is the electrical potential.

Typically, the full equation is simplified by relating mobility to diffusivity through the Einstein relation where u_i is mobility, F is the Faraday constant, D_i^o is the self-diffusion coefficient, R is the universal gas constant and T is absolute temperature:

$$u_i = \frac{F \cdot D_i^o}{R \cdot T} \tag{2.3}$$

Porosity, as used in the continuum transport equation, relates to capillary porosity. In cement pastes the capillary porosity size ranges between one nanometer and 10 μm (Menéndez 2010). Capillary porosity is used here as the volume available for salt precipitation. Typically, the capillary porosity of a hydrated cement paste is in the range 0.05–0.30 volume fraction. The corresponding value for concrete is the paste porosity multiplied by the volumetric paste content. Also, the interfacial transition zone between paste and aggregate may contribute significantly to the concrete porosity.

The capillary porosity ϕ is rather arbitrarily calculated from the mass of water lost from a saturated specimen upon heating it to constant mass at 105°C.

On the other hand, the ***formation factor*** Y is the ratio of the pore solution electrical conductivity σ_p to the bulk concrete electrical conductivity σ_b (Collins 1961):

$$Y = \frac{\sigma_p}{\sigma_b} \tag{2.4}$$

Technically, the formation factor is defined as a differential quantity ($d\sigma_p/d\sigma_b$), but since changing the pore solution can create commensurate changes in the pore structure, the simpler definition of the previous equation is used here.

Fortunately, determining the electrical conductivity is relatively easy. The bulk conductivity of concrete can be measured using readily available equipment. In fact, it has been shown that a simple modification of the ASTM C 1202 rapid chloride test can be used to determine the bulk concrete conductivity accurately (Snyder et al. 2000).

The pore solution conductivity poses a slightly more difficult problem. In principle, one could express the pore solution (Barneyback and Diamond 1981). Under practical conditions, however, the quantity expressed is likely to be smaller than can be measured by available conductivity meters.

Alternatively, one can dilute the expressed solution and use a technique such as ion chromatography to determine species concentrations in the original specimen.

The conversion from concentration to conductivity must be carried out with care because, at the concentrations usually found in the pore solution of cement pastes (Andersson et al. 1989; Larbi et al. 1990; Christensen 1993; Christensen et al. 1994), the relationship between concentration and conductivity is not linear. A solution to this problem has been proposed (Snyder et al. 2003) and has been claimed to be accurate to within 10% at concentrations up to 2 mol/l.

2.2.2 Absorption

Sulfate will diffuse through the concrete structure and may react with cement pastes. To measure the rate of ingress of water or a solution in a concrete specimen, there are two approaches; in-situ testing and laboratory testing. In the in-situ approach, the concrete cannot be conditioned and it is impossible to know the water content of the concrete. In the laboratory approach, the concrete might not be exactly the "same" as in-situ concrete, i.e., it may differ in curing, in hydration degree, etc., if, for example, substantial time has elapsed between obtaining cores and testing them. On the other hand, the concrete can be conditioned in the laboratory and thereafter the water content is known.

2.2.3 Wet-Dry Cycles

Penetration of sulfate into a concrete, either immersed or partially saturated, may need to be measured. But in the case of cycles alternating between immersion and exposure to air (drying conditions), additional factors need to be included in the analysis. In the field, it is not unusual to have a structure that is not constantly immersed or is only partially immersed in a sulfate solution. For instance, on pier columns, the portion totally and constantly immersed in the water is observed not to deteriorate while the portion in the tidal zone may be completely destroyed (Hime et al. 2001). It is essential to determine the mechanisms of deterioration in this case, which clearly differs from total immersion, to predict the sulfate resistance of a concrete subjected to wet-dry cycles or to partial immersion.

Since concrete is a porous material, the ingress of water is rarely completely prevented. Temperature cycles together with the presence of water and dissolved salts are also a prerequisite for several deterioration mechanisms: freezing and thawing of water, chemical attack, crystallization pressure, and thermal stresses. In a concrete structure under external sulfate attack, with fluctuating temperature and wet-dry cycles or partially immersed, the potential deterioration mechanisms are varied (Scherer et al. 2001). Several often overlooked factors include (i) mismatch of coefficients of thermal expansion between the components of the concrete, such as the various hydration products or salt precipitated in the pores; and (ii) polymorphic changes or changes in hydration state of salts precipitated in pores.

It is generally accepted that in the case of wet-dry cycles with exposure to a sulfate solution, the pores fill slowly with salts by precipitation during the dry cycle. During wet cycles, the pores are refilled generally with lower concentration of salts. Each cycle starts with more fluid ingressing pores. But in successive cycles the remaining volume of pores decreases until the pore is just filled and no more aqueous phase can enter during the wet stage. The pore size distribution, the type of salt, and the environmental conditions (RH, temperature, and duration of the cycles) control the type and extent of the deterioration observed. Depending on the environmental conditions, efflorescence can be observed on the surface of the concrete, or subfluorescence could occur below the surface, or both. Efflorescence is not aesthetically pleasing and can often be easily eliminated by washing the surface, while subfluorescence is more likely to generate stresses that will deteriorate the structure. On the other hand, the efflorescence and subfluorecence on brickwork (for example) often arises by precipitation of calcite as a consequence of reaction with atmospheric carbon dioxide. Normally this process, while unsightly, is not expansive (Dow and Glasser 2003).

Crystallization pressure. It is generally agreed (Flatt 2002) that the main cause of spalling under exposure to sulfate solution and wet-dry conditions is due to crystallization pressure. There are several theories as to how crystallization pressure develops, for example those of Scherer (1999) and Flatt (2002).

To summarize, a crystal will grow in all directions until its surface attains a local weighted mean curvature (Flatt 2002) that is in equilibrium with the concentration of the solution. Because of the interaction of large repulsive forces between the crystal and the surface of the pore, the crystal will stop growing toward the pore surface. Direct contacts will not however be possible (Scherer 1999) because of the large forces needed to overcome the surface tension. A thin film of solution remains at the interface of the crystal and the pore wall (Correns 1949; Rodriguez-Navarro and Doehne 1999). The concentration of the solution in this thin film will not be the same as that in the rest of the pore because this concentration is not in equilibrium with the radius of the pore but with the tip of the crystal, which is free to grow into the area not in contact with the wall.

Depending on the cation associated with sulfate, the maximum crystallization pressure varies as shown in Table 2.1 (Winkler and Singer 1972).

It is noteworthy that, in some cases, highly hydrated solid salts cannot generate as much pressure as water-poor salts, thus not much water may be required to generate a potentially destructive pressure.

2.3 Sulfate Attack Processes

External sulfate attack is a common form of concrete deterioration. It occurs when concrete comes in contact with water containing sulfates (SO_4). Sulfates can be found in some soils, in seawater, in wastewater treatment plants, etc.

Table 2.1 Properties of sodium and magnesium sulfate salts

					Crystallization pressure [MPa]					
					C/Cs = 2		C/Cs = 10		C/Cs = 50	
Salt	Formula	Density [g/cm^3]	Molecular weight [g/mol]	Molar volume [cm^3/mol]	0°C	50°C	0°C	50°C	0°C	50°C
Mirabilite	$Na_2SO_4.\ 10H_2O$	1.46	322.19	220	7.2	8.3	23.4	27.7	39.7	47.3
Thenardite	Na_2SO_4	2.69	142.04	93	29.2	34.5	97.0	115.0	165.0	196.5
Epsomite	$MgSO_4.\ 7H_2O$	1.68	246.40	147	10.5	12.5	35.0	41.5	59.5	70.8
Hexahydrate	$MgSO_4.\ 6H_2O$	1.75	228.45	130	11.8	14.1	39.5	46.9	67.1	30.0
Kieserite	$MgSO_4.\ H_2O$	2.45	138.39	57	27.2	32.4	91.0	107.9	154.3	184.0

C Solute concentration in bulk solution (supersaturation) [mol/cm^3]
Cs Solute concentration at the crystal solution interface [mol/cm^3]
After Winkler and Singer (1972)

Waterborne sulfates react with the hydration products of tri-calcium aluminate (C_3A) and, in the longer term, with hydration products of tetra-calcium ferrite-aluminate as well as with calcium hydroxide ($Ca(OH)_2$) to form expansive crystalline products (gypsum, monosulfate (AFm) or ettringite (AFt)). Expansion resulting from ettringite formation generally causes tensile stresses to develop in the concrete. When these stresses become greater than the concrete's tensile capacity, the concrete begins to crack. These cracks allow easy ingress for more sulfates and the deterioration accelerates. Sulfates also cause the chemical disintegration of some of the cement hydration products.

The main factors that affect the rate and severity of sulfate attack are the:

1. Nature of the reaction products formed with the sulfate solution.
2. Impermeability of the cement matrix which provides a barrier against penetration by sulfate ions.
3. Permeability of the concrete as a whole.
4. Concentration of sulfates in the waterborne solution.
5. Mobility of sulfate in groundwater or other source.
6. Nature of the accompanying cations, e.g., Na^+, Mg^{2+}, Ca^{2+}, etc.
7. pH of the sulfate bearing ground water/solution.
8. Presence and concentration of other dissolved salts such as chloride and carbonate.
9. Temperature of the exposure.
10. C_3A content of the cement.
11. $Ca(OH)_2$ content.
12. Extent to which stresses resulting from mass gains can be relieved by dissolution and leaching of cement components.

Sulfate attack in concrete occurs in the presence of water and can be encountered in a range of environments:

- External sulfate attack
  - Waters and soils with high sulfate content
  - Sea water
- Crystallization of sulfate salts
- Acid attack by sulfuric acid
  - Wastewater treatment plants
  - Industrial wastewater
- Internal sulfate attack
  - Additional ettringite formation in the post-hardened state
  - Aggregates with sulfide inclusions (pyrite and pyrrhotite, which slowly oxidize to sulfate)

In some of these processes the same chemical reactions are involved, although processes may differ in the origin of the sulfate, or in boundary conditions, or both. In particular, internal sulfate attack is characteristically isochemical, or nearly so, while external attack – as the term suggests – involves both gain and loss of mass.

2.4 External Sulfate Attack

External sulfate attack is one of the more common alterations observed in concrete, and has been well known for over a century. This is also one of the most studied degradation processes in concrete.

The deterioration of concrete is produced when sulfates react in the hardened concrete, causing an excessive expansion, giving rise to cracks, surface spalling, delamination and loss of mechanical properties.

The origin of the sulfate is usually from ground or surface water, often at high concentrations. Also, the source may be from natural soils or those treated with sulfates that can be dissolved in water and ingress concrete.

The ground waters and soils can be classified in relation to their potential aggressiveness to the hardened concrete, depending on the sulfate content.

The EHE – Spanish regulation for structural concrete – includes criteria about environmental aggressivity. These criteria relate the quantity of the sulfates in waters and soils and hence define a degree of aggressiveness. Table 2.2 shows a classification of potential damage to the concrete as a function of the sulfate content in water or total sulfates in soils according to EHE-08 (2008).

Ground water can contain different types and quantities of sulfates, the composition of which is influenced by the flow of water through soils. In turn, the soil is influenced by land use, industrial activity and effluents, acid rain, etc.

The main constituents of aggressive water and soils are the calcium, magnesium, sodium and potassium charge-balancing sulfate. The action of the different ions must be taken into account, because they can produce secondary ionic interchange reactions. If these secondary processes occur, there will generally be negative consequences to the hydrated cement paste and therefore for hardened concrete.

Sulfate ions in aqueous solution can react with hydrated aluminates from the cement clinker or supplementary cementitious materials to form ettringite (Candlot's salt). Ettringite is approximately $3CaO{\cdot}Al_2O_3{\cdot}3CaSO_4{\cdot}32H_2O$ (abbreviated, $C_6A\overline{S}_3 \cdot H_{32}$). It is very poorly soluble in water and produces a large increase in volume in relation to the original solid reactants. But ettringite

Table 2.2 Potential aggressiveness of waters and soils

		Aggressiveness of attack		
Aggressive medium	(SO_4^{2-}) content	Slight	Medium	High
Water	mg/l	200–600	600–3,000	>3,000
Soil	mg/kg	2,000–3,000	3,000–12,000	>12,000

will not form unless water is available and in that sense, water is as important a precursor to ettringite formation as sulfate.

Primary ettringite is one of the products formed normally during Portland cement hydration through the reaction of sulfate ions, which come mainly from the gypsum (or other soluble sulfate-rich phases) added as a set retarder. This primary ettringite forms while the concrete is in its plastic state and does not induce expansion. Indeed, some ettringite is useful in compensating for chemical shrinkage in the course of hardening.

In most commercial cements, soluble sulfate (gypsum, anhydrite) is exhausted within the first few hours of hydration. However ongoing hydration continues to liberate aluminium from clinker solids. In this period, when free calcium sulfate is exhausted, previously formed ettringite can be transformed into calcium monosulfoaluminate, ideally $4CaO{\cdot}Al_2O_3{\cdot}CaSO_4{\cdot}12H_2O$. Thus ettringite is not necessarily expansive and plays an important role in controlling the dimensional stability of concrete. However, when ettringite forms in hardened concrete, significant expansions can be induced that generate damage.

In the hydrated matrix of the cement the source of reactive alumina is normally monosulfate (AFm) according to the reaction:

$$\underbrace{C_4Al_2(OH)_{12} \cdot SO_4 \cdot 6H_2O}_{\textit{Monosulfate}} + 2Ca^{2+} + 2SO_4{}^{2-} + 20H_2O \rightarrow \underbrace{C_4Al_2(OH)_{12}(SO_4)_3 \cdot 26H_2O}_{\textit{Ettringite}} \quad (2.5)$$

Unreacted C_3A or ferrite, ideally C_4AF, are also potential sources of aluminium to form ettringite that can react with the external sulfates.

Monosulfate will normally also be present in composite cements containing blast furnace slag, fly ash or natural pozzolana but in the hydrated matrix of these cements, alumina is partly present in phases such as hydrotalcite or hydrogarnet or substituted in C-S-H. The result is that aluminate does not appear to be available to form an expansive reaction product (Gollop and Taylor 1996a).

Strength loss and disintegration are also associated with decalcification of C-S-H, which is an important mechanism during attack by $MgSO_4$ solutions but which also occurs to a lesser extent in Na_2SO_4 solution (Gollop and Taylor 1996b).

Current standards for sulfate resistant cements in CEN member countries are still based on national specifications and can be divided into two categories (CEN/TR 15697 2008).

- Portland cements, type CEM I according to EN 197-1, 2000, with a maximum permitted C_3A content.
- Portland composite cements containing appropriate levels of glassy blast furnace slag, fly ash or natural pozzolan.

Low C_3A cements are believed to provide chemical resistance to sulfate attack. The products of reaction with sulfates are not expansive and consequently the matrix is not disrupted, inhibiting further reaction. The un-reactive nature of the hydration products of low C_3A cements is attributed to a low level of monosulfate (AFm) and/or the formation of an iron-rich form which is slow reacting and is claimed to produce a "non-expansive" form of ettringite (Odler 1984).

Portland composite cements (i. e. CEM II, III, IV and V types, according to EN 197-1 (2000)) provide resistance to sulfate attack which is attributed to the microstructure (Mehta 1988; Soroushian and Alhozimy 1992; Stephens and Carrasquillo 2000; Hughes 1985). This arises from the significantly lower permeability of the hydrated matrix. Additional positive factors are:

- Reduced level of free calcium hydroxide in the matrix which reduces calcium availability for leaching, for ettringite formation and also the formation of gypsum when the matrix is exposed to concentrated sulfate solutions.
- Formation of hydrates containing alumina which are non-reactive to sulfate solutions.

The reduced availability of calcium may also result in the formation of ettringite with a morphology and distribution throughout the hydrated matrix which is claimed to be non-expansive (Brown and Taylor 1999). Much argument persists about the distinction between "expansive" and "non-expansive" ettringite.

Another, and often overlooked factor, is that resistance to external sulfates is normally positively influenced by the level of SO_3 in the binder; the higher the level in a range from ≈ 1 to 4%, the greater is the resistance. This applies to concrete produced from CEM I cements (Frigione and Zenone 1992) and also particularly to slag and fly ash containing concretes. Where the ash or slag is added during the concrete mixing process (Gollop and Taylor 1996b; Kollek and Lumly 1990; Prusinski and Carrasquillo 1995) the SO_3 level is lowered by dilution and the hydrated matrix becomes more vulnerable to attack by penetrating sulfates in comparison with a binder having an optimized SO_3 content. The improved resistance can be attributed to the increased level of sulfated phases, such as ettringite, formed during initial hydration, which remain stable in the presence of an elevated sulfate level.

The morphology of ettringite crystals often takes the form of needle-like hexagonal prisms. Prusinski and Carrasquillo (1995) define the structure of the ettringite formed from parallel columns of cationic units $\{Ca_3[Al(OH)_6].12H_2O\}^{3+}$. It can be seen in Fig. 2.1 that the octahedral alumina ($Al(OH)_6{}^{3+}$) is bound to the adjacent CaO_8-polyhedra through sharing of hydroxyl (OH^-) ions, while the channels between the columns are occupied by tetrahedral sulfate ($SO_4{}^{2-}$) ions and much loosely bound and disordered water. The decomposition of ettringite upon exposure to elevated temperatures is a result of the easy removal of the loosely-bound water (Taylor 1997).

The morphology of ettringite often takes the form of long hexagonal crystals such as those in Fig. 2.2 but other morphologies are well known such as stout prisms and prisms with hollow re-entrant ends (Menéndez 1999, 2010). Ettringite morphology is much influenced by sorption of ions and molecules, for example of plasticizers.

In ordinary Portland cement, sulfate attack is produced mainly by reaction of sulfates with the aluminates and portlandite phases. The main chemical reactions

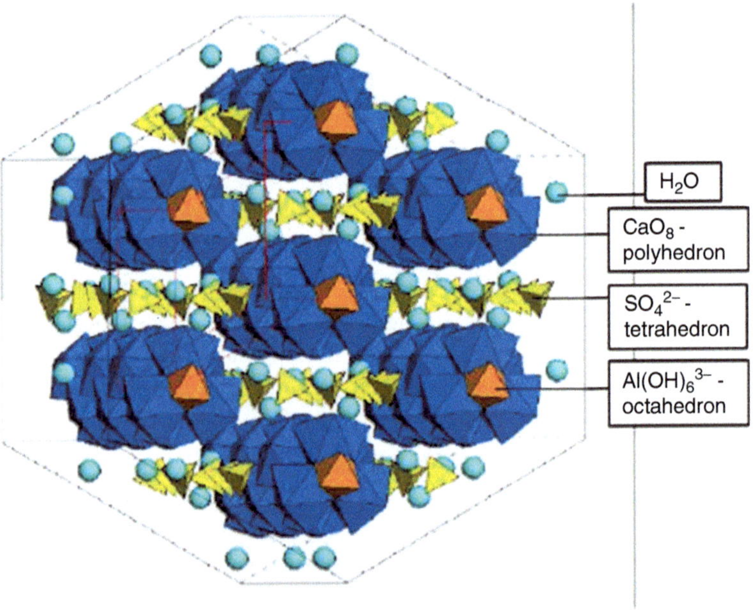

Fig. 2.1 Structure of ettringite (Stark and Bollman 1999)

Fig. 2.2 Typical microstructure of ettringite

produced are described in the following sections. Several authors have described the chemical reactions that are produced in the hardened concrete when sulfates penetrate (ASTM C 1012 2004; Skalny et al. 2002; Taylor 1997; Biczok 1967; Mehta 1986, 2005).

Fig. 2.3 Scheme of the sulfate attack from the surface to the interior

In external chemical attack on adequately compacted concrete, the sulfate-cement paste interaction zone develops at the surface, moving gradually inwards as the process continues. If reaction consists of more than one step, the region in which the first step takes place is located farthest away from the original surface, followed by the region in which step two takes place, etc. The process is accompanied by the migration of some of the products of these reactions in the opposite direction. The rate of migration will depend on the paste porosity and its permeability. The driving force for this complex migration is based on the requirement that each point in every zone has to remain electrically neutral. It is not sufficient just to gain sulfate ions: these require charge-balancing cations and, moreover, the permeating solution, while changing in composition, still has to remain electrically neutral at each point. Figure 2.3 shows a schematic of the general process.

The permeability of concrete is often determined from laboratory specimens. However, these are normally made under carefully controlled conditions and the resulting product approaches a uniform state whereas concrete made in the field is often inhomogeneous. For example, it may not achieve uniform compaction. Thus, field concretes often perform less well than their laboratory equivalents. This distinction between laboratory and field concretes is difficult to quantify but is very real. If the concrete undergoing chemical attack is insufficiently compacted, the corrosive agent may penetrate and cause damage at a greater depth and the damage to the concrete may be more extensive.

Different products are formed during sulfate attack, the more common being ettringite and monosulfoaluminate. Secondary gypsum, which is gypsum neo-formed within concrete that has already hardened, is also found.

2.4.1 Calcium Sulfate Solutions ($CaSO_4$)

The action of calcium sulfate is limited by its solubility in water, 15.2 mmol/l at 20°C (≈1.46 g SO_4^{2-}/l). However, when these ions enter concrete, the pH is different and there is a change in the reactivity and solubility with respect to ground water. Sulfate ions can react with the tricalcium aluminates of the cement or with the tetracalcium aluminate hydrate to form ettringite. Reaction begins in the areas

Table 2.3 Increase of volume of different primary reactant

Primary reactant	Product	Volumetric change ($\Delta V_p/V_p$)
$3CaO{\cdot}Al_2O_3$ *Tricalcium aluminate*	$3CaO{\cdot}Al_2O_3{\cdot}3CaSO_4{\cdot}31\text{-}32H_2O$ *Ettringite*	1.26
$4CaO{\cdot}Al_2O_3{\cdot}SO_3{\cdot}12H_2O$ *Monosulfate*	$3CaO{\cdot}Al_2O_3{\cdot}3CaSO_4{\cdot}31\text{-}32H_2O$ *Ettringite*	0.51
$4CaO{\cdot}Al_2O_3{\cdot}13H_2O$ *Tetracalcium aluminate hydrate*	$3CaO{\cdot}Al_2O_3{\cdot}3CaSO_4{\cdot}31\text{-}32H_2O$ *Ettringite*	0.48

After Tixier and Mobasher (2003)

closest to exposed surfaces. In this case, the Ca ions needed for reaction can be supplied by the cement and/or ground water.

The three different species can react with ingressing sulfate according to the following chemical reactions:

$$\underset{\text{Tricalciumaluminate}}{3CaO \cdot Al_2O_3} + 3\underset{\text{Gypsum}}{(CaO \cdot SO_3 \cdot 2H_2O)} + 26H_2O \rightarrow \underset{\text{Ettringite}}{6CaO \cdot Al_2O_3 \cdot 3SO_3 \cdot 32H_2O} \tag{2.6}$$

$$\underset{\text{Monosulfate}}{4CaO \cdot Al_2O_3 \cdot SO_3 \cdot 12H_2O} + 2\underset{\text{Gypsum}}{(CaO \cdot SO_3 \cdot 2H_2O)} + 16H_2O \rightarrow \underset{\text{Ettringite}}{6CaO \cdot Al_2O_3 \cdot 3SO_3 \cdot 32H_2O} \tag{2.7}$$

$$\underset{\text{Tetracalcium aluminate hydrate}}{4CaO \cdot Al_2O_3 \cdot 13H_2O} + 3\underset{\text{Gypsum}}{(CaSO_4 \cdot 2H_2O)} + 13H_2O \rightarrow \underset{\text{Ettringite}}{3CaO \cdot Al_2O_3 \cdot 3CaSO_4 \cdot 31 - 32H_2O} + \underset{\text{Portlandite}}{Ca(OH)_2} \tag{2.8}$$

Tixier and Mobasher (2003) calculated the volumetric change for the individual reactions of tricalcium aluminate, monosulfate and tetracalcium aluminate hydrated using the density, molar mass and molar volume of each compound. The volumetric changes are as shown in Table 2.3.

An initial increase in mechanical strength is produced through filling of the pores (capillary pores and air voids) with ettringite, but a continuing production of ettringite generates pressure and induces additional microcracking with expansion.

The microstructure of monosulfate as dense tabular crystals, and of ettringite as needle- like crystals, is shown in Figs. 2.4 and 2.5, respectively (Menéndez 1999, 2010). Usually monosulfate appears inside the paste, and ettringite appears in pores and inside the paste, with less expansion caused by monosulfate. Figures 2.4 and 2.5 are images of polished surfaces of concrete from field concretes exposed to external sulfate attack and microcracks form especially due to ettringite development. Figure 2.4 shows crystallizations of monosulfate with tabular morphology but without formation of microcracks. On the other hand, Fig. 2.5 shows very compact ettringite in the cement paste inducing the formation of microcracks.

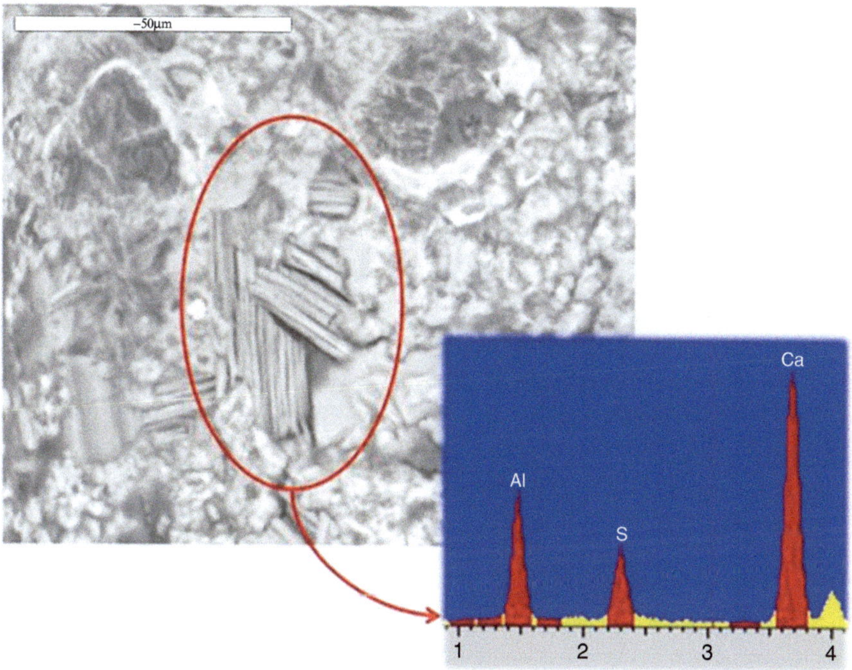

Fig. 2.4 Aspect of the monosulfate crystals growing in cement paste with a tabular aspect, the EDAX obtained from the circled region (shown in the *inset*) confirms the identity of the crystals

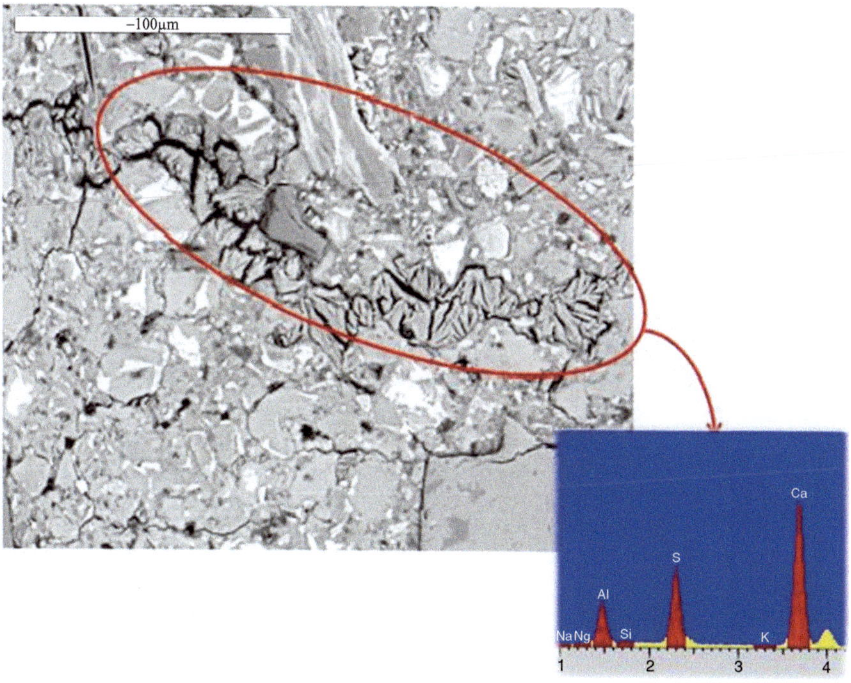

Fig. 2.5 Appearance of the ettringite crystals and associated microcracks developing inside cement paste (Menéndez 2010)

2.4.2 Sodium and Potassium Sulfate Solutions (Na_2SO_4 and K_2SO_4)

Alkali sulfates in solution also react with cement paste. Both sulfate ions and alkali ions (sodium and potassium) react, albeit in different ways. Alkalis are partly absorbed and affect the solubility of cement components while sulfate reacts with formation of low-solubility solids. In the first step of reaction, portlandite is dissolved and reacts with the sulfate ions to form ettringite and, at higher concentrations above 1–2 g/l, secondary gypsum. On the other hand, alkalis may react, especially at concentrations above ~0.5 M, to form a sodium-containing variety of AFm, termed the "U-phase". The reaction to form gypsum is as follows:

$$\underset{\textit{Soild}}{Ca(OH)_2} \rightarrow Ca(OH)_2 + \underset{\textit{Solution}}{2\,Na^+ + SO_4^{2-}} \xrightarrow{H_2O} \underset{\substack{\textit{Solid secondary gypsum}\\ \textbf{\textit{Increase of volume}}}}{CaSO_4 \cdot 2H_2O} + \underset{\textit{Solution}}{2\,Na^+ + 2\,OH^-} \quad (2.9)$$

The secondary gypsum has solubility in water of about 2.2 g/l at 0°C and its precipitation leads to an increase in volume, uncorrected for leaching, of about 17.7% in relation to the initial products. Gypsum formation is usually most pronounced near the exposed surface, as is the degree of decalcification.

The alkali sulfates can react with the monosulfate to form ettringite, while the Ca^{2+} necessary for its formation is provided by portlandite or, after portlandite is consumed, by C-S-H gel. As the calcium of the C-S-H gel is consumed, decreasing its CaO/SiO_2 ratio, the mechanical properties are adversely affected.

$$2SO_4^{2-} + \underset{\textit{Monosulfate}}{Ca_4Al_2(OH)_{12} \cdot SO_4 \cdot 6H_2O} + 2Ca^{2+} \rightarrow \underset{\substack{\textit{ettringite}\\ \textbf{\textit{Expansion}}}}{Ca_6Al_2(OH)_{12}(SO_4)_3 \cdot 26H_2O} \quad (2.10)$$

$$Ca(OH)_2 \rightarrow Ca^{2+} + 2OH^- \quad (2.11)$$

When concrete is altered by alkali sulfates, the sequence of zones has many similarities to those developed by calcium sulfate attack:

- Zones with ettringite formation developed by transformation of monosulfate are associated with local decreases of portlandite.
- Zones with crystallization of gypsum are also marked by a decrease in portlandite and partial decalcification of C-S-H gel (reduction in the CaO/SiO_2 ratio in the C-S-H).
- Sulfate is adsorbed in the decalcified C-S-H gel.
- Concrete which has partially reacted may exhibit little or no change apparent in the microstructure.

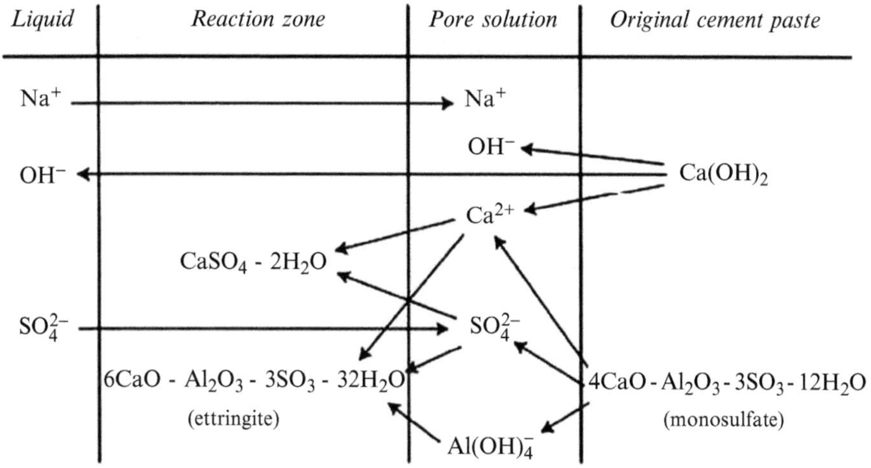

Fig. 2.6 Possible reactions with alkali sulfate solutions (Skalny et al. 2002)

As the alkali ions are relatively small, they can penetrate into the concrete and tend to migrate towards the interior, bringing about an increase in the alkalinity of the aqueous phase. As a consequence, and under adverse conditions, an alkali-silica reaction can occur, as well as a migration of OH^- ions outwards.

The possible reactions are shown in the diagram in Fig. 2.6.

Similar to attack by calcium sulfate, the initial action of the alkali sulfates may result in an increase in strength because of the filling of pores with newly-formed ettringite. However, the process is physically expansive and stress is created: once the stress exceeds the tensile strength of the concrete, cracking and spalling occur.

A surface scaling is often produced because of local ettringite formation or crystallization of gypsum, or both. Microcracking also occurs as a result of the decalcification and shrinkage of the C-S-H gel.

In case of high temperature and a high concentration of sulfates in the pore solution it is possible that the "U-phase" [$4CaO{\cdot}0.9Al_2O_3{\cdot}1.1SO_3{\cdot}0.5Na_2O{\cdot}16H_2O$] is formed instead of ettringite. In laboratory experiments formation of U-phase was observed in samples that were immersed in 1 M NaOH solution at 20 and 80°C (Moranville and Li 1999). Concrete deterioration may result from excessive formation of U-phase or by its transformation to ettringite. Studies of the system CaO-Al_2O_3-$CaSO_4$-H_2O in the presence of Na_2O show the stability of the U-phase and loss of stability of ettringite (Damidot and Glasser 1996).

Usually gypsum appears as dense crystals formed in parallel columnar growths, as shown in Fig. 2.7 (Menéndez 1999, 2010).

The above summary is rather qualitative. However immersion in alkali sulfate solutions is commonly used in tests of the sulfate resistance of cements and it is therefore of considerable practical importance to quantify reactions. Quantification is difficult for several reasons (i) an infinite number of possibilities exists and, presumably, lacking a valid system of interpolation and extrapolation, an infinite

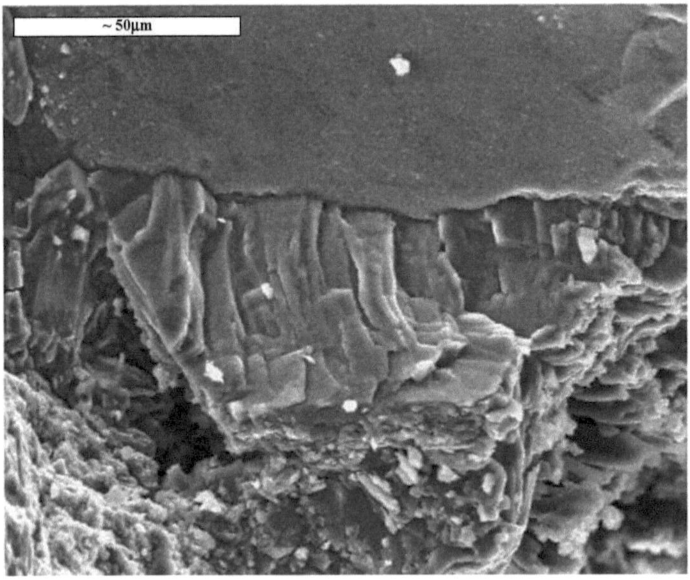

Fig. 2.7 Crystallizations of gypsum at the aggregate-paste interface. The aggregate is shown at the *top* and bladed columnar gypsum (*centre*) has filled the annulus between paste (*bottom*) and aggregate (Menéndez 2010)

number of experiments would be necessary to provide quantification, (ii) existing test methods do not adequately control all relevant variables and (iii) even where generic trends emerge from prescriptive testing, the nature of the attacking solution and selection of test conditions may be far removed from reality.

Consequently, we include in the following Sect. 2.4.3, details of an in-depth study into the influence of sodium sulfate concentration on cement paste – with a view to assist in quantifying the effects and thereby draw conclusions for the way forward in designing tests for sulfate attack. This is very important for standardization of future testing of sulfate resistance.

2.4.3 Influence of Sodium Sulfate Concentration on the Degradation of Portland Cement: Implications for the Design of Tests for Sulfate Attack

We have undertaken a quantitative treatment of the reaction between cement paste and sodium sulfate. Results are presented herein as a series of "snapshot" diagrams. They give a truer picture of volume changes relative to the simplistic calculations which are widely used, inasmuch as the mass of solids is automatically corrected for dissolution and recalculated to mineral composition.

Sulfate induced deterioration of Portland cement pastes, mortars and concretes in aqueous sulfate solutions is a well-known problem and numerous standard tests have been adopted in which cement performance is determined by immersion in aqueous sodium sulfate solutions. However, solution concentrations and other test conditions vary between specifications and there does not seem to be agreement between standards about what variables, temperature and initial solution concentration excepted, require to be controlled. Features and processes occurring in the course of degradation have been well established, as have performance criteria, albeit on a largely empirical basis. But doubts persist on the applicability of the various standard tests: how effectively do they predict the real-life performance? And how can test results be quantified? Can test results be extrapolated to other, non-standard conditions?

To answer these and other questions we have analyzed the physical chemistry of reactions between sodium sulfate and cement. A single model cement composition has been selected to calculate the impacts of changing solution volumes and concentrations of sodium sulfate, using a fixed mass of cement at 25°C. The results confirm that, in general, ettringite and gypsum are favoured initial products of attack. But the trends are not always progressive: crossing phase boundaries, the presence of which is indicated by the spontaneous development of mineralogically-zoned microstructures, can result in abrupt changes as the sodium sulfate concentration increases. Mineralogical changes and the balance between precipitation and dissolution affect the specific volume, and hence the potential for expansion, of the solids. In general, no one set of test conditions will have universal applicability and suggestions are made for matching test conditions to anticipated service environments. It is shown how the realism and applicability of present tests can be improved and deviations between laboratories using supposedly identical tests reduced, by more rigorous selection and definition of the test conditions.

Skalny et al. (2002) reviewed standard tests for the sulfate resistance of Portland cement and also described a number of non-standard tests. It is apparent that sodium sulfate is the preferred medium in which to determine the resistance of cements to sulfate attack as a wide range of prescriptive standards based on immersion in sodium sulfate exist: for example ASTM C150, C1157, C1012, E632, and in the U.K., as described in BRE Digest 363 (1996). A uniform Eurocode for sulfate-resistant cements is also at the discussion stage but thus far, without agreement. An important distinction exists between “internal” and “external” sulfate attack (Skalny et al. 2002). Internal sulfate attack is isochemical: the sulfate content – and that of other chemical species – remains constant. But in external sulfate attack the sulfate content changes, usually increasing, as a consequence of reactions in the service environment. The aggregates used in making mortars and concretes are normally unaffected: attack is concentrated on the cement paste. We therefore also exclude the presence of alkali-susceptible aggregates from consideration (but see discussion for comments on the relevance to alkali-aggregate attack).

The normal features of sodium sulfate attack on cements and concretes are, as described previously, (i) physical expansion with cracking of cement paste as attack proceeds with (ii) gain in sulfate by the solids (iii) loss of strength and coherence and (iii) changes in local chemistry and mineralogy, with development of a chemically and mineralogically zoned microstructure marked by loss of calcium. Simple diffusion theory, cited in (Skalny et al. 2002), has been applied to the propagation of attack with time but, on account of the chemically and mineralogically graded structure in attacked zones, with concomitant local changes in porosity and microstructure which impact on diffusibility, simple diffusion theory is inadequate to represent the progress of attack and quantify its kinetics.

We approach the problems associated with the design and interpretation of tests from a different perspective. The increasing performance lifetime expected for constructions based on Portland cement require quantification: how can assurances of performance lifetime be demonstrated, given that the existing body of empirical tests and prescriptive standards lack a scientific basis for extrapolation to longer time scales? The prediction of performance lifetimes is not simple, as is evidenced from the literature as well as the general paucity of quantitative approaches; Skalny et al. (2002) review the more constructive of these. Moreover, as limestone and blending agents are increasingly added to cement, the chemistry of the cement matrix changes and accompanying granulometric and microstructural changes affect its properties. These additional factors increase the number and range of compositions requiring to be assessed in respect of future performance. From being, very nearly, a material of constant composition, the cement paste binder of modern cement matrices has grown very complex indeed.

The generic approach used in the title study seeks to embrace a broad range of compositions (although for brevity only one example is presented, of a plain Portland cement). Data are used to make an analysis of the features and processes occurring in the course of tests and we consider the presence or absence of CO_2.

We start with the following generalizations.

1. Generic approaches to durability are preferable to composition-specific approaches, as the latter are normally of limited applicability.
2. Methodologies that permit calculation of sulfate impacts are preferable to those requiring lengthy experimental programs.
3. The many microscopic and chemical studies of deterioration mechanisms reveal that diffusion and mineralogical change are driven by a system of chemical potentials: a re-equilibration occurring between cement and aqueous solution drives and directs the course of reaction.
4. The attack process only rarely goes to completion in the course of either typical test conditions or field exposures, with the result that the resulting compositionally – and mineralogically – graded structures provide a snapshot, or series of snapshots, of how degradation progresses with time.
5. Calculation using standard iterative programs can be applied to solve many of the problems arising in the interpretation of the results of laboratory tests made under controlled conditions as well as, but with less certainty, to field observations.
6. While equilibrium conditions may define the final state or condition of the system, reaction kinetics may mediate the attainment of this state. However calculation, even if based on equilibrium, can also highlight many of the kinetic restraints. But tests and field observations also potentially reveal useful clues about rate-limiting processes.

By "calculation" we have in mind the methodologies developed so successfully by geochemists over the past few decades, using applied thermodynamics. Computer codes are freely available with which to minimize the free energy of a defined system and distribute chemical mass of substance between states (gas, liquid and solid) and, in complex systems, between coexisting solids of essentially fixed composition. We use the code GEMS (Kulik et al. 2003) for calculations. The code needs to couple with an extensive database. Until recently only some of the specialized database needed for calculation of cement equilibrium has been available but many missing data have now been obtained (Matschei et al. 2008). Calculations made in the course of the title study couple these data for cement substances with other data available in standard compilations. The actual calculations use an iterative process to converge on the state of minimum free energy and require a computer to implement; a PC is adequate to perform the required calculations. The problem we have encountered is (i) to determine and shape what calculations are required (ii) to formulate the problem, as well as the required calculation, in physicochemical terms and, finally, (iii) to appreciate the significance of the calculated results. In this respect, the computer is still servant, not master, and the need for a thorough understanding of cement science is strengthened, not reduced, by the advent of computer-based calculations.

We apply this approach to analysis of the reaction between Portland cement and sodium sulfate. The intention is to demonstrate how changing the initial sodium sulfate concentration and conditions of reaction affect the outcomes.

2.4.3.1 Input Data and Scope of Calculation

Table 2.4 shows the composition of the model cement used in subsequent calculations. This composition is believed to be reasonably representative of modern cement clinker except for the omission of iron oxide, omitted because in its present state of development the database is inadequate to model iron. But it is believed that iron oxide, initially present in cement mainly in the ferrite phase, remains relatively un-reactive in the course of early hydration, so a fully hydrated cement paste will contain hydrated iron oxide, *i.e.*, much of the total iron will not be incorporated in the principal hydrated cement solids. This is a useful simplification because in the range of sodium sulfate concentrations assessed (0–0.5 molal), hydrated iron oxides behave as inert solids and are essentially un-reactive with sulfate. On that account, iron need not be included in the assessment except to take into account the slight dilution which will occur in calculations of phase amounts.

The strategy for calculation is to interface the hydrated Portland cement, as defined in Table 2.4, with aqueous sodium sulfate, varying either the mass of aqueous solution relative to mass of cement, or the concentration of sodium sulfate, or both. The units of calculation require comment. The computer conserves mass in the course of calculations and it is therefore required to define the input masses of all components, including water. This necessitates use of the molal (wt/wt, symbol m) concentration scale. Where practicable we have also given more

Table 2.4 Composition of the model cement

		CaO	SiO_2	Al_2O_3	$CaSO_4$
Cement 1 (calcite free)	wt.%	68	22	4.5	5.5 (~3.2% SO_3)

Table 2.5 Calculated solid composition of hydrated model cement pastes (initially 100 g cement, table 4) at 25°C (initial w/c = 0.5)

Mineral	Formula weight [g/mol]	Density [g/cm³]	Cement [mol]	[g]
C-S-H ~$1.6CaO{\cdot}SiO_2{\cdot}2H_2O$[a]	185.84	2.43	0.366	68.01
Portlandite $Ca(OH)_2$	74.09	2.24	0.499	36.97
SO_4-AFt $Ca_6Al_2(SO_4)_3(OH)_{12}{\cdot}26H_2O$	1255.11	1.77	0	0
SO_4-AFm $Ca_4Al_2(SO_4)(OH)_{12}{\cdot}6H_2O$	622.52	2.01	0.044	27.39

[a]no gel-water included

familiar units, *e.g.*, weight %, and molar aqueous concentrations. However for many purposes, and particularly in dilute solutions, the differences between molar and molal scales are slight. It is also necessary to define a mole unit for phases such as C-S-H; Table 2.5 gives the numerical values used. Calculations were done for two sets of conditions, selected to reflect the different conditions imposed on test specifications. Either (i) a fixed mass of cement was reacted with a fixed mass of aqueous solution, the sodium sulfate concentration of which was varied in small increments up to 0.5 m, or (ii), the same fixed mass of cement was reacted with varying mass of aqueous sodium sulfate, the initial composition of which (with a few exceptions) was fixed at either: 0.1 or 0.25 m. Supplementary calculations are also presented for illustrative purposes and a discussion section deals with the presence of other sulfates, especially magnesium sulfate, and the role and influence of atmospheric carbon dioxide on tests conducted in aqueous sodium sulfate solutions. The calculations assume that the cement is 100% hydrated prior to contacting sodium sulfate. However some of the kinetic features associated with incomplete reaction are treated in discussion by assuming that local equilibrium is attained.

2.4.3.2 Results and Interpretation

Input Data

The composition of the prehydrated model cement is shown in Table 2.5 in both mole and mass proportions. Basically, calculations were done on one cement composition, without calcium carbonate; but a second composition, using the

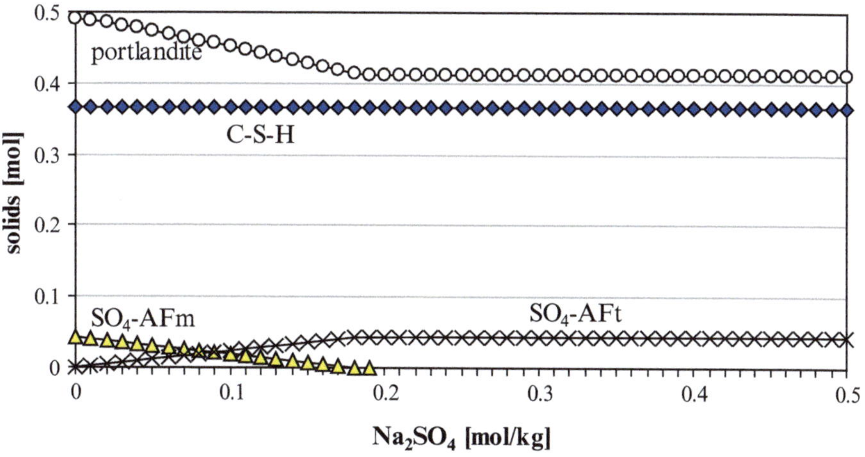

Fig. 2.8 Calculated changes on the solid mass balances at 25°C (See text for definition of the conditions)

same cement formulation but modified by replacement of cement with the molar equivalent of 5 wt.% calcite ($CaCO_3$) was also included.

In each calculation set the sodium sulfate mass, M, is equal to the product of the volume of the aqueous solution, V and its concentration, C (using molar units: for molal units the solution mass is used instead). Thus a large volume of dilute solution may be formulated to contain the same *total* number of moles of sodium sulfate as a smaller mass of more concentrated solution. However the amount of water will differ between the two cases and, on that account, because the soluble masses are affected by mass of water, the impacts on cement will not necessarily be identical; we return to this point subsequently. An important characterization parameter, used extensively in the course of this section is the mass ratio, defined as:

$$\text{mass ratio} = \frac{\text{mass of solution}}{\text{mass of cement}} \tag{2.12}$$

We now interface a known mass of the carbonate-free hydrated cement paste, Table 2.5, with a known mass and concentration of sodium sulfate. C-S-H remains essentially inert and persists in almost unchanged mass throughout the calculations, for example decreasing from 0.41 to 0.39 mol as a result of minor dilution. In this initial calculation, the water/solid weight ratio was fixed at 0.5 and the excess aqueous solution shown in Figs. 2.8, 2.9, and 2.10 is that which is calculated to remain after the hydration demand of the solids has been satisfied. We commence by examining the results of calculation using a fixed mass of aqueous phase.

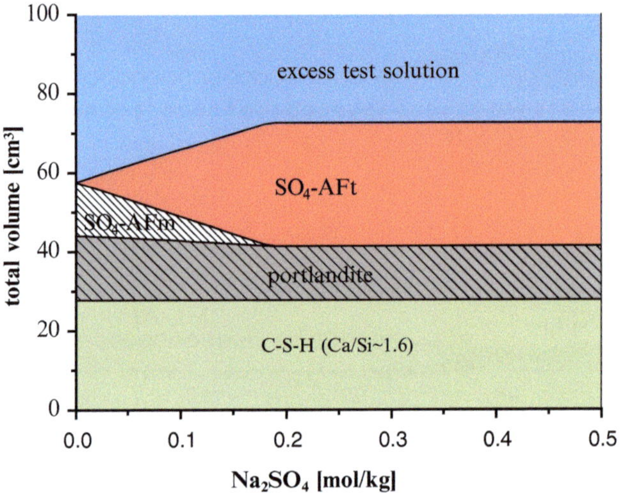

Fig. 2.9 Calculated solid volume changes in dependence of the initial Na_2SO_4 concentration (Using data from Fig. 2.8)

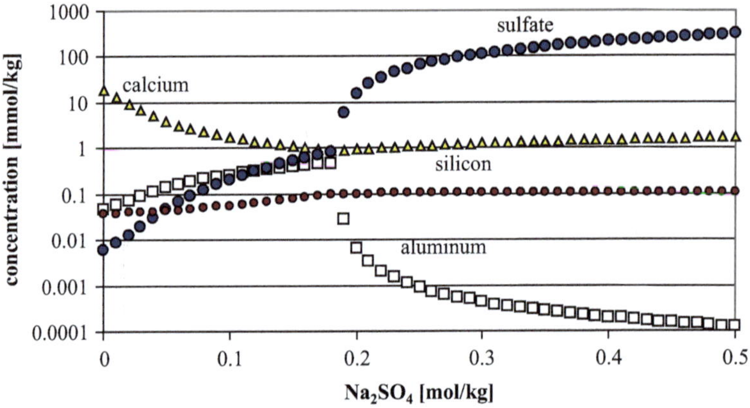

Fig. 2.10 Composition of the aqueous phase in dependence of the initial Na_2SO_4-concentration (Data from Fig. 2.8)

Results and Significance

Fixed Mass of Cement and Changing Aqueous Composition

Figures 2.8, 2.9, and 2.10 show the calculated impact of adding a fixed mass of sodium sulfate solution (500 g) to a total 100 g of model cement substance (defined in Table 2.5) i.e., a mass ratio of 5.0, while varying the initial concentration of sodium sulfate between zero and 0.5 m. Thus the composition of "pure" water (no added sodium sulfate) lies at the left-hand side of the Figures. The initial paste mineralogy has been calculated on the basis that (i) the cement has fully hydrated in the presence of water in a previous step and (ii) the resulting hydration products have subsequently been equilibrated with sodium sulfate under the defined conditions. The resulting mineralogical constitution can be determined graphically by constructing the appropriate vertical line (including zero, which predicts the assemblage formed in the course of initial hydration; see Table 2.5) and noting the intercepts, numerical values of which can be obtained from the vertical scale.

Figure 2.8 shows that as the aqueous concentration of sodium sulfate increases – corresponding to sweeping a vertical line from left to right across the diagram and noting its intercept with the phases formed – the mass amounts of portlandite and sulfate – AFm at first decrease rapidly while the mass of ettringite (AFt) increases. C-S-H is not significantly involved in reaction and, as a result, its mass is essentially unaffected.

The corresponding volume relationships, calculated from phase amounts and the known physical densities of the phases, are shown in Fig. 2.9; to facilitate comparison, the scale of concentrations is identical with that used for Fig. 2.8. The formation of ettringite consumes much water and, as a consequence of its low physical density relative to its solid precursors, the total volume occupied by solids at first increases rapidly as ettringite forms at the expense of AFm, etc. Note that the calculated volume increase is unambiguous but its translation into dimensional change is uncertain because some, possibly all, the volume increase of the solids is accommodated in pores. Hence the calculation discloses only a potential for volume increase (PVI). In this example the PVI is essentially exhausted after addition of 0.19/2 = 0.095 moles of sodium sulfate per 100 g of cement and reaches a 27 vol. % increase in solids volume, shown by a construction line and double-headed arrow in Fig. 2.9; that is, under the conditions of calculation, 100 g cement will react with a maximum of 0.095 mol (or 13.5 g of sodium sulfate, equivalent to 7.6 g of SO_3). Of course this value applies only to this particular composition but the calculation could readily be re-computed for other cement compositions and the trends thus established will have a generic similarity. For example, other factors being equal, increasing cement alumina contents will increase the amount of sodium sulfate required to complete reaction, also increasing the amount of ettringite and hence the potential for volume increase, whereas increasing the initial sulfate content of the cement will decrease the mass of sodium sulfate solution required to reach the point at which the specific volume of solids reaches a maximum. Under the conditions selected for calculation, a plateau in the potential for expansion occurs; increasing the concentration of sodium sulfate beyond 0.19 m does not change either the phase distribution or potential for expansion significantly, at least up to 0.5 m initial sodium sulfate concentration. The maximum upper limit of the calculations corresponds approximately to 71 g sodium sulfate/kg solvent.

A large body of evidence from empirical tests broadly supports this conclusion: the physical expansion of cement will increase rapidly at first with increasing sodium sulfate concentration but thereafter, expansion will slow. This observation is often cited as justification for preferring to recommend a relatively dilute sodium sulfate solution in standard tests. Superficially the present set of calculations supports this approach but caution is required in reaching generic conclusions because, as is shown subsequently, the maximum potential for expansion is influenced by other factors.

Figure 2.10 shows the impact of changes to the composition of the aqueous phase; it should be read in conjunction with Fig. 2.9. Note that Gibbs phase rule considerations apply: as long as the system retains one or more degrees of freedom, aqueous compositions must change continuously and abrupt changes in slope are only possible at invariant points. While the point at ~0.19 molal sodium sulfate appears to be invariant it is not; although compositions ~0.19 m mark a transition between different solubility regimes. Between 0 and 0.19 molal aqueous sodium sulfate, concentrations of calcium decrease while those of silicon, sulfate and aluminium rise (note that a log concentration scale has to be used to accommodate the large numerical ranges of species solubilities). But at molalities above 0.19 and, as the mass of sodium sulfate required to convert AFm to AFt is exceeded, the aqueous concentrations of aluminium decrease rapidly while those of sulfate increase; calcium and silicon change much less. Note that in regimes of low sodium sulfate molalities, the extent of sulfate removal from the aqueous phase is often high; solution concentrations may decrease from the original value as depletion approaches completion. However, for a constant mass ratio, sulfate depletion tends to be much less complete at higher molalities mainly because the binding capacity of the cement solids for sulfate is limited and approaches saturation. The ratio of $Na:SO_4$, initially 2.0, changes as a consequence of reaction and the changing numerical value of this ratio affects the solubility of other species.

Aluminium, often thought of as "insoluble" in cement undergoing sodium sulfate attack, is in fact almost as soluble as calcium at low sodium sulfate concentrations. Despite the elevation of pH arising from reaction of sodium sulfate with cement solids, the aluminium solubility is controlled primarily by the nature of the solubility-limiting solid, not primarily by pH. This has consequences for leaching as well as for transport processes leading to formation of ettringite: at low Na_2SO_4 molalities, all the components of ettringite – calcium, aluminium and sulfate – are readily soluble but at higher sodium sulfate concentrations, the components of ettringite, particularly aluminium, are much less soluble and, presumably, become less available to participate in mineralogical conversions requiring through-solution transport. Thus kinetic barriers to ettringite formation arising from vastly differing component solubilities are reduced in dilute sodium sulfate, <0.19 m approximately: the accelerated transport thus achieved is an argument in favour of using low concentrations, <0.19 m.

A characteristic feature of the example calculations, Figs. 2.8, 2.9, and 2.10, is that the mass of aqueous solution remains relatively small compared to mass of cement. On that account the overwhelming mass of cement species, water (and possibly sulfate) excepted, occur in solids, not in the aqueous solution. The highest species solubility, of sulfate only approaches the initial sulfate concentration at the highest molality and highest mass ratios included in the calculations. But, while the aqueous phase is depleted to varying extents in sulfate, its initial sodium concentration remains little changed. Therefore, an apparent deficit of negative charge develops over most of the aqueous sodium sulfate concentration range. Of course the aqueous solution must at all time conserve charge.

In the sodium-free situation, corresponding to the left hand edge of the diagram, the necessary aqueous charge-balancing anions, mainly required to charge balance soluble calcium, are provided by hydroxide ions with the result that the 25°C pH of the aqueous phase is not that of water but is about 12.6 as a consequence of the solubility of cement substances. However, as sodium sulfate is added, this simple explanation of the origin of charge balance is no longer adequate. Calcium solubility is rapidly reduced and removal of sulfate ions from solution, by reaction with cement solids, results in progressively more hydroxide requiring to be added to maintain charge balance of the remaining sodium. The pH therefore rises beyond the saturation value achieved by portlandite, reaching 13.8 in the sodium sulfate solution initially 0.5 molal. Thus the near-neutral pH of sodium sulfate notwithstanding, the pH is driven upwards as a consequence of the reaction between sodium sulfate and cement solids and the need to conserve electroneutrality. This pH, ~13.8, is of course at the high end of values normally encountered in the pore fluids of commercial cements and has potential consequences to the interpretation of test results: see additional discussion. However, an important conclusion must be that sodium is not "inert"; by affecting pH, it alters the solubilities of other species which, in turn, impacts on transport mechanisms and the kinetics and equilibrium of formation of potentially expansive zones.

Fixed Mass of Cement, But Changing Aqueous Composition and Mass of Solution

In this calculation set we varied both the mass of aqueous solution and its concentration. To depict the results of calculations in two dimensions, we restrained the initial aqueous concentration to one of two initial molalities with respect to sodium sulfate: either 0.1 or 0.25 molal. Calculations were commenced at a mass ratio of aqueous phase/cement ~2.0 and extended to mass ratio 50. The cement content, 100 g of substance, was unchanged from the previous calculation. However the mass of aqueous phase is no longer constant but instead is allowed to vary continuously between 200 and 5,000 g (0.2–5 kg). It will be appreciated that, especially at high molalities and larger volumes of aqueous phases, it is possible to have much wider mass ranges, mainly extending to higher numerical values, than used in the previous example.

Results for variable mass at an initial sodium sulfate concentration of 0.1 m are shown in Figs. 2.11, 2.12, and 2.13 for 25°C. Figure 2.11 relates to Fig. 2.8 as follows: if a vertical line is constructed in Fig. 2.8 at 0.1 m, it will reproduce the data shown in Fig. 2.11 along the vertical line at an aqueous phase/cement mass ratio equal to 5.0. Similar relations obtain for other pairs of figures, for example between Figs. 2.9 and 2.12, or 2.10 and 2.13. However, as noted, conditions in Fig. 2.11 extend to greater total mass of sodium sulfate than in the first set of calculations, made using a low mass ratio. Thus in Figs. 2.8, 2.9, and 2.10, despite increasing the concentration of sodium sulfate at fixed total mass of aqueous phase,

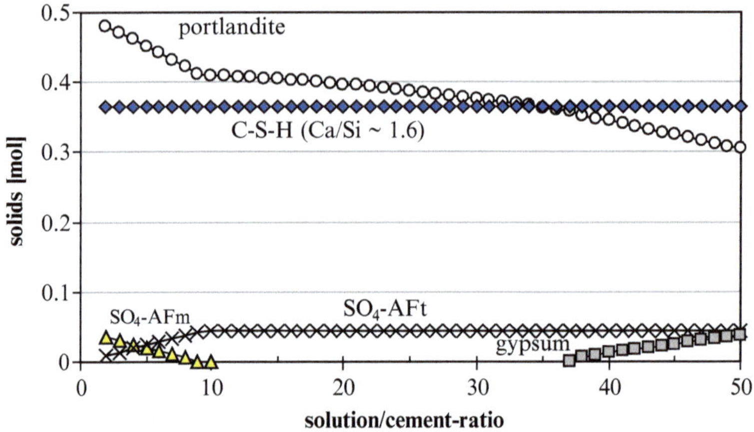

Fig. 2.11 Calculated changes on the solid mass balance as a function of changing solution/cement-ratio (initial concentration 0.1 mol Na_2SO_4/kg) 25°C

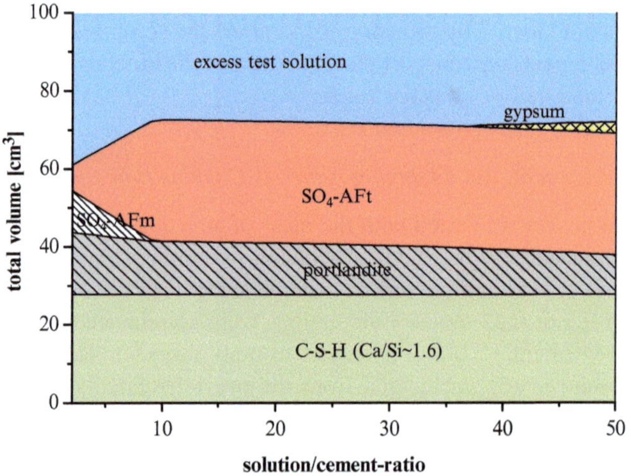

Fig. 2.12 Calculated solid volume changes as a function of changing solution/cement-ratios (initial concentration 0.1 mol Na_2SO_4/kg) (See text and Fig. 2.11 for conditions)

gypsum is not predicted to crystallize. On the other hand, if the mass ratio is allowed to vary, even at low initial molality, 0.1 m, gypsum crystallization is attained at high mass ratios, >5. Thus in Figs. 2.11 and 2.12 the phase boundary for gypsum is not attained until solution/cement ratios exceed ~37. Thus mass ratios, as well as initial sulfate concentration, control the appearance of gypsum and explain in part different threshold values for gypsum precipitation reported in the literature.

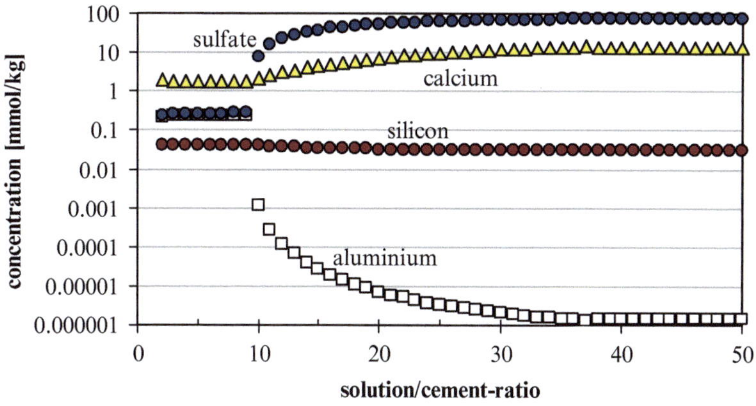

Fig. 2.13 Composition of the aqueous phase as a function of changing solution/cement-ratios (initial concentration 0.1 mol Na_2SO_4/kg) (Other data as for Figs. 2.11 and 2.12)

As a consequence of the changing solid phase distribution, Fig. 2.12, two accelerated regions of change in species solubility occur: one at low aqueous solution/cement ratios, about 10, at which point sulfate AFm is completely converted to AFt, ettringite, and another, at aqueous solution/cement ratio about 37, above which gypsum (and other products of attack) develop. Note that the amount of C-S-H remains essentially constant throughout the range of mass ratios, except for minor dilution effects, but that all other phases, portlandite included, must change in amount in response to (i) the redistribution of mass attending reaction and (ii) changing mass ratio, with higher mass ratios increasing the importance of dissolution. Since dissolution removes mass from the solids, it is potentially an important mechanism for relieving stress resulting from potentially expansive formation of additional solid. Calculation shows the interaction between phase reconstitution and solid volume: as the aqueous solution/solid ratio increases, the total volume of solids, shown in Fig. 2.12, at first increases, passing through a maximum at the point where sulfate AFm is completely converted to ettringite. Thereafter, as the aqueous solution/solid volume ratio continues to increase, the dissolution effect dominates with the result that the solid volume contracts very slightly although, as gypsum appears, the total solids volume again increases.

The literature contains many conflicting statements regarding gypsum: is it expansive or not? Bearing in mind that the diagrams we generate here do not predict expansion, only a potential for expansion, gypsum must without doubt be regarded as giving rise to a potential for expansion under the defined conditions. Note also that, in reaching this conclusion, it is first necessary to obtain a true mass balance, including a correction for the amount of dissolved substance, before predicting the specific volume of the remaining solids. Many estimates in the literature, by failing to correct for the aqueous solubility of solids, generally utilize insufficient data to sustain quantitative calculations of the potential volume changes: simple consideration of the molar volumes of reactants or their densities is inadequate.

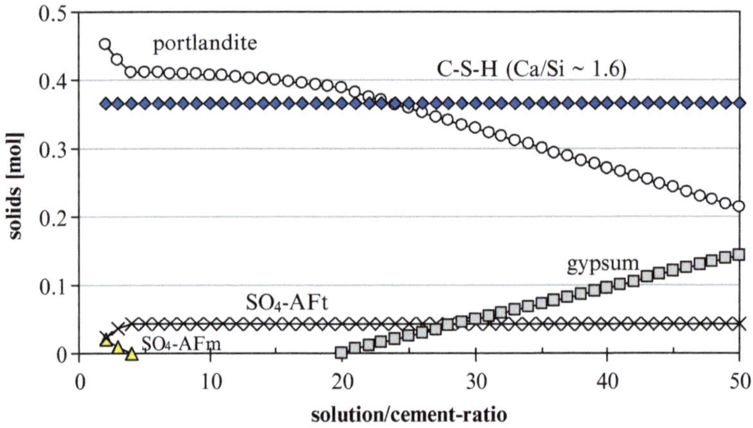

Fig. 2.14 Calculated changes on the solid mass balance as a function of changing solution/cement-ratio (initial concentration 0.25 mol Na_2SO_4/kg)

The changing aqueous phase compositions are shown in Fig. 2.13 as a function of mass ratio. The point at which AFm is reacted introduces an abrupt inflection in the aqueous phase composition: as long as AFm is present, aluminium and sulfate solubility remains relatively high, in the range 0.1–1·10 exp^{-3} molal. However once sufficient sulfate has been supplied to convert AFm totally to AFt, sulfate solubility increases by about two orders of magnitude while aluminium solubility decreases by two to four (or more) orders of magnitude. The solubility of calcium and silicon also change but less drastically. Nevertheless the comparatively high solubility of calcium throughout the higher range of mass ratios, *ca* 10·10 exp^{-3} molal, coupled with increasing solution volume, will increase the contribution of leaching to the mass balances as the mass ratio increases. In theory, another set of changes could occur as the threshold for gypsum precipitation is attained but results presented on a log scale are insensitive to the relatively small changes occurring in this region.

Data for a higher concentration of sodium sulfate, 0.25 molal, are shown in Figs. 2.14, 2.15, and 2.16 as functions of mass ratio up to ~50. These may be compared with Figs. 2.11, 2.12, and 2.13, respectively.

The differences arising from an increase in molality of the sodium sulfate solution result mainly from the greater mass of sodium sulfate available for reaction per gram of cement. Thus sulfate AFm is converted to ettringite, and gypsum first appears, at lower ratios of aqueous solution/solid with increasing initial sulfate molality, other factors remaining constant. Likewise, rapid changes in slope of the total solids volume are shifted to progressively lower mass ratios. Sulfate solubility is generally increased at the higher molality although the differences appear slight when presented on a log concentration scale. Calcium concentrations as a function of changing initial molality of sodium sulfate are discussed subsequently.

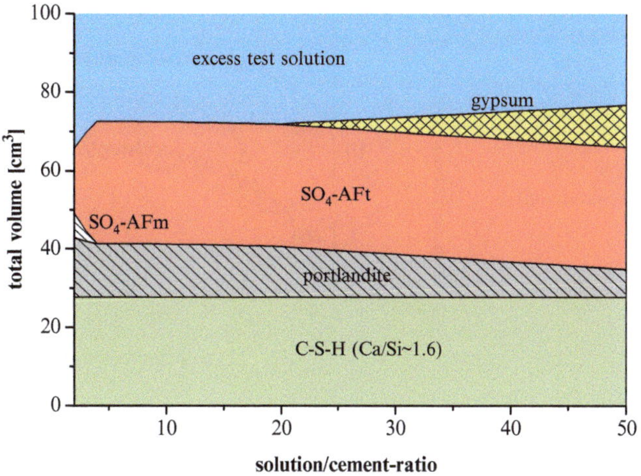

Fig. 2.15 Calculated solid volume changes as a function of changing solution/cement-ratios (initial concentration 0.25 mol Na_2SO_4/kg)

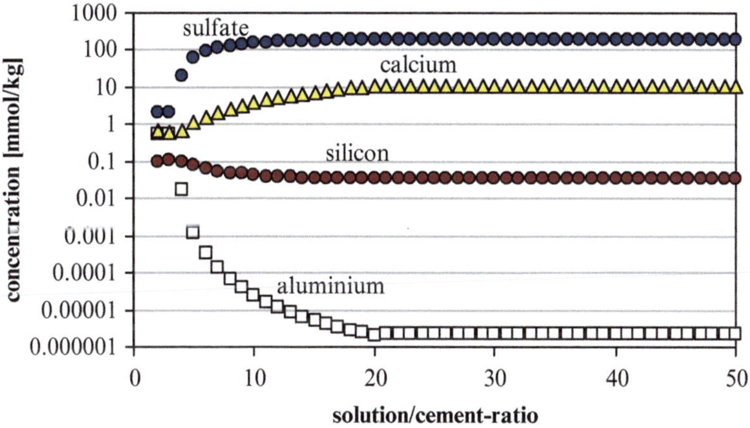

Fig. 2.16 Composition of the aqueous phase as a function of changing solution/cement-ratios (initial concentration 0.25 mol Na_2SO_4/kg)

The combined impacts of changing solution molality and aqueous solution/ cement mass ratio are shown in Fig. 2.17. Because of large differences in solubility, two diagrams are desirable to display the results, one for relatively soluble species, calcium and sulfate, Fig. 2.17a, and another for less soluble species, aluminium and silicon, Fig. 2.17b. Under these conditions, leaching is largely a function of calcium solubility and it is instructive to note how this varies as a function of aqueous solution/cement mass ratio. Figure 2.17a discloses that at low numerical mass ratios, calcium is more soluble in the 0.5 m solution but the opposite situation obtains at higher mass ratios, with a crossover occurring at about mass ratio 20.

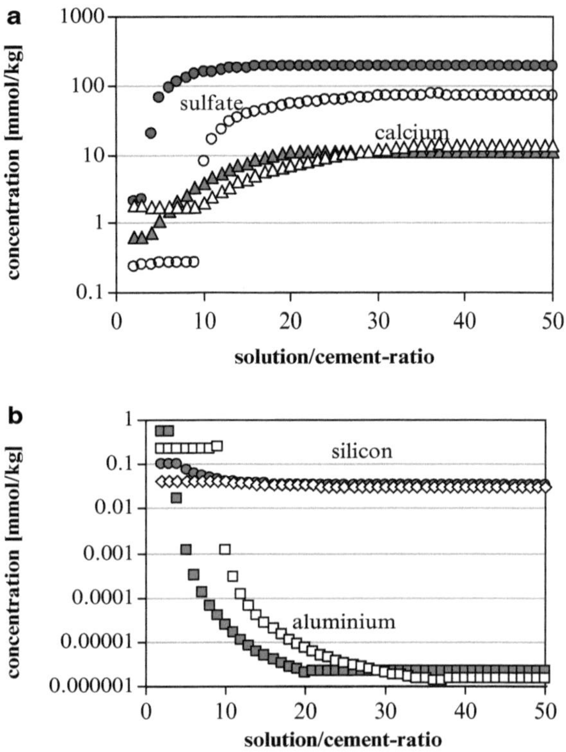

Fig. 2.17 Comparison of the aqueous phase compositions (**a**) sulfate and calcium (**b**) silicon and aluminium (0.1 m (*blank markers*) and 0.25 m (*filled markers*) Na_2SO_4/kg) (See text for conditions imposed on the calculations)

Another notable trend is the striking reversibility of the aqueous ratios of soluble Al/Si ratios, shown in Fig. 2.17b.

At low aqueous solution/solid ratios and lower sulfate molality, a regime exists in which aluminium is much more soluble than silicon. However this regime is gradually truncated by increasing the sodium sulfate molality, as well as by increasing the aqueous solution/solid ratio, so that over much of the range of compositions included in the scope of calculations, aluminium solubility is about four orders of magnitude less than that of silicon. While mobility is not a direct function of solubility, the two usually correlate in dilute solutions and it is noteworthy that aluminium – often regarded as having low mobility in alteration processes – can be quite mobile under the conditions outlined above.

The above calculations assume that certain numerical values remain constant. But one of the most striking features of tests made in dilute sodium sulfate solutions is the extent to which depletion of sulfate from the aqueous phase may occur. Of course depletion significantly affects the interpretation of test results because throughout much of the test duration, the actual aqueous solution composition will be significantly less than the initial composition. How much less will be a function

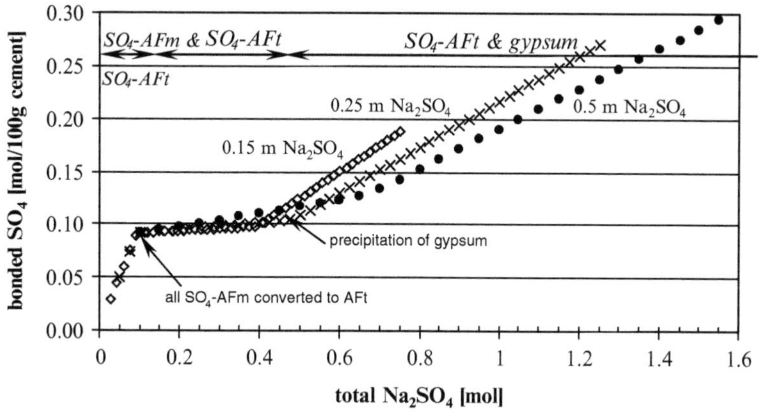

Fig. 2.18 Calculation of chemically bound sulfate and sulfate bearing phases of model cement in dependence of the total available amount of Na_2SO_4

of mass ratio and the extent to which reaction is completed. However, depending on the mass of added solution, the total available moles of Na_2SO_4 will also change as reaction progresses.

To evaluate the aqueous sulfate depletion effect, Fig. 2.18 shows the aqueous sulfate balance for three calculated cases; for solutions initially 0.1, 0.25 and 0.5 m Na_2SO_4/kg. To keep the calculations as generic as possible we also calculate the total number of moles of sulfate bonded in solids, neglecting any sorption by C-S-H (Fig. 2.18). Note that at low total number of moles of sodium sulfate, and irrespective of concentration in the range 0.10–0.50 m, the initial slopes are essentially coincident and very steep: this corresponds to the strong partition of sulfate into cement solids while mainly ettringite (AFt) is forming, largely at the expense of SO_4-AFm. Once AFt is maximized (*i.e.*, all SO_4-AFm has reacted to form AFt), addition of more sulfate no longer increases the amount of chemically bonded sulfate significantly until the solution becomes saturated with respect to gypsum. Once gypsum can precipitate the content of chemically-bonded sulfate again increases. Thus the rate of increase of bonded sulfate depends on the initial aqueous sodium sulfate concentration and mass.

An important conclusion from Fig. 2.18 is that the mineralogical changes due to the addition of sodium sulfate up to the point at which gypsum precipitates are largely dependent on the *total mass* of sodium sulfate, as distinct from its initial concentration, ~13 g of Na_2SO_4 per 100 g of cement being required to achieve the maximum content of ettringite. Thus in tests where a defined volume of aqueous phase is used, a high mass of 0.1 m sodium sulfate solution has roughly the same damage potential as a low mass of more concentrated (0.5 m) sodium sulfate solution; under these restrictions the total available mass of sulfate tends to limit the maximum degree of possible deterioration of the cement paste.

Based on the relations shown in Fig. 2.18, a mathematical expression can be derived to calculate the theoretical sulfate buffering capacity of carbonate-free

Portland cement. Monosulfoaluminate and related solid solutions will react with cement components and sulfate according to Eq. 2.13 to form ettringite:

$$\begin{aligned}&Ca_4Al_2(SO_4)_x(OH)_{14-2x} \cdot 6H_2O + (3 - x)Na_2SO_4 + (3 - x)Ca(OH)_2\\&\quad \rightarrow Ca_6Al_2(SO_4)_3(OH)_{12} \cdot 26H_2O + (6 - 2x)NaOH\end{aligned} \tag{2.13}$$

A correction has to be introduced for the amount of sulfate present in the cement. The theoretical maximum amount of ettringite formed in the course of sulfate attack is thus mainly dependent on the amount of reactive $Al_2O_3(n_{Al_2O_3})$, mainly as C_3A) and the SO_3/Al_2O_3-ratio of the cement. Thus with $x = n_{SO_3}/n_{Al_2O_3}$ (molar ratios) and from the known amount (moles) of reactive Al_2O_3 which corresponds to the theoretical maximum molar quantity of monosulfoaluminate formed in the hydrated cement, it is possible to calculate the maximum molar amount of sulfate (n_{SO4}) which can be bonded by the cement paste in the course of sulfate attack (Eq. 2.14):

$$n_{SO_4^{2-}(bound)} = 3n_{Al_2O_{3(cement)}} - n_{SO_{3(cement)}} \tag{2.14}$$

The linear relation between chemically bound and total added sulfate (Fig. 2.18) enables the mass of Na_2SO_4 required to be removed from the test solution sufficient to convert all monosulfoaluminate to ettringite to be calculated from the known weight fractions of reactive mAl_2O_3 and m_{SO3}, g per 100 g cement, according to Eq. 2.15:

$$m_{Na_2SO_4} = 4.18m_{Al_2O_{3(cement)}} - 1.77m_{SO_{3(cement)}} \tag{2.15}$$

The estimation is valid up to a mass ratio of ~300 and can be theoretically improved by multiplication of m_{Na2SO4} with a correction factor (k = 1/0.995 = 1.005, see Fig. 2.18) to correct for the amount of SO_4 dissolved in the test solution.

The following example should help to select test conditions and to demonstrate the relations between mass ratio and the amount of added total sodium sulfate: suppose it was desired to conduct tests in a solution containing initially 3.0 g Na_2SO_4/kg and, furthermore, that this concentration should not be allowed to decrease in the course of testing to <2.0 g Na_2SO_4/kg in order to avoid a strong increase of the pH of the test solution. We therefore have available for reaction 1.0 g Na_2SO_4/kg of aqueous phase. Suppose also that the test specimen initially contains 100 g cement which contains 4 wt.% reactive Al_2O_3 and 2.5 wt.% SO_3. Reference to Eq. 2.15 shows that to convert all AFm to AFt, we require approximately 12.3 g Na_2SO_4. This amount would be furnished by 36.9 g of total Na_2SO_4 and the mass ratio should therefore be not less than 123. Of course results will vary with different cements: the calculation is illustrative rather than definitive but illustrates that large volume of test solutions may be required to simulate performance in isochemical conditions, as might occur in conditions where the supply of sulfate was effectively infinite.

The above calculations assume complete reaction. However as reaction proceeds, local equilibria will develop in a cement monolith: the zonation as observed in field concretes is described elsewhere in this chapter. Thus formation of zones with high specific volumes, *e.g.* due to temporary formation of gypsum as a result of incomplete reaction, may increase the potential for expansion during early stages of the immersion of the test specimen (see subsequent text). Another problem of these estimations is the definition of reactive aluminium, as distinct from total Al. Most available studies confirm a strong dependence of the sulfate resistance upon cement C_3A contents, but with large scatter in the correlation. However the ferrite phase provides another potential source of alumina; and, although ferrite generally hydrates slower than C_3A, it will eventually contribute to ettringite formation in the presence of sufficient sulfate.

Marchand et al. (2002) (cited in Skalny et al. 2002) applied similar mathematical relations to compute the content of chemically bound sulfate to estimate the amounts of ettringite and gypsum formed in the course of sulfate attack based on the cement clinker composition. But in contrast to the title study, conversion of hydrogarnet to ettringite was assumed to occur with marked reduction of the potential for positive (expansive) volume changes. However the eventual products of the sulfate attack reactions used here accord with field observations. Moreover, more refined thermodynamic calculations show that hydrogarnet is destabilized by sulfate, so the situation observed in tests and in concretes removed from service correspond closely to those used for calculation.

Development of Zoned Microstructures

As is well known, the approach to equilibrium is often mediated by kinetics. At present we lack accurate coupled reaction-transport models with which to simulate diffusion profiles. However, to enable a simulation of zones involving different local equilibria developed in the course of sodium sulfate attack, we proceed as follows. Envisage a physical monolith of hydrated cement immersed in sodium sulfate solution and undergoing progressive attack at 25°C. The mineral zonation is simulated using the model cement paste, Table 2.5, and applying a gradient simulated by changing the mass ratio of aqueous solution/paste with depth, as shown schematically in Fig. 2.19. The attack profile thus deduced for 0.1 m sodium sulfate (~14.2 g/l) is shown in Fig. 2.20. The mass ratio is of course zero for fresh cement and initially projects at the left-hand side of the diagram: the progression of attack is simulated by sweeping a vertical construction line across the diagram, from left to right and noting its intersections and lengths of line segments on the diagram.

This approach allows us to apply a gradient of increasing amounts of total sulfate, with simultaneous correction for leaching which is assumed to be confined to the decalcification zone. The calculation well simulates the reactions of a cement paste immersed in sodium sulfate solution. Thus, at the beginning of the reaction the mass-ratio is very high ($\gg$1,000) as only a small thickness of cement paste interacts with the sulfate solution. But as we move into the solid, away from the

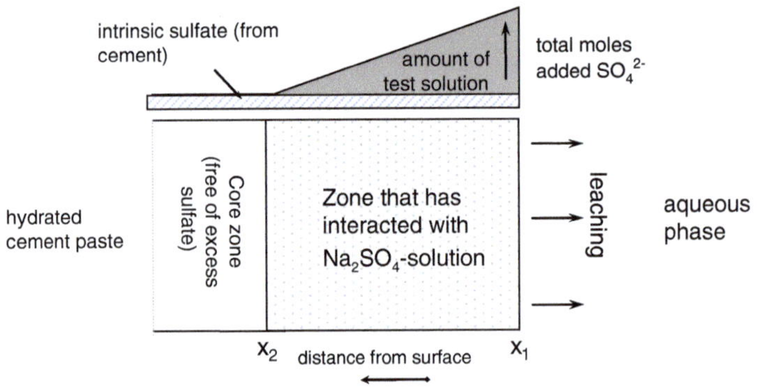

Fig. 2.19 Schematic showing imaginary cross-section of hydrated cement exposed to increasing amounts of sodium sulfate test solution

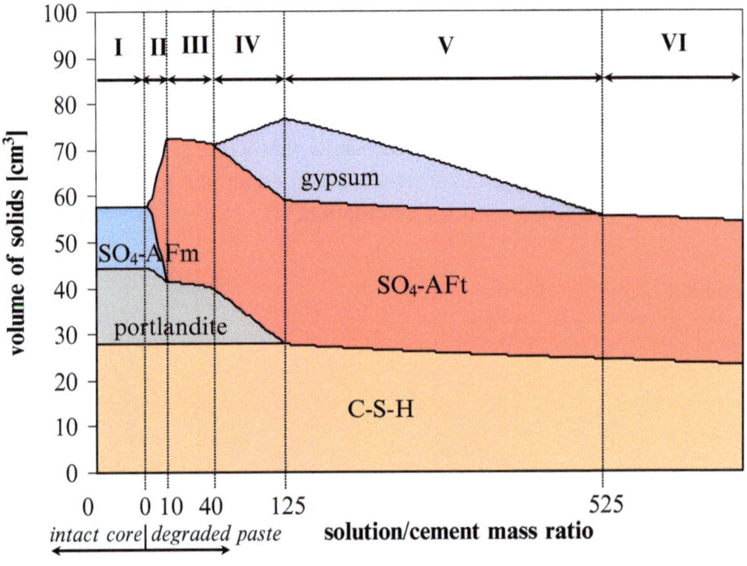

Fig. 2.20 Predicted mineral zonation in the course of sulfate attack in 0.1 m Na_2SO_4 at 25°C. The model shows specific volume of solids corrected for leaching, as a function of mass ratio. *I* – SO_4-AFm, CH, C-S-H, *II* – SO_4-AFm, SO_4-Aft, CH, C-S-H (Ca/Si ~ 1.6), *III* – SO_4-AFt, CH, C-S-H (Ca/Si ~ 1.6), *IV* – SO_4-AFt, CH, C-S-H (Ca/Si ~ 1.6), gypsum, *V* – SO_4-AFt, C-S-H (Ca/Si↓), gypsum, *VI* – SO_4-AFt, C-S-H (Ca/Si↓)

surface layer, the mass ratio decreases and local equilibria dictate the minerals and mineral amounts present. The mass ratio necessary to achieve selected milestones along the reaction path are shown at the bottom of Fig. 2.20. Thus, for example, a combination of reaction and dissolution result in the calculated disappearance of portlandite at mass ratio ~125 while the maximum potential for expansion occurs at the same ratio. At mass ratios <125, the Ca/Si ratio of C-S-H is ~1.6 (portlandite

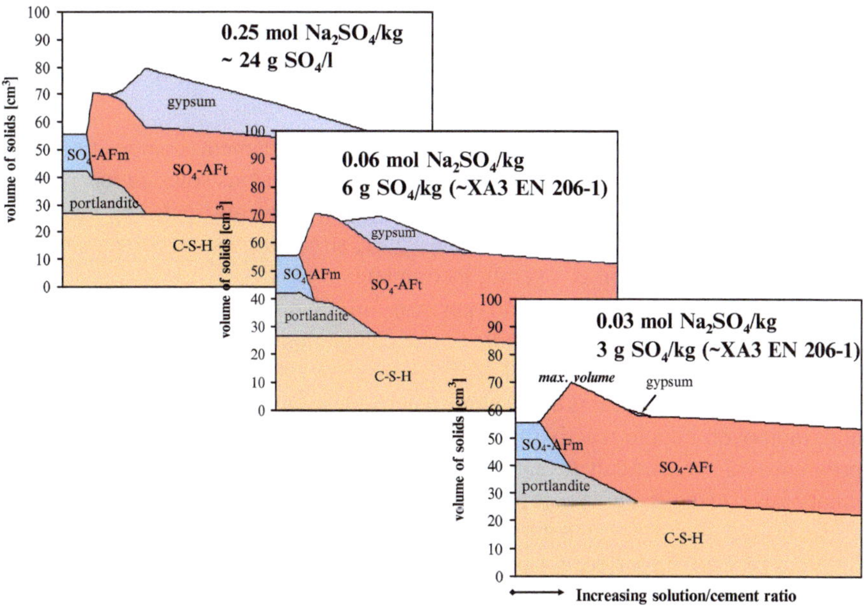

Fig. 2.21 Comparison of predicted mineralogical zonation in the course of sulfate attack as function of increasing initial Na_2SO_4 concentrations at 25°C. The model shows phase % and specific volume of solids, corrected for leaching, as a function of mass ratio

saturated) while at mass ratios >125, its Ca/Si ratio decreases (numerical values are not shown but are predicted in the course of the calculation). The predicted zonation is in good agreement with the experimental observations of Gollop and Taylor (1992), Planel et al. (2006) and Maltais et al. (2004).

Note that, due to combined leaching processes and sulfate ingress, the maximum specific solid volume is predicted to occur in zones IV and V, not in the outermost zone VI. Precipitation of gypsum leads to a significant specific volume increase. The prediction for higher concentrations agrees with experimental observations reported by Gollop and Taylor (1992) and Bellmann et al. (2006). But, as shown in Fig. 2.21, at low initial Na_2SO_4 concentrations (sulfate concentration similar to that proposed in EN 206-1 for aqueous solutions) the impact of gypsum on potential for expansion is significantly diminished. In partial support, Bellmann et al. (2006) reported that experimentally, no significant gypsum formation occurred at initial sulfate concentrations <3 g SO_4/l: we predict that gypsum will form but only in traces and so may escape detection. Thus at low initial Na_2SO_4 concentrations the maximum total solid volume is computed to occur in zone III corresponding to maximum ettringite formation but with no significant leaching of portlandite and C-S-H and no (or very little) gypsum. Calculation thus agrees with experiment and shows how the potential for expansion changes according to the concentration selected for test.

2.4.3.3 Additional Discussion

General Applicability of the Model Calculations

The calculations presented here relate to only one cement composition and, moreover, assume that reaction between cement and aqueous solution has gone to completion at least locally; Figs. 2.20 and 2.21 have been constructed on this basis. This latter case, depicted in Figs. 2.20 and 2.21, simulates test situations and field exposures where reaction is, overall, incomplete.

Historically, a range of cement types and formulations have been used in tests and in construction. To apply results of the calculations presented here, it will be necessary either to repeat the calculations or to develop extrapolations. We are aware of computer programs that include reactive transport in one or two dimensions and, moreover, couple reactive transport with equilibrium calculations. But these programs can be difficult to handle with confidence: assumptions about kinetic inputs have to be made. So for the present we restrict discussion to evaluating those factors that emerge naturally, from comparison between the calculations reported here and experience, from testing and field observations.

Conclusions Arising from the Calculations

It can be seen from the data presented that the most sensitive variables in respect of the mineralogical composition are the initial alumina and sulfate contents of the cement: the calculations are relatively insensitive to the amounts of calcium and silicon oxides, including C-S-H, in the composition ranges usually encountered in Portland cement. For this reason the potential for volume change is also much less than in commercial cements than might be inferred from the calculations presented: "real" pastes are perhaps 60% or more C-S-H and this dilutes the consequences of volume changes amongst the minor hydrate phases. Nevertheless we agree that increasing cement alumina content magnifies the importance of specific volume changes upon reaction with aqueous sodium sulfate solution; in general, low chemical alumina contents will lessen the potential for expansion and conversely, high chemical alumina contents will increase that potential. This accords well with conventional wisdom, although it should be noted that (i) we perform calculations on the basis of total alumina whereas in commercial cements, alumina appears in both tricalcium aluminate and ferrite, yet sulfate resistance is often correlated only with tricalcium aluminate content and (ii) we assume a fixed but relatively low sulfate content in cement, additional sulfate requiring to be supplied from a percolating aqueous phase. However, the initial sulfate content of the cement is an important but somewhat neglected variable as it has an important influence on the volume of cement solids achieved in isochemical hydration (Matschei et al. 2007a). Thus the amount of sulfate in the clinker is also an important factor. We return to this as well as the alumina availability in clinker in subsequent discussion.

Attention is directed to several trends occurring in response to changing aqueous solution mass and volume. We previously described the role of gypsum in controlling the potential for expansion and how the amount of gypsum is affected by the sodium sulfate concentration. Other test parameters also require to be defined if laboratory tests are to be reproducible. For example, using a fixed mass of sodium sulfate but increasing its initial molality in the range 0.1–0.5 m, sulfate is removed from the aqueous phase and increasingly transferred to solids, with the result that the mean ratio of SO_3/Al_2O_3 in cement solids is enhanced. Calculations have been done for mass ratio equal to 5, but inspection of the underlying physical chemistry discloses that similar trends will be encountered across all concentrations encountered in standard tests known to us. There are also changes in the mass of aqueous phase occurring in the course of reaction, arising from the differing water content of solids, but these changes are relatively minor over the range of calculations and as a good first approximation, changes to the mass of the aqueous phase at high mass ratios can be neglected. However (i) the high water content required for ettringite formation might, in some circumstances, create a local shortage of water and thus inhibit ettringite formation and (ii) the mass of sulfate transferred to the solids may in some circumstances be limited by too low a mass ratio. A simple way of explaining the consequences of this mass ratio effect is to note that, as the sulfate concentration of the aqueous phase decreases as a consequence of its reaction with cement solids, the aqueous phase becomes progressively less able to transfer sulfate to the solid. Thus test conditions in which (i) the mass of aqueous solution is small and (ii) the aqueous concentration is initially low are predicted to encounter the greatest depletion of sulfate and have the greatest tendency towards premature cessation of reaction. If tests are to be reproducible from one laboratory to another, it is therefore essential that the ability of the aqueous phase to deliver sulfate should remain constant, or nearly so. The easiest way of ensuring this condition is achieved, that the concentration of the aqueous phase should remain substantially unchanged throughout the test duration, is to specify a high mass ratio.

The calculations were also broadened to include two additional scenarios: a fixed initial mass ratio of 5, and a variable mass ratio, ranging between approximately 2 and 50, both calculation sets being repeated for different initial sodium sulfate concentrations. The initial mass ratio does not of course remain constant throughout the test partly because water is consumed/liberated in the course of reaction (although, as noted, this is generally a relatively minor factor) but mainly because the cement binds sulfate (a mass gain) as well as experiencing mass loss through dissolution. Note, however, that as sulfate redistributes between solid and aqueous phases, we cannot necessarily ignore changes in its concentration in the aqueous phase. This is especially so at (i) low mass ratios and (ii) low initial sulfate concentrations, with a worst case scenario in the course of a test being achieved by application of both sets of conditions. However at mass ratios above approximately 10, and in the range of sulfate concentrations used here, changes in aqueous sulfate concentration occurring in the course of reaction are predicted to remain relatively small. Certainly one conclusion arising from calculations done here is that realistic tests of sulfate attack conducted in the laboratory should always use a mass ratio greater than 10. In this way, the concentration of sulfate will remain reasonably close to the initial value, even when reaction with cement solids approaches completion. Note that this restriction also simulates the constancy of the sulfate concentration in certain natural conditions, *e.g.* sea water, which comprises an essentially infinite reservoir of sulfate at constant composition. Thus the mass ratio affects the course of reaction and, if reproducibility is important, the mass ratio requires to be specified preferably to be >10. An example of how to fix the mass ratio was given in Sects. 2.4.3.2 and 2.4.3.3 (see sub-section "Conclusions Arising from the Calculations").

Limitations of the Calculations

By assuming that the reaction of clinker components goes to completion, the calculations performed implicitly include all the alumina, irrespective of its form in clinker. Alumina occurs in commercial cements mainly in two phases: tricalcium aluminate and ferrite, the latter being a solid solution of variable composition between C_2F and a maximum of ~70% C_2A. In practice, while C_3A is reactive in early hydration, ferrite reactivity is variable and dependent on clinker microstructure, Al/Fe ratio, and minor element content, especially of Mg and Ti. But in general the reactive fraction; that is, the fraction of total alumina actually available for incorporation into the hydrated paste minerals, will be somewhat less than the total chemical alumina. The extent of the reduction of alumina available for formation of hydrates will become better known for several representative commercial cements in the course of Nanocem projects, currently in progress at EPFL Lausanne. We await results but in the meantime, encourage application of these data only to hydration at longer ages, typically greater than 28 days, by which time the necessary factor to correct for un-reactive alumina is likely to reduce and approach more closely the actual bulk alumina content. Ongoing projects in the Nanocem network will, as noted, enable definition of realistic alumina availability factors, numerical values of which can, if desired, be used to modify the alumina content disclosed by analysis in subsequent calculations of the type undertaken here.

The calculations do not permit actual physical expansion to be determined, only a change in specific volume, which we term a *potential* for expansion. The magnitude of the potential for expansion is not necessarily a linear function of the progress of attack and is certainly not a linear function of the sodium sulfate concentrations: mass ratio is another and very important characterization parameter. The literature has, however, concluded that potential for expansion does tend to be associated with the onset and development of certain phase changes, with notable potential for expansion being associated with formation of AFt and gypsum. This conclusion accords with established wisdom but note qualifications given in the text, particularly in respect of the role of gypsum. Expansion is not uniquely associated with the *presence* of a particular phase or phases: rather it is associated with the *amounts* of the phases and the *specific volume change* attending their formation reaction as well as the *mechanism* whereby recrystallization occurs. For example, many cement pastes undergoing expansive sulfate attack contain empty or partially empty pores: ettringite precipitates in the matrix yet expansion occurs even though some pores remain unfilled. But the potential for expansion will always be greatest when leaching is minimized. Thus when AFm is converted to AFt (the balanced reaction also involves consumption of portlandite and conversion of liquid water to structural water), the process can be made thermodynamically favourable by relatively small incremental changes in aqueous sodium sulfate *concentrations* but much mass transport is nevertheless required to *complete* the transformation. This kinetic restriction, on the need for mass transport, is likely to be of considerable importance in creating local potentials for expansion, especially in the low concentration range, up to about 0.15 m where solubilities of components common to both AFm and AFt are roughly (within an order-of-magnitude) equal. Note that mass fluxes do not depend on sulfate mobility alone: other ions are required to diffuse in order to satisfy mass balance and electroneutrality and we do not at present know which transport processes are rate-limiting.

In this context we also remark on the role of dissolution. Cement components are somewhat soluble in pore fluid as well as in the much greater volume of aqueous sodium sulfate solution added in the course of testing. Other factors being equal, a low mass ratio will result in less dissolution than a high mass ratio. Since the mineralogical changes occurring in the course of sodium sulfate attack generally lead to an increase in specific volume of the cement solids, dissolution may be regarded as a "safety valve" for relief, or partial relief, of expansive stresses that occur in response to mineralogical changes including gain in mass of solids. In the range of mass ratios up to 10, approximately, the specific volume of solids typically undergoes an initial rapid increase at low sulfate concentrations, arising from the sharp increase of ettringite content attending addition of sulfate. This potentially expansive effect is observed at all concentrations for which calculations were made. Note that the volume change in calculations presented here automatically corrects the specific volume of solids for dissolution, the extent of which is unique for each calculation set. Simple considerations of reaction stoichiometry are often used in the literature to determine expansive potential, but these will not give identical potentials for expansion to those calculated here unless they are corrected for dissolution. The magnitude of the correction will increase as the mass ratio increases and will not be constant even for a particular sodium sulfate concentration. Thus the methodology used in the title paper, by correcting for dissolution, must (i) inevitably be more realistic and (ii) explains poor reproducibility of data reported in the literature: not all relevant variables have been controlled.

Linkages Between Laboratory and Field Studies

Field conditions are not always sufficiently well defined to construct simulations. For example, leaching is in part a function of coupling between the cement and its service environment. Thus leaching is maximized in water or in strata which are permeable and/or permit rapid advective flow. Some regimes, as in fractured granites, may concentrate attack at certain points or zones where flow channels impinge on the cement. Others, such as impermeable shale, may have such low diffusion rates that leaching may be limited to that necessary to achieve local saturation. In the following treatment we will deal with situations where leaching is not significantly limited by the service environment as these environments, permitting ready ionic movement to or from the cement, are also of concern. One of the classical correlations between cement mineralogy and resistance to sulfate attack, as measured by immersion in sodium sulfate solutions and as inferred from field studies, concerns the relationship between clinker C_3A content and sulfate resistance. Tikalsky et al. (2002) have recently reviewed correlations between sulfate-induced expansivity at 6 months and clinker mineralogy: a positive correlation was obtained with increasing C_3A content and a negative correlation with ferrite concentration. However the justification for including the cement C_3S and MgO contents in the calculation, which somewhat improved the fit between calculated and observed data, lacks scientific justification. The generally large scatter of data obtained in this and other studies, with many outlier points, makes it inadvisable to rely on this approach dependably to engineer clinker mineralogy for optimum sulfate resistance. In the following discussion we highlight some reasons for the poor correlations. Additionally, we highlight the role of carbonate in calculation of the potential for expansion.

In standard tests of sulfate attack based on immersion, sulfate can only be added to cement solids from an external source. A concentration gradient develops where the cement matrix is of poor quality, perhaps because of formulation to a high w/c ratio, with the result that the inward diffusion profile is essentially unimpeded by solid and may closely match a Fick's law profile. But in well-made concrete, a sharply zoned structure develops. The concentration of sulfate cannot be described by smooth Fickian profiles. Zoning arises for several reasons; (i) in inhomogeneous solids, the requirement of obeying the phase rule has to be superimposed on diffusion profiles and (ii) the complex changes in solubility conditioned by the local composition of the aqueous phase. Inspection of the solubility data presented here helps explain this "chemical" effect: many of the diffusing species have solubility minima and as a result, tend to become "trapped" at or near points of minimum solubility. The resulting mass gains and losses not only are reflected in mineralogical changes but lead to zones, some of which tend to densify while others experience net loss of material without compensatory formation of space-filling phases and, as a result, become porous. Zones of ettringite precipitation are particularly easy to densify because ettringite formation converts much liquid water to solid. It is difficult to compress either water or ettringite but, while liquid water can move under pressure gradients, it is difficult to suppress ettringite crystallization in stressed regions. Thus space filling considerations, as used here, identify zones where stress is liable to accumulate and explain why, although porous zones may also be forming, overfilling and cracking can still occur in close spatial proximity to porous regions. Thus, for reactions occurring along a sulfate concentration gradient, it is predicted that a maximum potential for expansive stresses will be generated within subsurface layers or zones of the cement matrix specifically (i) when the first derivative of the specific volume-composition function is a maximum and (ii) when species solubility are low. In practical terms, these conditions are likely to occur in the course of sulfate attack at the point or zone at which sulfate concentrations attain or exceed the threshold necessary to convert AFm to AFt.

In typical test conditions, the potential for mass dissolution will also vary locally, as a function of depth, but will always be low at low numerical values of the mass ratio, such that the aqueous phase will become saturated in cement components without much mass of cement having to dissolve. Thus in most practical tests made at low mass ratio, leaching is largely confined to near-surface layers with the result that our model, allowing as it does a "safety valve" to relieve stress by dissolution, tends not to operate to any significant extent except in the near-surface layers. A third factor also needs to be introduced, concerning the rapid changes in aqueous phase composition encountered in certain ranges of sodium sulfate molality and its impact on kinetics. Both dissolution and precipitation of a given phase will be facilitated if (i) the aqueous solubility of the relevant phases is relatively high, *e.g.* in the millimolar range, and (ii) the species solubilities of the forming or dissolving phases are approximately in the congruent ratio. Inspection of the species solubility diagrams discloses that for the model cement, the conditions are met at total sodium sulfate concentrations close to 0.09 m sulfate (corresponding to about 12.8 g Na_2SO_4 per l). While this value is not necessarily the same for all cements, scoping studies suggest that its value is similar and that concentrations in the range 10–15 g/l have generic significance.

The *mass* of sulfate required to complete the process of converting AFm to AFt can be estimated from Eqs. 2.14 and 2.15. Thus a narrow range of compositions close to this calculated sulfate concentration is predicted to be especially favourable for recrystallization and transport of species necessary to alter the balance between AFm and AFt, while at the same time, reducing the impact of kinetic barriers resulting from transport processes. These arguments favour test specifications requiring relatively moderate concentrations of sodium sulfate particularly as at higher Na_2SO_4 concentrations, $\gtrsim$13 g/l, solubilities change drastically and, moreover, gypsum formation is artificially enhanced with respect to all but a few natural environments. Note that the importance of maintaining high mass ratios, to avoid depletion of the solution in sulfate, is increased when using dilute sodium sulfate solutions.

Renewal of the test solution is sometimes used to avoid problems arising from the change in composition of the aqueous phase. With renewal, the composition of the solution fluctuates with only the concentration of the aqueous phase at the commencement of a cycle being known. Thus an additional element on non-reproducibility is built into such tests and by attempting to avoid one problem, renewal introduces others. The microstructure of cements undergoing attack also reveal reduction of portlandite content extending from the surface in contact with sodium sulfate to significant depth, and this depletion is often interpreted as evidence for leaching as was concluded from many of the interpretations described earlier.

As shown Fig. 2.20, leaching certainly occurs and is an important mechanism for release of expansive stress. But calculations also show that portlandite is consumed in the course of ettringite formation from cement precursors. We suggest that the facile explanation often given, that the reduction or disappearance of portlandite in selected zones at depth arises from leaching, is only partially correct. However modern Portland cement contains sufficient portlandite that it is unlikely *totally* to be depleted in the course of AFt formation unless significant leaching also occurs.

Analysis of dynamic situations is much assisted by the development of summary diagrams such as Figs. 2.20 and 2.21. However the links between chemical models and diffusion/reaction kinetics; can also be developed by application of models developed by Marchand et al. (2002). Their equilibrium calculations, although less developed in terms of the thermodynamic database than those presented here, are broadly in agreement. However, by also incorporating porosity-permeability relationships, they are better able to develop and evaluate the influence of w/c ratio of the concrete formulation in controlling sulfate penetration.

For the interpretation of field tests, the procedures described here can be reversed: if the profile of the mineralogical assemblage can be established, perhaps through microscopic/petrographic studies, and if the approximate local pore fluid composition can be determined by calculation. In general the sequence of mineralogical zoning can be predicted by choosing the diagram that most nearly applies to the conditions imposed by the test specification and, as described above, sweeping a vertical line across the appropriate diagram from the point of maximum sulfate concentration (most likely to be achieved at the near-surface layers of the cement) to the left. The left-hand side of the relevant diagram thus defines the chemistry and mineralogy of the interior portions, those as yet unaffected by sulfate uptake. The prediction will not be exact for a theoretical reason – we have not developed diagrams for all conditions – as well as for practical reasons: (i) sulfate is more strongly removed than sodium, so that the ratio of sodium to sulfate in the pore fluid, initially fixed at 2.0, tends to increase with depth: this inequality has potential to affect phase balances (ii) the pH varies as a consequence of the amount of hydroxide needed to maintain charge balance for sodium and, as a consequence, phase solubility and stability are somewhat affected and may deviate from the model predictions and (iii), some sulfate may be sorbed in C-S-H. However studies in progress at Aberdeen (Skapa 2009) show that the relevant correction for sorbed sulfate is small. So, ignoring these complications, we argue as follows. As sulfate penetrates, its binding affects the paste mineralogy. Because each phase change requires a certain minimum sulfate concentration to achieve, as well as a continuing supply of sulfate to complete, the operation of the phase rule, superimposed on diffusive processes, give rise to a zoned mineralogical structure. With time and continued diffusion, the zones will move inward. The process of migration is driven by thermodynamic forces so it is inevitably progressive until all parts of the cement have attained equilibrium. But the aluminate portion of the paste has a large capacity to bind sulfate with the result that a theoretical expectation, of Fickian concentration gradients of sulfate, may become distorted. The smooth Fickian concentration profiles anticipated from simple diffusion theory are instead converted into stepwise concentration profiles, each step marking the onset or completion of a specific mineralogical conversion involving loss or gain (usually the latter) of sulfate. In other words, phase rule considerations have to be superimposed onto diffusion theory.

A Portland cement will thus exhibit a series of buffering reactions for sulfate, each successive reaction being initiated in response to the permeating aqueous phase attaining a certain critical sulfate concentration. The buffering reactions are conditioned by successive mineralogical transformations, each requiring a change in the nature of one or more of the sulfate-containing solids. The steps automatically become distributed throughout the cement as a function of time and depth, each step marking the attainment of a critical threshold sulfate concentration. Typically, these reaction features will be observed microscopically (or perhaps even macroscopically) extending with time to greater depths and spatially orientated roughly parallel with the exposed surface in a sequence corresponding to decreasing sulfate pore fluid contents. The faithful reproduction of the predicted sequence of zoning in field specimens provides indirect but powerful evidence for the attainment of a local equilibrium at each point along the diffusion profile.

The calculated diagrams also shed much qualitative understanding on reaction kinetics. For example, if we accept that some proportionality exists between the kinetics of formation of a phase and the concentration in excess of saturation of its constituent ions in solution, we can venture predictions. Take as an example conditions favouring ettringite formation: reference to Fig. 2.10 shows that under the conditions depicted, the order-of-magnitude solubilities of calcium, aluminium and sulfate are not only relatively high but also nearly equal over a range of low concentrations, extending up to about 0.09 mol total sodium sulfate (~13 g Na_2SO_4 per 100 g cement). Since ettringite is also stable over much of this range, it is predicted that ettringite will form readily; all the kinetic indicators are favourable. If, however, we were to increase the molality of the sodium sulfate, for example, to 0.4 m, the solubility of alumina will have declined by two to three orders of magnitude relative to that of sulfate, making it difficult to mobilize the Al necessary to form ettringite; the supply of aluminium species may therefore become rate-limiting. Thus if it is desired to accelerate reaction, kinetic factors favour the use of low sulfate concentrations, calculated according to sulfate buffer capacity Eq. 2.15, thereby giving a more realistic representation of internal ion transport occurring during the recrystallization which accompanies sulfate attack.

Finally, we remark briefly on the reaction of mortars and concretes with sodium sulfate. If siliceous aggregates are used, a potential for alkali-aggregate reaction (AAR) also exists. We note that sodium reacts less strongly with cement substances than sulfate. The preferential removal of sulfate requires that another anion be supplied to balance the positive change on sodium: in this instance, the ion is mainly OH, hydroxide. This additional hydroxide accounts for the calculated rise in pH occurring in the course of reaction of cement with sodium sulfate. This pH elevation is unique to sodium (or potassium) sulfate; magnesium and calcium sulfate do not exhibit this elevation of pH. The practical consequences of this are that, if the aggregate is potentially susceptible to AAR, the sodium sulfate test may well exacerbate the sensitivity and introduce an artifact, namely an additional contribution arising from the reactivity of the aggregate.

Additional data have to be introduced to determine the role of dissolved carbon dioxide. However indications are that the sodium sulfate test is strongly influenced by access of air to the test system. If access cannot be controlled, it is better to exclude air and carbon dioxide, as the ingress of carbonate may significantly alter the phase assemblages of the hydrated cement paste (Matschei et al. 2007b).

2.4.3.4 Summary

Colleagues have commented on the rather academic approach we have taken to modeling the role of sodium sulfate. But an important conclusion of the title study is that measuring the resistance of cement to sulfate attack by immersion in sodium sulfate solutions is itself academic: sodium sulfate solutions are rare in nature. The sodium component is not, as is often asserted, inert but contributes to a sharp rise in pH which affects test results significantly. We conclude that tests of the resistance of cement to sodium sulfate attack do exactly what they claim: namely, to measure the resistance of the matrix to sodium sulfate. But the nature and course of attack differs from that encountered in most natural conditions. It is therefore pointless to develop an accelerated test based on sodium sulfate and expect it to apply to other chemical conditions. However, as presently specified, tests do not necessarily give either good repeatability or reliable indication of how different cements will perform under realistic service conditions such that attack occurs by mixed sulfates. These failings are due to a combination of circumstances: failure to specify completely the test conditions, including access of carbon dioxide and control of mass ratios, and the simplicity of the tests relative to realistic conditions, where natural waters and soils range widely in sulfate concentrations, as well as the ability of the surrounding medium to transport sulfate and the complex cation and anion populations encountered in many realistic exposures.

The present results do however show how limited improvement in reproducibility can be obtained by appropriate design of the sodium sulfate test. Even dilute CO_2, as occurs in air, should be excluded from the test cell as air affects the mechanisms and interacts strongly with both the aqueous sodium sulfate and cement solids.

In CO_2-free conditions much evidence suggests that reaction mechanisms undergo change at solution concentrations above about 13–14 millimolar and, since dilute solutions are often encountered in nature, simulations should not use high concentrations of sodium sulfate. In particular, high concentrations exaggerate the role of gypsum. However an intractable problem remains: empirical test results cannot dependably be extrapolated or even interpolated to other conditions (temperature, sodium sulfate concentrations, mass ratios, etc.). Calculation can, however, be used for this purpose. On the other hand, a basic difficulty with calculations of the type undertaken here is that they do not necessarily reflect the kinetics of the situation. But this difficulty is not insuperable: numerous codes exists which couple transport to reaction, see for example Marchand et al. (2002). The problem in applying these codes is twofold: of understanding and evaluating the physical and kinetic factors and of data acquisition, necessary to implement reactive transport models. Both these difficulties are being slowly overcome and we predict the way ahead is to combine calculation with focused experimental work. A trial calculation, comparing the predicted mineralogical zoning of cement attacked by 0.15 m Na_2SO_4, shows excellent correlation with experimentally-determined profiles (Planel et al. 2006; Maltais et al. 2004). Moreover, calculation enables the change in specific volume of solids, and hence potential for expansion, to be predicted, including correction for dissolution which in some conditions, provides a stress relief mechanism.

Subsequent studies must extend these calculations to explore the role of carbonate. Carbonate and sulfate interact strongly and, until these interactions have been quantified and the evidence presented, it is not possible to depict many realistic scenarios for "sodium sulfate" attack. For this reason we recommend that carbon dioxide be excluded from tests.

However a further concern about reliance on calculation has been identified to us: to what extent can the potential for expansion be correlated with real physical expansion? At present we do not know the relationship, although it is likely that relationships can be developed. Future studies must therefore have as an objective the linking of *physical* parameters, most importantly dimensional change, with *chemical/mineralogical* changes.

2.4.4 Magnesium Sulfate Solutions ($MgSO_4$)

In magnesium sulfate solutions, both cation and anion react strongly with calcium hydroxide from the paste; magnesium hydroxide crystallizes as brucite while sulfate ions react with the portlandite to form secondary gypsum, which may react with the calcium aluminates to form ettringite.

The reactions involved are as follows:

$$\underset{\textit{Soild}}{Ca(OH)_2} + \underset{\textit{Solution}}{Mg^{2+}} \rightarrow \underset{\textit{Solid-Brucite}}{Mg(OH)_2} + \underset{\textit{Solution}}{Ca^{2+}} \tag{2.16}$$

$$\underset{\textit{Soild}}{Ca(OH)_2} + \underset{\textit{Solution}}{SO_4{}^{2-}} \rightarrow \underset{\substack{\textit{Solid} \\ \textbf{\textit{Increase in volume}}}}{CaSO_4 \cdot 2H_2O} + \underset{\textit{Solution}}{2\ OH^-} \tag{2.17}$$

$$\begin{aligned} &\underset{\textit{Calcium aluminate}}{4CaO \cdot Al_2O_3 \cdot 13H_2O} + \underset{\textit{Gypsum}}{3(CaSO_4 2H_2O) + 13H_2O} \\ &\rightarrow \underset{\substack{\textit{Ettringite} \\ \textbf{\textit{Expansion}}}}{3CaO \cdot Al_2O_3 \cdot 3CaSO_4 \cdot 31 - 32H_2O} + \underset{\textit{Portlandite}}{Ca(OH)_2} \end{aligned} \tag{2.18}$$

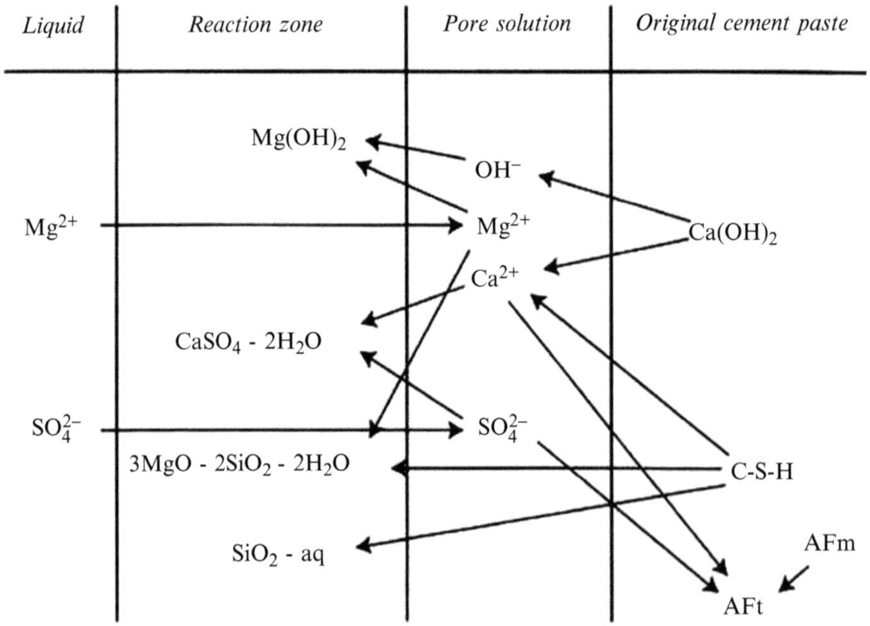

Fig. 2.22 Possible reactions with magnesium sulfate solutions, according to Skalny et al. (2002)

Magnesium ions do not directly replace calcium in C-S-H but form hydrated magnesium silicate. The Mg-silicates apparently do not have binding properties and lead to disintegration of the bulk paste. The rate of reaction by replacement of calcium by magnesium is slow for C-S-H with respect to the kinetics of reaction with portlandite. Thus two processes usually occur simultaneously: a gradual decomposition of the C-S-H gel, forming amorphous hydrous silicate [SiO_2 aq.] or magnesium silicate hydrate phase, the approximate composition of which is [$3MgO{\cdot}2SiO_2{\cdot}2H_2O$], while the crystallization of gypsum, brucite and hydrous magnesium silicates still occurs.

$$\begin{aligned} &Mg^{2+} + SO_4{}^{2-} + CaO \cdot SiO_{2aq} + 3H_2O \\ &\quad \rightarrow CaSO_4 \cdot 2H_2O + Mg(OH)_2 + 13H_2O + SiO_{2aq} \end{aligned} \tag{2.19}$$

$$\begin{aligned} &4Mg^{2+} + 2SO_4{}^{2-} + 2[CaO \cdot SiO_2aq] + 8H_2O \\ &\quad \rightarrow 3MgO \cdot 2SiO_2 \cdot 2H_2O + 2[CaSO_4 \cdot 2H_2O] + Mg(OH)_2 \end{aligned} \tag{2.20}$$

Skalny et al. (2002) analyzed the possible reactions between magnesium ions and the hydrated cement compounds, as shown in Fig. 2.22. They explained the immediate conversion of the calcium hydroxide into magnesium hydroxide and later the degradation of the C-S-H phase in terms of the low solubility of the compounds formed and the pH equilibrium between the solid and aqueous phases.

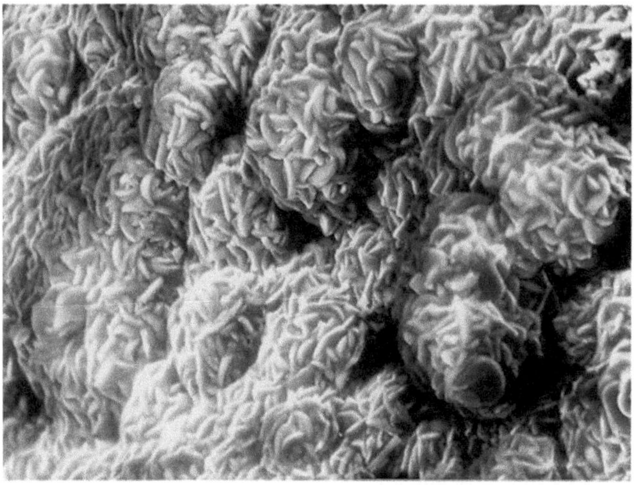

Fig. 2.23 Crystallizations of brucite on the surface of concrete (Menéndez 2010)

The main products are insoluble brucite and later, magnesium silicate. Also, calcium aluminate sulfate phases decompose in regions with low pH.

There are different degrees of alteration in different areas of the concrete with various CaO/SiO_2 ratios in the C-S-H gel. In concrete, the ratio CaO/SiO_2 varies from normal in the interior to near zero with the virtual disappearance of calcium in the most affected zones.

In concretes exposed to magnesium sulfate, brucite typically crystallizes on the exposed surfaces as small, compact crystals that clog pores. This produces a coating effect that can partially protect the solids from further external attack. A typical microstructure of brucite crystallization is shown in Fig. 2.23. In the interior, crystallization of gypsum and, to a lesser extent, ettringite, occurs.

2.4.5 *Ammonium Sulfate Solutions ($(NH_4)_2SO_4$)*

Attack by ammonium sulfate solutions is a combination of acid attack, with formation of gaseous ammonia and, at the same time, reaction with C-S-H gel. This initially extracts Ca with gypsum formation and leads to a decrease in the gel CaO/SiO_2 ratio and eventually its conversion to amorphous hydrous silica. The reactions are as follows:

$$Ca(OH)_2 + (NH_4)_2SO_4 \rightarrow CaSO_4 \cdot 2H_2O + 2NH_3 \quad (2.21)$$

$$CaO \cdot SiO_{2aq} + (NH_4)_2SO_4 + H_2O \rightarrow CaSO_4 \cdot 2H_2O + 2NH_3 + SiO_{2aq} \quad (2.22)$$

2.4.6 Sulfuric Acid (H_2SO_4)

Cement paste is strongly altered by sulfuric acid. Sulfuric acid can arise in a variety of ways: from industrial waste waters, within sewage systems, and also from ground water. This acid may be formed due to the oxidation of sulfides, especially pyrite present in soils or concrete aggregates. The reaction or oxidation of pyrites is as follows:

$$2FeS_2 + 7O_2 + 2H_2O \rightarrow 2FeSO_4 + 2H_2SO_4 \quad (2.23)$$

or

$$4FeS_2 + 15O_2 + 2H_2O \rightarrow 2Fe_2(SO_4)_3 + 2H_2SO_4 \quad (2.24)$$

Sewage is also a source of sulfuric acid: bacterial attack on sulfates in the sewage leads to formation of sulfide, typically hydrogen sulfide, which is further oxidized to sulfuric acid on the sewer walls.

Ferrous sulfate and sulfuric acid react with the alkaline constituents of the concrete pore solution to form alkaline and calcium sulfates, and ferric hydroxide. This process causes expansion, as a result of the higher volume of oxidation products relative to the original solids; the volume increase is about 15%. Subsequently a secondary expansion occurs due to the ettringite formation in the paste-aggregate interface in the same way as in conventional sulfate attack.

$$FeSO_4 + Ca(OH)_2 + 2H_2O \rightarrow CaSO_4 \cdot 2H_2O + Fe(OH)_2 \quad (2.25)$$

$$H_2SO_4 + Ca(OH)_2 \rightarrow CaSO_4 \cdot 2H_2O \quad (2.26)$$

$$\begin{aligned} &3(CaSO_4 \cdot 2H_2O) + 3CaO \cdot Al_2O_3 \cdot 6H_2O + 19H_2O \\ &\rightarrow 3CaO \cdot Al_2O_3 \cdot 3CaSO_4 \cdot 31H_2O \end{aligned} \quad (2.27)$$

Unlike other types of sulfate attack, chemical corrosion by sulfuric acid is not a pure sulfate attack, but a combined attack by the proton (an acid) and sulfate attack in which the acid components enhance dissolution and thereby play a significant role in the corrosion mechanism.

In sulfuric acid attack, calcium sulfate is typically formed in the reaction of the acid with calcium hydroxide and the C-S-H gel, being converted ultimately into amorphous hydrous silica.

$$Ca(OH)_2 + H_2SO_4 \rightarrow CaSO_4 \cdot 2H_2O \quad (2.28)$$

$$CaO \cdot SiO_2aq + H_2SO_4 + H_2O \rightarrow CaSO_4 \cdot 2H_2O + SiO_2aq \quad (2.29)$$

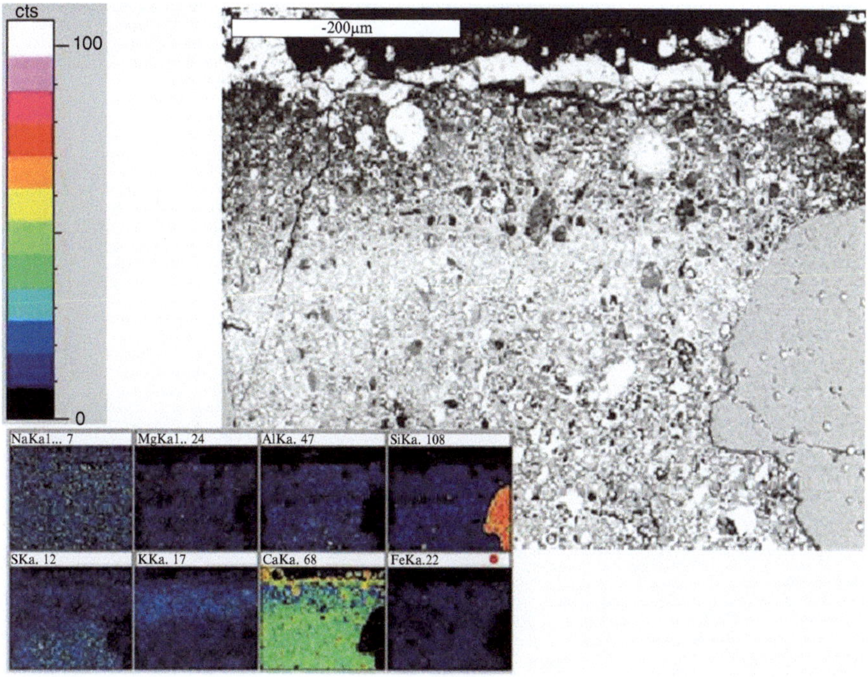

Fig. 2.24 Decalcification of cement paste and re-crystallization of calcium carbonate on surface of concrete in contact with air, at *top* of image. The *inset* shows element mapping at false color scale depicting concentration from the higher values to the lowest (*red*, *orange*, *yellow*, *green*, *blue* and *black* respectively, according to the *key bar* shown in the *left hand side* of the image)

As the pH of the service environment decreases because of the presence of acid, ettringite and monosulfate are destabilized and solubilised with leaching and conversion to gypsum and aluminium hydroxide. At the same time, ettringite crystallization may continue in the interior where moderately high pH still persists; calcium sulfates can also form in this zone. This process produces a degradation of the cement paste and a decrease in strength.

The types of damage described are observed when concrete pipes are exposed to acidic water with high sulfate content, as in sewers. Near the exposed surface a strong decalcification is observed in the cement paste, Calcium carbonate may also form on these surfaces in contact with air and not exposed to high pH. Also, some re-crystallization of calcium sulfates can occur on exposed surfaces.

This form of degradation is shown in Figs. 2.24 and 2.25. Figure 2.24 shows a field concrete from a sewage pipe. The top of the image shows formation of a thin layer of calcium carbonate covering its surface (white color).

Figure 2.25 shows another area of the same pipe with formation of large crystals of gypsum on the surface exposed to acid solution (Biczock 1967; Menéndez and Andrade 2009; Menéndez 1999).

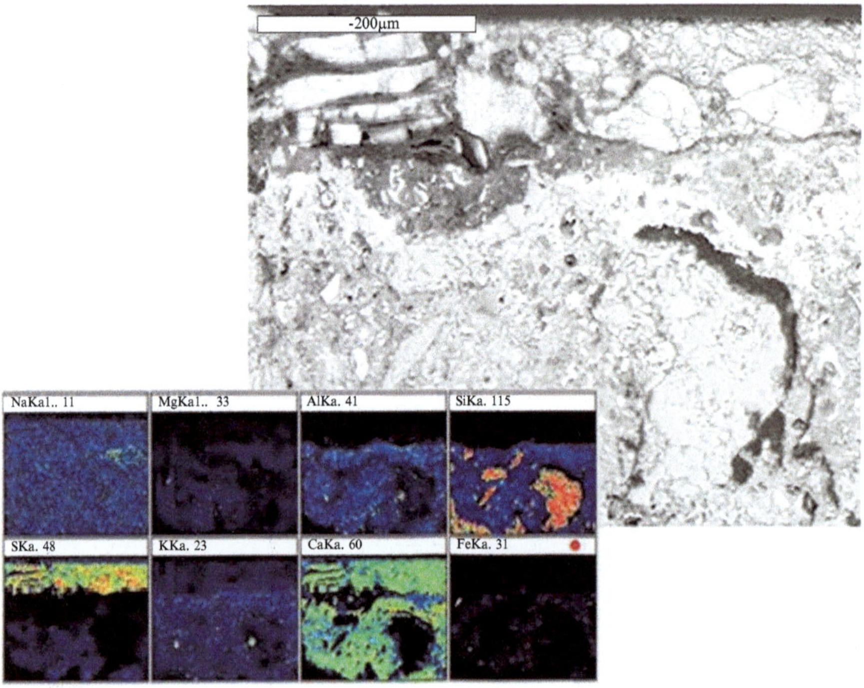

Fig. 2.25 Crystallizations of calcium sulfate on the exposed concrete surface (See caption of Fig. 2.24 for additional explanation)

2.4.7 Sea Water

Sea water has a complex chemistry including magnesium, sulfate, sodium, chloride and dissolved CO_2 species. Despite the potentially high aggressiveness of sea water towards cement, there are some processes that mitigate this aggressiveness. The presence of chloride lowers the aggressiveness of the sulfate while magnesium precipitates in form of magnesium hydroxide (brucite), clogging pores and sealing the surface of the concrete. Precipitation of calcium carbonate also adds to and enhances the formation of a semi-protective surface layer. (Menéndez 1999, 2010). Attack by magnesium species in the context of sea water is addressed specifically in Chap. 3 in this Report.

2.4.8 Thaumasite Formation

Thaumasite is formed when sulfate, silicate and calcium react in the presence of calcium carbonate. The humidity must be high and typically, temperatures are 0–10°C, but possibly up to ~25°C; thaumasite itself is stable up to about 28°C.

Thaumasite, like ettringite – to which it is structurally similar – has an expansive effect, but can also provoke more damage to concrete than ettringite because its formation is associated with depletion of C-S-H which is necessary to supply silicon. These processes simultaneously produce expansion and degradation of the cement paste that can lead to complete disintegration of hardened concrete.

Thaumasite is thus formed by the reaction of the calcium silicate of the hydrated paste with sulfates and carbonates. These sulfates and carbonates can derive from water in the service environment, but according to Matschei et al. (2007b) it is not necessary that other components come from the service environment. However their thermodynamic calculations show that thaumasite is unlikely to form in fresh, unaltered cement because of the low internal silica activity even though a supply of carbonate may be available internally. Thus $CaCO_3$ added to cement will not of itself initiate subsequent thaumasite formation. Leaching may, however, lower the Ca content and thereby increase silica activity to the point at which thaumasite becomes stable. Thus thaumasite attack does not necessarily involve transport into the cement: leaching may be sufficient to stabilise thaumasite.

$$\underset{\textit{Ettringite}}{3CaO \cdot Al_2O_3 \cdot 3CaSO_4 \cdot 31 - 32H_2O} + \underset{\textit{Carbonate}}{CO_2} + C - S - H$$
$$\rightarrow \underset{\textit{Thaumasite}\ \textbf{\textit{Expansion}}}{CaSiO_3 \cdot CaCO_3 \cdot CaSO_4 \cdot 15H_2O} \quad (2.30)$$

The morphology of thaumasite crystals is similar to ettringite, but thaumasite frequently grows in less elongated prismatic needles and of course has different elemental composition, as shown in Fig. 2.26 (Menéndez 1999, 2010).

2.5 Internal Sulfate Attack

Internal sulfate attack is associated mainly with delayed ettringite formation. But aggregates containing sulfides can also induce alteration arising from sulfate generated by oxidation of sulfides to sulfate and, as a consequence, increased sulfate concentration. The precursor contaminations are related to inclusions of pyrites or pyrrhotite. These are found in a wide range of aggregate types including igneous, metamorphic and sedimentary types, the latter group including limestone with calcite and dolomite.

2.5.1 Delayed Ettringite Formation

Delayed ettringite formation (DEF) refers to an isochemical process whereby the distribution of sulfate changes as a consequence of a thermal excursion. Most commonly, but not exclusively, the thermal excursion occurs early in the

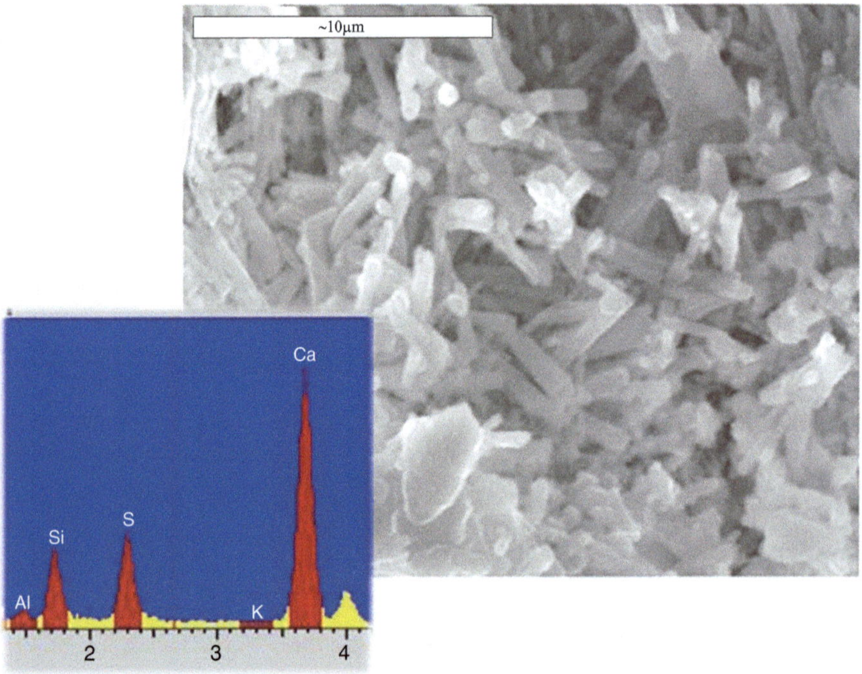

Fig. 2.26 Crystallization of thaumasite and its composition. The *inset* shows an elemental microanalysis with diagnostic signals for Ca, Si and with minor Al (Menéndez 2010)

history of cement while it is still hardening. For example, the thermal excursion may result from the heat of hydration of clinker not being dissipated, or the concrete may be cured in hot water, or both. A typical symptom of DEF is that, following return to normal service temperature, the concrete initially behaves normally but sometime thereafter it undergoes spontaneous physical expansion. The expansion may be delayed by weeks, months or even years. In theory, DEF can occur entirely without side reactions although in many real examples, DEF is complicated by the occurrence of other types of degradation, most notably alkali-aggregate reaction. However "pure" DEF can be induced in the laboratory and has been much studied.

There are conflicting views regarding the expansion mechanism of DEF. Two different theories have been proposed to explain the DEF expansion mechanism in heat-cured cementitious systems (Okurut and Ekolu 2004):

- *Ettringite crystal growth* occurs in preformed cracks and at the aggregate-paste interface. The theory proposes that the growth of ettringite crystals at, for example, the interfacial transition zone between aggregates and pastes can generate sufficient expansive pressure to cause disruption in cementitious matrices (Heinz and Ludwig 1987; Sylla 1988; Moranville-Regourd 1995; Diamond 1996; Fu and Beaudoin 1996a, b; Yang et al. 1999a, b; Zhang 1999; Zhang et al. 2002).

- The *homogeneous paste expansion theory* was proposed based on two features that were observed in heat-cured mortars: (i) that the width of gaps formed around aggregate particles upon DEF expansion were proportional to the aggregate size and (ii) that the gaps also replicated the shape of the particles (Johansen et al. 1993a, b; Johansen and Thaulow 1997). Several other later works (Shayan and Quick 1992; Lewis and Scrivener 1995; Skalny et al. 1996; Scherer 1999; Famy 1999) have subscribed to this theory, arguing that paste movements occur first, creating cracks and gaps around aggregates where ettringite subsequently crystallizes. The theory has been criticized on the grounds that paste expansion cannot occur uniformly due to the localized variability in materials and conditions within the microstructural surroundings of different expansion zones. Rather, the paste expansion should be heterogeneous (Collepardi 2003).

The two theories are not necessarily mutually exclusive. However it is well known that in cements subjected to warm or hot curing conditions, the amount of ettringite-type phase diminishes sharply. Sulfate liberated from ettringite substantially increases the amount of AFm, the S/OH ratio of which increases to its theoretical maximum. Upon cooling, this balance is reversed: hydroxyl increasingly replaces sulfate in the AFm phase and sulfate thus displaced reforms ettringite. Thus the mineralogical balances are well understood. However the microstructural expression of these phase changes is not defined by theory: the sites at which ettringite precipitates, and hence the exact reaction mechanism, are not predicted. However it appears probable that significant Al is incorporated in C-S-H in the course of warm cure and this Al is only accessible for ettringite formation by subsequent migration of sulfate into C-A-S-H. For reaction to continue, water must also migrate into C-A-S-H (the high water content of ettringite and thaumasite has been noted) and it is this imbibition of water into cement gel-like C-A-S-H that is responsible for much of the observed swelling. In this view, cracks are mainly a secondary feature and their subsequent infilling with solids is not intrinsically expansive.

The process of delayed ettringite formation results in gaps forming around the aggregates. In some cases re-crystallization of ettringite in the gaps, crystallizing perpendicularly to the surfaces of the aggregates and the paste, have been observed; these spaces are shown in Fig. 2.27 (Menéndez 1999, 2010).

2.5.2 *Presence of Sulfides in Aggregates*

Oxidation of sulfides in aggregates results in additional sulfate being produced that can induce formation of ettringite in the post-hardening stage. The suspect aggregates have particles of pyrite or pyrrhotite, or both, that are slowly oxidized, liberating sulfate that reacts with the cement components, including any remaining calcium monosulfoaluminate, to form ettringite.

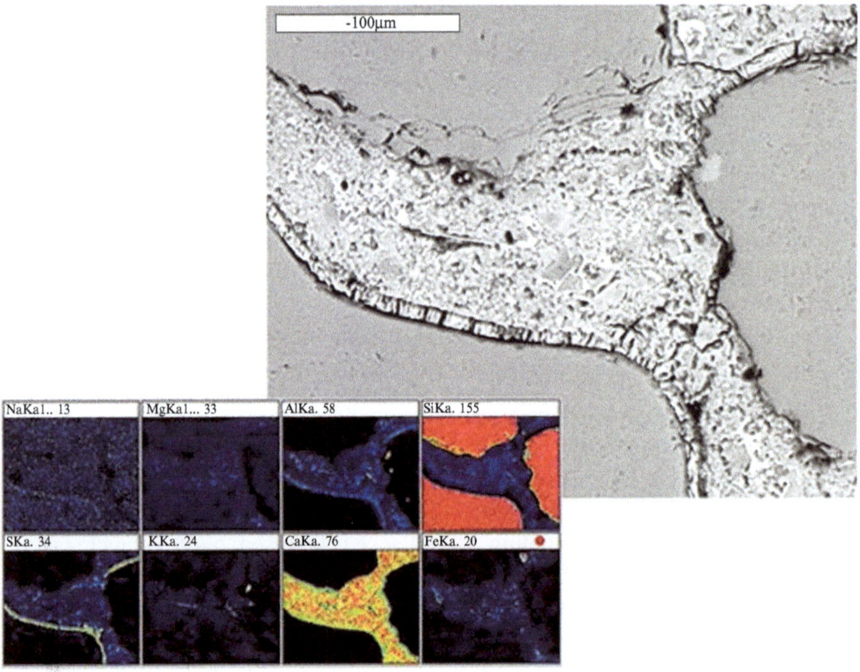

Fig. 2.27 Gaps around the aggregates with secondary ettringite inside as a consequence of the delayed ettringite formation. The false color (*inset*) enable siliceous aggregate (*orange*) to be distinguished from regions high in sulfur (*yellow-green* fringe) and Ca

The pyrite and pyrrotite reactions of oxidation are as follows:

Pyrite oxidation

$$FeS_2 + 7/2O_2 + H_2O \rightarrow Fe^{2+} + 2SO_4{}^{2-} + 2H^+ \quad (2.31)$$

$$4Fe^{2+} + O_2 + 4H^+ \rightarrow 4Fe^{3+} + 2H_2O \quad (2.32)$$

$$FeS_2 + 14Fe^{3+} + 8H_2O \rightarrow 15\ Fe^{2+} + 16H^+ + 2SO_4{}^{2-} \quad (2.33)$$

Pyrrotite oxidation

$$FeS_{1-x} + 7/2H_2O \rightarrow (1 - x)Fe^{3+} + SO_4{}^{2-} + H^+ + 3H_2O \quad (2.34)$$

The oxidation products may also react to form gypsum once the potential for ettringite formation is exceeded. In the presence of alkaline constituents dissolved in the concrete pore solution, ferric ions may also react to form ferric hydroxides. This process is also associated with expansion, due to the increase in volume of the final products, about 15%. Thus secondary expansion is initiated by formation of expansive products including both ettringite and ferric hydroxide.

$$\text{Iron sulfides} + O_2 + H_2O \rightarrow FeSO_4 + H_2SO_4 \tag{2.35}$$

$$FeSO_4 + Ca(OH)_2 + 2H_2O \rightarrow CaSO_4 \cdot 2H_2O + Fe(OH)_2 \tag{2.36}$$

$$H_2SO_4 + Ca(OH)_2 \rightarrow CaSO_4 \cdot 2H_2O \tag{2.37}$$

$$\begin{aligned} &3(CaSO_4 \cdot 2H_2O) + 3CaO \cdot Al_2O_3 \cdot 6H_2O + 19H_2O \\ &\rightarrow 3CaO \cdot Al_2O_3 \cdot 3CaSO_4 \cdot 31H_2O \end{aligned} \tag{2.38}$$

In the presence of calcium carbonate, formation of gypsum is also possible, if the pH decreases to near neutral, according to the reaction.

$$CaCO_3 + H_2SO_4 \rightarrow CaSO_4 + H_2O + CO_2 \tag{2.39}$$

Figure 2.28 shows this alteration in a field concrete with limestone aggregate and inclusions of pyrite, exposed to a temperature of 5–30°C and high relative humidity for more than 20 years (Menéndez 1999, 2010).

2.6 Crystallization of Salts

When concrete is exposed to elevated temperatures it is possible to form hydrated sulfates from anhydrous salts in the course of wet-dry cycling.

In case of alkali sulfates, if water evaporates from the surface of concrete such that the pore solution contains high amounts of alkali sulfates, crystallization of these salts may take place in regions close to the surface. Under these conditions the crystallization pressure generated in the process may disrupt the matrix. For example, in the case of sodium sulfate, a particularly unfavourable situation may occur because of a phase transition between sodium sulfate and its decahydrate with the anhydrous phases being stable above ~32°C. If the anhydrous solid subsequently comes into contact with water at lower temperatures, conversion of anhydrous salt to its hydrous phase, the decahydrate, may take place (Skalny et al. 2002; Pettifer and Nixon 1980).

For sodium sulfates the reaction is as follows:

$$\underset{\text{Thenardite}}{Na_2SO_4} + 10H_2O \rightarrow \underset{\text{Mirabilite}}{Na_2SO_4 \cdot 10H_2O} \tag{2.40}$$

In this case there is an increase in volume of about 315% between the anhydrous product (thenardite) and the hydrate (mirabilite).

There are different types of sulfates that can re-crystallize to form efflorescence (at the surface and generally harmless) or subflorescence, occurring in the concrete. Table 2.6 shows the types of salts and their solubility (Gaspar 2007).

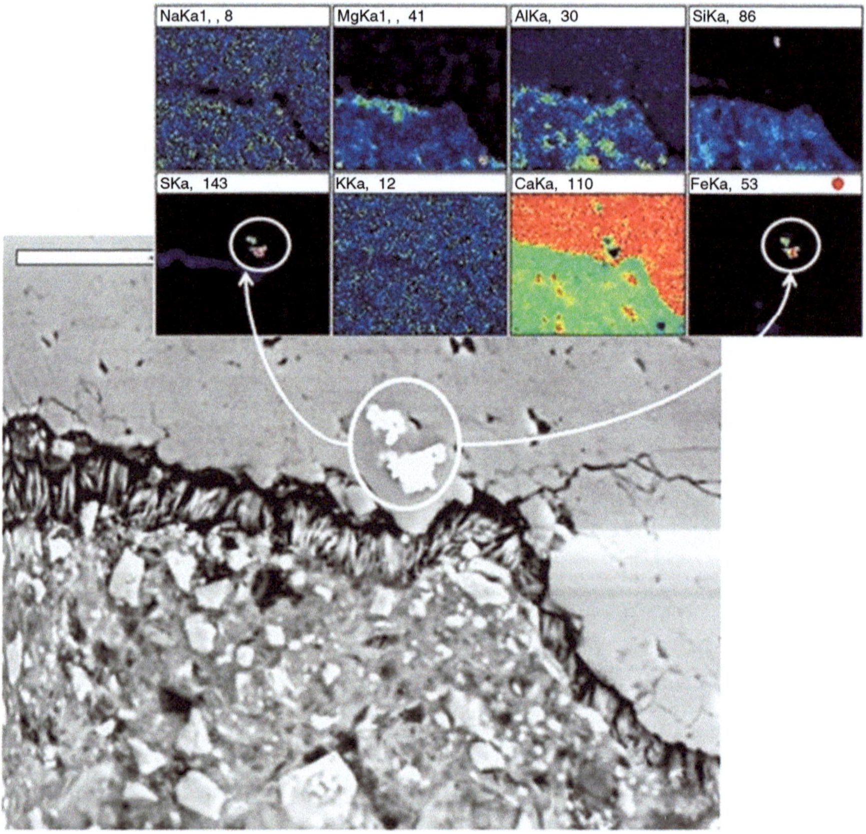

Fig. 2.28 Crystallization of ettringite in the aggregate-paste interface, as a consequence of the oxidation of pyrite in the aggregate. The optically bright grains (*circled*) are iron sulfide in the aggregate which has not yet reacted (Menéndez 2010)

Table 2.6 Types of sulfate salts and their solubility

Type of sulfate	Chemical formula	Water solubility
Calcium sulfate anhydrous	$CaSO_4$	3,0 g/l (20°C)
Calcium sulfate dihydrate	$CaSO_4{\cdot}2H_2O$	2,1 g/l (30°C)
Magnesium sulfate heptahydrate	$Mg_2SO_4{\cdot}7H_2O$	409 g/l (30°C)
Potassium sulfate	K_2SO_4	130 g/l (30°C)
Sodium sulfate anhydrous	Na_2SO_4	400 g/l (30°C)
Sodium sulfate decahydrate	$Na_2SO_4{\cdot}10H_2O$	Soluble in water

Figures 2.29 and 2.30 show examples of salt crystallization in concrete. These salts were formed in a field concrete exposed to wet-dry cycles of water containing alkali sulfates in the course of more than 10 years exposure (Menéndez 1999, 2010). The identity of the crystals is confirmed by EDAX.

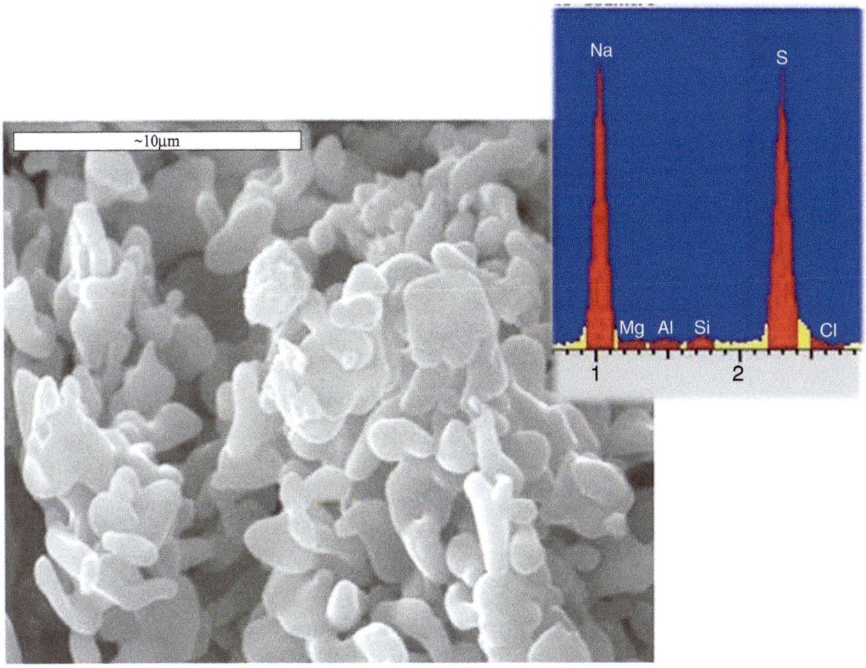

Fig. 2.29 Crystallization of sodium sulfate

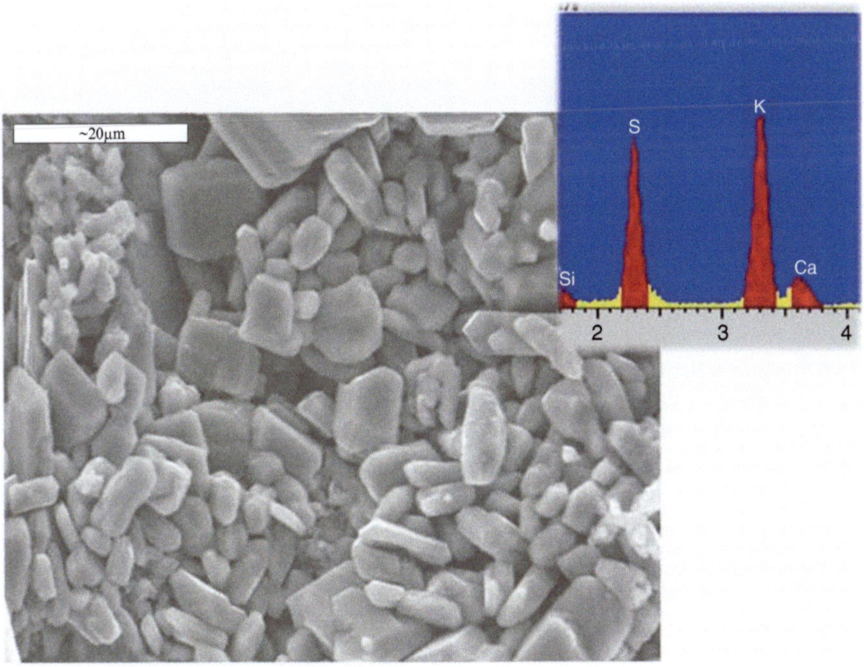

Fig. 2.30 Crystallization of potassium sulfate

2.7 Conclusions

While Portland cement composites are durable in a wide range of natural environments, some service environments, notably those rich in sulfate, are aggressive to concretes made with conventional Portland cement. Performance is affected by a process termed *sulfate attack*. This is an umbrella term describing a wide range of conditions and content, and different types of cations associated with sulfate. Sulfate attack has both chemical and physical aspects but the two aspects are difficult to deconvolute.

To some extent, the term *sulfate attack* is a misnomer because the aspects involving chemical change have to maintain local as well as overall electroneutrality. Thus cations are a part of the process and the nature of the cation (s) associated with sulfate strongly influences the mechanism and kinetics of reaction, as well as its physical consequences. For example, a species containing protons (H^+ or H_3O^+) such as aqueous sulfuric acid, is strongly acidic and tends both to dissolve calcium from cement and precipitate solids such as ettringite and gypsum as neutralisation products. In some conditions, where accumulation of solids – especially with low physical densities – exceeds loss by dissolution, physical expansion may occur.

However, when attack occurs by aqueous magnesium sulfate, dissolution is much less important because magnesium reacts strongly with the cementitious phases resulting in formation of insoluble substances such as brucite (magnesium hydroxide). As a result, both magnesium and sulfate appear as essential solids forming as a consequence of attack. Thus the mechanism of attack, as well as changes in specific volume of solids, differs from attack by other sulfates or by sulfuric acid.

Examination of concrete removed from service or reacted under controlled laboratory conditions has been invaluable to understand the sequence of mineralogical reactions and relate the findings to cement formulation and exposure conditions. Conventional petrographic analysis and electron optical methods, both coupled with X-ray diffraction and chemical mapping, have been used to characterize reaction. They define the course of reaction and its consequences, although they only indirectly measure dissolution. On the other hand, accelerated testing of sulfate resistance has an important role in assessing future performance. Testing is usually done in aqueous solutions of sodium sulfate. Historically sodium sulfate was chosen on the basis that sodium interacts weakly, or not at all, with cementitious phases so that attack is due mainly or entirely to "sulfate". Although sodium is not significantly incorporated in the solid products of attack, sodium is shown profoundly to affect the solubilisation of cement species and the pH of the pore solution. Moreover, sodium affects the relative importance of dissolution relative to precipitation. These changes are shown by calculation to be functions of sodium sulfate concentration, temperature and ratio of mass of solution to mass of solid. Thus the results from sodium sulfate immersion testing are not generic but depend on the exact choice of conditions. No rational system exists whereby tests based on sodium sulfate immersion can be extrapolated to other sulfate conditions

The relevance of testing in sodium sulfate has long been debated but without firm and well-found conclusions. In fact, sodium sulfate is only rarely an important component of most natural waters and is therefore unrepresentative of all but a few service environments. Calculation supports this view. Moreover, sodium sulfate tests, as commonly specified, fail to control all relevant variables, notably access of atmospheric carbon dioxide and, as a result, are non-reproducible. By adding to the test protocols, the reproducibility can be improved although it is still impossible to extrapolate data obtained from sodium sulfate immersion to other conditions and other geochemical environments.

Examples of concrete removed from tests or from service conditions can, however, establish data about the nature and consequences of reaction and unequivocally determine the presence of sulfate attack. Microscopy discloses that a leached and relatively weak near-surface layer develops, underlain by a zone of sulfate enrichment. This zone is characterized by formation of secondary ettringite and/or gypsum. The mass balance in the sulfate deposition zone, coupled with the large molar volume particularly of ettringite, frequently gives rise to the observed expansion and cracking. Thus, in practice, the balance between leaching and deposition, as well as the densities of the solids, are important considerations in determining the rate of attack and its consequences to the integrity of concrete. Arguments about whether a given phase is "expansive" or "non-expansive" are of marginal relevance because of the balancing effect of dissolution; the latter acting as a "safety valve".

The flux of sulfate (as well as other ions) is diffusion controlled but not in any simple way because of the need to maintain electroneutrality, locally as well as globally. Moreover, superimposition of the phase rule on diffusion profiles predicts that locally discontinuous changes in mineralogy will occur, and hence variations in local matrix permeability will arise. Microscopy usefully identifies, at least qualitatively, the sequence and evolution of these attack zones. Comparison between the observed mineral zonation and microchemistry and phase equilibrium studies discloses that under normal, *i.e.*, non-accelerated conditions, local phase equilibrium is often maintained, albeit with denser relicts of less-altered cements sometimes being left behind the advancing fronts.

Many of the features of sulfate attack – provided the associated cations can be specified – can also be modelled, giving rise to expectations that modelling, guided by observation and testing, may represent a way forward which obviates need for immersion testing in sodium sulfate. Modelling is rapid and capable of great flexibility in respect to choice of the input variables. However, modelling still requires guidance about (i) transport: the transport properties of the altered zones, including the role of cracks in the transport process, needs to be placed on a more quantitative basis and (ii), the links between what models calculate, namely space filling and a potential for expansion, with the actual physical and engineering properties of concrete.

Thus the present system for prescriptive specification of cements, favouring low tri-calcium aluminate clinker contents, combined with low permeability of the matrix, remains the best guidance for the design of sulfate-resistant Portland cement concretes. However these empirical guidelines, based on long experience, are probably inadequate for Portland cement and are almost certainly inadequate for specification of concrete made with high replacement of cement by supplementary materials such as slag and fly ash. Paradoxically, these supplementary materials are often characterised by having high chemical alumina contents yet, when blended with Portland cement, are also reported to enhance sulfate resistance. It is probable that a radical rethink of existing predictive methods and performance standards is required to take into account the character of these complex matrices.

Acknowledgment Many people have stimulated our interest in this problem, many of whom are members of RILEM TC-211 PAE. We have slowly gone up a learning curve and we thank all those who have assisted our progress. We also thank Dr G.K. Moir for information on the progress of European Standardization for sulfate-resistant cement specifications. The support of the NANOCEM network has also been invaluable in support of T.M's PhD studies. Nanocem is a not-for-profit consortium of leading European academics and manufacturers.

References

Andersson, K., Allard, B., Bengtsson, M., Magnusson, B.: Chemical composition of cement pore solutions. Cem. Concr. Res. **19**(3), 327–332 (1989)

ASTM C 1012-04: Standard Test Method for Length Change of Hydraulic-Cement Mortars Exposed to a Sulfate Solution. Annual Book of ASTM Standards, vol. 04.01. ASTM International, West Conshohocken (2004)

Barneyback Jr., R.S., Diamond, S.: Expression and analysis of pore fluid from hardened cement pastes and mortars. Cem. Concr. Res. **11**(2), 279–285 (1981)

Bellmann, F., Moeser, B., Stark, J.: Influence of sulfate solution concentration on the formation of gypsum in sulfate resistance test specimen. Cem. Concr. Res. **36**(2), 358–363 (2006)

Biczok, I.: Concrete Corrosion-Concrete Protection. Chemical Publishing Inc., New York (1967). 543 p

BRE Digest 363: Sulfate and Acid Resistance of Concrete in the Ground. Building Research Establishment, Garston (1996)

Brown, P.W.: An evaluation of the sulfate resistance of cements in a controlled environment. Cem. Concr. Res. **11**(5–6), 719–727 (1981)

Brown, P.W., Taylor, H.F.W.: The role of ettringite in external sulfate attack. In: Marchand, J., Skalny, J. (eds.) Materials Science of Concrete, Special Volume. Sulfate Attack Mechanisms, pp. 73–77. American Ceramic Society, Westerville (1999)

CEN/TR 15697.: Cement – performance testing for sulfate resistance – state of the art report (2008)

Christensen, B.J.: Microstructural studies of hydrating Portland cement-based materials using impedance spectroscopy. Ph.D. thesis, Northwestern University, Evanston, IL (1993)

Christensen, B.J., Coverdale, R.T., Olson, R.A., Ford, S.J., Garboczi, E.J., Jennings, H.M., Mason, T.O.: Impedance spectroscopy of hydrating cement-based materials: measurement, interpretation, and application. J. Am. Ceram. Soc. **77**, 2789–2804 (1994)

Clifton, J., Frohnsdorff, G., Ferraris, C.: Standards for evaluating the susceptibility of cement-based materials to external sulfate attack. In: Marchand, J., Skalny, J.P. (eds.) Materials Science of Concrete: Sulfate Attack Mechanisms, Special Volume. American Ceraic Society/Wiley, New York (1999)

Collepardi, M.: A state-of-the-art review on delayed ettringite attack on concrete. Cem. Concr. Compos. **25**, 401–407 (2003)

Collins, R.E.: Flow of Fluids Through Porous Materials. Reinhold Publishing, New York (1961)
Correns, C.W.: Growth and dissolution of crystals under linear pressure. Discuss. Faraday Soc. **5**, 267–271 (1949)
Crank, J.: The Mathematics of Diffusion, 2nd edn. Oxford University Press, London (1975)
Damidot, D., Glasser, F.P.: Thermodynamic investigation on the $CaO-Al_2O_3-CaSO_4-H_2O$ system at 25°C and the influence of Na_2O. Cem. Concr. Res. **23**, 221–238 (1996)
Diamond, S.: Delayed ettringite formation-processes and problems. Cem. Concr. Compos. **18**(3), 205–215 (1996)
Dow, C., Glasser, F.P.: Calcium carbonate efflorescence on Portland cement and building materials. Cem. Concr. Res. **33**, 147–154 (2003)
EHE-08.: Ed. Instruction of Concrete Structures. Ministry of Foment, Spain (2008)
Famy, C.: Expansion of heat-cured mortars. Ph.D. thesis, University of London, 256 p (1999)
Ferraris, C.F., Stutzman, P.E., Snyder, K.A.: Sulfate resistance of concrete: a new approach. PCA R&D Serial No. 2486, Published by PCA without copyright, Illinois, USA. (2006)
Flatt, R.J.: Salt damage in porous materials: how high supersaturation is generated. J. Cryst. Growth **242**, 435–454 (2002)
Frigione, G., Zenone, F.: Influence of gypsum content in Portland cements on sulfate resistance. In: Proceedings of the 9th International Congress on Chemistry of Cement, New Delhi, IV, pp. 60–86 (1992)
Fu, Y., Beaudoin, J.J.: Microcracking as a precursor to delayed ettringite formation in cement systems. Cem. Concr. Res. **26**(10), 1493–1498 (1996a)
Fu, Y., Beaudoin, J.J.: Letter to the Editor on the distinction between delayed and secondary ettringite formation in concrete. Cem. Concr. Res. **26**(6), 979–980 (1996b)
Gaspar, D.: Manual of Efflorescence. Masonry Mortars. ANEFHOP, Madrid (2007)
Gollop, R.S., Taylor, H.F.W.: Microstructural and microanalytical studies of sulfate attack. I. Ordinary Portland cement paste. Cem. Concr. Res. **22**, 1027–1038 (1992)
Gollop, R.S., Taylor, H.F.W.: Microstructural and microanalytical studies of sulfate attack V. Comparison of different slag blends. Cem. Concr. Res. **26**, 1029–1044 (1996a)
Gollop, R.S., Taylor, H.F.W.: Microstructural and microanalytical studies of sulfate attack IV. Reactions of a slag cement paste with sodium and magnesium sulfate solutions. Cem. Concr. Res. **26**, 1013–1028 (1996b)
Heinz, D., Ludwig, U.: Mechanism of secondary ettringite formation in mortars and concretes subjected to heat treatment. In: Scanlon, J.M. (ed.) Concrete Durability, Katharine and Bryant Mather International Conference. SP-100, vol. 2, pp. 2059–2071. American Concrete Institute, Detroit (1987)
Helfferich, F.: Ion Exchange. McGraw-Hill, New York (1962)
Hime, W.G., Martinek, R.A., Backus, L.A., Marusin, S.L.: Salt hydration distress: observations on an unidentified or misidentified cause of concrete distress. Concr. Int. **23**(10), 43–50 (2001)
Hughes, D.C.: Sulfate resistance of OPC/fly ash and SRPC pastes: pore structure and permeability. Cem. Concr. Res. **15**, 1003–1012 (1985)
Johansen, V., Thaulow, N.: Heat curing and late formation of ettringite. In: Erlin, B. (ed.) Ettringite-The Sometimes Host of Destruction, ACI SP-177, 1999. ACI, Seattle, pp. 47–64 (1997)
Johansen, V., Thaulow, N., Skalny, J.: Simultaneous presence of alkali-silica gel and ettringite in concrete. Adv. Cem. Res. **5**(17), 23–29 (1993a)
Johansen, V., Thaulow, N., Jacobsen, U.H., Palbφl, L.: Heat cured induced expansion-size of gaps around aggregate. In: Proceedings of the 3rd Beijing International Symposium on Cement and Concrete, China, vol. 3, pp. 144–156. International Academic Publishers. Also in RH&H Bulletin, No. 47 (1993b)
Kollek, J.J., Lumley, J.D.: Comparative sulfate resistance of SRPC and Portland slag cements. Durability of building materials and components. In: Proceedings of the 5th International Conference, E & FN. Spon, Brighton, UK, paper 43 (1990)

Kulik, D., Berner, U., Curti, E.: Modelling chemical equilibrium partitioning with the GEMS-PSI code. PSI Scientific Report IV, pp. 109–122, http://les.web.psi.ch/Software/GEMS-PSI/ (2003)

Larbi, J.A., Fraay, A.L.A., Bijen, J.M.: The chemistry of the pore fluid of silica fume-based cement systems. Cem. Concr. Res. **20**, 506–516 (1990)

Lewis, M.C., Scrivener, K.L.: Micromechanical effects of elevated temperature curing and delayed ettringite formation. In: Proceedings of the Materials Research Society's Symposium on Mechanisms of Chemical Degradation of Cement-Based Systems, Boston, MA, USA, pp. 243–250 (1995)

Maltais, Y., Samson, E., Marchand, J.: Predicting the durability of Portland cement systems in aggressive environments – laboratory validation. Cem. Concr. Res. **34**, 1579–1589 (2004)

Marchand, J., Samson, E., Maltais, Y., Beaudoin, J.J.: Theoretical analysis of the effect of weak sodium sulfate solutions on the durability of concrete. Cem. Concr. Compos. **24**, 317–329 (2002)

Matschei, T., Lothenbach, B., Glasser, F.P.: The AFm-phase in Portland cement. Cem. Concr. Res. **37**(2), 118–130 (2007a)

Matschei, T., Lothenbach, B., Glasser, F.P.: The role of calcium carbonate in cement hydration. Cem. Concr. Res. **37**(4), 551–558 (2007b)

Matschei, T., Lothenbach, B., Glasser, F.P.: Thermodynamic properties of Portland cement hydrates in the system CaO-Al_2O_3-SiO_2-$CaSO_4$-$CaCO_3$-H_2O. Cem. Concr. Res. **38**(1), 1–18 (2008)

Mehta, P.K.: Concrete. Structure, Properties and Materials. Prentice-Halls, Inc., Englewood Cliffs (1986)

Mehta, P.K.: Concrete in the Marine Environment. Taylor & Francis, New York (2005)

Mehta, P.K.: Sulfate resistance of blended cements. In: Proceedings of the International Workshop on the Use of Fly Ash, Silica Fume and Other Siliceous Materials in Concrete. Concrete 88, W.G. Ryan, Sydney, pp. 337–351 (1988)

Menéndez, E.: Electron microscopy: a tool to study degradation process in field concretes like sulfate attack and ettringite formation. In: Pietersen, H.S., Larb, J.A., Janssen, H.H.A. (eds.) Proceedings of the 7th Euroseminar on Microscopy Applied to Building Materials, pp. 171–180, Delft (1999)

Menéndez, E.: Análisis del hormigón en estructuras afectados por reacción árido-álcali, ataque por sulfatos y ciclos de hielo-deshielo (Concrete analysis in structures affected by alkali-silica reaction, sulphate attack and freezing-thawing cycles). IECA, Madrid (2010)

Menéndez, E., Andrade, C.: Analysis of deterioration of the concrete of a sewer for industrial waste water with sulfates. In: Alexander, M.G., Bertron, A. (eds.) Proceedings of the International RILEM TC 211-PAE Final Conference on Concrete in Aggressive Aqueous Environments. RILEM Publications S.A.R.L., vol. 1, pp. 45–53 (2009)

Moranville, M., Li, G.: The U-phase – formation and stability. In: Marchand, J., Skalny, J. (eds.) Materials Science of Concrete, Sulfate Attack Mechanism, pp. 175–188. The American Ceramic Society, Westerville (1999)

Moranville-Regourd M (1995) Implications of curing temperatures for durability of cement-based systems. In: Proceedings of the Materials Research Society's Symposium on Mechanisms of Chemical Degradation of Cement-Based Systems, Boston, USA, pp. 211–218 (2010)

Odler, T.: A discussion of the paper "Mechanism of sulfate attack on Portland cement and concrete – another look" by P.K. Mehta. Cem. Concr. Res. **14**, 147–148 (1984)

Okurut Ekolu, S.: Role of Heat Curing in Concrete Durability Effects of Lithium Salts and Chloride Ingress on Delayed Ettringite Formation. PCA R&D serial no. 2869. Doctor of Philosophy, Department of Civil Engineering, University of Toronto (2004)

Pettifer, K., Nixon, P.J.: Alkali metal sulphate – a factor common to both alkali aggregate reaction and sulphate attack on concrete. Cem. Concr. Res. **10**, 173–181 (1980)

Planel, D., Sercombe, J., Le Bescop, P., Adenot, F., Torrenti, J.M.: Long term performance of cement paste during combined calcium leaching-sulfate attack: kinetic and size effect. Cem. Concr. Res. **36**, 137–143 (2006)

Prusinski, J.J., Carrasquillo, R.L.: Using medium – to high volume fly ash blended cements to improve the sulfate resistance of high-lime fly ash concrete. In: Proceedings of the Fifth-International Conference on Natural Pozzolans in Concrete, Milwaukee, ACI SP-153, pp. 43–65 (1995)

Rodriguez-Navarro, C., Doehne, E.: Salt weathering: influence of evaporation rate, supersaturation and crystallization pattern. Earth Surf. Process. Landf. **24**, 191–209 (1999)

Rubinstein, I.: Electro-Diffusion of Ions. SIAM, Philadelphia (1990)

Samson, E., Marchand, J.: Numerical solution of the extended Nernst-Planck model. J. Colloid Interface Sci. **215**, 1–8 (1999)

Samson, E., Marchand, J., Robert, J.L., Bournazel, J.P.: Modeling ion diffusion mechanisms in porous media. Int. J. Numer. Methods Eng. **46**, 2043–2060 (1999)

Scherer, G.: Crystallization in pores. Cem. Concr. Res. **29**(8), 1347–1358 (1999)

Scherer, G.W., Flatt, R., Wheeler, G.: Materials science research for the conservation of sculpture and monuments. MRS Bull. **26**(1), 44–50 (2001)

Shayan, A., Quick, G.W.: Relative importance of deleterious reactions in concrete: formation of AAR products and secondary ettringite. Adv. Cem. Res. **4**(16), 149–157 (1992)

Skalny, J.P., Johansen, V., Thaulow, N., Palomo, A.: DEF as a form of sulfate attack. Mater. Constr. **46**(244), 5–29 (1996)

Skalny, J., Marcharnd, J., Odler, I.: Sulfate Attack on Concrete. Spon Press/Taylor & Francis, London (2002)

Skapa, R.; Optimum sulfate content of Portland cement. Ph.D. thesis, University of Aberdeen (2009)

Snyder, K.A., Ferraris, C.F., Martys, N.S., Garboczi, E.J.: Using impedance spectroscopy to assess the viability of the rapid chloride test for determining concrete conductivity. J. Res. Natl. Inst. Stand. Technol. **105**(4), 497–509 (2000)

Snyder, K.A., Feng, X., Keen, B.D., Mason, T.O.: Estimating the electrical conductivity of cement paste pore solutions from OH-, K+ and Na+ concentrations. Cem. Concr. Res. **33**, 793–798 (2003)

Soroushian, P., Alhozaimy, A.: Correlation between fly ash effects on permeability and sulfate resistance of concrete. In: Proceedings of the 9th International Congress Chemistry of Cement, New Delhi, V, pp. 196–202 (1992)

Standard EN 197-1: Cement. Composition, specifications and conformity criteria for common cements, Committee of European Normalization (2000)

Stark, J., Bollman, K.: Delayed ettringite formation in concrete. Proc. Nordic. Concr. Res. Mtg. Island, pp. 4–28 (1999)

Stephens, J.B., Carrasquillo, R.L.: Evaluating performance-based test and specifications for sulfate resistance in concrete. Texas Department of Transportation, Report No 0-1706-3 (2000)

Sylla, H.M.: Reactionen im zementstein durch wärmebehandlung (Reactions in cement paste due to heat treatment). Beton **38**(11), 449–454 (1988)

Tang, L.: Concentration dependence of diffusion and migration of chloride ions, part 1: theoretical considerations. Cem. Concr. Res. **29**, 1463–1468 (1999a)

Tang, L.: Concentration dependence of diffusion and migration of chloride ions, part 2: experimental evaluations. Cem. Concr. Res. **29**, 1469–1474 (1999b)

Taylor, H.F.W.: Chemistry of Cement, 2nd edn. Thomas Telford Publishing, London (1997)

Tikalsky, P., Roy, D., Scheetz, B., Krize, T.: Redefining cement characteristics for sulfate-resistantPortland cement. Cem. Concr. Res. **32**, 1239–1246 (2002)

Tixier, R., Mobasher, B.: Modeling of damage in cement-based materials subjected to external sulfate attack. I: formulation. J. Mater. Civil Eng. **15**(4), 305–313 (2003)

Truc, O., Ollivier, J.P., Nilsson, L.O.: Numerical simulation of multi-species transport through saturated concrete during a migration test – MSDIFF code. Cem. Concr. Res. **30**, 1581–1592 (2000a)

Truc, O., Ollivier, J.P., Nilsson, L.O.: Numerical solution of multi-species diffusion. Mater. Struct. **33**, 566–573 (2000b)

Tyrrell, H.J.V., Harris, K.R.: Diffusion in Liquids. Butterworths, London (1984)
Winkler, E.M., Singer, P.C.: Crystallization pressure of salts in stone and concrete. Geol. Soc. Am. Bull. **83**(11), 3509–3513 (1972)
Yang, R., Lawrence, C.D., Sharp, J.H.: Effect of type of aggregate on delayed ettringite formation. Adv. Cem. Res. **11**(3), 119–132 (1999a)
Yang, R., Lawrence, C.D., Lynsdale, C.J., Sharp, J.H.: Delayed ettringite formation in heat-cured Portland cement mortars. Cem. Concr. Res. **29**(1), 17–25 (1999b)
Zhang, Z.: Delayed ettringite formation in heat-cured cementitious systems. Ph.D. thesis, Dept of Civil Engineering, Purdue University (1999)
Zhang, Z., Olek, J., Diamond, S.: Studies on delayed ettringite formation in earlyage, heat-cured mortars, I. Expansion measurements, Changes in dynamic modulus of elasticity, and weight gains. II. Characteristics of cement that may be susceptible to DEF. Cem. Concr. Res. **32**(11), 1729–1742 (2002)

Chapter 3
Magnesium Attack of Cementitious Materials in Marine Environments

Manu Santhanam

Contents

Keywords Sea water • Magnesium sulfate • Chloride • Degradation Mechanisms • Binder • Brucite • Water/cement

3.1 Introduction

Concrete undergoes numerous chemical reactions with the chemicals in a marine environment. While sodium chloride is the dominant salt in marine waters, there is also a substantial amount of magnesium sulfate. The attack by the Mg ion is particularly dangerous for the system, as it can cause a complete disintegration of the C-S-H in the long term. However, in the presence of a strong chloride environment such as in seawater, there are some differences in the mechanism of magnesium attack as

M. Santhanam (✉)
Department of Civil Engineering, IIT Madras, Chennai 600 036, India
e-mail: manus@iitm.ac.in

M. Alexander et al. (eds.), *Performance of Cement-Based Materials in Aggressive Aqueous Environments*, RILEM State-of-the-Art Reports 10,
DOI 10.1007/978-94-007-5413-3_3, © RILEM 2013

opposed to a typical sulfate attack. This chapter explores the chemistry of magnesium and sulfate attack in a chloride environment.

3.2 Reaction Mechanisms

3.2.1 Magnesium Sulfate Attack

The attack by sulfate solutions produces diverse effects, and the associated mechanisms can best be described chemically by the individual chemical reactions that take place. It must, however, be understood that combination of two or more effects could produce mitigating or accelerating factors, whereby the attack due to the combination of chemicals cannot be predicted from the results of testing the individual chemical. The role of magnesium sulfate in seawater, or concentrated seawater, cannot be predicted from magnesium sulfate solutions. There is significant research that differentiates the attack of sulfates in soils or ground water where there are no chlorides to the attack of similar sulfate concentrations in seawater where chlorides are present.

The primary products of sulfate attack reactions are gypsum and ettringite. Secondary effects include the decalcification of the C-S-H gel, and the loss of cementitious structure. The reactions involved in sulfate attack are well understood by researchers (Cohen and Bentur 1988). Magnesium sulfate ($MgSO_4$) reacts with the hydrated cement paste to form gypsum ($CaSO_4.2H_2O$), ettringite ($6CaO.Al_2O_3.3SO_3.32H_2O$), and brucite ($Mg(OH)_2$) according to the following approximate equations:

$$Ca(OH)_2 + MgSO_4 + 2H_2O \rightarrow \underset{\text{Gypsum}}{CaSO_4 \cdot 2H_2O} + \underset{\text{Brucite}}{Mg(OH)_2} \tag{3.1}$$

$$C_4AH_{13} + 3C\overline{S}H_2 + 14H \rightarrow \underset{\text{Ettringite}}{C_6A\overline{S}_3H_{32}} + CH \tag{3.2}$$

$$C_4A\overline{S}H_{12-18} + 2C\overline{S}H_2 + (10-16)H \rightarrow \underset{\text{Monosulfate}}{C_6A\overline{S}_3H_{32}} \tag{3.3}$$

$$C_3A + 3C\overline{S}H_2 + 26H \rightarrow C_6A\overline{S}_3H_{32} \tag{3.4}$$

The formation of ettringite causes expansion (Mehta 1983; Brown and Taylor 1999), although the mechanism of expansion is still debated by researchers. The brucite layer that forms on the surface has a very low solubility. Thus, the formation of brucite according to Eq. (3.1) continues until the $Ca(OH)_2$ becomes depleted. With the consumption of $Ca(OH)_2$, a gradual decrease of pH occurs. This renders the calcium silicate hydrate (C-S-H) gel unstable, and causes it to liberate further $Ca(OH)_2$ into the surrounding solution to increase the pH. This can once again go into reaction with the attacking solutions. Thus, a progressive reduction of the CaO/SiO_2 ratio occurs in the C-S-H. At advanced stages of the attack, the Mg^{2+} ion can replace Ca^{2+} in the C-S-H to form magnesium silicate hydrate (M-S-H), which is reported to be non-cementitious (Bonen 1992). The formation of M-S-H is the last stage of magnesium ion attack. A direct attack of the C-S-H by $MgSO_4$ can also cause the formation of M-S-H and secondary gypsum (Bonen and Cohen 1992). The M-S-H and gypsum are precipitated in bands parallel to the surface of the specimen (Bonen and Cohen 1992; Gollop and Taylor 1992; Diamond and Lee 1999). Approximate reactions are given below:

$$C_3S_2H_x + 3M\overline{S} + (x + 21)H \rightarrow 3C\overline{S}H_2 + 3MH + 3SH_y + (12 + X - Y)H \quad (3.5)$$

$$4MH + SH_y \rightarrow M_4SH_{8.5} + (4.5 - y)H \quad (3.6)$$

Bonen and Cohen (1992) investigated the effect of magnesium sulfate solutions on Portland cement pastes. They suggested that the attack by magnesium ion primarily leads to the formation of a brucite layer at the surface. A layer of gypsum underneath the brucite layer is also formed, by a two-way diffusion of ions. The SO_4^{2-} ions from the solution diffuse inwards, while the OH^- ions from the cement paste move outwards. Because of the low solubility of brucite, the penetration of Mg^{2+} beneath the brucite layer into the interior of the paste specimen is restricted.

Summary

1. The mechanisms described here for sulfate attack are purely the individual mechanisms and are not representative of the more holistic reactions that occur when the sulfates are present as seawater or concentrated seawater (for example, in desalination outfall tunnels).
2. Magnesium sulfate deteriorates concrete by a direct attack on the C-S-H; presence of CH slows down this attack, unlike in the case of sodium sulfate, where better resistance is achieved with lower CH content.
3. A brucite layer forming on the concrete surface lends temporary protection to the concrete.
4. Under the action of flowing water, the layer of brucite may be removed or eroded, and the attack front can move into the cement paste.

3.2.2 *Role of the Magnesium Ion*

In sulfate attack research, the role of the cation is often not addressed adequately. Considerable influence is brought about in the attack when the cation is Mg^{2+}. In this section, the role of the magnesium ion is assessed by doing a comparative analysis of the attack of concrete by magnesium sulfate and magnesium chloride. This is more to address the mechanism which gives rise to the concerns when magnesium is included in a solution rather than to address the effect of concentrated seawater.

A comparative study of attack of Portland cement mortars by magnesium sulfate and magnesium chloride solutions was conducted by Moukwa (1990). Pore structure analysis by Mercury Intrusion Porosimetry (MIP) showed that the attack by magnesium sulfate actually led to a densification of the pore structure at early ages, and a corresponding increase in strength. On the other hand, attack by magnesium chloride caused an increase in the porosity of the mortar, as well as in the threshold diameter. (In MIP, the threshold diameter refers to a critical pore size, beyond which there is a sudden increase in the intrusion – similar to a 'bottleneck').

The primary difference in the pattern of attack by magnesium sulfate and magnesium chloride is seen in the rates of deterioration. The attack by $MgSO_4$ starts slowly, owing to the limited penetration of the sulfate ions into the mortar because of the formation of a surface layer of brucite. The reaction at this stage becomes diffusion controlled. The sulfate ions diffusing through the brucite layer react with the hydrated cement products to produce gypsum and ettringite. The reduction in pH due to the consumption of calcium hydroxide (CH) can ultimately lead to the decalcification of the C-S-H. The formation of M-S-H occurs when the C-S-H is directly attacked by the magnesium sulfate solution. The brucite layer tends to break down leading to an increased penetration at later ages. This is reflected in the increased rate of attack as seen from the sudden jump in mass, and a sudden drop in the compressive strength seen at later ages with $MgSO_4$ attack. The presence of M-S-H is primarily in those areas within the specimen where the brucite layer is cracked. This implies that the damaging effects of the attack can be delayed considerably if a thick brucite layer is formed.

No real evidence of the time for removal or abrasion of the brucite layer, or a change in the rate of attack after removal of the layer, exists in the literature. Generally, the softening of the cement paste under this layer, because of the formation of gypsum, and later on M-S-H, would lead to a gradual weakening of the layer, causing it to break or separate, allowing further ingress of the solution. Typically, in seawater, thickness of the brucite layer is around 30 μm (Skalny et al. 1999).

$MgCl_2$ attack occurs very rapidly at first, but then slows down ultimately. The high quantity of brucite formed in $MgCl_2$ attack, along with the rapid consumption of CH indicates that the primary effect of the Mg ion is to drive the reaction towards brucite formation. The rate of attack, as seen from the mass change and compressive strength measurements, slows down at later ages, possibly because of a decreased formation of brucite. $MgCl_2$ attack depletes the CH very fast to form brucite, and triggers the decalcification of C-S-H, thus rendering the paste porous. Due to the absence of sulfate ions, the only secondary product that forms as a result of decalcification is M-S-H. Because no other secondary products are deposited, the structure is rendered porous, causing a rapid drop in strength at early ages. The chlorides enter into reaction with C_3A to produce chloroaluminate compounds such as Friedel's salt. These chloroaluminate compounds are reported to be similar in appearance to ettringite but do not cause any expansion (Verbeck 1975). Chlorides also become incorporated within the C-S-H. The $CaCl_2$ that is formed as a result of the reaction between $MgCl_2$ and CH is highly soluble and does not precipitate.

Table 3.1 Composition of seawater from around the world

	Concentration (mg/l)					
Major ions	Mediterranean sea	North sea	Atlantic ocean	Baltic sea	Arabian gulf	Red sea
Na	12,400	12,200	11,100	2,190	20,700	11,350
Mg	1,500	1,110	1,210	260	2,300	1,867
Cl	21,270	16,550	20,000	3,960	36,900	22,660
SO_3	2,596	2,220	2,180	580	5,120	3,050
TDS	38,795	33,060	35,370	7,110	66,650	40,960

Based on Biczok (1967) and Mehta (1999)

Summary

1. The brucite layer forming on the surface of concrete is highly insoluble, and provides protection to the underlying concrete.
2. For concrete with low permeability, the brucite layer could be highly stable, as the penetration of the sulfate ions would be reduced.
3. Mechanical damage to the brucite layer (such as erosion by flowing water) can initiate direct attack on C-S-H

3.2.3 Combined Sulfate and Chloride Attack

In the case of combined magnesium chloride and magnesium sulfate attack, such as in seawater and concentrated seawater, the sulfate-bearing phases formed include ettringite, gypsum, and chloroaluminates such as Friedel's salt (Brown and Badger 2000). Compositions of seawater from around the world (based on Biczok 1967; Mehta 1999) are presented in Table 3.1.

It can be seen that for the most part, seawater consists of sodium chloride and magnesium sulfate. An example of a PC mortar immersed in a concentrated seawater solution is shown in Fig. 3.1. The relative distribution of these phases in the concrete varies. While ettringite and gypsum are found closer to the top surfaces of concrete, Friedel's salt exists more in the interior zones (Brown and Badger 2000). Sulfate resisting cements (low C_3A type) could be less beneficial than expected in protecting the concrete, as C_3A has the capability of binding the chlorides into non-expansive products such as Friedel's salt.

Deterioration of concrete in seawater primarily manifests in the form of erosion or loss of constituents from the concrete surface (Hewlett 1998). According to Lea (1970), CH present in cement paste is considerably more soluble in seawater than in plain water. Furthermore, the presence of CO_2 and $MgSO_4$ in seawater renders the ettringite unstable. The combined effect of permeable concrete and presence of chloride, magnesium, and CO_2 in seawater accounts for a reduction in alkalinity of the paste, resulting in loss of adhesion and strength. Surface zones of concrete in seawater often show deposits of aragonite, in addition to the brucite that forms during magnesium sulfate attack. The formation of these compounds on the surface provides some temporary barrier to the entry of the solution.

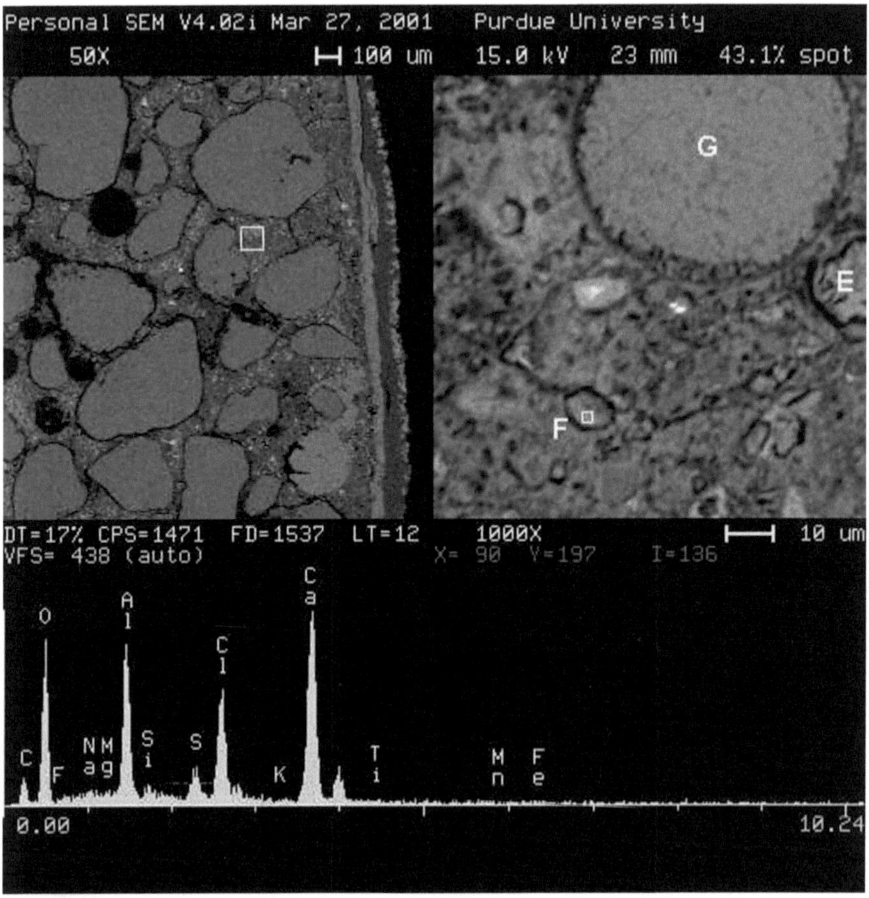

Fig. 3.1 PC mortar stored in seawater showing large air void completely filled with gypsum (*G*); a small inclusion of Friedel's salt (*F*) and a deposit of ettringite (*E*) are also visible (Santhanam 2001)

The formation of ettringite in seawater and concentrated seawater typically does not lead to expansion and cracking; it is reported to be non-expansive in the presence of high concentration of chlorides (Mehta 1999). It should be noted that there is considerable research on the chemical attack mechanisms of seawater but this is seldom considered in practice as experience shows that the concretes used in practice suffer very little deterioration. The lack of expansion due to the chlorides and the formation of a protective brucite layer are both factors that explain this.

In seawater, the layer of brucite would be enveloped by a layer of aragonite (Skalny et al. 1999), because of the high levels of dissolved CO_2. This insoluble layer of aragonite would serve as an additional barrier. Brucite has a hardness of 2.5–3, while aragonite has a hardness between three and four. Section loss from the concrete would be promoted when the seawater erodes these two layers, followed by the gypsum layer underneath. While there is no evidence in literature as to the time this would take, it can be presumed that the process would be slow.

Tumidajski and Chan (1996) opine that the brucite formed from interaction of Mg^{2+} with CH can also react with chloride in a magnesium brine solution (akin to seawater), forming magnesium hydroxychloride. Their studies indicated that longer initial curing coupled with the use of mineral admixtures was good for developing resistance to brine. The best overall durability to brine was exhibited by a concrete mix incorporating slag and silica fume in a ternary blend.

Kurdowski (2004) studied the attack of a simulated seawater solution on C-S-H paste, and determined that the diffusion of Cl^- was much faster than for Mg^{2+}. C-S-H was seen to become quickly decalcified in strong chloride solutions (such as seawater), and formation of M-S-H was also observed. The formation of a surface layer of brucite and basic magnesium chloride was reported. Under the surface skin, the paste contained phases such as Friedel's salt.

Summary

1. Ettringite formation is restricted in the surface zones of concrete in seawater; brucite and aragonite form protective layers on the surface
2. Presence of chloride in large amounts also makes the ettringite non-expansive.
3. The brucite layer on surface is semi-stable

3.3 Role of Concrete Mixture Composition

3.3.1 Type of Cement

The conventional approach to sulfate resistance has been the use of low C_3A cements. While this approach is good from the viewpoint of avoiding ettringite formation in the hardened concrete systems subjected to external sulfate attack, it does not address the formation of gypsum, or the direct attack on C-S-H that is possible when magnesium sulfate is the attacking solution. It is well understood in cement chemistry (Taylor 1997) that C_3S hydration produces almost three times as much CH as C_2S. Thus, the potential for gypsum formation is higher in concrete that is prepared with a high C_3S cement (such as OPC). Conversely, when the quantity of CH is low in the paste, conditions for direct attack on C-S-H can be created in magnesium sulfate attack.

Furthermore, when the attacking solution is a mixture of various aggressive species (chlorides and sulfates, for instance, as in seawater), C_3A can be beneficial in binding some of the chlorides into non-expansive products, as explained previously.

The above issues indicate that the role of cement composition is not very clear, and the choice of cement type would not be crucial in producing a sulfate resistant concrete in seawater. Rather, the control of permeability of concrete is the single most important factor affecting its sulfate resistance. Permeability of concrete depends on its water to cementitious materials ratio. The lower the w/c, the lower is the interconnected porosity, and thus the permeability.

It must be stated here, however, that the discussion above is based on Portland cement systems. Complete resistance to sulfate attack can be achieved by using other systems, such as Phosphate Cement, Alkali Silicate Cement, and Geopolymer Cement.

3.3.2 Water to Cement Ratio

Long term studies by the Portland Cement Association (PCA) indicate clearly that irrespective of the C_3A content of the cement, all concretes with w/c of less than 0.45 did not fail within the design service life of 40 years when exposed to a sulfate environment. A study of the concrete standards around the world clearly shows that w/c lower than 0.45 is recommended for concrete in sulfate bearing soil and groundwater.

While there is no direct relation between the permeability and the rate of attack, as far as magnesium sulfate attack is concerned, it is stated in a number of published works that the brucite layer on the surface is able to provide protection only in the case of dense impermeable concretes (Mehta and Monteiro 1993; Brown and Badger 2001). In fact, Brown and Badger explicitly state that "As hydrous silica and brucite are not compatible, their eventual conversion to a magnesium silicate should be anticipated. The fact that these compounds did not form as a glaze on the concrete surfaces facing the Mg source is an indication of the low quality of the concretes investigated. Surface coatings of these compounds are well-documented in marine structures. However, concretes for marine applications are typically produced at low water-to-cement ratios and, as a consequence, exhibit much lower permeabilities than the concretes we analyzed. Thus, brucite/hydrous silica/ magnesium silicate deposits are typically observed near the concrete surfaces."

The penetration of magnesium is minimal as compared to that of sulfate (or chloride) ions, since the creation of the insoluble brucite layer restricts the further penetration of magnesium ions. However, the decrease in permeability can be expected to decrease the ingress of the sulfate and chloride ions. This in turn would reduce the weakening of the underlying concrete (because of gypsum formation in sulfate attack, or increase in porosity in the case of chloride attack). As stated earlier, the surface layer of brucite, which offers protection, can be damaged or removed when the concrete underneath becomes softened, or when it is removed by abrasion or erosion.

According to Mehta and Haynes (1975), "Concrete submerged for 67-year in seawater has shown that low permeability concrete is highly durable to sulfate attack; however, more permeable concrete is susceptible to sulfate attack. Large concrete blocks were retrieved off the Los Angeles Harbour breakwater. Concrete blocks of size 69 × 69 × 42 in. which were retrieved were part of a test program initiated by the Corps of Engineers in 1905. Cores were tested for compressive strength and material deterioration to determine if the concrete was attacked by seawater. Results from other reports of sulfate attack on concrete are summarized. It is concluded that for long-time durability of concrete exposed to seawater, a reduced permeability and reduced alkalinity of concrete appear to be as important as low C_3A content of cement."

Irassar (2009) summarized the results of sulfate performance in laboratory and field tests where limestone is used as a constituent of cement (PLC). For $MgSO_4$ solution at temperatures higher than 15°C, specimens with w/c < 0.6 and moderate sulfate resistant cement ($C_3A < 8\%$) showed no deterioration in this environment for periods of more than a year. This could result from the evolution of a protective layer of brucite and carbonates, as occurs in seawater environments. At low temperatures, decalcification of C-S-H by Mg can promote formation of thaumasite.

3.3.3 Mineral Admixtures

When mineral admixtures such as silica fume are used, the pozzolanic reaction consumes calcium hydroxide. Complete consumption of the CH would occur for large contents of silica fume (>10%). Thus the attack by magnesium sulfate solutions can proceed quickly to the stage where decalcification of the C-S-H gel starts occurring. This leads to a poor performance of the systems containing silica fume compared to Type I cement (Al-Amoudi et al. 1995; Cohen and Bentur 1988). Ganjian and Pouya (2005) reported that silica fume concrete performed poorly compared to OPC concrete even in seawater – the lack of a brucite layer was stated as the reason for the poor performance.

It should be noted that most data in the literature report the performance of silica fume pastes or mortars where the replacement level is 10% or more. For intermediate contents of silica fume (5–7%), the pozzolanic reaction will not consume all the CH.

The reduction in permeability and refinement of pore structure with the use of mineral admixtures can often overcome this negative effect. Thus, it is essential to determine critical dosage levels of the pozzolans and slag cements to maximize their benefits, and minimize the deleterious effects of magnesium sulfate attack.

Roy et al. (2001) showed that both metakaolin and silica fume, at replacement levels less than 10%, showed good performance against chemical attack. In the case of fly ash, satisfactory performance was observed at 15–20% replacement levels. Lee et al. (2005a) suggested that metakaolin replacement actually reduces the resistance against magnesium sulfate attack, because of gypsum formation and C-S-H decalcification. Lee et al. (2005b) showed that the performance of cement mortars with silica fume was poor with respect to magnesium sulfate attack; a high degree of compressive strength loss was exhibited by cement mortars containing 5–10% silica fume, for w/c of 0.45.

Lawrence (1990) states with respect to magnesium sulfate attack that "there is well-documented evidence of considerable variability of performance for OPC/pulverized fuel ash and OPC/ground granulated blast furnace slag combinations, and the ability of composite cements to provide a guaranteed performance is suspect." However, according to Mehta (1999), the performance of fly ash or slag with respect to sulfate resistance depends on the composition of the mineral admixtures. In mixes where the hydrated phases – before sulfate exposure – had monosulfate, performance was poor, but in cases where the hydrated calcium sulfoaluminate phase was ettringite, the mixes performed well in a sulfate exposure. Additional evidence of positive performance of slag-based systems was obtained by Frigione and Sersale (1989), in a $MgSO_4$ percolation test. Hekal et al. (2002) found that hardened cement pastes with 40% slag showed higher resistance to 10% $MgSO_4$ compared to OPC mixes, while replacement of cement by 10–15% silica fume did not result in a major improvement.

Field investigations of sulfate attack on columns half buried in soil (Irassar et al. 1996) indicate that mineral admixtures (fly ash, natural pozzolan, and slag) improved the sulfate resistance of concrete buried under the soil, but were more susceptible to salt crystallization based attack in the concrete above soil. Concrete with 20% fly ash was seen to be good overall.

According to ASTM C989 (2008), good sulfate resistance was exhibited in blends with 60–65% slag, irrespective of the alumina content of the slag. Low alumina slags (11%) are seen to increase sulfate resistance independent of the C_3A content of the cement. However, high alumina slags adversely affect the resistance.

In Dehwah's (1999) study, deterioration was not noted in plain and blended cement concrete specimens exposed to sodium chloride plus sodium sulfate solutions even when the SO_4^{2-} concentration was as high as 4%. However, in the case of magnesium sulfate solutions, deterioration, in the form of surface damage, was noted in the plain and fly ash cement concrete specimens when the SO_4^{2-} concentration was 2.5% and above. In silica fume and blast furnace slag cement concrete specimens exposed to magnesium sulfate solutions, deterioration was noted when the SO_4^{2-} concentration was as low as 1%.

Al-Amoudi et al. (1994) attributed the exacerbated sulfate deterioration in silica fume and blast furnace slag cements when exposed to both the sulfate and sulfate–chloride environments, to the magnesium-oriented type of sulfate attack that was more operative in blended cements due to the absence of calcium hydroxide. In contrast to the gypsum and ettringite-oriented sulfate attacks, this type of attack was not inhibited in the presence of chloride ions. Due to the effect of the magnesium ions on concrete deterioration in blended cements, the authors recommended the usage of additional protective measures, such as the application of a water-resistant, epoxy-based coating to the exterior surface of the structure to protect the reinforced concrete structural utilities in these exposures.

With respect to most of the above studies, it must be added that the cementitious systems (paste/mortar/concrete) were prepared with relatively high water to cementitious materials ratios, in order to study the mechanism in a short duration.

Furthermore, the lower rate of pozzolanic reaction deems it necessary to cure theblended cement systems for a longer duration than OPC mixtures. Mehta (1999) also feels that anomalous behaviour of concretes with mineral admixtures could be due to use of improper test methods and mix proportions. In summary, it is essential to maximise the lowering of permeability due to the mineral admixtures, in order to improve the resistance to sulfate attack.

Summary

1. Type of cement is not as important as control of permeability in the case of magnesium sulfate attack; however, it is beneficial to use cement with relatively low C_3A and C_3S.
2. Using concrete with lowest possible water to cement ratio is the best protection against sulfate attack.
3. Mineral admixtures are known to reduce resistance to magnesium sulfate attack; but their performance in a mixed attack as in seawater is good.
4. Reduction in permeability by using mineral admixtures can provide a good approach to resisting sulfate attack.

3.4 Effect of Solution Concentration

According to Biczok (1967), the mechanism of reaction changes when the concentration of the solution changes. In magnesium sulfate attack, ettringite production is observed at a low concentration (<4,000 ppm SO_4^{2-}), a combined ettringite and gypsum formation is observed at an intermediate concentration (between 4,000 and 7,500 ppm SO_4^{2-}), and magnesium corrosion dominates at high concentrations (>7,500 ppm SO_4^{2-}).

Studies by Santhanam (2001) on attack by mixed sulfate and chloride solutions showed that expansion of samples in 2.5x seawater concentrations was not much higher than in seawater. Details of the solutions used, based on data provided by Biczok (1967), are given in Table 3.2. The expansion and compressive strength of mortars in the different solutions are shown in Figs. 3.2 and 3.3.

The mortars in SWL and SWH show the same ultimate expansion. For all the solutions, there is an abrupt initial increase of mass, which probably reflects the formation of brucite. Beyond this, the mass of the mortars in the high concentration solutions increases at a steady rate. Compressive strength behaviour is similar for all the mortars although strength loss was higher in the SWH than the SWL.

Scanning electron microscopy of the PC mortars in SWL and SWH (Figs. 3.4 and 3.5 respectively) indicate that the quantity of ettringite is greater for the low concentration solution compared to SWH. On the other hand, both mortars show a well formed layer of brucite.

Table 3.2 Details of solutions used by Santhanam (2001)

Designation	Solution	Strength	Equivalent SO_3 (ppm)	Equivalent Cl^-/Mg^{2+} (ppm)
SWL	Sea water (typical)			
	NaCl	2.70%	2,233	19,090/1,250
	$MgCl_2$	0.32%		
	$MgSO_4$	0.22%		
	$CaSO_4$	0.13%		
	$CaCl_2$	0.06%		
SWH	Sea water (High concentration)	2.5 times SWL	5,583	47,725/3,125
GWL	Ground water (typical)			
	$MgSO_4$	0.30%	2,233	1,043/950
	$CaSO_4$	0.04%		
	$MgCl_2$	0.14%		
GWH	Ground water (High concentration)	5.0 times GWL	11,167	5,215/4,750

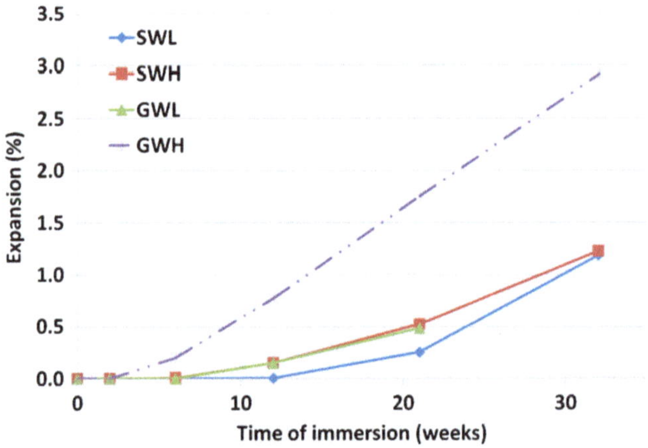

Fig. 3.2 Expansion of PC mortars in different storage solutions

Summary

1. Deterioration of mortars in seawater is not as severe as in groundwater – this indicates the benefits of the presence of chloride ions.
2. Expansion of mortars is not affected by increasing the seawater concentration. The loss in strength is increased and is accompanied by a mass loss consistent with the expected reactions. Hence, the control of permeability is increasingly important at higher concentrations.
3. Increasing the seawater concentration by 2.5x seems to increase the rate of deterioration by a factor of 1.5–2x in permeable mortars. In terms of section

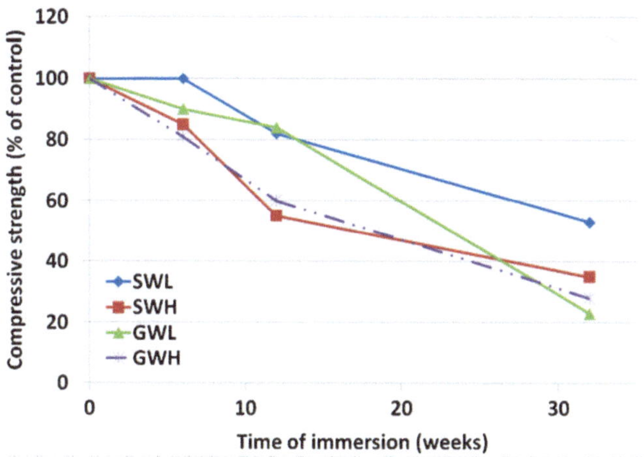

Fig. 3.3 Compressive strength for PC mortars in different storage solutions

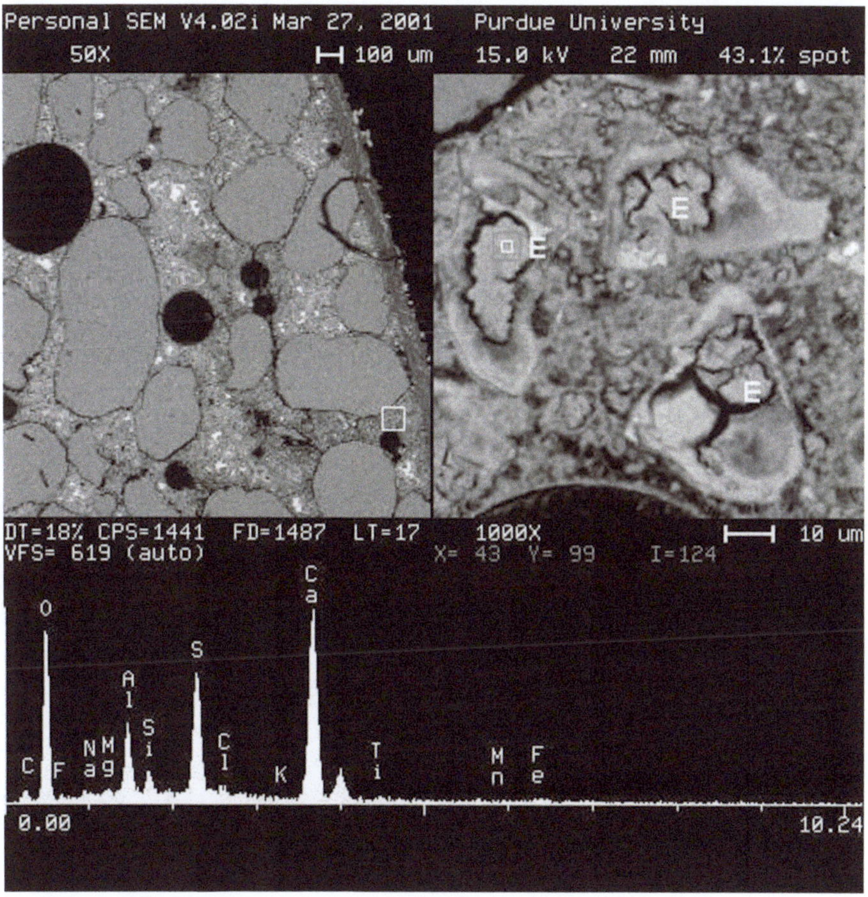

Fig. 3.4 PC mortar in SWL showing the formation of ettringite (*E*)

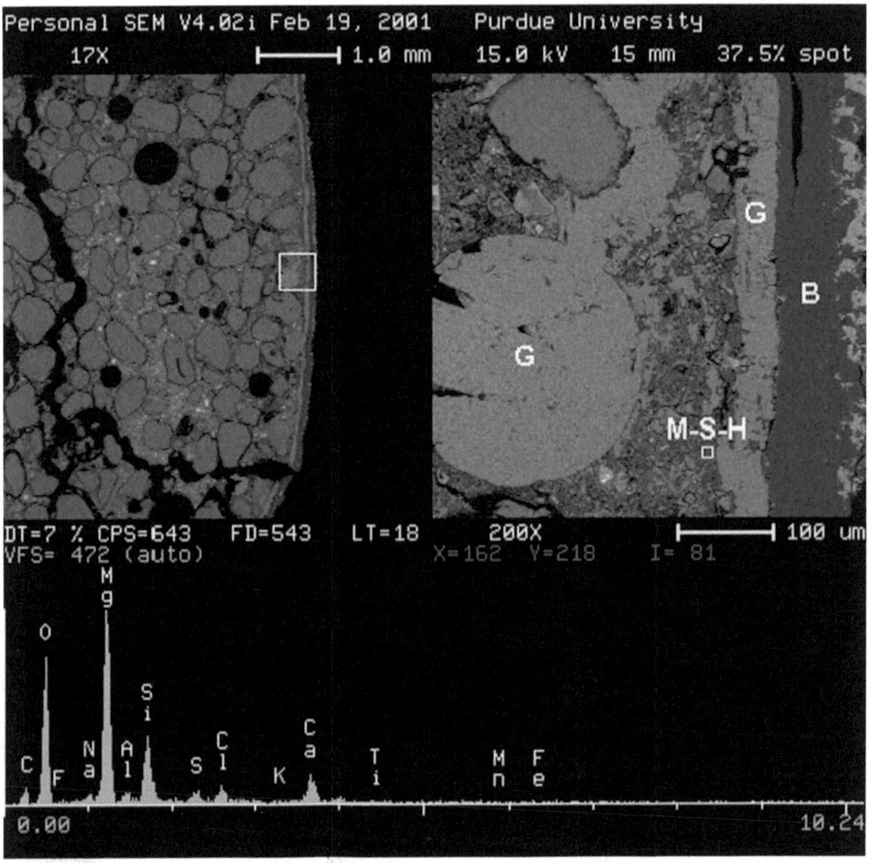

Fig. 3.5 PC mortar in SWH (the extensive cracking on the low magnification image on the *left* is due to specimen preparation; *G* gypsum, *B* brucite, *M-S-H* magnesium silicate hydrate)

loss, the loss of compressive strength would be a better indicator – thus, it is probably safe to say that the estimated section loss in 2.5x concentration seawater would be about 1.5 times that in normal seawater.

4. The primary attack in seawater appears to be the direct attack on C-S-H, once the protective brucite film is destroyed.

3.5 Lessons from Literature

The available literature on magnesium sulfate attack, or sulfate attack in general, indicates that:

1. Control of permeability is more important than cement composition.
2. Sulfate attack usually occurs in combination with other processes of deterioration; particularly, in seawater, erosion and weathering accelerate the progress of chemical attack.

3. Sulfate attack damage in real structures manifests in the form of loss of adhesion and strength, rather than expansion and cracking that is commonly observed in laboratory tests; furthermore, salt crystallization due to drying and wetting is also a common form of deterioration.

Literature clearly indicates that Mg ion attack is highly detrimental to the long term performance of concrete. The attack can also be roughly divided into two phases: (1) when the underlying concrete is protected by the brucite layer formed on the surface – this prevents further entry of the solution, and (2) final stage when the C-S-H is gradually converted to M-S-H.

References

Al-Amoudi, O.S.B., Rasheeduzzafar, M.M., Abduljauwad, S.N.: Influence of chloride ions on sulfate deterioration in plain and blended cements. Mag. Concr. Res. **46**(167), 113–123 (1994)

Al-Amoudi, O.S.B., Maslehuddin, M., Saadi, M.M.: Effect of magnesium sulfate and sodium sulfate on the durability performance of plain and blended cements. ACI Mater. J. **92**(1), 15–24 (1995)

ASTM C989.: Standard specification for ground granulated blast-furnace slag for. Use in concrete and mortars. Annual book of ASTM Standards, ASTM International, West Conshohocken, Pennsylvania (2008)

Biczok, I.: Concrete Corrosion Concrete Protection. Chemical Publishing, New York (1967)

Bonen, D.: Composition and appearance of magnesium silicate hydrate and its relation to deterioration of cement based materials. J. Am. Ceram. Soc. **75**(10), 2904–2906 (1992)

Bonen, D., Cohen, M.D.: Magnesium sulphate attack on Portland cement paste – I. Microstructural analysis. Cem. Concr. Res. **22**, 169–180 (1992)

Brown, P.W., Badger, S.: The distributions of bound sulphates and chlorides subjected to mixed NaCl, $MgSO_4$, Na_2SO_4 attack. Cem. Concr. Res. **30**, 1535–1542 (2000)

Brown, P.W., Badger, S.: Reply to the discussion by William G. Hime and Stella Ll. Marusin of the paper "The distribution of bound sulfates and chlorides in concrete to mixed NaCl, $MgSO_4$, Na2SO_4 attack". Cem. Concr. Res. **31**, 1117–1118 (2001)

Brown, P.W., Taylor, H.F.W.: The role of ettringite in external sulphate attack. In: Skalny, J., Marchand, J. (eds.) Material Science of Concrete – Sulphate Attack Mechanisms, pp. 73–98. American Ceramic Society, Westerville (1999)

Cohen, M.D., Bentur, A.: Durability of Portland cement – silica fume pastes in magnesium sulphate and sodium sulphate solutions. ACI Mater. J. **85**, 148–157 (1988)

Dehwah, H.A.F.: Influence of cement composition on concrete durability in chloride–sulfate environments. PhD thesis, Loughborough University (1999)

Diamond, S., Lee, R.J.: Microstructural alterations associated with sulphate attack in permeable concretes. In: Skalny, J., Marchand, J. (eds.) Material Science of Concrete – Sulphate Attack Mechanisms, pp. 123–173. American Ceramic Society, Westerville (1999)

Frigione, G., Sersale, R.: The action of some aggressive solutions on Portland, pozzolanic, and blast furnace cement mortars. Cem. Concr. Res. **19**, 885–893 (1989)

Ganjian, E., Pouya, H.S.: Effect of magnesium and sulfate ions on durability of silica fume blended mixes exposed to the seawater tidal zone. Cem. Concr. Res. **35**, 1332–1343 (2005)

Gollop, R.S., Taylor, H.F.W.: Microstructural and microanalytical studies of sulphate attack: I. Ordinary Portland cement paste. Cem. Concr. Res. **22**, 1027–1038 (1992)

Hekal, E.E., Kishar, E., Mostafa, H.: Magnesium sulphate attack on hardened blended cement pastes under different circumstances. Cem. Concr. Res. **32**, 1421–1427 (2002)

Hewlett, P.C.: Lea's Chemistry of Cement and Concrete, 4th edn. Arnold Publishing, London (1998)
Irassar, E.F.: Sulfate attack on cementitious materials containing limestone filler – a review. Cem. Concr. Res. **39**, 241–254 (2009)
Irassar, E.F., Di Maio, A., Batic, O.R.: Sulphate attack on concrete with mineral admixtures. Cem. Concr. Res. **26**, 113–123 (1996)
Kurdowski, W.: The protective layer and decalcification of C-S-H in the mechanism of chloride corrosion of cement paste. Cem. Concr. Res. **34**, 1555–1559 (2004)
Lawrence, C.D.: Sulphate attack on concrete. Mag. Concr. Res. **42**(153), 249–264 (1990)
Lea, F.M.: Chemistry of Cement and Concrete. Wiley, New York (1970)
Lee, S.T., Moon, H.Y., Hooton, R.D., Kim, J.P.: Effect of solution concentrations and replacement levels of metakaolin on the resistance of mortars exposed to magnesium sulphate solutions. Cem. Concr. Res. **35**, 1314–1323 (2005a)
Lee, S.T., Moon, H., Swamy, R.N.: Sulphate attack and role of silica fume in resisting strength loss. Cem. Concr. Compos. **27**, 65–76 (2005b)
Mehta, P.K.: Mechanism of expansion associated with ettringite formation. Cem. Concr. Res. **13**, 401–406 (1983)
Mehta, P.K.: Sulfate attack in a marine environment. In: Skalny, J., Marchand, J. (eds.) Material Science of Concrete – Sulphate Attack Mechanisms. American Ceramic Society, Westerville (1999)
Mehta, P.K., Haynes, H.H.: Durability of concrete in seawater. J. Struct. Div. **101**(8), 1679–1686 (1975)
Mehta, P.K., Monteiro, P.J.M.: Concrete. Structure, Properties, and Materials, 2nd edn. Prentice Hall, Englewood Cliffs, New York (1993)
Moukwa, M.: Characteristics of the attack of cement paste by $MgSO_4$ and $MgCl_2$ from the pore structure measurements. Cem. Concr. Res. **20**, 148–158 (1990)
Roy, D.M., Arjunan, P., Silsbee, M.R.: Effect of silica fume, metakaolin, and low-calcium fly ash on chemical resistance of concrete. Cem. Concr. Res. **31**, 1809–1813 (2001)
Santhanam, M.: Studies on sulphate attack: mechanisms, test methods, and modeling. PhD dissertation, Purdue University, W. Lafayette (2001)
Skalny, J., Marchand, J., Odler, I.: Sulfate Attack on Concrete. E&FN Spon, London (1999)
Taylor, H.F.W.: Cement Chemistry, 2nd edn. Thomas Telford, London (1997)
Tumidajski, P.J., Chan, G.W.: Durability of high performance concrete in magnesium brine. Cem. Concr. Res. **26**, 557–565 (1996)
Verbeck, G.J.: Mechanisms of corrosion of steel in concrete. Corros. Metals Concr., ACI Publication SP-49, 21–38 (1975)

Chapter 4
Leaching of Cementitious Materials by Pure Water and Strong Acids (HCl and HNO_3)

Josée Duchesne and Alexandra Bertron

Contents

J. Duchesne (✉)
Département de géologie et de génie géologique, CRIB, Université Laval, CRIB (Centre de recherche sur les infrastructures en béton), Québec, QC, Canada G1V 0A6
e-mail: josee.duchesne@ggl.ulaval.ca

A. Bertron
LMDC (Laboratoire Matériaux et Durabilité des Constructions), Université de Toulouse; UPS, INSA, 135, avenue de Rangueil, F-31 077 Toulouse Cedex 04, France
e-mail: Alexandra.Bertron@insa-toulouse.fr

M. Alexander et al. (eds.), *Performance of Cement-Based Materials in Aggressive Aqueous Environments*, RILEM State-of-the-Art Reports 10,
DOI 10.1007/978-94-007-5413-3_4, © RILEM 2013

Keywords Pure water • Nitric acid • Hydrochloric acid • Leaching • Decalcification • Degradation mechanisms • Chemical and mineralogical zonation • Stability of hydrates

4.1 Introduction

Ordinary Portland cement (OPC) is highly alkaline with pH values normally above 12.5, and is easily attacked by pure water and acidic solutions (Chandra 1988; Revertegat et al. 1992; Pavlik 1994a; Faucon et al. 1996, 1998; Israel et al. 1997; Allahverdi and Skvara 2000a; Kamali et al. 2008; Glasser et al. 2008). In fact, all hydrated cement compounds are stable only in solutions with well defined ranges of concentrations for Ca^{2+} and OH^{-} ions. It is well known that in contact with a solution with low mineral content and/or as the pH of the solution decreases, hydrolytic decomposition of the hydrated cement compounds disturbs the equilibrium of the cement matrix, and causes severe degradation of the material, leading to a deterioration in its technical properties (Allahverdi and Skvara 2000a; Pavlik 1994a; Adenot and Buil 1992; Reardon 1990; Biczok 1967; Bertron et al. 2007; Haga et al. 2005a, b; Mainguy et al. 2000; Kamali et al. 2008; Jain and Neithalath 2009; Faucon et al. 1996, 1997, 1998). Leaching involves a deterioration of many physical and mechanical properties of cement-based materials such as porosity, elastic modulus, compressive strength, internal friction angle and creep (Burlion et al. 2006; Haga et al. 2005b; Constantinides and Ulm 2004; Bernard et al. 2008; Heukamp et al. 2001; Stora et al. 2009; Sellier et al. 2011).

The attack by acidic waters represents a topic of growing significance due to increasing damages to concrete structures globally. This undesirable situation is a consequence of increasing sources of acidic media due to growing urban and industrial activities, and because cementitious materials are highly vulnerable in acidic environments due to their high alkalinity (Zivica and Bajza 2001; Pavlik 1994a; Allahverdi and Skvara 2000a, b).

Moreover, most studies carried out for the investigation of the safety of radioactive waste disposal concrete facilities involve tests and simulations based on attack by pure water. This is a very severe scenario that is implemented, notably to model the long term behaviour of the concrete exposed to subterranean waters. The durability of water pipes or dams is also the focus of such studies.

This review focuses on the attack of concrete by strong acids producing soluble calcium salts and on the attack by pure water. In both cases, the degradation is due to the leaching of cementitious materials, responsible for higher porosity and loss of strength. The intensity of the attack depends mainly on the pH conditions.

This chapter presents a description of the media, and the degradation mechanisms of cementitious materials by pure water and strong acids. Finally, factors influencing kinetics and durability are presented.

4.2 Characterization of the Media

The spectrum of aggressive acidic media is broad. Acidic media originate usually from industrial processes but can be due to urban activity and natural occurrences. Free acids in natural waters are rare. Exceptions are CO_2 waters and sulfurous and sulfuric acids found as a result of the oxidation of sulphide minerals such as pyrites, a process which is catalyzed by the presence of the aerobic bacterium, thiobacillus ferro-oxidans (Fattuhi and Hugues 1988; Pavlik 1994a; Zivica and Bajza 2001). As the scope of this section is limited to the influence of strong acid producing soluble calcium salts (HCl and HNO_3 mainly) and pure water, the characterization will be limited to these.

Significant quantities of free acids can occur in industrial environments in the form of leakage and random spillage. In these cases, the concentration of acid which comes in the contact with concrete structures may reach high values (Harrison 1987; Fattuhi and Hugues 1988; Allahverdi and Skvara 2000a). Further, acidic media can result from air pollution due to the presence of atmospheric carbon dioxide, sulfur dioxide and nitrogen oxides (Zivica and Bajza 2001). In fact, industrial emissions of sulphur and nitrogen compounds which later can be converted to sulphuric and nitric acids are encountered as the origin of such acid precipitations with a pH level ranging from 5.0 to 3.0 (Pavlik 1994a; Sersale et al. 1998). Acidic rain or mist is then another aggressive agent for concrete structures.

Strong acids (HNO_3, H_2SO_4) can also be produced by microorganisms, namely bacteria, as end-products of their metabolisms (Sand 1997). Attack by strong acid HNO_3 can for example occur in nitrifying basins of waste water treatment plants. In such basins, the oxidation of ammonia into nitrate ions by autotrophic nitrifying bacteria leads to the production of acid, the pH at the concrete surface – under the biofilm – being lower than the pH of the water in the nitrifying tank (in the range 6.6–7.5) (Siegrist and Gujer 1987; Leemann et al. 2010a). Basically, the attack results in a combination of cement paste hydrate dissolution and the carbonation of the matrix because of the bicarbonate anions naturally occurring in the wastewaters (Leemann et al. 2010a, b).

Water is essential to chemical attack, since it acts as a solvent for aggressive agents, and it helps to transport these aggressive agents and their reaction products. Thus, water is always present in chemical deterioration processes. Pure, salt free, distilled or deionised waters along with natural glacier water, snow-melt water, rain water, waters from lakes, creeks and rivers, are all detrimental to concrete with pH close to neutral conditions (Zivica and Bajza 2001, 2002; Adenot and Buil 1992). Leaching of concrete by flowing water has caused severe damage in dams, pipes or conduits, and is potentially important for the long-term storage of nuclear wastes (Taylor 1997; Moranville et al. 2004). The French Atomic Energy Commission (CEA) has for many years studied the leaching of concrete by water; their concern is the long-term behaviour of nuclear waste storage facilities and the modelling thereof (Faucon et al. 1996).

4.3 Degradation Mechanisms of Cementitious Materials by Pure Water and Strong Acids

4.3.1 General Mechanisms of Leaching

4.3.1.1 Physical and Chemical Characterization

In this section the degradation mechanisms of leaching common to pure water and strong acids are detailed. Mechanisms specific to strong acids (with soluble salts, such as HNO_3 and HCl), which mainly concern the characteristics of the outer layer of leached cementitious matrices, will be described in Sect. 4.3.2.

Leaching by pure water involves two types of phenomena: (i) diffusion of ionic species (Ca^{2+} and OH^- mainly) in the pores of the cementitious matrix because of concentration gradients between the highly alkaline and basic interstitial solution and the external aggressive solution and (ii) dissolution of the Ca-bearing cementitious phases by hydrolysis to supply Ca ions to maintain equilibrium.

The principle of the attack of concrete by strong acids with soluble salts such as HCl or HNO_3 is very similar to that by pure water except that the dissolution of the hydrates and/or the anhydrous residual grains of the cement paste is the consequence of acid-base reactions. The high solubility of the calcium salts (namely $CaCl_2$ and $Ca(NO_3)_2$) leads to a strong calcium leaching by the aqueous solution. This effect contributes to the progress of the acidic front. Due to the action of acidic solutions, the pH value of pore solution in the concrete is decreased, and the equilibrium in the cement matrix is disturbed. This effect also causes the gradual hydrolytic decomposition of all the hydration products.

The alteration of cement-based materials subjected to leaching manifests as a mineralogical zonation of the matrix, the zones being delimited by quite abrupt dissolution/precipitation fronts (Heukamp et al. 2001; Bernard et al. 2008; Kamali et al. 2008; Carde et al. 1996; Adenot and Buil 1992). The mineralogical zonation can be explained by assuming local equilibrium (Adenot and Buil 1992). Figure 4.1 presents an example of different zones observed macroscopically on laboratory samples exposed to leaching in pure water.

Apart from hydroxyl ions, calcium ions are the main ions of the cement paste released during leaching. Fig. 4.2, from Adenot and Buil (1992), presents the ratio of amount of leached ions to amount of added acid as a function of time for cement pastes subjected to pure water.

The dissolution of the various hydrates during leaching is directly dependent on their solubility properties, which have been established by various authors (Reardon 1990; Berner 1992; Matschei et al. 2007). Figure 4.3 is a schematic description of the solid ($Ca(OH)_2$ and C-S-H)-liquid equilibrium in relation to the calcium ion concentration in the pore solution. The dissolution of portlandite occurs when the Ca ion concentration in the pore solution drops below 22 mmol/L. C-S-H gels are stable between liquid Ca ion concentrations of 22 and 2 mmol/L, depending on the Ca/Si molar ratio of the gel. At Ca ion concentration less than 2 mmol/L in the

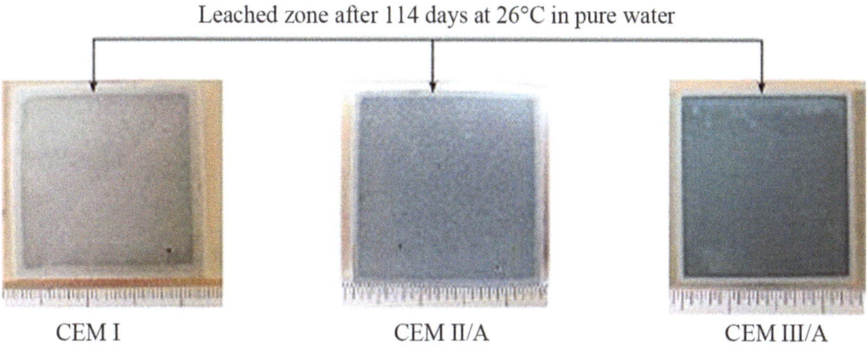

Fig. 4.1 Altered zones observed macroscopically on cross-sections of leached cement paste samples with W/C = 0.4 (Kamali et al. 2008)

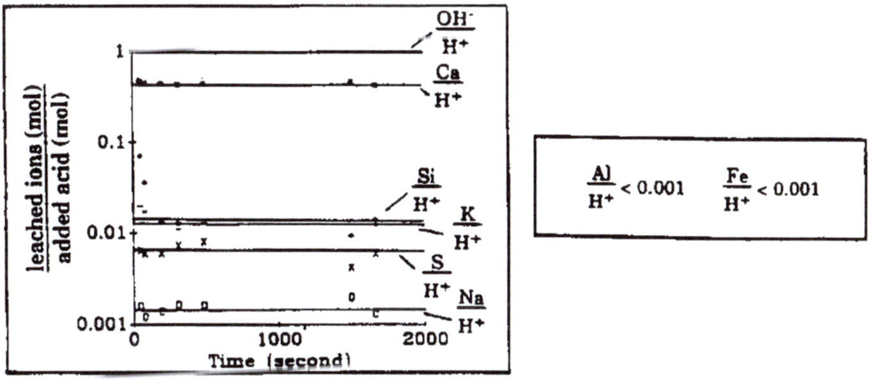

Fig. 4.2 Ratio of amount of leached ions to amount of added acid (Adenot and Buil 1992)

pore solution, the C-S-H phase is no longer stable and only the silicon rich gels are stable. A value of 1.7 mmol/L for the Ca ion concentration in the pore solution was reported by Henocq (2005) as the transition between C-S-H and silica gels.

Figure 4.4 illustrates the profile of calcium content of a matrix leached by decarbonated ion-exchanged water (Haga et al. 2005a). As schematized by the authors, this release of calcium is linked with the successive dissolutions of the various hydrates.

In the cementitious matrix, portlandite is the first hydrate to be dissolved, followed by AFm and AFt. Then, the C-S-H is progressively decalcified (Faucon et al. 1996, 1997, 1998; Moranville et al. 2004). At pH values lower than 12.5, portlandite is the first constituent to start dissolution, then ettringite (10.6), C-S-H (~10.5) and finally calcium aluminate and ferrite hydrates decompose successively until a silica gel residue is obtained (Reardon 1990; Pavlik 1994a; Myneni et al. 1998; Moranville et al. 2004; Beddoe and Dorner 2005). There are variations in reported values for pH stability limit of C-S-H, a pH value of 8.8 was given by Reardon (1990) to describe the C-S-H – amorphous silica phase transition.

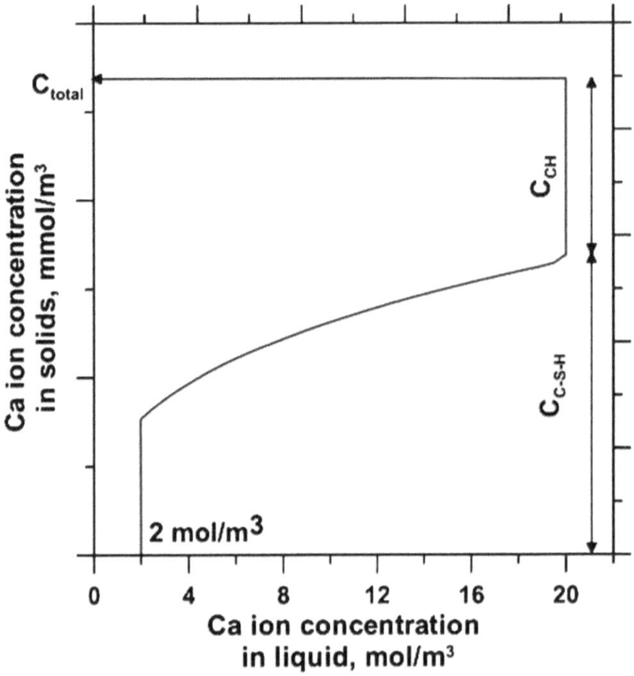

Fig. 4.3 Pore solution-Solid equilibrium curve for calcium (Jain and Neithalath 2009)

As mentioned above, this results in a well-marked chemical and mineralogical zonation of the cementitious matrix. Figure 4.5 (from Bernard et al. 2008) presents a schematic model describing the different zones and their composition. Figure 4.6 (from Faucon et al. 1998) presents changes in crystallised hydrates composition from the surface to the sound zone of a cement paste sample leached by mineralised water for 6 months.

The outer layer of leached cementitious matrices is of interest. Many authors report the presence of this layer acting as a protective barrier at the interface of the aggressive solution and point out the importance of Al and Fe as stabilizing elements in the protective layer (Adenot and Buil 1992; Faucon et al. 1996, 1998; Pavlik 1994a). According to Faucon et al. (1998), the superficial layer creates a barrier and diffusion front moves toward the core of the cementitious material.

Faucon et al. (1996) have extensively studied this surface layer formed on contact with demineralised water. According to these authors, during dissolution of AFm and AFt phases, both iron and magnesium ions pass through the pore solution and the gradients of concentration enhance diffusion of those species to the surface where they precipitate in the form of hydrotalcite (a magnesium-bearing phase with brucite structure) and iron-substituted C-S-H. Iron-substituted hydrogarnet (C_3AH_6) was also observed in the surface layer as this phase presents

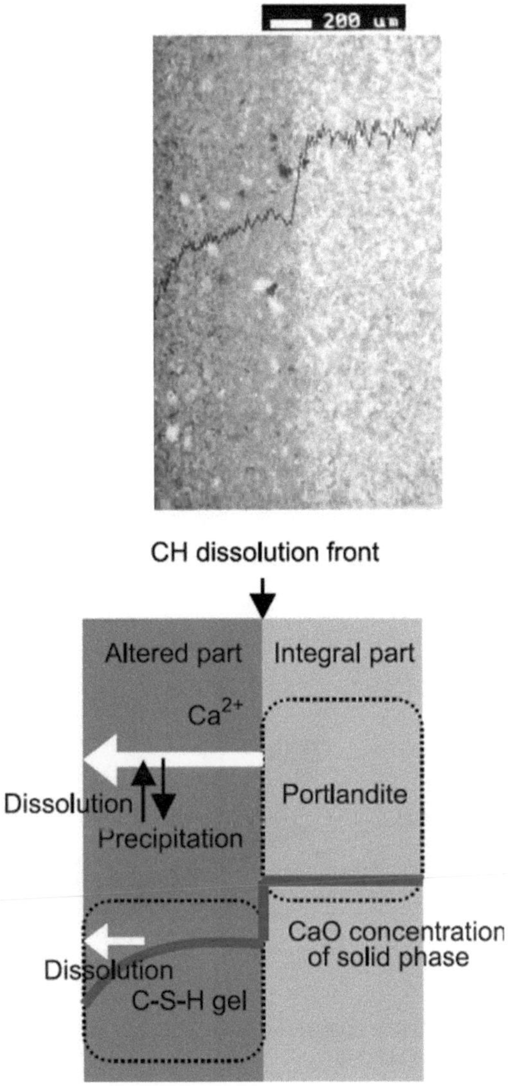

Fig. 4.4 CaO concentration on the cross-section of a 5 weeks leached sample and schematic reaction model of the dissolution (Haga et al. 2005a). The observations and analyses were obtained using EPMA

slow dissolution kinetics. These secondary precipitations can explain the Fe, Al and Si enrichment observed in the protective surface layer of cementitious material subjected to low pH solution. Hidalgo et al. (2007) also investigated the outer layer of leached cement paste and revealed the formation of a C-A-S-H gel.

The characteristics of this outer layer change when the pH of the leaching solution decreases. This will be detailed in Sect. 4.3.2.

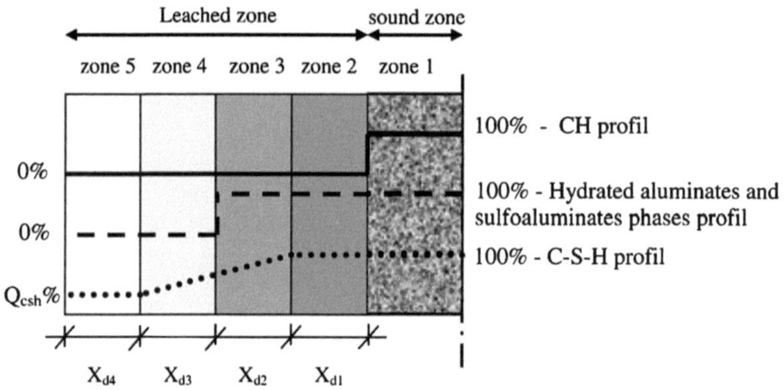

Fig. 4.5 Schematic model of cement-based material leaching (Bernard et al. 2008). Zone 1 is the sound zone; zone 2 is the zone where portlandite is totally depleted; zone 3 corresponds to the zone where portlandite is totally dissolved and C-S-H begin to be decalcified; zone 4 represents the zone where portlandite, hydrated aluminates and sulfoaluminates phases are totally dissolved and C-S-H continue to be decalcified; finally zone 5 is the much altered zone

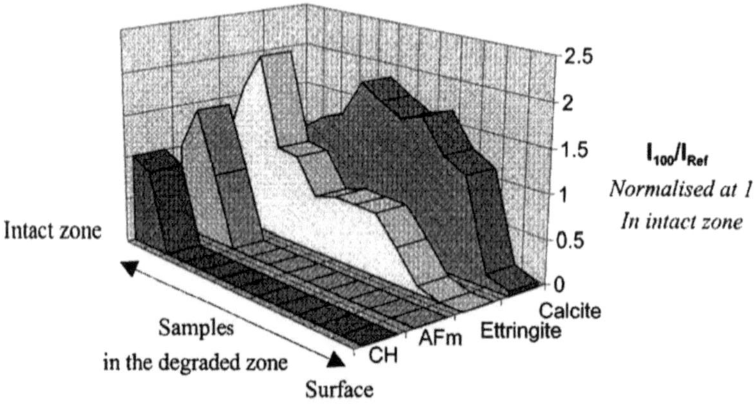

Fig. 4.6 Mineralogical composition of cement paste leached for 6 months by demineralised water (Faucon et al. 1998)

4.3.1.2 Classification of the Attack and Macroscopic Consequence of Leaching

Figure 4.7 presents a general scheme of deterioration of concrete by chemical reaction proposed by Regourd (1981). In this scheme three reactions are described:

A – Reactions involving hydrolysis and leaching of the components of hardened cement paste

B – Exchange reactions between the aggressive fluid and components of hardened cement paste

C – Reactions involving formation of expansive products.

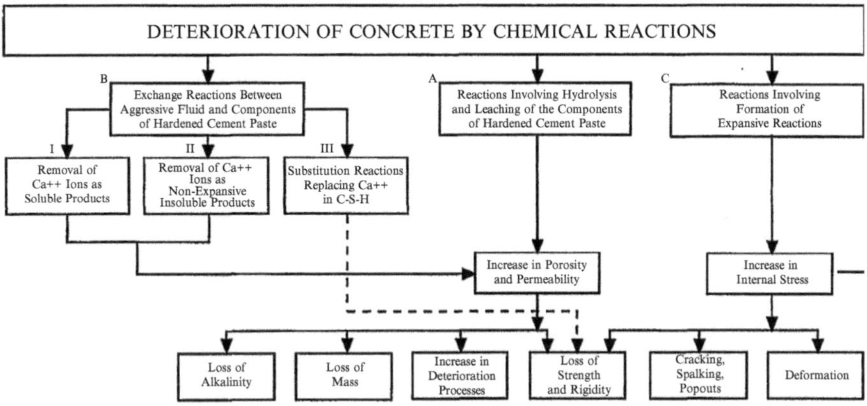

Fig. 4.7 Scheme of deterioration of concrete by chemical reaction (Regourd 1981)

Different examples are given by the authors including:

B_I: Acidic solutions forming calcium chloride, calcium acetate, etc.
A: Soft water attack on calcium hydroxide and C-S-H

Following this classification, leaching by pure water and strong acid with soluble salts matches the A and B_I deterioration pathways. Both lead to the increase in porosity and permeability of the matrix. Gradual degradation of engineering properties of concrete can be observed. Among the signs of leaching are the loss of alkalinity, loss of mass, loss of strength and rigidity. More specifically, consequences of the leaching phenomena included the peeling-off of the concrete skin, with the presence of a pale rusty brown outer layer.

The changing mineralogy due to the leaching processes causes a loss of strength dependant on the mortar type (Allahverdi and Skvara 2000b). The leaching of CH creates greater porosity than the leaching of C-S-H, the leaching of the former being identified as responsible for the increase in the volume of pores larger than 0.2 μm, while the decalcification of the latter increases the volume of pores smaller than 0.2 μm (Haga et al. 2005b). Nevertheless, according to Carde and François (1999) the leaching of C-S-H is mostly responsible for the loss of strength during cement-based material leaching. Young's Modulus in the leached zones can decrease by more than 80% (Stora et al. 2009). The increase in creep of cementitious specimens subjected to leaching has also been investigated and quantified by several authors (Sellier et al. 2011; Le Bellégo et al. 2000).

4.3.1.3 Kinetics

The corrosion processes are very fast at the beginning and controlled by the diffusion of the aggressive substance through solution to the surface and/or by the chemical processes at the surface itself (Pavlik 1994a; Moskvin et al. 1980; Adenot

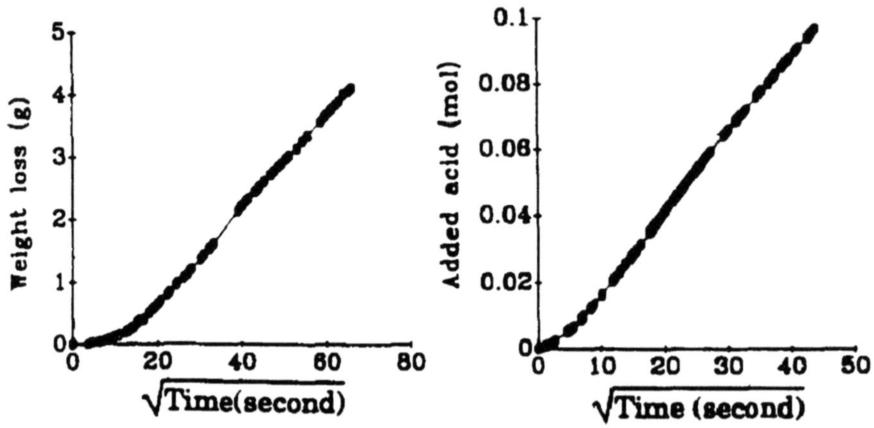

Fig. 4.8 Linear relationship between weight loss and added acid as a function of the square root of time for cement-based material subjected to leaching (Adenot and Buil 1992)

and Buil 1992). The leaching causes an opening of the porosity and thus an increase of the permeability of the concrete. Thus, the corrosion process is controlled by the thickness and the diffusion resistance of the protective layer. The degradation rate can be described by Fick's law relating the degraded thickness to the square root of immersion time in aggressive solution (Carde et al. 1996). According to Adenot and Buil (1992), the various dissolution/precipitation fronts are proportional to the square root of time as long as an unleached zone exists as shown in Fig. 4.8.

The pH of the water has a strong influence on the leaching kinetics. The degradation kinetics will be influenced by ion transfers caused by concentration gradients between the corrosive water and the pore water of the cement paste and by the dissolution/precipitation due to the variation in concentrations of the various elements in the pore water of the cement paste. The presence of calcium salts has a retarding effect on the rate at which the calcium hydrate is leached. However, the solubility of portlandite is increased in the presence of alkali salts which consequently tend to accelerate leaching (Duchesne and Reardon 1995).

Moving water is more detrimental to a concrete structure than stagnant water. If water moves at a low velocity along the concrete surface, the calcium concentration in the adjacent layer attains a high value, so that the diffusion of calcium from concrete is greatly retarded and may even be stopped entirely in stagnant groundwater. However, where concrete is surrounded by flowing groundwater, an increase of calcium concentration is prevented by the steady water supply, and the diffusion of calcium from concrete is maintained at a uniform rate. In this manner, a layer poor in calcium forms on the concrete surface and acts as a barrier between groundwater and sound concrete, preventing the diffusion of calcium from the concrete (Biczok 1967).

The following sections detail the degradation mechanisms specific to strong acids. The differences are minor and mainly relate to the properties of the outer layer of the degraded cementitious matrix.

4.3.2 Specificity of the Degradation Mechanisms by Strong Acids

This section is concerned with the attack of concrete by strong acids producing soluble calcium salts.

According to Allahverdi and Skvara (2000a), the chemical reactions of OPC with a mono-proton acid of the general formula HA can be written as follows:

Direct attack of acid on portlandite:

$$Ca(OH)_2 + 2HA \rightarrow Ca^{2+} + 2A^- + 2H_2O \tag{4.1}$$

Decomposition of hydration products e.g. C-S-H:

$$x\text{CaO} \cdot y\text{SiO}_2 \cdot n\text{H}_2\text{O} + 2x\text{HA} \rightarrow x\text{Ca}^{2+} + 2x\text{A}^- + y\text{Si(OH)}_4 + (x + n - 2y)\text{H}_2\text{O} \tag{4.2}$$

The corresponding reactions for high alumina cement (HAC) are as follows:

The first reaction is the aluminate hydrate decomposition by the acid:

$$x\text{CaO} \cdot y\text{Al}_2\text{O}_3 \cdot n\text{H}_2\text{O} + 2x\text{HA} \rightarrow x\text{Ca}^{2+} + 2x\text{A}^- + 2y\text{Al(OH)}_3 + (x + n - 3y)\text{H}_2\text{O} \tag{4.3}$$

The first reaction is followed by attack on the aluminium hydroxide if the pH value of the acidic solution is below 4:

$$2y\text{Al(OH)}_3 + 6y\text{HA} \rightarrow 2y\text{Al}^{3+} + 6y\text{A}^- + 3y\text{H}_2\text{O} \tag{4.4}$$

All the phases present in OPC and HAC react with acidic solutions and dissolution of phases occurs either congruently such as portlandite or incongruently such as C-S-H phase causing residual layers on the exposed surface of material. The severity of the acid attack is significantly dependent on solubility of the calcium salt of the acid and stability of the corroded layer.

4.3.2.1 Preamble on Chemical Parameters and Aggressiveness of Strong Acids

According to the reviews presented by Zivica and Bajza (2001, 2002), the degree of aggressiveness of an acid is dependent on the chemical character of the acid-forming anion present. The severity of the acid attack is notably dependent on the solubility of the calcium salt (Biczok 1967; Bertron et al. 2007; Bertron and Duchesne 2012; Pavlik 1994a; Beddoe and Dorner 2005), other influencing parameters, such as the molar volume of the salt, its mesoscopic shape or its affinity for the cementitious matrix being also mentioned (Larreur-Cayol et al. 2011).

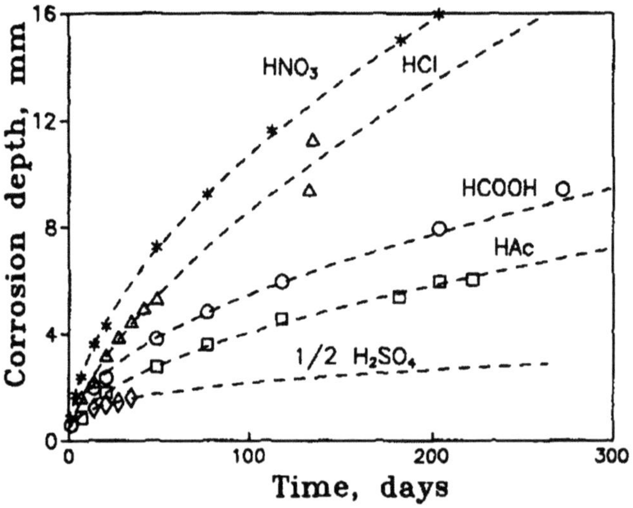

Fig. 4.9 Effect of different acids on the corrosion depth (Pavlik 1994a)

Acids with soluble Ca-salts such as nitric or hydrochloric acid have a strongly decalcifying effect and can form a porous corrosive layer with relatively weak diffusion resistance (Revertegat et al. 1992; Pavlik 1994b; Pavlik and Uncik 1997). Because the acid attack is based on the processes of decomposition and leaching of the constituents of the cement matrix, the conditions of transport phenomena, such as the supply of aggressive acidic solutions and removal of the products of the attack is of specific significance for the intensity of attack (Lea 1971; Mehta 1985, 1986).

Figure 4.9 presents the effect of different acids on the corrosion depth measured as a function of time. Here, nitric (HNO_3 – strong acid) and hydrochloric (HCl – strong acid) acids present more aggressive behaviour compared to sulphuric (H_2SO_4 – strong acid), acetic (CH_3COOH – weak acid, denoted HAc on Fig. 4.9 by the author) and formic (HCOOH – weak acid) acids. Among acids producing soluble Ca salts, strong acids are more aggressive than weak acids for a given concentration of acid.

pH and Concentration of Acid

According to many authors (Zivica and Bajza 2001, 2002; Bertron et al. 2007; Pavlik 1994a), the pH value gives no correct information about the real concentration of acid in the solution and its aggressiveness, because pH values are dependent on the dissociation degree of the acid. Therefore, a strong acid (e.g. HCl, HNO_3) may already cause, in small quantities in solution, significantly lower pH values than a weak acid in multiple quantities in the solution, owing to its low dissociation degree. For the aggressiveness of acidic solutions, the real concentration of acid is more significant than pH value. Figure 4.10 from Pavlik (1994a) presents the effect of the concentration of nitric acid on the corrosion depth measured on cement paste.

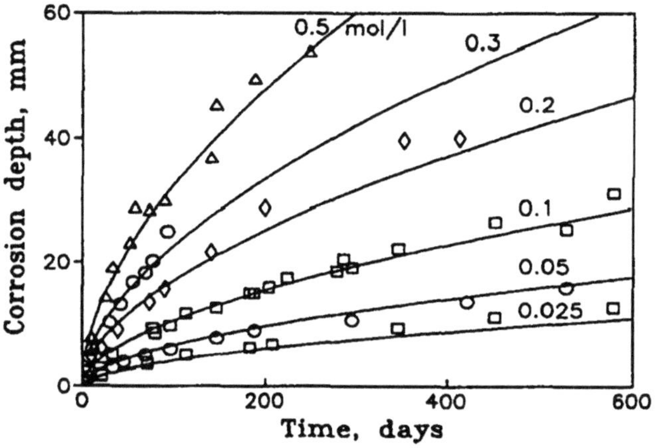

Fig. 4.10 Effect of acid concentration on corrosion depth (Pavlik 1994a – data for nitric acid)

Fig. 4.11 Effect of acid concentration on corrosion depth for a strong (nitric) acid and a weak (acetic) acid (Pavlik 1994a)

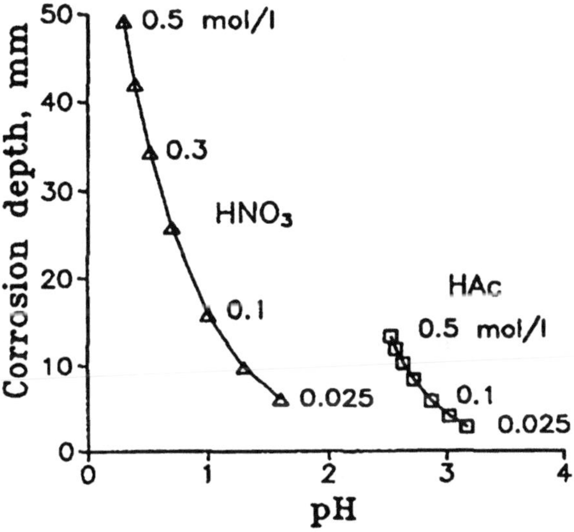

In general, the intensity of aggression increases with an increase in acid concentration (Fig. 4.11), except for some anomalous behaviour observed in the case of sulphuric acid (Pavlik 1994a; Fattuhi and Hugues 1988; Kong and Orbison 1987).

Solubility of Calcium Salts

Table 4.1 presents the solubility in water of calcium salts of some acids. The calcium, from the reaction between the acid and portlandite ($Ca(OH)_2$) and C-S-H of the cement paste, is leached from the concrete by water if the Ca-salt of

Table 4.1 Solubility and other properties of calcium salts of some strong acids

Calcium salt of acid	Formula	Molar mass (g/mol)	Density	Molar volume (cm^3/mol)	Water solubility (g/100 cc) (Lide 1994)
Calcium sulfate	$CaSO_4{\cdot}2H_2O$	172.17	2.32	74.21	0.241 (0°C)
Calcium chloride	$CaCl_2{\cdot}6H_2O$	219.08	1.71	128.12	536 (20°C)
Calcium nitrate	$Ca(NO_3)_2{\cdot}4H_2O$	236.15	1.896	124.55	266 (0°C) 660 (30°C)

Table 4.2 Chemical composition of corroded layers (Pavlik 1994b)

	SiO_2 + i.r.	Fe_2O_3	Al_2O_3	CaO + MgO	Fe_2O_3/SiO_2
Outer zone – corroded layer – white part	96.7	0.2	1.1	2.0	0.22
Corroded layer – brown part	79.1	15.7	2.7	2.45	19.85
Uncorroded cement paste	23.1	3.7	3.7	66.5	15.94

i.r. insoluble residue

the acid is soluble. Soluble calcium salts are, for example, $CaCl_2$ and $Ca(NO_3)_2$ which can form under the action of HCl and HNO_3 respectively (Biczok 1967; Pavlik 1994a).

4.3.2.2 Attack by Nitric Acid

The corrosion of hardened cement paste by nitric acid has been studied in detail by Pavlík (1994a, b), who reported that the corroded layer developed by the action of nitric acid solution with concentrations ranging between 0.025 (pH = 1.6) to 0.5 mol L^{-1} (pH = 0.3) is soft, and porous with visible cracks. The colour of such a corroded layer is white with the exception of a relatively thin brown zone adjacent to the uncorroded cement paste.

The final reaction products between the hydrates and the acid are the corresponding calcium salt of the acid (if not soluble) as well as hydrogels of silicon, aluminium, and ferric (hydr)oxides (Reardon 1990; Bertron et al. 2007; Chandra 1988; De Ceukelaire 1992; Delagrave et al. 1996). The solubility of aluminium and ferric (hydr)oxides depends on the pH value of the acting solution, while SiO_2 is relatively insoluble in acidic solutions. Aluminium (hydr)oxide dissolves at pH values lower than 3–4 and the iron (hydr)oxides at values lower than 1.5–2 (Pavlik 1994a). Pavlik (1994b) analyzed the chemical composition of the two main zones in the corroded layer (Table 4.2). The white zone of the corroded layer was almost totally composed of SiO_2, i.e. 96.7 wt.%, while the brown zone contained a markedly increased amount of ferric (hydr)oxide, i.e. 15.7 wt.% Fe_2O_3 besides a high content of SiO_2 , i.e. 79.1 wt.%.

Nitric acid attack is a typical acidic corrosion causing volume reduction (shrinkage) of the corroded layer due to leaching of highly soluble calcium nitrate. Pavlík and Unčík (1997) reported values of approximately 13 and 7% of shrinkage per length of the corroded layer for OPC paste with a w/c = 0.5 exposed for a period of nearly 300 days to solutions of nitric and acetic acids respectively. Such volume contractions of the corroded layer, especially for the case of nitric acid, can result in the formation of visually observable cracks across the corroded layer. In the presence of these cracks, the transport rate of acid and corrosion products to and from the corrosion front increases and this accelerates the process of deterioration.

Moudilou et al. (2002) used a dynamic leaching device which is a modified Soxhlet-type leaching device to assess trace metals released by pure cement pastes. Nitric acid kept at a constant pH of 5 was used to maintain acidity in the leaching cell. The calcium content leached from the cement paste was monitored over time and it was confirmed that the release of Ca ions was governed by a diffusion process. Effective diffusion coefficients were calculated and values of 1.58×10^{-9} and 3.22×10^{-9} $cm^2\ s^{-1}$ were obtained for temperature of 20 and 40°C, respectively.

4.3.2.3 Attack by Hydrochloric Acid

The effect of hydrochloric acid on cement based materials has been studied or reported by various authors (Allahverdi and Skvara 2000a; Chandra 1988; De Ceukelaire 1992; Israel et al. 1997). Hydrochloric acid attacks the cement based material and allows the leaching of Ca. The specimen loses its binding ability and its porosity is increased. Si, Al, and Fe gels are produced. The mixture of these hydroxides concentrates in a layer covering the undamaged concrete. It is also possible that some F-S, A-S or S-A-S complexes are formed (Chandra 1988). As reported by Pavlik (1994b), Moskvin et al. (1980) divide the corroded layer formed in hydrochloric acid into 2 or 3 zones. The first zone (innermost) is the zone where the components of the cement paste react with the aggressive medium. This thin zone lies at the boundary of the uncorroded cement paste delineated by pH values between 12 and 12.5 on the side of the cement paste and by pH between 4 and 5 on the other side. The second and the third zones are composed of corrosion products. The second zone is composed mainly of the hydrated oxides of Si, A1 and Fe; the third and outermost zone is composed mainly of hydrated silica. Aluminium (hydr)oxides dissolve at pHs less than 3, and ferric (hydr)oxides at pHs less than 1.5 (Moskvin et al. 1980). Other authors predict the presence of $Fe(OH)_3$ in the layer at pH above 2 and $Al(OH)_3$ at pH above 4 (Zivica and Bajza 2001; Pavlik 1994a; Chandra 1988).

The detrimental action of chloride ions on the deterioration mechanisms of the hydrated cement paste is not solely limited to the decalcification process. Chloride ions can also react with the aluminates to form new chloride-bearing compounds – calcium monochloroaluminates or Friedel's salt (3 $CaO{\cdot}Al_2O_3{\cdot}CaCl_2{\cdot}10H_2O$) and chloroferrite hydrates ($C_3F{\cdot}CaCl_2{\cdot}10H_2O$) (Israel et al. 1997; Midgley and Illston 1984). Upon crystallisation, chloroaluminates can form expansive compounds, and generate significant tensile pressures in the cement paste which may lead to microcracking causing disintegrating stresses in concrete, which falls apart. Delagrave et al. (1996) have observed formation of Friedel's salt only in mixtures with high water/cement ratio. These authors observed that the decrease of the water to binder ratio reduces significantly the rate of deterioration. Chloroaluminate crystals were observed only in the cement pastes having a water to binder ratio of 0.38. For w/c lower than 0.25, no chloroaluminates were found (Delagrave et al. 1996; Delagrave 1996).

Although the amount of chloroaluminates seems to be influenced by the chloride concentration of the solution, the formation of this compound appears to be mainly affected by the amount of aluminates in the hydrated cement paste (Midgley and Illston 1984). When present in very high concentrations (>15%), chloride ions can also form expansive Ca-oxychloride compounds which can cause severe deterioration to concrete.

4.4 Factors Influencing Kinetics and Durability

This section is mainly based on the extensive review published by Allahverdi and Skvara (2000a, b) to which was incorporated the works of other authors among whom those of Pavlik (1994a, b).

It is well known that the corrosion process is very rapid at the start of acid attack. At this stage, the process is controlled by the diffusion of aggressive agents through solution to the surface, and/or by the chemical processes taking place at the surface itself. As corrosion proceeds, the corroded layer increases in thickness, and thus it becomes more protective. This results in the process ultimately becomes controlled by diffusion of aggressive agents through the protective layer (Pavlik 1994b; Pavlik and Uncik 1997; Grube and Rechenberg 1989; Kamali et al. 2003). By contrast, there are cases where the corroded layer is removed continuously; the corrosion rate then becomes controlled by transport processes across the solid-liquid phase boundary, or by the rate of the corrosion reactions (Grube and Rechenberg 1989).

According to Allahverdi and Skvara (2000b), the rate of deterioration of acid attack may be influenced by a number of factors which can be classified into three groups:

1. Factors related to the leaching solution: pH of the solution (linked with the initial pH but also with experimental procedure characteristics such as frequency of

renewal if the pH is not maintained by the addition of acid), strength of acid, concentration of acid or pH value, diffusivity of acid, and mobility of the acidic solution.

2. Factors related to cementitious material: type of cement (chemical composition, mineralogy, thermodynamic stability, stability and thickness of the corroded layer, permeability, cement content, curing time, neutralizing capacity of the material, and influence of aggregate).
3. Other factors including temperature and capillary suction forces acting under the alternative cycles of wetting and drying.

The main factors are described hereafter.

4.4.1 Diffusivity of Acid

Different acids have different diffusivities in the pore structure of the cement matrix. In the presence of acids such as acetic, nitric, and hydrochloric, the formation of a core layer was observed (Pavlik 1994a; Israel et al. 1997). Diffusivity of dissolved calcium has an important role and is dependent on the gradient between the pore solution and the aggressive solutions and on the microstructure of the cement matrix. It was observed that a reduction in w/c ratio reduces the diffusivity and slows down the process of degradation. The role of supplementary cementing materials on the porosity and thus the diffusivity is not clear; conflicting results are available in the scientific literature (Pavlik and Uncik 1997; Fattuhi and Hugues 1988; Sersale et al. 1998; Zivica 2004; Bertron et al. 2007; Oueslati 2011; Oueslati and Duchesne 2009, 2012).

4.4.2 Stability of the Corroded Layer

The stability of the corroded layer is another important factor which influences the rate of deterioration. The resistance of the corroded layer is influenced by its chemical composition and by the presence of external mechanical effects including the flow rate and the properties of the aggressive fluid. The corroded layer may be stable due to its high Si and Al contents and/or to the absence of any external mechanical effects and acts as a protective layer reducing the transport of the aggressive solution to the corrosion front. In the presence of a stable protective layer, the degradation depth or weight loss follows a parabolic function with time (Allahverdi and Skvara 2000b; Pavlik 1994a, 1996; Revertegat et al. 1992; Adenot and Buil 1992; Oueslati 2011). The corroded layer can also be removed continuously. In this case, the parabolic function becomes to a linear one (Allahverdi and Skvara 2000b).

Table 4.3 Empirical equations to model the corrosion depth (Allahverdi and Skvara 2000b)

Equation	Acid – concentration	Cement	Authors
$d(mm) = 0.174C^{0.579}\ t^{0.583}$ d = corrosion depth C = acid concentration (vol. %) t = time (days)	0.5–5 vol.% formic acid	OPC w/c = 0.4	Bajza (1992)
$d(mm) = 1.232C^{0.525}\ t^{0.52}$ d = corrosion depth C = acid concentration (mol/L) t = time (days)	0.025–0.5 mol L^{-1} acetic acid	OPC w/c = 0.4	Pavlik (1994a)
$d(mm) = 4.11C^{0.7}\ t^{0.56}$ d = corrosion depth C = acid concentration (mol/L) t = time (days)	0.025–0.5 mol L^{-1} nitric acid	OPC w/c = 0.4	Pavlik (1994a)
$m = kC^{0.4074}\ t^{0.5572}$ m = dissolved CaO ($g\ m^{-2}$) C = acid concentration (mol/L) t = time (min) k = a constant depending on pH	Hydrochloric acid	OPC w/c = 0.5	Herold (1997)

4.4.3 Capillary Suction Forces

When materials are subjected to alternate wetting and drying cycles, the degradation process can be accelerated by the action of capillary suction forces (Allahverdi and Skvara 2000b; Attiogbe and Rizkalla 1988). These capillary forces cause acid to penetrate into the material during wetting by suction, while during drying microcracking can occur due to the accompanying shrinkage, which results in a higher penetrability of the material (Israel et al. 1997).

Some authors have published empirical equations to express the corrosion depth of hardened cement paste as a function of the acid concentration and time. Table 4.3 presents the empirical equations reported by Allahverdi and Skvara (2000b).

4.5 Conclusion

Cement based materials are highly alkaline and easily attacked by pure water and strong acids. The mechanism of the acid attack is mainly a decalcification of the cement hydration products. More precisely, calcium hydroxide will dissolve at a pH value of 12.5, followed by ettringite at pH value of 10.7, C-S-H at pH between ~10.5 and 8.8 (according to different authors) and finally calcium aluminate and ferrite hydrates will decompose successively until a C-A-S-H gel, in the case of leaching at pH above 7, or a silica gel residue containing Al and Fe, is obtained at pH between 1 and 6. In all cases the outer layer shows higher porosity and significant loss of strength.

References

Adenot, F., Buil, M.: Modelling of the corrosion of the cement paste by deionized water. Cem. Concr. Res. **22**, 489–496 (1992)

Allahverdi, A., Skvara, F.: Acidic corrosion of hydrated cement based materials. Part 1: Mechanism of the phenomenon. Ceram. Silik. **44**(3), 114–120 (2000a)

Allahverdi, A., Skvara, F.: Acidic corrosion of hydrated cement based materials. Part 2: Kinetics of the phenomenon and mathematical models. Ceram. Silik. **44**(4), 152–160 (2000b)

Attiogbe, E.K., Rizkalla, S.H.: Response of concrete to sulphuric acid attack. ACI Mater. J. **85**, 481–488 (1988)

Bajza, A.: Corrosion of hardened cement pastes by formic acid solutions. In: Proceedings of the 9th International Conference on the Chemistry of Cement, New Delhi, V, pp. 402–408 (1992)

Beddoe, R.E., Dorner, H.W.: Modelling acid attack on concrete: Part I. The essential mechanisms. Cem. Concr Res. **35**, 2333–2339 (2005)

Bernard, F., Kamali-Bernard, S., Prince, W.: 3D multi-scale modelling of mechanical behaviour of sound and leached mortar. Cem. Concr. Res. **38**(4), 449–458 (2008)

Berner, U.R.: Evolution of pore water chemistry during degradation of cement in a radioactive waste repository environment. Waste Manag. **12**, 201–219 (1992)

Bertron, A., Duchesne, J.: Attack of cementitious materials by organic acids in agricultural and agrofood effluents. In: Alexander, M., Bertron, A., De Belie, N. (eds.) RILEM 211-PAE State-of-the-Art Report: Concrete in Aggressive Aqueous Environments (2012)

Bertron, A., Duchesne, J., Escadeillas, G.: Degradation of cement pastes by organic acids. Mater. Struct. **40**(3), 341–354 (2007)

Biczok, I.: Concrete Corrosion and Concrete Protection. Chemical Publishing Company/Publishing House of the Hungarian Academy of Sciences, Budapest/Hungary (1967). 543 pp

Burlion, N., Bernard, D., Chen, D.: X-ray microtomography: application to microstructure analysis of acementitious material during leaching process. Cem. Concr. Res. **36**(2), 346–357 (2006)

Carde, C., François, R.: Modelling the loss of strength and porosity increase due to the leaching of cement pastes. Cem. Concr. Compos. **21**, 181–188 (1999)

Carde, C., François, R., Torrenti, J.M.: Leaching of both calcium hydroxide and C-S-H from cement paste: modeling the mechanical behaviour. Cem. Concr. Res. **26**, 1257–1268 (1996)

Chandra, S.: Hydrochloric acid attack on cement mortar – an analytical study. Cem. Concr. Res. **18**, 193–203 (1988)

Constantinides, G., Ulm, F.J.: The effect of two types of C-S-H on the elasticity of cement-based materials: results from nanoindentation and micromechanical modeling. Cem. Concr. Res. **34** (1), 67–80 (2004)

De Ceukelaire, L.: The effects of hydrochloric acid on mortar. Cem. Concr. Res. **22**, 903–914 (1992)

Delagrave, A.: Mécanismes de pénétration des ions chlore et de dégradation des systèmes cimentaires normaux et à haute performance. Ph.D. thesis, Université Laval, Québec, 362 p (1996)

Delagrave, A., Pigeon, M., Revertégat, E.: Influence of chloride ions and pH level on the durability of high performance cement pastes. Cem. Concr. Res. **24**(8), 1433–1443 (1996)

Duchesne, J., Reardon, E.J.: Measurement and prediction of portlandite solubility in alkali solutions. Cem. Concr. Res. **25**, 1043–1053 (1995)

Fattuhi, N.I., Hugues, B.P.: The performance of cement paste and concrete subjected to sulphuric acid attack. Cem. Concr. Res. **18**, 545–553 (1988)
Faucon, P., Le Bescop, P., Adenot, F., Bonville, P., Jacquinot, J.F., Pineau, F., Felix, B.: Leaching of cement: study of the surface layer. Cem. Concr. Res. **26**(11), 1707–1715 (1996)
Faucon, P., Adenot, F., Jorda, M., Cabrillac, R.: Behaviour of crystallised phases of Portland cement upon water attack. Mater. Struct. **30**, 480–485 (1997)
Faucon, P., Adenot, F., Jacquinot, J.F., Petit, J.C., Cabrillac, R., Jorda, M.: Long-term behaviour of cement pastes used for nuclear waste disposal: review of physico-chemical mechanisms of water degradation. Cem. Concr. Res. **28**, 847–857 (1998)
Glasser, F.P., Marchand, J., Samson, E.: Durability of concrete – degradation phenomena involving detrimental chemical reactions. Cem. Concr. Res. **38**(2), 226–246 (2008)
Grube, H., Rechenberg, W.: Durability of concrete structures in acidic water. Cem. Concr. Res. **19**, 783–792 (1989)
Haga, K., Shibata, M., Hironaga, M., Tanaka, S., Nagasaki, S.: Change in pore structure and composition of hardened cement paste during the process of dissolution. Cem. Concr. Res. **35**, 943–950 (2005a)
Haga, K., Sutou, S., Hironaga, M., Tanaka, S., Nagasaki, S.: Effects of porosity on leaching of Ca from hardened ordinary Portland cement paste. Cem. Concr. Res. **35**, 1764–1775 (2005b)
Harrison, W.H.: Durability of concrete in acidic soils and waters. Concrete **21**, 18–24 (1987)
Henocq, P.: Modélisation des interactions ioniques à la surface des silicates de Calcium Hydratés. Ph.D. thesis, Université Laval, Québec (2005)
Herold, G.: In: Scrivener, K.L., Young, J.F. (eds.) Mechanisms of Chemical Degradation of Cement-based Systems, pp. 98–105. E & FN Spon, London (1997)
Heukamp, P.H., Ulm, P.J., Germaine, J.T.: Mechanical properties of calcium leached cement pastes: triaxial stress and the influence of the pore pressure. Cem. Concr. Res. **31**, 767–774 (2001)
Hidalgo, A., Petit, S., Domingo, C., Alonso, C., Andrade, C.: Microstructural characterization of leaching effects in cement pastes due to neutralisation of their alkaline nature: Part I: Portland cement pastes. Cem. Concr. Res. **37**(1), 63–70 (2007)
Israel, D., Macphee, D.E., Lachowski, E.E.: Acid attack on pore-reduced cements. J. Mater. Sci. **32**, 4109–4116 (1997)
Jain, J., Neithalath, N.: Analysis of calcium leaching behavior of plain and modified cement pastes in pure water. Cem Concr. Compos. **31**, 176–185 (2009)
Kamali, S., Gérard, B., Moranville, M.: Modelling the leaching kinetics of cement-based materials – influence of materials and environment. Cem. Concr. Compos. **25**(4–5), 451–458 (2003)
Kamali, S., Moranville, M., Leclercq, S.: Material and environmental parameter effects on the leaching of cement pastes: experiments and modelling. Cem. Concr. Res. **38**, 575–585 (2008)
Kong, H.L., Orbison, J.G.: Concrete deterioration due to acid precipitation. ACI Mater. J. **84**, 110–116 (1987)
Larreur-Cayol, S., Bertron, A., Escadeillas, G.: Alteration of cement-based materials by various organic acids in agro-industrial waste-waters. Cem. Concr. Res. **41**(8), 882–892 (2011)
Le Bellégo, C., Gerard, B., Pijaudier-Cabot, G.: Chemo-mechanical effects in mortar beams subjected to water hydrolysis. J. Eng. Mech. **126**(3), 266–272 (2000)
Lea, F.M.: The Chemistry of Cement and Concrete. Chemical Publishing Company Inc., New York (1971)
Leemann, A., Lothenbach, B., Siegrist, H., Hoffmann, C.: Influence of water hardness on concrete surface deterioration caused by nitrifying biofilms in wastewater treatment plants. Int. Biodeter. Biodegr. **64**(6), 489–498 (2010a)
Leemann, A., Lothenbach, B., Hoffmann, C.: Biologically induced concrete deterioration in a wastewater treatment plant assessed by combining microstructural analysis with thermodynamic modeling. Cem. Concr. Res. **40**(8), 1157–1164 (2010b)
Lide, D.R.: CRC Handbook of Chemistry and Physics, 75th edn. CRC Press Inc., Boca Raton (1994)
Mainguy, M., Tognazzi, C., Torrenti, J.M., Adenot, F.: Modelling of leaching in pure cement paste and mortar. Cem. Concr. Res. **30**, 83–90 (2000)

Matschei, T., Lothenbach, B., Glasser, F.P.: The AFm phase in Portland cement. Cem. Concr. Res. **37**, 118–130 (2007)

Mehta, P.K.: Studies on chemical resistance of low water/cement ratio concretes. Cem. Concr. Res. **15**, 969–978 (1985)

Mehta, P.K.: Concrete, Properties and Materials. Prentice-Hall Inc., Englewood Cliffs (1986)

Midgley, H.G., Illston, J.M.: The penetration of chlorides into hardened cement pastes. Cem. Concr. Res. **14**, 546–558 (1984)

Moranville, M., Kamali, S., Guillon, E.: Physicochemical equilibria of cement-based materials in aggressive environments—experiment and modelling. Cem. Concr. Res. **34**(9), 1569–1578 (2004)

Moskvin, V.M., Ivanov, F.M., Alexejev, S.N., Guzeev, E.A.: Corrosion of Concrete and Reinforced Concrete, Methods of Their Protection. Strojlzdat, Moscow (1980)

Moudilou, E., Bellotto, M., Defosse, C., Serclerat, I., Baillif, P., Touray, J.C.: A dynamic leaching method for the assessment of trace metals released from hydraulic binders. Waste Manag. **22**, 153–157 (2002)

Myneni, S.C.B., Traina, S.J., Logan, T.J.: Ettringite solubility and geochemistry of the Ca $(OH)_2$–$Al_2(SO_4)_3$–H_2O system at 1 atm pressure and 298 K. Chem. Geol. **148**, 1–19 (1998)

Oueslati, O.: Durabilité des matériaux cimentaires soumis aux acides organiques – Résistance chimique, mécanique et de corrosion. Ph.D. thesis, Université Laval, Quebec (2011)

Oueslati, O., Duchesne, J.: Acetic attack of blended cement paste. In: Alexander, M.G., Bertron, A. (eds.) Proceedings of the RILEM TC 211-PAE Final Conference, Toulouse, vol. 1, pp. 108–115 (2009)

Oueslati, O., Duchesne, J.: The effect of SCMs and curing time on resistance of mortars subjected to organic acids. Cem. Concr. Res. **42**(1), 205–214 (2012)

Pavlik, V.: Corrosion of hardened cement paste by acetic and nitric acids; Part I: Calculation of corrosion depth. Cem. Concr. Res. **24**, 551–562 (1994a)

Pavlik, V.: Corrosion of hardened cement paste by acetic and nitric acids; Part II: Formation and chemical composition of the corrosion products layer. Cem. Concr. Res. **24**, 1495–1508 (1994b)

Pavlik, V.: Corrosion of hardened cement paste by acetic and nitric acids; Part III: Influence of water/cement ratio. Cem. Concr. Res. **26**, 475–490 (1996)

Pavlik, A., Uncik, S.: The rate of corrosion of hardened cement pastes and mortars with additive of silica fume in acids. Cem. Concr. Res. **27**, 1731–1745 (1997)

Reardon, E.J.: An ion interaction model for the determination of chemical equilibria in cement/water systems. Cem. Concr. Res. **20**, 175–192 (1990)

Regourd, M.: 32-RCA: resistance of concrete to chemical attack. Mater. Struct. **14**(80), 130–137 (1981)

Revertegat, E., Richet, C., Gegout, P.: Effect of pH on the durability of cement pastes. Cem. Concr. Res. **22**, 259–272 (1992)

Sand, W.: Microbial mechanisms of deterioration of inorganic substrates—a general mechanistic overview. Int. Biodeter. Biodegr. **40**(2–4), 183–190 (1997)

Sellier, A., Buffo-Lacarrière, L., El Gonnounia, M., Bourbon, X.: Behavior of HPC nuclear waste disposal structures in leaching environment. Nucl. Eng. Des. **241**(1), 402–414 (2011)

Sersale, S., Frigione, G., Bonavita, L.: Acid depositions and concrete attack: main influences. Cem. Concr. Res. **28**, 19–24 (1998)

Siegrist, H., Gujer, W.: Demonstration of mass transfer and pH effects in a nitrifying biofilm. Water Res. **21**, 1481–1487 (1987)

Stora, E., Bary, B., He, Q.C., Deville, E., Montarnal, P.: Modelling and simulations of the chemo–mechanical behaviour of leached cement-based materials: leaching process and induced loss of stiffness. Cem. Concr. Res. **39**(9), 763–772 (2009)

Taylor, H.F.W.: Cement Chemistry, 2nd edn. Thomas Telford, London (1997)

Zivica, V.: Acidic attack of cement based materials – a review. Part 3: Research and test methods. Construct. Build Mater. **18**, 683–688 (2004)
Zivica, V., Bajza, A.: Acidic attack of cement based materials: a review. Part 1. Principle of acidic attack. Construct. Build Mater. **15**, 331–340 (2001)
Zivica, V., Bajza, A.: Acidic attack of cement based materials: a review. Part 2. Factors of rate of acidic attack and protective measures. Construct. Build Mater. **16**, 215–222 (2002)

Chapter 5
Ammonium Nitrate Attack on Cementitious Materials

Gilles Escadeillas

Contents

Keywords Ammonium nitrate • Leaching • Accelerated testing • Degradation mechanisms • Test methods • Kinetics • Dissolution • Crystallization

G. Escadeillas (✉)
LMDC (Laboratoire Matériaux et Durabilité des Constructions), Université de Toulouse; UPS, INSA, 135, avenue de Rangueil, F-31 077 Toulouse Cedex 04, France
e-mail: Gilles.Escadeillas@insa-toulouse.fr

M. Alexander et al. (eds.), *Performance of Cement-Based Materials in Aggressive Aqueous Environments*, RILEM State-of-the-Art Reports 10,
DOI 10.1007/978-94-007-5413-3_5, © RILEM 2013

5.1 Introduction

Of the factors that can quickly degrade concrete, ammonium salts are among the most aggressive (Lea 1965; Biczok 1972). Ammonium nitrate, a very commonly used fertilizer, is at the origin of severe degradation of concrete structures, especially in factories manufacturing these fertilizers or in storage silos.

In these environments and with very high concentrations of ammonium nitrate solutions, it is important to note that even high-performance concretes, either with or without silica fume, can also be susceptible to corrosion (Schneider and Chen 1998). Moreover, calcium aluminate cement concrete is not durable (Escadeillas and Hornain 2008).

Recently, ammonium nitrate has been used in the laboratory to accelerate the kinetics of degradation of cementitious materials for durability studies (in place of pure water leaching) (Carde et al. 1997; Moranville et al. 2004; Bernard et al. 2008; Kamali et al. 2008; Stora et al. 2009). The test is even used as a general durability indicator in a performance-based approach (Assie et al. 2007).

The aim of this section is to present the chemical reactions underlying the degradation of concrete subjected to ammonium nitrate, then the kinetics of these reactions, and finally some tests to evaluate the durability of concrete in this environment.

5.2 Damage Observed on Structures

Damage observed on concrete in contact with ammonium nitrate solution is of two types, depending on the environment:

1. A very large increase in porosity with a weakening of mechanical properties in concrete continuously immersed in a solution of ammonium nitrate.
2. Very notable swelling with occurrence of cracks linked to the formation of expansive crystals in concrete in contact with the air or subjected to wet-dry cycles.

In ammonia factories, the signs of damage (swelling and cracks) are generally observed very early on, within 3 years after the start of production in the plant (Baxi and Patel 1998).

5.3 Chemical Reactions

Chemical attack by ammonium nitrate leads to a very soluble calcium nitrate salt, a slightly soluble calcium nitro-aluminate salt, and ammonia NH_3 emanation. This process mainly induces total leaching of $Ca(OH)_2$ and progressive decalcification of C-S-H.

The chemical reactions can be divided into three types that are, in sequence: (1) dissolution of the cementitious compounds and then, depending on the environment, (2) simple diffusion or (3) crystallization.

5.3.1 Dissolution Process

Ammonium nitrate salts are highly hygroscopic and soluble in water (24 mol l^{-1} at 20°C). They decompose in water according to (5.1):

$$NH_4NO_3 \leftrightarrow NH_4^+ + NO_3^- \tag{5.1}$$

In a basic environment, ammonium ions NH_4^+ (weakly acid) are transformed into aqueous ammonia NH_{3aq} with the formation of H^+ ions according to (5.2):

$$NH_4^+ \leftrightarrow NH_{3Aq} + H^+ \quad (pKa = 9.25) \tag{5.2}$$

Ammonium ions react according to an exchange reaction $2NH_4^+ \rightarrow Ca^{2+}$. The calcium salt produced ($Ca(NO_3)_2$) is highly soluble in water, and therefore this exchange reaction leads to the progressive dissolution of all the calcium-bearing cementitious phases. Moreover, in the presence of cementitious materials (pH between 12 and 13.5), the equilibrium of (5.2) is strongly moved to the right. NH_{3aq} ammonia is the predominant form and hydronium ions (H^+) are released. This production of H^+ decreases pH and accelerates the successive dissolution of the Ca-bearing phases. Portlandite, C-S-H and aluminates (phases AFm and AFt) are dissolved according to (5.3), (5.4), (5.5) and (5.6).

$$Ca(OH)_2: \quad Ca(OH)_2 + 2NH_4NO_3 \rightarrow Ca^{2+} + 2NO_3^- + 2(NH_3, H_2O) \tag{5.3}$$

C-S-H :

$$\begin{aligned} &(CaO)_x(SiO_2)_y(H_2O)_z + 2wNH_4NO_3 \\ &\rightarrow (CaO)_{(x-w)}(SiO_2)_y(H_2O)_{(z-w)} + wCa^{2+}2wNO_3^- \\ &+ 2w(NH_3,\ H_2O) \end{aligned} \tag{5.4}$$

AFm :

$$\begin{aligned} &3CaO \cdot CaSO_4 \cdot Al_2O_3 \cdot 12H_2O + 12NH_4NO_3 \\ &\rightarrow 4Ca^{2+} + SO_4^{2-} + 2Al^{3+} + 12NO_3^- + 12(NH_3, H_2O) + 6H_2O \end{aligned} \tag{5.5}$$

AFt :

$$\begin{aligned} &3CaO \cdot 3CaSO_4 \cdot Al_2O_3 \cdot 32H_2O + 12NH_4NO_3 \\ &\rightarrow 6Ca^{2+} + 3SO_4^{2-} + 2Al^{3+} + 12NO_3^- + 12(NH_3, H_2O) + 26H_2O \end{aligned} \tag{5.6}$$

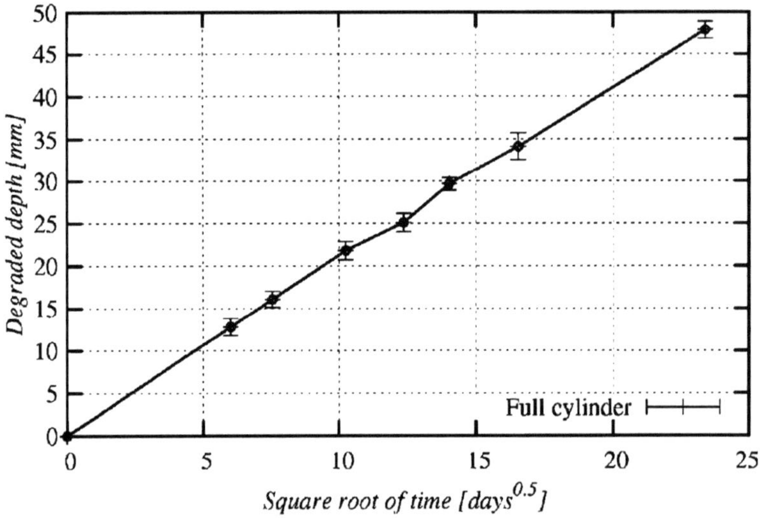

Fig. 5.1 Degraded depth evolution vs. the square root of time for concrete specimens (Nguyen et al. 2007)

The role of gaseous ammonia in this reaction is controversial. Torrenti et al. (2008) assume that the gaseous ammonia released is in no way responsible for the dissolution of portlandite and can be neglected in the material balance for the degradation reaction. For Escadeillas and Hornain (2008), the gaseous ammonia released contributes to a reduction of the basicity, prevents the reaction from reaching equilibrium and permits total leaching without a change of solution.

5.3.2 Diffusion Process

When cementitious materials are immersed in ammonium nitrate solutions, calcium ions diffuse to the exterior without any formation of calcium salts in the material, as the solubility of portlandite increases significantly in these solutions (from 21 mmol l^{-1} in pure water to 2.7 mol l^{-1} for a concentration of 6 mol l^{-1} of ammonium nitrate) (Berner 1988).

In this case, the degradation is governed by a diffusion mechanism and its kinetics can be described by Fick's law, relating the degraded thickness to the square root of the immersion time in the aggressive solution. Figure 5.1 shows this evolution in concrete, but it is interesting to note that the degraded depth develops in the same way for cement paste and mortars (Nguyen et al. 2007).

Figure 5.2 shows that, during this diffusion process, the pH of the solution increases very quickly in the first few days and then progressively reaches a quasi-static value of about 9. In accordance with the thermodynamic equilibrium curve, such as the one established by Berner (1992), the rapid increase of pH in the

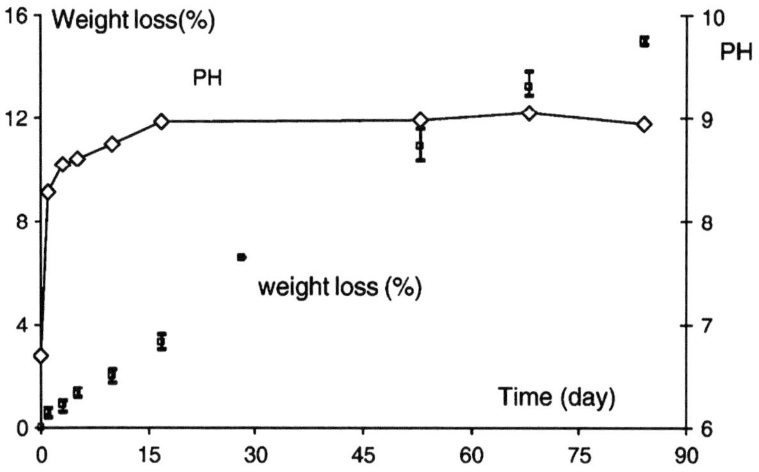

Fig. 5.2 Relative weight loss of cement paste samples and variation of pH of the solution with time (Xie et al. 2008)

first stage may correspond to the fast dissolution of portlandite crystals. The final pH is around the pKa value of the acid-base pair NH_4^+/NH_3 (9.25) for concentrated NH_4NO_3 solution. On the other hand, Fig. 5.2 shows that the weight loss of cement samples increases continuously with time during the leaching process.

The removal of calcium ions was also highlighted by Carde and Francois (1999) with electron microprobe analysis. Figures 5.3 and 5.4 give profiles of calcium concentration in the solid phase obtained on cement pastes cast without or with silica fume and then immersed in the aggressive solution. The results indicate a difference between the two profiles:

- In the presence of portlandite, the profile is almost constant in the inner half of the degraded zone (the release of calcium is attributed to the dissolution of $Ca(OH)_2$, which is confirmed by XRD analysis) and linear in the outer half of the degraded zone (progressive decalcification of C-S-H), and
- In the absence of portlandite, the variation of Ca is a linear function of the depth of the degraded zone (progressive decalcification of C-S-H).

At the same time, leaching in concentrated solutions of ammonium nitrate removes only calcium from the solid while largely preserving silicate and other ions (Chen et al. 2006).

5.3.3 *Crystallization Process*

When concrete is subjected to drying or to wet-dry cycles, salts may crystallize in the partially dry concrete.

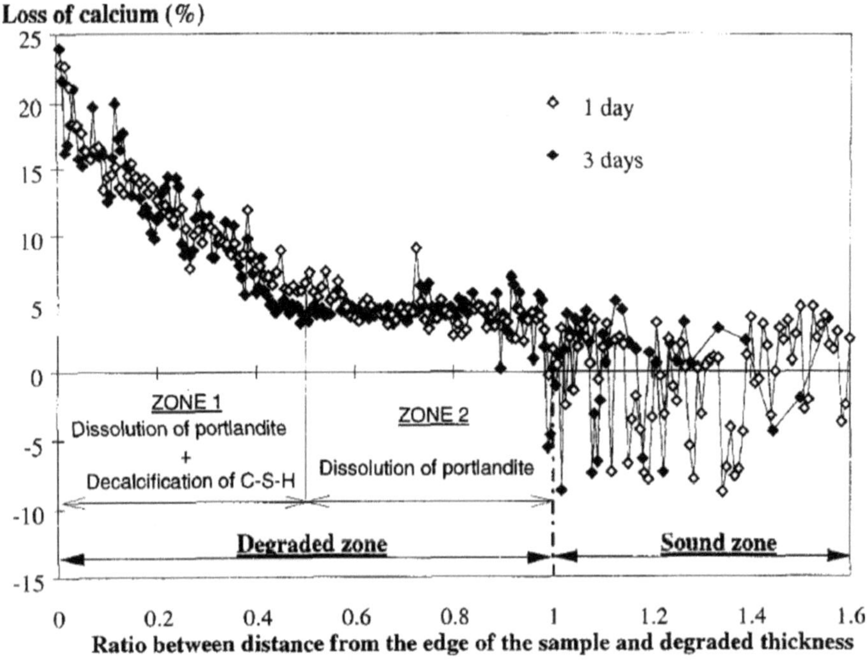

Fig. 5.3 Profiles of loss of calcium in the solid phase for a cement paste without silica fume. Two zones of the same thickness are observed in the degraded zone: (1) both dissolution of portlandite and progressive decalcification of C-S-H, (2) only dissolution of portlandite (Carde and François 1999)

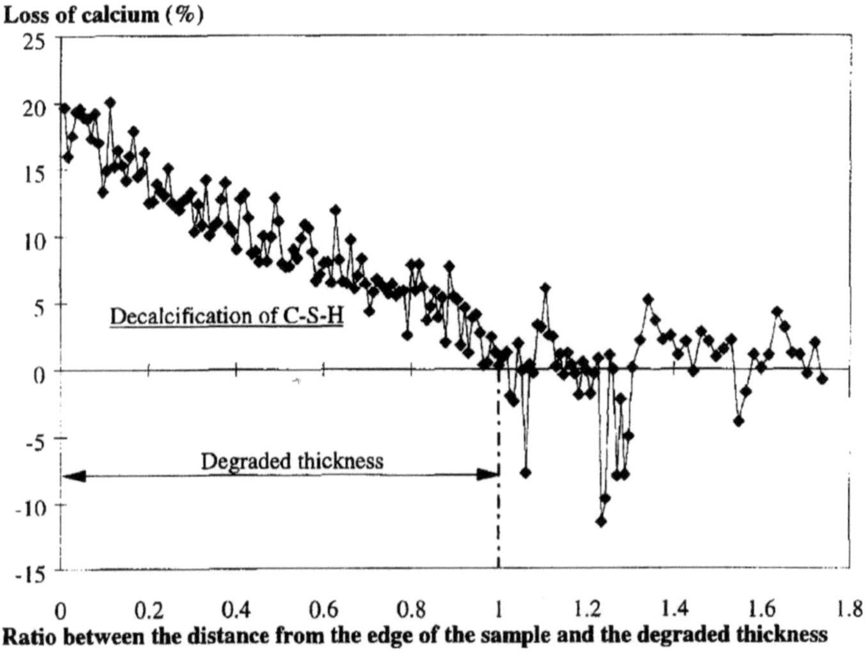

Fig. 5.4 Profile of loss of calcium in the solid phase for paste containing silica fume. The linear profile observed in the whole degraded zone shows that the loss of calcium is due only to the progressive decalcification of C-S-H (Carde and François 1999)

Nitrate ions can react with calcium ions to form highly soluble calcium nitrate $Ca(NO_3)_2$. The general reaction equation is (5.7):

$$Ca(OH)_2 + 2NH_4NO_3 \rightarrow Ca(NO_3)_2 + 2NH_3 + 2H_2O \quad (5.7)$$

Aluminates can also react with nitrate ions to form a less soluble calcium nitro-aluminate salt, $3CaO{\cdot}Al_2O_3{\cdot}Ca(NO_3)_2{\cdot}10H_2O$ (Ukraincik et al. 1978). Similarly, in the presence of sulphates, the possibility of formation of a double salt $CaSO_4{\cdot}(NH_4){\cdot}2SO_4{\cdot}2H_2O$ has also been mentioned (Mohr 1925).

Under these conditions, such crystallization leads to high volume expansion, the formation of cracks and, very quickly, total destruction of the material.

5.3.4 *Kinetics of Degradation*

The kinetics of degradation of cementitious materials by ammonium nitrate have been well studied both through industrial applications (Mohr 1925; Lea 1965; Schneider et al. 1992; Escadeillas 1993; Baxi and Patel 1998) and in accelerated tests to characterize the leaching of these materials for applications related to the storage of radioactive wastes (Carde 1996; Carde et al. 1997; Tognazzi 1998; Tognazzi et al. 2000; Le Bellego et al. 2000; Le Bellego 2001; Kamali 2003; Kamali et al. 2003, 2008; Nguyen et al. 2007; Bernard et al. 2008).

Most tests were conducted by total immersion of the samples in saturated ammonium nitrate solutions (480 g l^{-1} or 6 mol l^{-1}) (Carde et al. 1997; Tognazzi et al. 2000; Heukamp et al. 2001; Loosveldt 2002; Moranville et al. 2004; De Larrard et al. 2009) or in dilute solutions (0.12–1.2 mol l^{-1}) (Bajza et al. 1986; Jauberthie and Rendell 2003; Schneider et al. 1992; Schneider and Chen 1999, 2005).

The degraded depth is generally determined by using phenolphthalein on sawn samples (Schneider and Chen 1998). Phenolphthalein turns from colourless in acidic solutions to pink in basic solutions with the transition occurring around pH 9.5 (Fig. 5.5). However, phenolphthalein does not give the exact position of the dissolution front of portlandite (pH = 12). By comparing the phenolphthalein measurement and SIMS microprobe analysis, Le Bellego et al. (2000) have shown that, for cement paste immersed in ammonium nitrate, the total degraded depth e_t can be determined by correcting the degraded depth e_{phenol} measured by phenolphthalein with the formula: $e_t = 1.17\ e_{phenol}$.

In paste, mortar or concrete made with CEM I cement, with a concentration of ammonium nitrate of about 6 mol l^{-1}, the rate of degradation varies between 0.6 mm $\sqrt{d}^{-1}$ and 2 mm $\sqrt{d}^{-1}$ depending on the authors (and for specimens with different w/c ratios). For lower ammonium nitrate concentrations (0.12–1.2 mol l^{-1}), the degraded thickness also increases more slowly, with values of 0.2 mm $\sqrt{d}^{-1}$ to 0.4 mm $\sqrt{d}^{-1}$, respectively. Thus, under saturated conditions, acceleration factors with respect to leaching by pure water vary between 30 and 300 (Perlot 2005; Perlot et al. 2007).

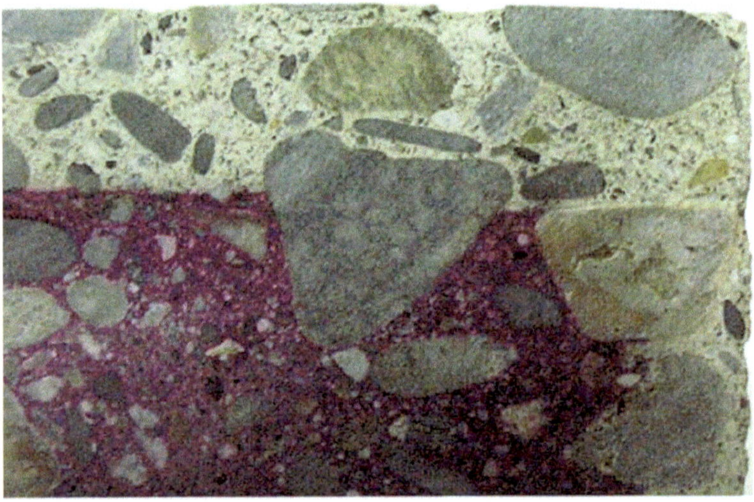

Fig. 5.5 Determination of degraded depth with phenolphthalein in concrete specimen immersed in saturated solution of ammonium nitrate for 6 months (Escadeillas 1993)

Some authors also assess the state of alteration of cementitious materials by their weight loss or their increase in water porosity (Bajza et al. 1986). These quantities, related to the dissolution of hydrates, also vary according to the square root of time, confirming the diffusive nature of degradation by ammonium nitrate.

Thus, Carde (1996) measured a mass loss rate of $2.8\%.\sqrt{}d^{-1}$. Goncalves and Rodrigues. (1991) measured mass losses of 8 and 11% after 14 years of immersion in solutions of 0.6 mol l^{-1} and 6 mol l^{-1} respectively. It should be noted that the mass losses were reduced with the use of slag (losses of 4 and 7% respectively).

These mass losses and increases in porosity are in direct relation with the decalcification of the cement hydrates. Figure 5.6 superposes the pore size distribution for sound and degraded mortar samples. The first group of pores (pore radius > 10 nm) observed on degraded mortars is associated with the progressive decalcification of calcium silicate hydrates while the second main pore class is essentially associated with the gaps formed when large portlandite crystals dissolve, the gaps becoming part of the connected macroscopic porosity (Perlot et al. 2006, 2007).

These results are in accordance with those of Ulm et al. (2003), which indicate that calcium leaching from the C–S–H structure changes the chemical composition of the solid phase without significantly changing the texture; the solid phase remains continuous.

However, results of open porosity and microstructural characterization have to be considered cautiously because the experimental determination accounts for only the average evolution of the microstructure in the degraded thickness (Perlot et al. 2006).

The evolution of degradation can also be investigated by determining the loss of strength. Carde and François (1999) connected loss of relative strength $\Delta\sigma/\sigma$ to the ratio between degraded area A_d and initial area A_t. Figure 5.7 shows a linear relationship between strength loss and degraded area. This figure also shows that mortar with silica fume (without portlandite) has the best behaviour.

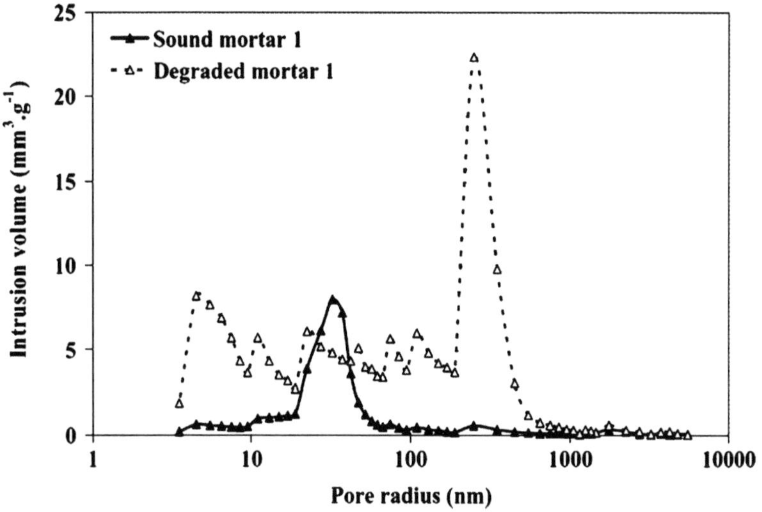

Fig. 5.6 Pore size distribution of sound and degraded zones of mortar samples immersed in ammonium nitrate solution 6 M (Perlot et al. 2006)

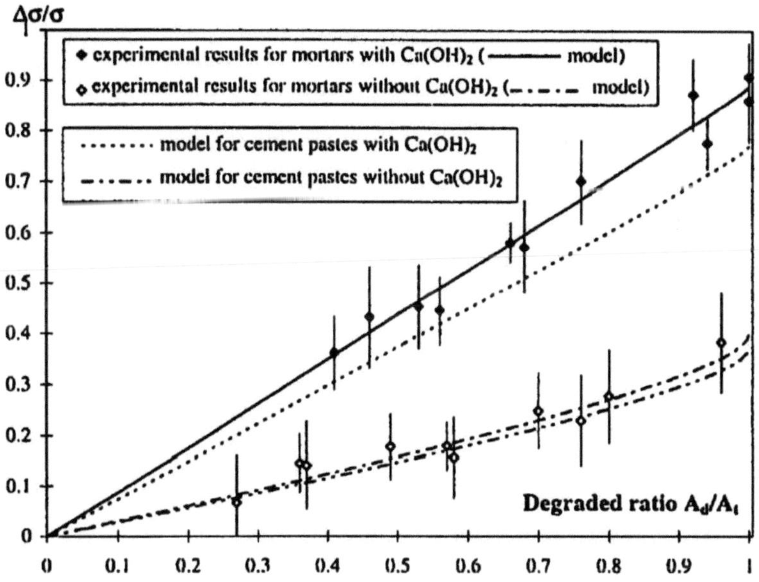

Fig. 5.7 Loss of relative strength vs. ratio of degraded area-to-initial area for mortars with and without silica fume immersed in ammonium nitrate solution (Carde and François 1997)

More recently, synchrotron X-ray computed microtomography, a powerful non-destructive observation technique, was applied to characterize the evolution of a mortar immersed in ammonium nitrate solution in the course of degradation (Burlion et al. 2006). The degradation kinetics, the leaching front position and the porosity increase were thus determined without interfering with the material. Analysis of the variations of the X-ray linear attenuation coefficient provided a quantitative determination of the evolution of the calcium content in the solid phase and the increase in porosity during leaching of a mortar. Additionally, the analysis demonstrated that the siliceous aggregates played no significant role in the leaching process, although the increase in porosity seemed to be high in the ITZ.

Finally, based on the experimental results for calcium leaching and coupled chemo-mechanical damage, material models formulated on a macroscopic level of observation have been developed for numerical predictions of the lifetime of concrete structures (Gérard 1996; Gérard et al. 2002; Le Bellego 2001; Le Bellego et al. 2003; Kamali et al. 2003; Kuhl et al. 2004).

5.4 Chemical and Mechanical Tests and Influencing Parameters

5.4.1 General Considerations

In the absence of standards and recommendations, conditions of degradation in ammonium nitrate tests are quite variable, especially regarding the ratio of sample surface area/volume of aggressive solution (Carde et al. 1997; Heukamp et al. 2001; Le Bellego et al. 2000; Schneider and Chen 1999; Perlot 2005; Perlot et al. 2007) or the temperature during the test. However, in general, the concentration of ammonium nitrate is about 6 mol l^{-1} (saturation), the solution being renewed approximately monthly. The pH may also be maintained with nitric acid between 6.5 and 8.2, which further accelerates the kinetics of degradation.

Degraded depths can reach 3 mm in 2–3 days (acceleration by a factor of at least 100 with respect to pure water) and exceed 15 mm in 3 months. It should be noted that the use of ammonium nitrate requires special precautions during storage (solid product of explosive nature), tests (production of ammonia) and disposal of solutions after testing (Perlot et al. 2007).

5.4.2 Accelerated Calcium Leaching Tests

An example of an experimental set-up for accelerated calcium leaching is presented in Fig. 5.8 (Nguyen et al. 2007). The pH and temperature are recorded by a data acquisition system. In this case, the solution was renewed each time the pH became higher than 8.2. In other cases, the solution was not renewed during the test (Assié et al. 2007). The duration of such tests varies from a few weeks to a year or more.

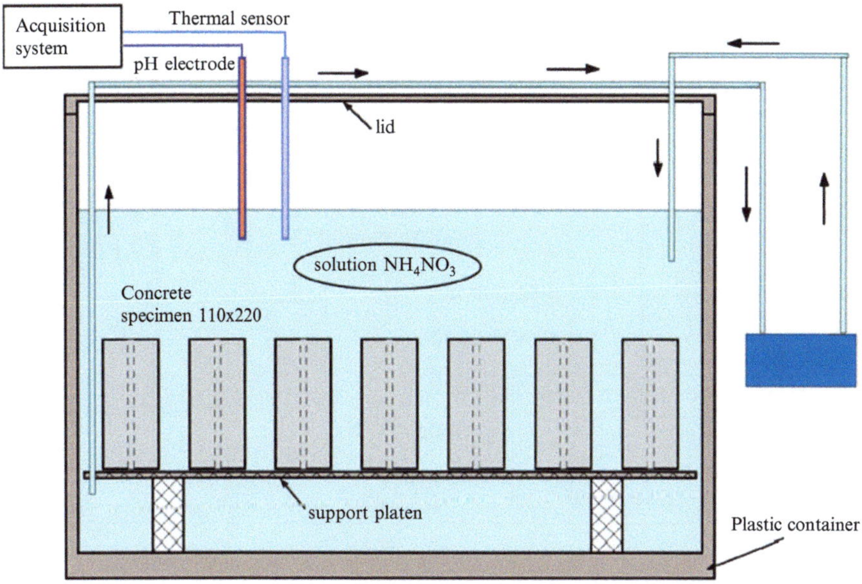

Fig. 5.8 Scheme of the calcium leaching test of concrete under 6 M ammonium nitrate solution (Nguyen et al. 2007)

5.4.3 Combined Chemical and Mechanical Tests with Ammonium Nitrate

5.4.3.1 Stress Corrosion Tests

Stress corrosion tests include two kinds of attack: chemical attack and mechanical attack. Such tests are performed on construction elements which are usually subjected to both chemical and mechanical attacks during long periods.

Generally, some specimens are loaded in 3- or 4-point-bending with a stress level of 30% of the initial flexural strength using apparatus like that shown in Fig. 5.9 (Schneider and Chen 1999, 2009). In order to make a comparison with the loaded specimens, unloaded specimens are also immersed in the same solutions. The results of Schneider and Chen (2005) show that external stresses on the specimens can significantly accelerate the chemical attack (Fig. 5.10).

5.4.3.2 Creep Tests

Creep tests coupled with chemical attack may also be conducted (Torrenti et al. 2008; Sellier et al. 2011). An example of this type of test is given in Fig. 5.11. In this case, a compressive test is performed. The force applied is kept constant using a nitrogen–oil oleopneumatic accumulator. A large vessel is used to renew the

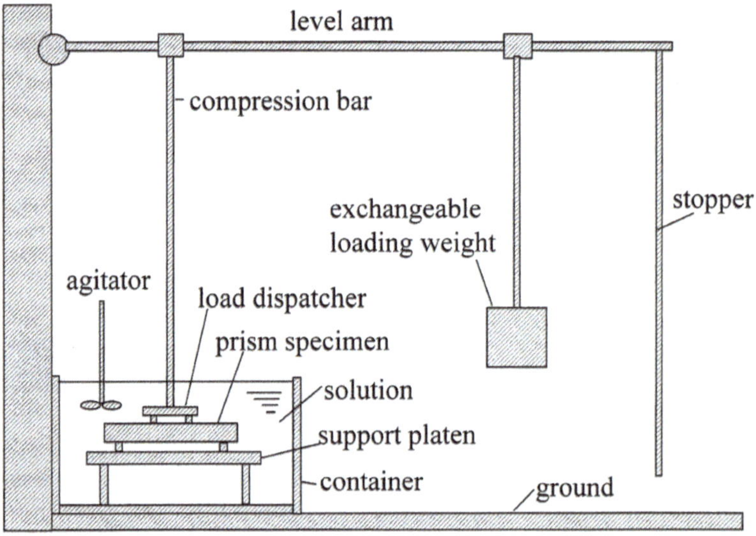

Fig. 5.9 Loading apparatus for stress corrosion tests (Schneider and Chen 1998)

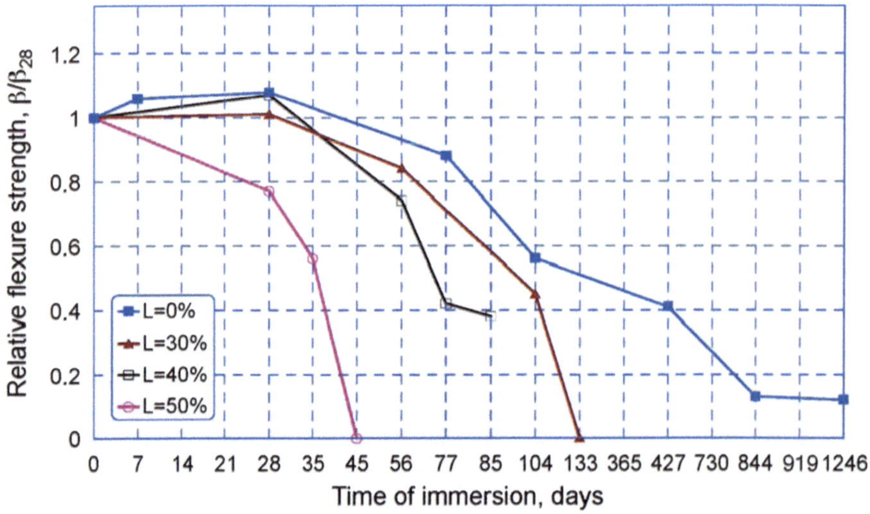

Fig. 5.10 Evolution of relative flexure strength vs. time for ordinary C40 concrete specimens immersed in 5% NH_4NO_3 solution under loading with different load levels L (Schneider and Chen 2005)

ammonium nitrate solution. Whenever the pH reaches 8.2, the solution is renewed. The set-up is fitted with an electrode and a thermal probe to monitor the pH and temperature of the solution. Movement between the plates is measured by an LVDT movement sensor. All the systems are connected to a computer for data acquisition.

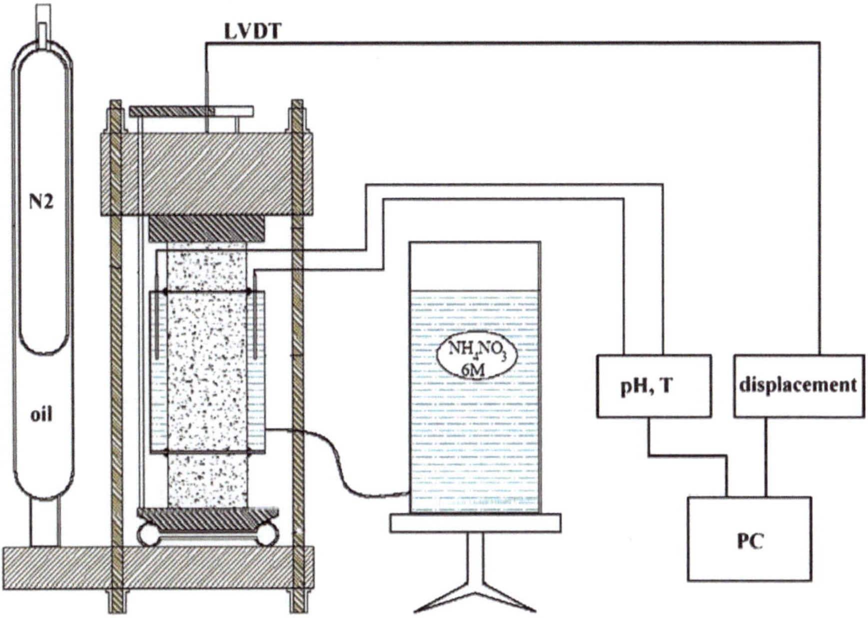

Fig. 5.11 Diagram of the test bench for creep combined with leaching (Torrenti et al. 2008)

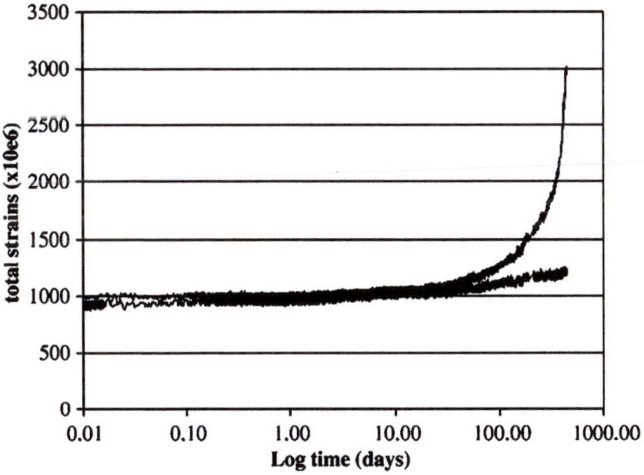

Fig. 5.12 Total strains in water (*lower curve*) and in the ammonium nitrate solution (*higher curve*) (Torrenti et al. 2008)

Deformation in the combined test is the same as under water at the beginning (for 20 days) but later deformation is observed to accelerate and a tertiary creep phase is reached, which leads to failure (Fig. 5.12). This demonstrates the effect of creep combined with leaching.

Bending tests were also carried out by Sellier et al. (2011), in similar conditions, to measure the variation of creep properties in compression and in tension.

Finally, a triaxial test, applying mechanical loading to a specimen previously subjected to ammonium nitrate attack, has also been used (Heukamp et al. 2003).

5.4.4 Influencing Parameters

5.4.4.1 Initial Strength of Materials

The test results indicate that high-strength concrete has a higher resistance to ammonium nitrate attack. Under loading with a load level of 30% of the reference strength, the service life of concrete C80 in 5% NH_4NO_3 solution was 2 months longer than that of concrete C40 (Schneider and Chen 2005). This relates to the high compactness and low permeability of this concrete, which confers high resistance to the penetration of aggressive ions.

5.4.4.2 Composition of Cementitious Materials

The kinetics of degradation increases with the w/c ratio. Carde (1996) indicated that the kinetics were higher for high w/c mixtures because the porosity of the degraded area was greater. This result was confirmed by Cassagnabère and Verdier (2009) and is illustrated in Fig. 5.13. The porous medium governs degradation by regulating the dissemination of the ionic species of the pore solution towards the aggressive environment.

The use of mineral additives in concrete, and more particularly pozzolanic additives, also limits the kinetics of degradation by lessening the diffusion in concrete through a more compact microstructure and low initial quantity of portlandite, which is the first hydrate to be attacked (Carde et al. 1997; Schneider and Chen 1998, 1999).

5.4.4.3 Initial Concentration of Ammonium Nitrate Solution

A higher concentration of the solution leads to a shorter lifetime of specimens with or without loading (Schneider and Chen 2005). The relations between the lifetime and concentration of the solutions can be described by exponential functions (Fig. 5.14).

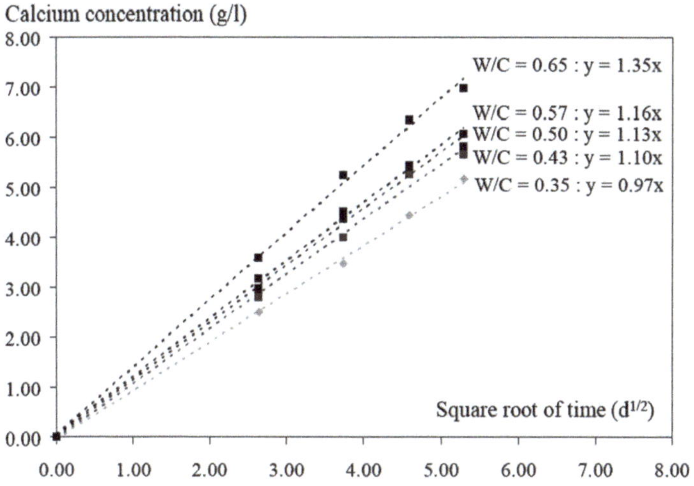

Fig. 5.13 Calcium leaching on mortars made with different w/c ratios (Cassagnabère and Verdier 2009)

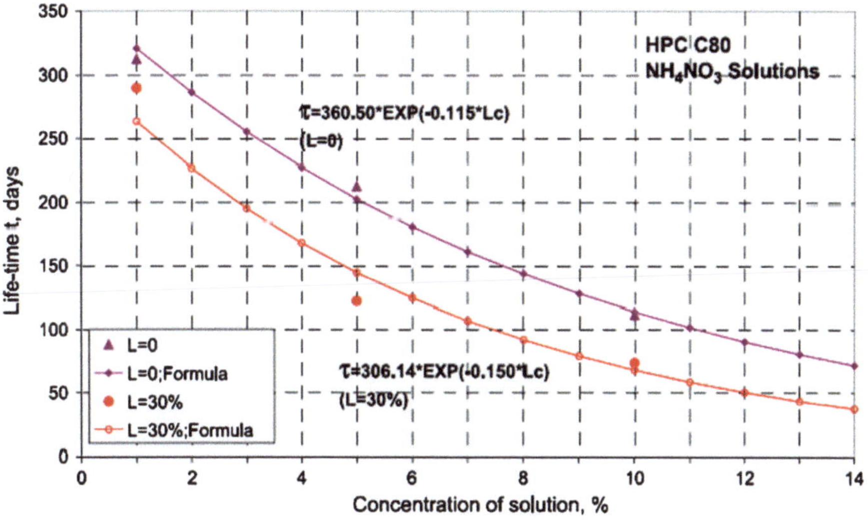

Fig. 5.14 Lifetime of HPC C80 immersed in ammonium nitrate solutions with different concentrations, with/without load (Schneider and Chen 2005)

5.5 Conclusion

Ammonium nitrate solutions produce rapid decalcification due to the high solubility of its calcium salts. In that sense, its action is very different from that of ammonium sulphate, which causes strong swelling of the sample due to the formation of needle-like gypsum crystals having low solubility. However, if the concrete is subjected to drying, swelling related to the precipitation of salts of calcium nitro-aluminates can be observed.

In all cases, degradation reactions related to ammonium nitrate solutions are swift and severe. Precautions should therefore be taken (high strength concrete, use of pozzolanic additives, low w/c) to increase the service life of structures placed in such environments. Fortunately, the number of structures affected is low and remains limited to plants manufacturing ammonium nitrate and to storage silos.

In recent years, ammonium nitrate has been especially used in laboratory tests to accelerate the phenomena observed in leaching by pure water and thus predict the very long-term behaviour of concrete materials (for the storage of radioactive waste). However, while many researchers have developed degradation models based on leaching of concretes by ammonium nitrate, it should be noted that there are currently no recommended tests. Results can thus differ greatly depending on experimental conditions (for example, the acceleration factor relative to pure water can range from 30 to 300).

References

Assié, S., Escadeillas, G., Waller, V.: Estimates of self-compacting concrete potential durability. Construct. Build Mater. **21**(10), 1909–1917 (2007)

Bajza, A., Rousekova, I., Vrana, O.: Corrosion of hardened cement pastes by NH_4NO_3 solutions. In: 8th International Congress on the Chemistry of Cement, Rio, pp. 99–103, (1986)

Baxi, C., Patel, S.K.: Concrete exposed to calcium ammonium nitrate. In: Gjorv, O.E., Sakai, K., Banthia, N. (eds.) Concrete under severe conditions 2: environment and loading, E&FN SPON London, pp. 76–83 (1998)

Bernard, F., Kamali-Bernard, S., Prince, W.: 3D multi-scale modelling of mechanical behaviour of sound and leached mortar. Cem. Concr. Res. **38**(4), 449–458 (2008)

Berner, U.R.: Modelling the incongruent dissolution of hydrated cement minerals. Radiochem. Acta. **44–45**, 387–393 (1988)

Berner, U.: Evolution of pore water chemistry during degradation of cement in a radioactive waste repository environment. Water Manage. **12**, 201–219 (1992)

Biczok, I.: Concrete Corrosion and Concrete Protection. Akadémiai Kiado, Budapest (1972)

Burlion, N., Bernard, D., Chen, D.: X-ray microtomography: application to microstructure analysis of a cementitious material during leaching process. Cem. Concr. Res. **36**(2), 346–357 (2006)

Carde, C.: Caractérisation et modélisation de l'altération des propriétés mécaniques due à la lixiviation des matériaux cimentaires. Ph.D. thesis, INSA Toulouse (1996) (in French)

Carde, C., François, R.: Effect of ITZ leaching on durability of cement-based materials. Cem. Concr. Res. **27**(7), 971–978 (1997)

Carde, C., François, R.: Modelling the loss of strength and porosity increase due to the leaching of cement pastes. Cem. Concr. Compos. **21**(3), 181–188 (1999)

Carde, C., Escadeillas, G., François, R.: Use of ammonium nitrate solution to simulate and accelerate the leaching of cement pastes due to deionized water. Mag. Concr. Res. **49**(181), 295–301 (1997)

Cassagnabère, F., Verdier, J.: Effect of W/C ratio on physical properties of totally decalcified mortars degraded with ammonium nitrate. In: Alexander, M.G., Bertron, A. (eds.) Proceedings of the RILEM TC 211-PAE Final Conference, vol. 1, pp. 65–77. Toulouse (2009)

Chen, J.J., Thomas, J.J., Jennings, H.A.: Decalcification shrinkage of cement paste. Cem. Concr. Res. **36**(5), 801–809 (2006)

De Larrard, T., Benboudjema, F., Colliat, J.B., Torrenti, J.M., Deleruyelle, F.: A probabilistic approach for assessing concrete degradation due to leaching. In: Alexander, M.G., Bertron, A. (eds.) Proceedings of the RILEM TC 211-PAE Final Conference, vol. 1, pp. 240–248. Toulouse (2009)

Escadeillas, G.: Méthodologie d'analyse de dégradation chimique accélérée des bétons par utilisation du nitrate d'ammonium. LMDC-Grande Paroisse contract report (1993)

Escadeillas, G., Hornain, H.: La durabilité des bétons vis-à-vis des environnements chimiquement agressifs. Chap. 12. In: Presses de l'ENPC (ed.) La durabilité des bétons, pp. 613–705, (2008)

Gérard, B.: Contribution of the mechanical, chemical and transport couplings in the long term behavior of radioactive waste repository structures. Ph.D. thesis, Laval University, Québec, Canada – ENS Cachan, France (1996)

Gérard, B., Le Bellego, C., Bernard, O.: Simplified modelling of calcium leaching of concrete in various environments. Mater. Struct. **35**(10), 632–640 (2002)

Goncalves, A., Rodrigues, X.: The resistance of cement to ammonium nitrate attack. In: Proceedings of the 2nd International Conference on CANMET/ACI Durability of concrete, Montreal, Canada (1991)

Heukamp, F., Ulm, F.J., Germaine, J.: Mechanical properties of calcium leached cement paste: triaxial stress states and the influence of pore pressure. Cem. Concr. Res. **31**(5), 767–774 (2001)

Heukamp, F.H., Ulm, F.J., Germaine, J.T.: Poroplastic properties of calcium-leached cement-based materials. Cem. Concr. Res. **33**(8), 1155–1173 (2003)

Jauberthie, R., Rendell, F.: Physicochemical study of the alteration surface of concrete exposed to ammonium salts. Cem. Concr. Res. **33**(1), 85–91 (2003)

Kamali, S.: Modélisation de la lixiviation de matériaux cimentaires, application à différents ciments dans différents environnements. Ph.D. thesis, ENS Cachan (2003) (in French)

Kamali, S., Gérard, B., Moranville, M.: Modelling the leaching kinetics of cement-based materials – influence of materials and environment. Cem. Concr Compos. **25**(4–5), 451–458 (2003)

Kamali, S., Moranville, M., Leclercq, S.: Material and environmental parameter effects on the leaching of cement pastes: experiments and modelling. Cem. Concr. Res. **38**(4), 575–585 (2008)

Kuhl, D., Bangert, F., Meschke, G.: Coupled chemo-mechanical deterioration of cementitious materials. Int. J. Solids Struct. **41**, 15–40 (2004)

Lea, F.M.: The action of ammonium nitrate salts on concrete. Mag. Concr. Res. **17**(52), 115–116 (1965)

Le Bellego, C.: Couplages chimie-mécanique dans les structures en béton attaquées par l'eau : étude expérimentale et analyse numérique. Ph.D. thesis, ENS Cachan (2001) (in French)

Le Bellego, C., Gerard, B., Pijaudier-Cabot, G.: Chemo-mechanical effects in mortar beams subjected to water hydrolysis. J. Eng. Mech. ASCE **126**(3), 266–272 (2000)

Le Bellego, C., Pijaudier-Cabot, G., Gerard, B., Dube, J.F., Molez, L.: Coupled mechanical and chemical damage in calcium leached cementitious structures. J. Eng. Mech. **129**(3), 333–341 (2003)

Loosveldt, H.: Etude expérimentale des comportements hydraulique et poromécanique d'un mortier sain ou dégradé chimiquement. Ph.D. thesis, Ecole centrale de Lille (2002) (in French)

Mohr, A.: Uber die Einwirkung von Ammoniumsalzlo¨sungen auf Beton. Der Bauingenieur **6**(8), 284–293 (1925) (in German)

Moranville, M., Kamali, S., Guillon, E.: Physicochemical equilibria of cement-based materials in aggressive environments – experiment and modeling. Cem. Concr. Res. **34**(9), 1569–1578 (2004)

Nguyen, V.H., Colina, H., Torrenti, J.M., Boulay, C., Nedjar, B.: Chemo-mechanical coupling behaviour of leached concrete – part 1: experimental results. Nucl. Eng. Des. **237**(20–21), 2083–2089 (2007)

Perlot, C.: Influence de la décalcification de matériaux cimentaires sur les propriétés de transfert. Application au stockage profond de déchets radioactifs. Ph.D. thesis, Université de Sherbrooke, Québec, Canada – Université Paul Sabatier, France (2005) (in French)

Perlot, C., Verdier, J., Carcasses, M.: Influence of cement type on transport properties and chemical degradation: application to nuclear waste storage. Mater. Struct. **39**(5), 511–523 (2006)

Perlot, C., Bourbon, X., Carcasses, M., Ballivy, G.: The adaptation of an experimental protocol to the durability of cement engineered barriers for nuclear waste storage. Mag. Concr. Res. **59**(5), 311–322 (2007)

Schneider, U., Chen, S.W.: The chemomechanical effect and the mechanochemical effect on high-performance concrete subjected to stress corrosion. Cem. Concr. Res. **28**(4), 509–522 (1998)

Schneider, U., Chen, S.W.: Behavior of high-performance concrete under ammonium nitrate solution and sustained load. ACI Mater. J. **96**(1), 47–51 (1999)

Schneider, U., Chen, S.W.: Deterioration of high-performance concrete subjected to attack by the combination of ammonium nitrate solution and flexure stress. Cem. Concr. Res. **35**(9), 1705–1713 (2005)

Schneider, U., Chen, S.W.: The leaching effect of concrete immersed in ammonium nitrate solution. In: Alexander, M.G., Bertron, A. (eds.) Proceedings of the RILEM TC 211-PAE Final Conference, vol. 1, pp. 54–64. Toulouse (2009)

Schneider, U., Nagele, E., Dumat, F., Holst, S.: Der Einfluss Mechanischer Spannungen auf den Korrosionwiderstand Zementgebundener Baustoffe. DAfStB Heft 429, Beuth Verlag GmbH, Berlin (1992) (in German)

Sellier, A., Buffo-Lacarrière, L., El Gonnounia, M., Bourbon, X.: Behavior of HPC nuclear waste disposal structures in leaching environment. Nucl. Eng. Des. **241**(1), 402–414 (2011)

Stora, E., Bary, B., He, Q.C., Deville, E., Montarnal, P.: Modelling and simulations of the chemo-mechanical behaviour of leached cement-based materials: leaching process and induced loss of stiffness. Cem. Concr. Res. **39**(9), 763–772 (2009)

Tognazzi, C.: Couplages fissuration – dégradation chimique dans les matériaux cimentaires: caractérisation. Ph.D. thesis, INSA Toulouse (1998) (in French)

Tognazzi, C., Torrenti, J.M., Carcasses, M., Ollivier, J.P.: Coupling between diffusivity and cracks in cement-based systems. Concr. Sci. Eng. **2**(8), 176–181 (2000)

Torrenti, J.M., Nguyen, V.H., Colina, H., Le Maou, F., Benboudjema, F., Deleruyelle, F.: Coupling between leaching and creep of concrete. Cem. Concr. Res. **38**(6), 816–821 (2008)

Ukraincik, V., Bjegovic, D., Djurekovic, A.: Concrete corrosion in a nitrogen fertilizer plant. In: Proceedings of the First International Conference on the Durability of Building Materials and Components, Ottawa, Canada, pp. 397–409, (1978)

Ulm, F.J., Lemarchand, E., Heukamp, F.H.: Elements of chemomechanics of leaching of cement-based materials at different scales. Eng. Fract. Mech. **70**(7–8), 871–889 (2003)

Xie, S.Y., Shao, J.F., Burlion, N.: Experimental study of mechanical behaviour of cement paste under compressive stress and chemical degradation. Cem. Concr. Res. **38**(12), 1416–1423 (2008)

Chapter 6
Attack of Cementitious Materials by Organic Acids in Agricultural and Agrofood Effluents

Alexandra Bertron and Josée Duchesne

Contents

A. Bertron (✉)
LMDC (Laboratoire Matériaux et Durabilité des Constructions), Université de Toulouse; UPS, INSA, 135, avenue de Rangueil, F-31 077 Toulouse Cedex 04, France
e-mail: alexandra.bertron@insa-toulouse.fr

J. Duchesne
Department of Geology and Geological Engineering, Centre de Recherche sur les infrastructures en Béton (CRIB), Université Laval Québec, Québec, QC, Canada G1V 0A6
e-mail: josee.duchesne@ggl.ulaval.ca

M. Alexander et al. (eds.), *Performance of Cement-Based Materials in Aggressive Aqueous Environments*, RILEM State-of-the-Art Reports 10,
DOI 10.1007/978-94-007-5413-3_6, © RILEM 2013

Keywords Acetic/oxalic/citric acid • Degradation mechanisms • Physicochemical zonation • Aggressiveness • PH • Calcium salt • Solubility • Molar volume

6.1 Introduction

Agricultural and agrofood industries such as the breeding, dairy or sugar industries produce large quantities of effluents. As these waste waters may be important sources of pollution of the environment, notably of natural water resources (Müller and Heil 1998; Martinez and Le Bozec 2000), they cannot be released into the environment without treatment. They must be collected, stored and eventually treated to comply with chemical, physical, and microbial standards before being released into public streams.

Such effluents may lead to serious degradation of structures – often made of concrete – for their production, collection, storage and treatment (cowsheds, housing structures, milking rooms, storage silos, collecting systems, treatment plant, dairy plants, ripening rooms, etc.) (De Belie 1997; De Belie et al. 1997a; 2000a, b; O'Donnell et al. 1995a, b; Bensted 1993; Bertron et al. 2005c). Progressive corrosion of concrete has been observed, which may be accelerated by mechanical action (high pressure cleaning, animal and machine traffic). In the long term, embedded reinforcement can become corroded and result in destruction of the facilities.

The effluents have complicated and variable compositions but share some common characteristics. The waste waters initially contain complex organic compounds in solution or in suspension, mineral compounds and a large population of micro-organisms – mainly bacteria. The metabolism of the micro-organisms produces organic acids, which are among the main final products of microbial fermentation together with CO_2 gas.

This chapter reviews the state of knowledge on the mechanisms of organic acid attack on cementitious matrices. Bacteria may also have some specific effects (carbonation linked with the production of CO_2 by bacterial respiration, consumption of some elements of the cementitious matrix, etc. (Bertron et al. 2005c; Magniont et al. 2011)). While bacterial growth and populations will not be explicitly developed, the consequences of microbial metabolism are treated.

Effluents contain a wide range of organic acids: mono-, di- and tri-carboxylic acids, among which are volatile fatty acids such as acetic or propionic acid, and also lactic, oxalic or citric acids. These acids have very varied chemical characteristics with significantly different aggressiveness to concrete. Also, the waste waters generally contain several acids and the resulting attack by the mixture of acids is not necessarily the superposition of the alteration caused by each acid taken separately (Bertron et al. 2009a; De Windt and Devilliers 2010; Larreur-Cayol et al. 2011b, c). The problem is complex and, to adequately understand the phenomenon, each acid should be considered separately before mixes of acids are studied.

From a general point of view, organic acids, which are weak acids, partly dissociate in water and react with hydrated and anhydrous compounds of the cement paste to give mainly calcium salts or complexes. Depending on their solubility, the salts may be dissolved in solution or they may form a solid deposit that can either protect the cementitious matrix or worsen its degradation. The solubility of the salts and the dissociation constant of the acids are known to influence the aggressiveness of the acids. But other parameters, such as the poly-acidity of the acids or the physical properties of the salts, the influence of which has been far less explored in the literature, must also be considered.

This chapter first considers the data available on the composition of agricultural and agrofood effluents in terms of organic acid concentrations and pH. It then gives some chemical data on organic acids, notably the dissociation constants of the acids and the solubility and some physical properties of salts.

Afterwards, we focus on the mechanisms and kinetics of the attack of the cementitious matrix by organic acids by considering, first, organic acids with soluble salts (acetic, propionic, butyric, lactic, etc. acids), and then acids with slightly soluble to insoluble salts (succinic, tartaric, oxalic, etc. acids). The degradation mechanisms are correlated with the various properties of the acids and their salts. The attack by acids with soluble salts is also compared, in terms of mechanisms and degradation kinetics, to those by strong acids such as HNO_3 or HCl. The influence of acid concentration and pH on the mechanisms and kinetics are also reviewed. Finally some studies considering mixes of acids are described.

This section focuses on the action of organic acids occurring at the surface of concrete, which is the most common mode of attack. However, reaction sometimes occurs inside concrete by liberation of the acid in the course of reaction with polymers which degrade under alkaline conditions. For example, in the nuclear industry, when cellulosic materials are immobilized with other low-and intermediate level radioactive wastes in cement matrices, cellulose may degrade to isosaccharinic acid mainly, and formic and acetic acids (Glaus et al. 1999). This type of organic acid – cementitious matrix reactions is not dealt with here.

6.2 Characterization of the Medium: Biochemical Compositions

In this section, the composition of agricultural and agrofood effluents will be given by briefly recalling the reactions that occur in the media. The micro-organisms responsible for these reactions will be mentioned only briefly and the paper will focus on the composition of the acidic part of the effluents.

6.2.1 Agricultural Effluents (From Livestock Rearing)

6.2.1.1 Liquid Manure

Liquid manure is the initial or diluted (hence mainly liquid) form of the mixture of animal excrement (faeces and urine). Complex organic compounds from animals (cattle, sheep and pigs) are naturally degraded by biological processes in anaerobic and aerobic reactions linked with the activity of specific bacteria. Three successive steps (acidogenesis, acetogenesis and methanogenesis) lead to the formation of the well known fermentation products: ethanol, volatile fatty acids, carbon dioxide (CO_2), methane (CH_4) and dihydrogen (H_2).

Liquid manure thus contains both mineral compounds (fertilizing elements such as N, P, K, and Ca, sulfate compounds, and metals such as Cu, Zn, Mn, Fe, Mo) and organic compounds, in various quantities, among which figure volatile fatty acids (VFA, short chain molecules with six or fewer atoms of carbon) such as acetic, propionic, butyric, iso-butyric and valeric acids, and also bacteria (enterobacter, sulphite-reducing *clostridium*, and faecal *streptococcus*).

There are many factors that cause variations in manure composition and pH: the animal food, the depth in the storage silo, the manure dilution rate, the sex and age of the animal (Levasseur 1998). The time (Ziegler and Heduit 1991) and conditions (temperature, aeration (Zhu et al. 2003)) of storage in the silo are also major factors as, during storage, manure is subjected to physical (migration of particles in suspension), chemical and bacterial transformations.

Consequently, the total quantities of VFA, and the proportions of the different acids at the beginning and during storage are variable. Table 6.1 lists some VFA contents that have been established from various manure samples.

Because of the presence of both VFA and the acid-base pair NH_4^+/NH_3, the pH of liquid manure is quite stable and lies between 6 and 8 (Table 6.2). Values of pH can vary according to the factors mentioned above, notably with the depth in the silo (pH increases with depth (Beline et al. 1998)). A few days of aeration causes an increase in slurry pH because of the rapid evaporation of VFA and a relatively smaller decrease in total ammonia concentration (Paul and Beauchamp 1989).

Table 6.1 Concentrations of the different organic acids in manure at initial state

	Concentrations (g/l)			Proportions (%)
Organic acids	Bernard and Heduit (1979)	De La Farge et al. (1983)	CEMAGREF data	Williams (1983)
Acetic	1.9	4.28	1.5–12.8	44
Propionic	0.8	1.86	0.4–2.8	22
Butyric	0.6	2.54	0.1–1.6	22
Iso-butyric	n.m.	n.m.	0–0.6	3
Valeric	n.m.	0.37	0.1–0.3	5
Iso-valeric	n.m.	n.m.	0.1–0.4	4

n.m. not measured

Table 6.2 Values of pH for different manure specimens

Authors	Zhang et al. (2006), Zhang and Zhu (2006)	Gelegenisa et al. (2007)	Bernard and Heduit (1979)	De La Farge et al. (1983)	Beline et al. (1998)
pH	6.0–7.7	7.4	7.0	7.2–7.6	6.8–7.2

6.2.1.2 Silage Effluent

During ensilage, soluble sugars in green food are transformed mainly into lactic acid and also into smaller quantities of volatile fatty acids (principally acetic acid) through lack of air and under the action of fermenting bacteria. The reactions are accompanied by an acidification that shows itself by the pH falling to below 4. This reduction limits undesirable fermentation (such as the development of butyric flora) and ensures the preservation of silage. In the silo, the formation of carbon dioxide (from plant respiration) leads to the progressive start of acid fermentation. The rarefaction of oxygen and the increasing acidification determine the succession of various populations of micro-organisms.

For pH higher than 4.5, acetic bacteria ferment sugars into lactic acid, acetic acid and CO_2. Then lactic bacteria develop, which acidify the medium so that its pH rapidly reaches about 3 and prevents butyric fermentation (for pH $>$ 4). The main bacteria are homofermentative bacteria (*Lactobacillus plantarum, Lactobacillus casei* and *Streptococcus faecium*). *Lactobacilli* represent 40–70% of silage bacteria. Other micro-organisms are found, such as butyric spores, moulds and yeasts, but they are contaminants and their quantities should remain low.

Green silage can release up to 400 l of effluent per ton of forage, mostly during the first 7 days (Savoie et al. 2000; Meinck et al. 1977). The quantity of leachate mainly depends on the forage water content and the pressure developed on silage (notably linked to the shape of the silo) (Savoie et al. 2000).

Silage effluent may have highly various compositions. The main factors are the nature of the silage (corn, grass, sugar cane, etc.), the silage mode (in balls, clamp, round or stack silos), the water and air contents, the temperature, the silage leaching linked with rainfall, and the time of ensilage. Among these parameters, the air-tightness of the silo has a great influence on the composition of acids.

Table 6.3 pH and VFA concentrations and lactic acid of silage effluents for different terms of ensilage

		Acids (g/l)						
Nature of silage	Term of analysis[a]	Lactic	Acetic	Propionic	Butyric	Iso-butyric	Iso-valeric	pH
Grass	A few days	5.73	1.70	0.23	0.35	0.09	0.11	4.5
	40 days	6.56	1.25	0.21	0.41	0.11	0.58	4.7
Grass	End of ensilage[b]	12.00	3.50	0.90	n.m.	n.m.	n.m.	4.4
		24.00	3.30	0.20	n.m	n.m	n.m	3.8
		20.00	5.50	0.80	n.m.	n.m.	n.m.	4.0
Grass	n.m.	44.6	n.m.	n.m.	2.7	n.m.	n.m.	3.9
Grass[c]	7 weeks	7.16	0.889	0.101	0.065	n.m.	n.m.	3.6
		9.20	0.909	0.158	0.074	n.m.	n.m.	3.6
		8.11	1.092	0.096	0.062	n.m.	n.m.	3.5
Maize	15 days	17.20	8.60	0.20	0.40	n.m.	n.m.	3.86

n.m. not mentioned
[a]After the beginning of ensilage
[b]Term not mentioned, generally 2–4 months
[c]Compositions for different types of fermentation inoculants
Galanos et al. (1995); O'Donnell et al. (1995a), (1995b); Patterson and Walker (1979); Weydert et al. (2001)

Less efficient sealing leads to lower concentrations of acids and higher pH (Amyot et al. 2003; Ouellet et al. 2001). Moreover, the treatments performed to improve the nutritive characteristics of silage influence the activity of lactic bacteria and thus modify the acid concentrations and the pH (Amyot 2005).

The pH at the end of ensilage, measured for different modes of silage, is generally between 3.5 and 4.5 (Galanos et al. 1995; O'Donnell et al. 1995a; Savoie et al. 2000; Ouellet et al. 2001; Amyot 2005; McDonald et al. 1991). Table 6.3 gives examples of silage effluent composition analysed for different terms of ensilage.

O'Donnell et al. (1995a, b) analysed grass silage juice in direct contact with concrete structures and silage juice at a distance of 1 m from the concrete inside the silage. They observed an increased pH (5.6 in contact versus 3.9 at 1 m from concrete) and a lower NaOH titratable acidity (58 mmol/l versus 380 mmol/l). Moreover, a higher concentration of lactic acid (44.6 g/kg vs 3.4 g/kg) and lower concentration of butyric acid (2.7 vs 11.7 g/kg) were found. According to the authors the higher pH linked with the alkalinity of the concrete influences the lactic and butyric acid concentrations and tends to favour butyric rather than lactic fermentation in the vicinity of the concrete.

6.2.2 *Agrofood Effluents*

6.2.2.1 Dairy Effluents: Whey and White Waters

Two dairy industry types exist: factories that only produce and distribute fresh milk, and factories that transform milk into dairy products (cheese, butter, etc.). The former produce effluents called white waters, which contain the constituents of milk

Table 6.4 pH of various dairy industry effluents

Effluent	Milk	Whey	White waters	Whey + white waters
pH	6.5–6.8	4.5–6.5	5.5–6.2	4.0–4.5

Table 6.5 pH and organic acid concentrations for cheese whey

	Acids (mg/l)					
Nature of whey	Lactic	Acetic	Propionic	Butyric	Citric	pH
Raw cheddar-cheese[a]	n.m.	836.3	74.2	52.4	n.m.	5.92
Cheddar-type cheese whey[b]	510	n.m.	n.m.	n.m.	1,770	6.3
Dutch-type cheese whey[b]	310	n.m.	n.m.	n.m.	1,840	6.3

n.m. not measured
[a]Yang et al. (2003)
[b]Johansen et al. (2002)

(water, casein, fats and lactose) and washing water (including residues of cleaning products), whereas the latter produce white waters and whey. Whey is the aqueous phase separated from the solid phase (the curd) during the making of cheese. It contains proteins, lactose, fat and mineral salts and a large population of bacteria.

Because of their high lactose contents, these effluents are rapidly subjected to acid fermentation. Lactose is transformed into lactic and butyric acids and gases (notably CO_2). The pH can quickly reach values of 3 or even 2. The acidification process is accelerated with increased temperature.

Table 6.4 gives ranges of pH for various dairy effluents. It shows that whey is the most aggressive dairy product for concrete structures in terms of pH. The type and composition of whey mainly depends upon the cheese processing techniques. Two types of whey can be distinguished: sweet whey (pH ~ 6.5) and acid whey (pH < 5).

Whey contains lactic ferments which are generally *streptococcus, lactobacillus* or *leuconostocs*. The activity of these bacteria produces lactic, acetic, propionic and butyric acids. Some whey may also contain citric acid. Table 6.5 gives some examples of whey composition for various types of cheese.

The treatments designed to reduce the pollution load of dairy effluents can change their chemical characteristics and notably lead to their acidification, for instance by increasing the acetic and butyric acid concentrations in whey (Yang et al. 2003; Saddoud et al. 2007).

6.2.2.2 Sugar Beet and Cane By-products: Molasses and Sugar Juices

Molasses is a by-product of sugar beet or cane factories. It consists of syrupy residue collected during the making or the refining of sugar. Cane molasses is notably recycled as livestock feed but is also used by fermenting industries to produce alcohol (rum from cane juice molasses, for example).

Table 6.6 Concentration of acids in cane juices and molasses (mg/g)

Acids		Oxalic	Glycolic	Malic	Lactic	Shikimic	Citric	Aconitic Cis-	Aconitic Trans-	Fumaric
Cane juice		42.5	62.1	103	ND	15.7	97	12.5	210	3
Molasses	G	2.27	2.32	2.64	ND	0.52	3.01	3.64	12.8	0.31
	L.	3.73	NM	NM	ND	0.47	NM	2.97	20.4	0.29
	RI	–	–	4.88	0.909	0.016	1.245	6.160	25.922	–

G Guadaloupe, *L* Louisiana, *RI* Reunion Island, *ND* not detectable *NM* not measurable, –:not provided
Celestine-Myrtil and Parfait (1988); Albet et al. (2009)

Molasses contains about 50% by mass of sugars (saccharose, or possibly other forms of sugar), nitrogenous compounds, organic acids and mineral matter (K, Na, Mg, Ca, P, etc.). It contains a broad range of organic acids. Sugar beet molasses contains mainly: citric, lactic, malic, acetic, formic and oxalic acids (Lodi and Rossin 1995), whereas sugar cane molasses contains aconitic, citric, malic, glycolic, oxalic and succinic acids. Fumaric and shikimic acids may also be found (Celestine-Myrtil and Parfait 1988). Examples of compositions are given in Table 6.6. The pH of raw beet molasses is around 8 (Coca et al. 2005), whereas cane molasses is more acidic, with pH between 4.5 and 6.

6.2.2.3 Effluents from Fermenting Industries (Winery Wastewaters, Malt Industry, Sugar Distilleries and Others)

Wine

There are six main steps in the wine-making process: grape harvesting, crushing and de-stemming; and juice preparation, fermentation and purification. Wine activities produce more or less seasonal polluting effluents, essentially composed of (i) pollution made of rough elements such as pulp and pips collected from washing during the wine-making period and (ii) organic pollution mainly composed of soluble alcohols, sugars and esters coming from the raw material (wort) or the finished product (wine).

Raw effluents have pH between 3.3 and 4.7, common values being around 4 (Orditz et al. 1998; Bories et al. 1998; Shepherd 1998).

Acetic, lactic, succinic, malic, citric and tartaric acids are the main organic acids in winery effluents. Three main organic acids occur in grapes: malic, tartaric and citric acid. Succinic acid is formed by yeast metabolism and so is found in wine but not in grapes (Wansbrough 1998). Acetic and lactic acids are also formed during fermentation. In grapes, tartaric acid content is between 2 and 15 g/l, malic 1–6 g/l, citric 0.5–1 g/l (Ribéreau-Gayon et al. 2006). Moosbrugger et al. (1993) give the typical organic acid composition of wine distillery effluent: tartaric acid 27%, malic acid 8%, lactic acid 29%, succinic acid 26% and acetic acid 10%. Table 6.7 gives the typical acid composition of wine.

Table 6.7 Typical acids in dry wine

Acid	Acetic	Lactic	Malic	Succinic	Tartaric	Citric	pH
Concentration (g/l)	0.3–0.7	0.3–5	0.0–6	1	5–10	0.2	3.2–3.3

Wansbrough (1998)

Table 6.8 Acid content of wine effluents at various steps of wine production

Sampling period		Beginning of grape harvest	Middle of grape harvest	Wine-making	Racking
pH		4.2	4.5	4.7	4.6
Concentration of acids (mg/l)	Acetic	48	170	49	387
	Tartaric	942	561	246	1724
	Succinic	45	37	15	287

From Bories et al. (1998)

Table 6.9 Mean composition of winery wastewater after treatment

Acid	Acetic	Lactic	Malic	Succinic	Tartaric	pH
Conc.(g/l)	0.19–0.30	0.16–0.20	0.06–0.07	0.04–0.06	0.18–1.3	4.99–6.11

Colins et al. (2005)

Bories et al. (1998) analysed the acid content and measured pH of wine residues at different steps of the wine-making process (Table 6.8).

Viaud et al. (1998) measured volatile fatty acids in various raw effluents and obtained concentrations between 0.16 and 6g/l of equivalent acetic acid.

After treatment (settling, aeration, filtration), pHs tend towards neutral (Orditz et al. 1998; Shepherd 1998; Viaud et al. 1998; Colin et al. 2005 – Table 6.9).

Vinasses (Distillery Residues of Molasses and Cane Juices)

Molasses can be fermented and then distilled to produce rum, for example. The distillery residue is named vinasses. The organic acid spectrum is the same as in molasses (see section 6.2.2.2), but with additional lactic acid, which is produced during rum fermentation (Celestine-Myrtil and Parfait 1988). Lactic acid is the main acid found in vinasses, with concentrations between 1.7 and 6.3g/l depending on the sample (Table 6.10). Citric, glycolic and aconitic acids are also well represented (up to 1.5 or 2g/l for each). The differences in compositions are probably linked with differences in the raw material.

6.2.3 Conclusion

Although agricultural and agrofood effluents have very variable compositions, it is possible to bring out some typical composition characteristics concerning their acidic part. These features are summarized below.

Table 6.10 Composition of acids of vinasses from cane juice and cane molasses (mg/l)

Acids		Oxalic	Glycolic	Malic	Shikimic	Lactic	Citric	Aconitic Cis-	Aconitic Trans-	Fumaric
Cane juice		7.5	NM	NM	1	1,680	ND	62.4	217	NM
Cane molasses	1.	66	1,560	NM	96.5	7,240	570	24	28	NM
	2.	97.5	1,900	62.5	81	6,260	2,000	480	845	14
	3.	29	930	NM	72.5	4,460	338	350	734	43
	4.	21	1,860	NM	72.3	3,700	383	345	784	41
	5.	–	–	1,068	–	3,880	353	2,101	5,896	–

ND not detectable, *NM* not measurable, –:not provided
From Celestine-Myrtil and Parfait (1988); Albet et al. (2009)

1. From the stock-rearing industry:
   - Liquid manure contains volatile fatty acids (VFA): mainly acetic and propionic acid but also butyric, iso-butyric and valeric acids, with a total concentration of the order of 10 g per litre. Because of the presence of the acid/base pair NH_3/NH_4^+in manure, pH is between 6 and 8.
   - Silage effluents contain mainly lactic acid (between 5 and 40 g/l) and acetic acid (in smaller quantities: less than the half the lactic acid concentration). Butyric acid is also found but in much lower quantities (if not, it is a sign of bad preservation). pH is between 3.5 and 4.5 (pH under 4 is a quality criterion).
2. From the dairy industry:
   The dairy industry produces whey and white waters. Whey is the most aggressive effluent, with a pH between 4.5 and 6.5 depending on the type of whey. Available values indicate that whey contains mainly lactic and acetic acids with lower quantities of other VFA, and sometimes citric acid. Diluted with white waters, whey remains very aggressive, with a pH between 4 and 4.5.
3. From the sugar production industry:
   Sugar beet molasses contains lactic, malic, acetic and oxalic acid, whereas sugar cane molasses contains aconitic, citric, malic and succinic acids. Their pH is between 4.5 and 8, cane molasses being more acidic than beet molasses.
4. From the distillery and fermentation industry:
   Winery raw wastewaters have pH around 4.5 and contain acetic, lactic, succinic, malic, citric and tartaric acid, whereas lactic, glycolic, malic, citric and aconitic acids are the main acids found in molasses vinasses.
   It is important to note that some treatments performed to reduce the pollution load of waste waters may, in some cases, lead to their acidification. These treatments and their consequences on the effluents were not developed here.
   The rest of the chapter will focus on the degradation of cementitious material by organic acids by first detailing the chemical parameters of the organic acids and then the mechanisms and kinetics of the attack.

6.3 Chemical Parameters and Aggressiveness of Organic Acids

6.3.1 Principle of Acid/Base Attacks

The cementitious matrix is a basic medium. The presence of an acidic medium in the environment of the cementitious material leads to an acid-base reaction between the acid and the hydrates and/or the anhydrous residual grains of the cement paste.

An acid/base reaction can be written in the following general form:

$$\text{Acid} + \text{Base} \rightleftarrows \text{Salt} + \text{Water} \tag{6.1}$$

Strong acids are completely dissolved in solution. Weak mono-acids, characterized by A^- anion are hydrolysed according to reaction (6.2):

$$AH + H_2O \rightleftarrows A^- + H_3O^+ \tag{6.2}$$

The dissociation constant of a weak acid is given by relation (6.3):

$$Ka = \frac{[H_3O^+][A^-]}{[AH]} \quad with\ pKa = -\log(Ka) \tag{6.3}$$

Bi-acids (general form AH_2) are dissociated according to reactions (6.4) and (6.5):

$$AH_2 + H_2O \rightleftarrows AH^- + H_3O^+ \quad with\ Ka_1 = \frac{[H_3O^+][AH^-]}{[AH_2]} \tag{6.4}$$

$$AH^- + H_2O \rightleftarrows A^{2-} + H_3O^+ \quad with\ Ka_2 = \frac{[H_3O^+][A^{2-}]}{[AH^-]} \tag{6.5}$$

Tri-acids (AH_3) are characterized by three dissociation constants and are hydrolysed according to three reactions of the same type as above.

During the reaction with the cement-based matrix, bi-and tri-acids are likely to release two or three times as much hydronium ion as a mono-acid, for the same concentration of acid.

It should also be noted that the predominant species in solution depend on the relative values of the pH and pKa_i as shown in Fig. 6.1.

A weak acid is characterized by a "buffer" zone: when the pH of the solution is around the value of the dissociation constant, an addition of base causes a very slight variation of the pH (see Chap. 15, Part III). Thus, for a weak acid, the acid concentration is more significant than the pH of the solution.

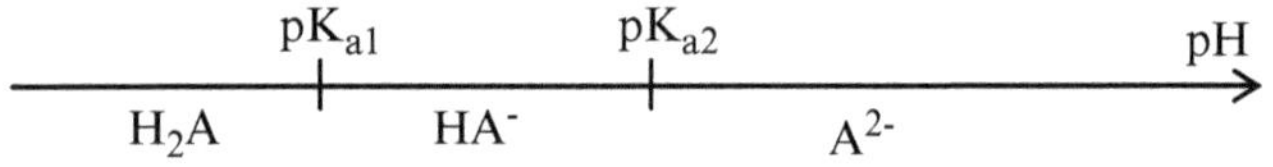

Fig. 6.1 Diagram of predominant species of a bi-acid H_2A, with dissociation constants pK_{A1} and pK_{A2} according to pH

Organic acids are likely to react with the different hydrated and anhydrous compounds of the cement paste (calcium hydroxide, C-S-H, hydrated aluminates, C_3S, C_2S) to give more or less soluble salts. For example, Bensted (1993) supposed that acetic acid was involved in at least three reactions (6.6), (6.7), and (6.8) during its action on the cementitious matrix:

Calcium hydroxide $2CH_3COOH + Ca(OH)_2 \rightarrow Ca(CH_3COO)_2 + 2H_2O$ (6.6)

C-S-H

$$6CH_3COOH + 3\ CaO.Al_2O_3.6H_2O \rightarrow 3Ca(CH_3COO)_2 + Al_2O_3.aq + nH_2O \tag{6.7}$$

Hexahydrate

$$6CH_3COOH + 3\ CaO.2SiO_2.3H_2O \rightarrow 3Ca(CH_3COO)_2 + 2SiO_2.aq + nH_2O \tag{6.8}$$

The aggressiveness of the acids depends on the chemical and physical data related to the acids and their salts. Some of them are listed in the following section.

6.3.2 Physicochemical Data of Organic Acids and Relative Aggressiveness Toward Cementitious Matrix

Various chemical and physical parameters have been shown to influence the aggressiveness of the acids toward the cementitious matrix: (i) the solubility of the acid salts in water (mainly calcium salts), (ii) the acids' dissociation constants, (iii) the mono or poly-acidity of the acids, (iv) the constants of the organometallic complexes formed in solution and (v) the physical properties of the salts (Medgyesi 1969; Bertron et al. 2005a, 2009a, 2011; Larreur-Cayol et al. 2011a, b, c; Larreur-Cayol 2012; De Windt and Devilliers 2010). This section provides a non-exhaustive list of these features and summarizes the correlations between these parameters and the aggressiveness of the acids, as highlighted by the various authors.

6.3.2.1 Inventory of Chemical and Physical Data of the Organic Acids and Their Salts

The chemical formulae and dissociation constants of the main organic acids in agricultural and agrofood effluents are given in Table 6.11. The pKa values of liquid manure mono-acids (acetic, propionic, butyric, iso-butyric and valeric acids) are very close: 4.81 ± 0.05, whereas the pKa of lactic acid, which is found in silage effluent for example, is lower: 3.86. Malic, tartaric, oxalic and succinic acids, found notably in molasses, are bi-carboxylic acids whereas citric acid, found in whey, is a tri-carboxylic acid. Oxalic acid found in wine is one of the strongest organic acids, with the lowest pKa (1.23 and 4.19).

When calcium salts are soluble or when they are slightly soluble but non-protective, the poly-acidity seems to influence the aggressiveness of the acids toward the cementitious matrix. The very high aggressiveness of citric acid, which dissolves the specimen's degraded zone and causes very high alteration kinetics as compared to the other acids studied, could be related, at least partially, to its tri-acidity (Bertron et al. 2009a; Larreur-Cayol et al. 2011a): for 1 mol of acid in solution, depending on the pH of the solution, citric acid is likely to release up to 3 mol of hydronium ion versus 1 mol for acetic acid.

Table 6.12 gives some values of solubility for Ca, Al, Mg and Fe salts of agricultural and agrofood organic acids.

Available values show that salts of organic acids, except citric acid, in liquid manure, silage effluent and whey: acetic, propionic, butyric, iso-butyric and lactic acids, are soluble to highly soluble in water, except for aluminium acetate and calcium butyrate and iso-butyrate, which are only slightly soluble at high temperatures. Given the concentrations in the media (Sect. 6.2 of this Chapter), these salts are unlikely to form for concrete structures in situations of immersion. This has been confirmed for acetic, propionic, butyric, iso-butyric and lactic acids by various authors (Pavlík 1994b; Bertron et al. 2004b, 2005a, b, 2007; Oueslati 2011) using various types of batch leaching tests with organic acid solutions at different concentrations (between 0.28 and 0.5 M).

In contrast, salts of oxalic, tartaric, citric and malic acids are slightly soluble to insoluble in water. These salts (especially calcium salt) are then likely to form during the action of these acids on concrete and either protect or damage the cementitious matrix (Larreur-Cayol et al. 2011a, b, c; Larreur-Cayol 2012). Also, calcium salts of succinic acid are not very soluble in water. Depending on the acid concentration, this salt may precipitate in solution. Calcium succinate salt was detected by Bertron et al. (2009a) during the attack of cement paste specimens immersed in succinic acid solutions 0.28 M at pH 4.

Moreover, complexation of cations, coming from cement paste dissolution, and anions, from acid dissociation, in solution may also have an influence on the aggressiveness of the acid on concrete. The complexation phenomena are governed by complexation constants. Some of these constants are given in Table 6.13. A more complete list is given in (De Windt and Devilliers 2010) and in (Larreur-Cayol et al. 2011b; Larreur-Cayol 2012), where equilibrium formation constants for Al–Ca–Mg organometallic and hydroxyl complexes are reported.

Table 6.11 pKa of different organic acids found in agricultural and agrofood effluents

	Acids					
	Acetic ($C_2H_4O_2$)	Propionic ($C_3H_6O_2$)	Butyric ($C_4H_8O_2$)	Iso-butyric ($C_4H_8O_2$)	Valeric ($C_5H_{10}O_2$)	Lactic ($C_3H_6O_3$)
pKa (25°C)	4.76	4.87	4.81	4.85	4.80	3.86
	Malic ($C_4H_6O_5$/$C_4H_5O_5^-$)	Tartaric ($C_4H_6O_6$/$C_4H_5O_6^-$)	Oxalic ($C_2H_2O_4$/$C_2HO_4^-$)	Succinic ($C_4H_6O_4$/$C_4H_5O_4^-$)	Citric ($C_6H_8O_7$/$C_6H_7O_7^-$/$C_6H_6O_7^{2-}$)	
$pK_1/pK_2/pK_3$ (25°C)	3.46/5.10	2.98/4.34	1.23/4.19	4.16/5.61	3.14/4.77/6.39	

Table 6.12 Solubility in water of calcium, aluminium, magnesium and iron salts of some organic acids (g/100 ml) (Seidell 1919; Weast et al. 1986; Lide 1991; Streit et al. 1998)

Acids		Acetic	Propionic	Butyric	Isobutyric	Lactic	Succinic	Citric	Oxalic	Malic	Tartaric
Ca salts	Formula	$Ca(C_2H_3O_2)_2$/ $Ca(C_2H_3O_2)_2.H_2O$/ $Ca(C_2H_3O_2)_2.2H_2O$	$Ca(C_3H_5O_2)_2.H_2O$	$Ca(C_4H_7O_2)_2.H_2O$	$Ca(C_4H_7O_2)_2.5H_2O$	$Ca(C_3H_5O_3)_2.5H_2O$	$CaC_4H_4O_4.H_2O$/ $CaC_4H_4O_4.3H_2O$	$Ca_3(C_6H_5O_7)_2.4H_2O$	CaC_2O_4/ $CaC_2O_4.H_2O$/ $CaC_2O_4.2H_2O$/ $CaC_2O_4.3H_2O$	$CaC_4H_4O_5.3H_2O$/ $CaC_4H_4O_5.2H_2O$	(d)$CaC_4H_4O_6.4H_2O$/(dl) $CaC_4H_4O_6.4H_2O$ /(Meso) $CaC_4H_4O_6.3H_2O$
	Sol. cold water g/100 cc (T°C)	37.4/43.6(0°C)/ 34.7(20°C)	42.8(0°C)	20.3(0°C) 18.2(20°C)	20.1(0°C) 22.3(20°C)	3.1(0°C)	0.517 (20°C)/ 0.193(0°C)	0.085(18°C)	0.00067(18°C)/ 0.00066*/ 0.0011*/ 0.0013* (25°C)	0.321/0.821(0°C)	0.0266(0°C)/ 0.0032(0°C)/i
	Sol. hot water g/100 cc (T°C)	29.7/34.3(100°C)/ 33.5(80°C)	48.4(100°C)	15.8(100°C)	28.7(62°C)	7.9(30°C)	0.475(40°C)/ 0.89(30°C)	0.096(23°C)	0.0014(95°C)/ 0.00072*/0.0014*/ 0.0018* (40°C)	0.451/1.224 (37.5°C)	0.0689 (87.5°C)/ 0.00078(87.5°C)/ 0.16(100°C)
Al salts	Formula	$Al(C_2H_3O_2)_3$	–	–	–	$Al(C_3H_5O_3)_3$	–	–	$Al_2(C_2O_4)_3.4H_2O$	–	–
	Sol. cold water g/100 cc (T°C)	vsls	–	–	–	vs	–	–	i	–	–
	Sol. hot water g/100 cc (T°C)	d	–	–	–	–	–	–	i	–	–
Mg salts	Formula	$Mg(C_2H_3O_2)_2$/ $Mg(C_2H_3O_2)_2.4H_2O$	–	–	–	$Mg(C_3H_5O_3)_2.3H_2O$	$MgC_4H_4O_4.5H_2O$	–	$MgC_2O_4.2H_2O$	–	$MgC_4H_4O_6.5H_2O$
	Sol. cold water g/100 cc (T°C)	vs/120(15°C)	–	–	–	3.3	24.4 (15°C)	–	0.07(16°C)	–	0.8(18°C)
	Sol. hot water g/100 cc (T°C)	vs or ∞	–	–	–	16.7(100°C)	66.4 (100°C)	–	0.08(100°C)	–	1.44(90°C)
Fe (III) salts	Formula	$FeOH(C_2H_3O_2)_2$	–	–	–	$Fe(C_3H_5O_3)_3$	–	$FeC_6H_5O_7.5H_2O$	$Fe_2(C_2O_4)_3.5H_2O$	–	–
	Sol. cold water g/100 cc (T°C)	i	–	–	–	s	–	sls	vs	–	–
	Sol. hot water g/100 cc (T°C)	–	–	–	–	vs	–	–	vs	–	–

Sol.: solubility; d: decomposes; vs: very soluble; s: soluble; sls: slightly soluble; vsls: very slightly soluble; i: insoluble -*: data calculated from Streit et al. (1998)

Table 6.13 Formation constants of some organic acid complexes: log (K_i) at 20°C

		Cations			
		Ca^{2+} (complex formula)	Al^{3+} (complex formula)	Mg^{2+} (complex formula)	Reference
Ligands	Acetate anion $(C_2H_3O_2)^-$	4.0 ($Ca(C_2H_3O_2)_2$)	4.3 ($Al(C_2H_3O_2)_2^+$)	1.3 ($MgC_2H_3O_2^+$)	Allison et al. (1991)
	Lactate anion $(C_3H_5O_3)^-$	1.42	–	1.37	Lange (1985)
	Oxalate anion $(C_2O_4^{2-})$	3.2 (CaC_2O_4)	13.4 ($Al(C_2O_4)_2^-$)	3.6 (MgC_2O_4)	Allison et al. (1991)
	Butyrate anion $C_4H_7O^{2-}$	0.9 ($CaC_4H_7O_2^+$)	2.2 ($AlC_4H_7O_2^{2+}$)	1.0 ($MgC_4H_7O_2$)	Allison et al. (1991)

Allison et al. (1991), listed by De Windt and Devilliers (2010), Lange and Dean (1985)

Regarding the values available for lactic, acetic and butyric acids, the salts of which are soluble in water, the reactions that lead to the formation of complexes are preponderant over the one leading to the formation of salts. In this case, complexation could tend to accelerate the kinetics of reactions between the acids and the cementitious matrix. In the literature, very few studies investigating the chelating effect of acid anions on the degradation mechanisms are to be found. De Windt and Devilliers (2010) and Larreur-Cayol (2012) modelled leaching tests of ordinary Portland cement pastes by acetic, butyric, succinic, citric and oxalic acids, using HYTEC® software. Their work indicated that cement paste degradation was essentially driven by acidic hydrolysis and not by complexation. Nonetheless, aluminium complexation by acetate and oxalate may have an effect on the stability of the alumina gel layer that is formed in the degraded layer, or on the proportion of aluminium inserted into the silica gel.

The physical characteristics of the calcium salts are also important parameters of the acid's aggressiveness. The salt's molar volume is particularly important since the comparison between this volume and that of the original compounds and porosity of the cement paste gives information about the possible impact, deleterious or not, of the formation of the salt on the matrix. Whether a salt is expansive or not depends in part on (i) where precipitation occurs and (ii) the balance between precipitation and dissolution. The molar volume yields a potential for expansion. Table 6.14 gives the density and molar volume of the slightly soluble to insoluble calcium salts listed in Table 6.12.

Calcium oxalate salts have the highest density and calcium citrate salts the lowest. All the calcium salts considered have higher molar volumes than $Ca(OH)_2$ (33 cm^3/mole (Tennis and Jennings 2000)). The molar volume of calcium oxalate is much lower than that of the other acids but it remains 1.9 times higher than that of $Ca(OH)_2$. The molar volume of calcium citrate is, by far, the highest among the salts considered. It is at least 4.8 times that of C-S-H ($C_{1.7}SH_z$), the molar volume of which is between 88 and 125cm^3/mol, depending on the scale of observation of

Table 6.14 Chemical formulae, density and molar volumes of calcium salts of some organic acids

Acid	Tartaric	Oxalic	Malic	Succinic	Citric
Formula	d-$CaC_4H_4O_6$ $4H_2O$/ dl-$CaC_4H_4O_6$. $4H_2O$/meso- $CaC_4H_4O_6$ $3H_2O$	CaC_2O_4/ CaC_2O. H_2O	$CaC_4H_4O_5$ $3H_2O$/ $CaC_4H_4O_5$ $2H_2O$/ $CaHC_4H_4O_5$ $6H_2O$	$CaC_4H_4O_4.H_2O$/ $CaC_4H_4O_4$. $3H_2O$	$Ca_3(C_6H_5O_7)_2$. $4H_2O$
Density (g/cm^3)	No data/1.817/no data	2.02/2.26	1.767/no data/no data	1.818–1.59/no data	1.1
Molar volume (cm^3/mol)	No data/143.137/no data	63.416/66.41	128/no data/no data	95.8–109.5/no data	518.4

Lange and Dean (1985), Kok and Khan (1994), Walsdorf et al. (1989), Mathew et al. (1994), Karipides and Reed (1980), Zabozlaev et al. (2007)

C-S-H and their state of saturation (Taylor 1990; Tennis and Jennings 2000). Calcium tartrate has intermediate molar volume (about 143cm^3/mol). Larreur-Cayol et al. (2011a) highlighted a correlation between the molar volume of the Ca-salt and the deleterious effect of the corresponding organic acid on the cement matrix. These authors noted that calcium oxalate and calcium tartrate, the molar volumes of which are moderate (66 and 143cm^3/mol), were very protective and moderately protective respectively, while calcium citrate salt, the molar volume of which is very high (518 cm^3/mol), is not protective (see Sect. 6.4.2). In the case of oxalic acid, the protective effect of the salt is directly linked with its ability to fill in the capillary porosity of the paste without any damage to the matrix (Larreur-Cayol et al. 2011a).

Other parameters are likely to influence the aggressiveness of organic acid toward the cementitious matrix, though these influences are not well understood and quantified yet.

Larreur-Cayol et al. (2011a) notably mention a correlation between the protective effect of salts and their affinity for the matrix. Depending on the acid, the authors noted very different adhesion of the salts onto the matrix. For example, calcium citrate precipitates but does not adhere to the matrix (no trace of salt is found in the degraded area of the sample, all the salt being found at the bottom of the tanks of the leaching tests), while calcium oxalate and tartrate seem to be strongly adherent. This difference between the three acids is not understood yet.

One can also mention: (i) the mesoscopic shape of the salt and (ii) the kinetics of formation of the salt, as other parameters influencing the aggressiveness of the acids. For example, it may be assumed that high kinetics of salt formation, compared to that of diffusion of chemical species, should enhance the protective effect of the salt, as rapid precipitation of salt inside the porosity of the matrix, could prevent the release of calcium and other dissolved chemical elements.

The influence of the kinetics of salt formation, the morphology of the salts and their affinity toward the cement matrix needs further investigation.

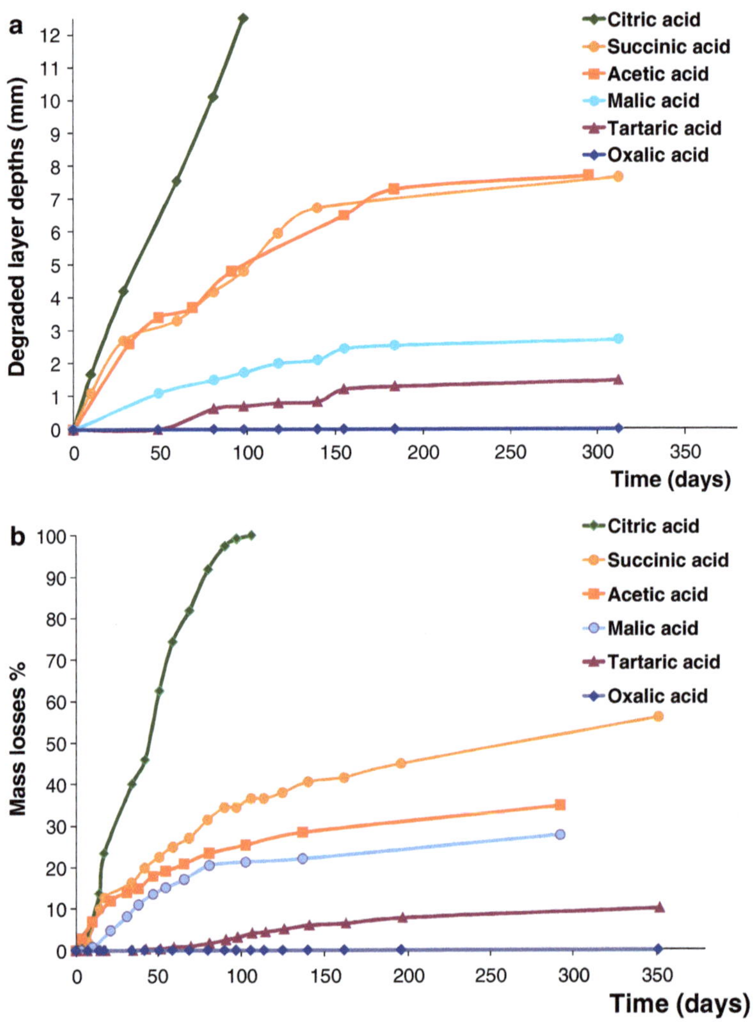

Fig. 6.2 Alteration kinetics of Portland cement paste specimens immersed in organic acids according to time of immersion (Bertron et al. 2011a) – (**a**) degraded layer depths and (**b**) relative mass losses

6.3.2.2 Relative Aggressiveness of the Acids

Larreur-Cayol et al. (2011a) and Bertron et al. (2011) performed immersion tests on cement paste specimens in six organic acids in order to evaluate the relative aggressiveness of the acids. Degradation depths and mass losses of the specimens were monitored over time (Fig. 6.2).

The increasing order of aggressiveness of the six acids on the cement matrix was as follows: oxalic acid < tartaric acid < malic acid < acetic acid < succinic acid < citric acid. As mentioned in the previous section, correlations were found by the authors between the aggressiveness of the acids on the one hand, and the physical properties of the salts, the polyacidity of the acids, etc. on the other. The low or null aggressiveness of oxalic acid was linked, according to the authors, with the formation of calcium oxalate in a thin outer layer of the cement matrix, and the low molar volume and the good adhesion of the salt to the cement paste. This resulted in a highly protective layer on the matrix. Tartaric acid was not very aggressive to the cement matrix. The higher molar volume of the salt (as compared to that of oxalic acid) is mentioned as a possible explanation for the higher aggressiveness of tartaric acid. Malic acid caused progressive but moderate alteration kinetics. Succinic and acetic acids were much more aggressive for the cement paste and had very similar behaviour in terms of altered depth kinetics. Mass losses were, however, higher for succinic acid, probably because of the progressive erosion of the outer parts of the specimens. No precipitation of salt was seen for acetic acid and small quantities of calcium succinate salts were observed for succinic acid attack. Citric acid was the most aggressive for the cement pastes. The slightly soluble citrate salt precipitated but, first, it was expansive (high molar volume as compared to that of $Ca(OH)_2$ or C-S-H) and, second, it did not adhere to the cement matrix and thus was not protective.

Previous results were obtained with organic acids, the calcium salts of which are soluble to insoluble in water. When considering organic acids with soluble salts, from the available data, the relative aggressiveness of the acids seems to be primarily influenced by the pKa of the acids. Bertron et al. (2005a) have shown through dissolution tests that the aggressiveness of the different acids present in liquid manure and silage effluents toward the cementitious matrix essentially depends on the dissociation constants. The main acids of liquid manure (acetic, propionic, butyric and iso-butyric acids), the pKa of which are close, were found to be equally aggressive. On the other hand, lactic acid, found with acetic acid in silage effluents, was more aggressive, in accordance with its pKa value.

6.4 Physical, Chemical, Mineralogical and Microstructural Mechanisms of the Attack by Organic Acids at pH Between 2.5 and 6

The mechanisms of alteration by the acids are primarily determined by the solubility of the calcium salts. In this section, two groups of acids are thus considered:

1. Acids having soluble calcium salts: there is no formation of salt in the usual conditions of tests (i.e. regular renewal of the aggressive medium and/or aggressive conditions periodically maintained by additions of concentrated acid (see Part III, Chap. 15)). This group includes acetic acid and other volatile fatty acids, such as propionic or butyric acids and lactic acid.
2. Acids having calcium salts that are slightly soluble to insoluble in water. Tartaric, oxalic, succinic, malic, citric acids are part of this group.

This chapter will first consider data regarding each acid taken separately and then the few papers dealing with mixes of acids.

6.4.1 Attack by Acids with Soluble Calcium Salts

The physical and chemical attack by organic acids with soluble calcium salts (acetic, propionic, butyric, iso-butyric and valeric acids and also lactic acid) has notably been investigated by Pavlìk (1994a, b); Bertron et al. (2005a, b, 2007, b); Israel et al. (1997), Oueslati (2011) and Oueslati and Duchesne (2009, 2011a, b, 2012).

6.4.1.1 Mechanisms at pH Around 4

Three types of tests were performed to expose cement pastes to mixes of several acids found in manure at pH between 2.5 and 6: immersion, semi-immersion and leaching tests (Bertron et al. 2004, 2005b, 2007, b; Oueslati 2011; Oueslati and Duchesne 2011a, b, 2012; Pavlík 1994a, b; Pavlik and Uncik 1997).

It was found that calcium salts and other salts coming from the reaction of these organic acids had neither protective effects in immersion tests nor destructive effects on the matrix in semi-immersion tests, in conditions favourable to their formation (Bertron 2004, Bertron et al. 2004; Oueslati 2011). Actually, in immersion tests, the salts dissolved in accordance with the solubility of these salts in water (Table 6.12) and in semi-immersion tests simulating the tidal zone of structures intended for effluent storage, the salt precipitation did not have any supplementary damaging effect.

The attack by organic acids found in liquid manure, investigated by immersion tests, results in a well-marked mineralogical and chemical zonation of the specimen (Figs. 6.3, 6.4 and 6.5).

The altered zone, distinguished from the sound zone by the change in colour of phenolphthalein (Bertron et al. 2005b (Fig. 6.3) and Oueslati and Duchesne 2012), showed high porosity (Table 6.15) (Bertron et al. 2007; Oueslati 2011; Oueslati and Duchesne 2009, 2012) and very low mechanical strength. This altered zone is easily removed if mechanical actions are applied (De Belie et al. 1996, 1997b; Pavlík 1994a, b). The altered zone of OPC pastes was not dissolved in tests performed at pH 3, 4 and 6 with 0.05 and 0.28 M acetic and other volatile fatty acid solutions with or without the addition of calcium hydroxide to bring the pH up to the target value (Bertron et al. 2004, 2005b; Israel et al. 1997). Nevertheless, for higher concentrations of acids (0.5 M acetic acid solutions with pH 4), Oueslati (2011) reported dissolution of the outer part of immersed OPC paste specimens cured for 1 month prior to the tests. This dissolution was no longer observed with specimens cured for 3 months.

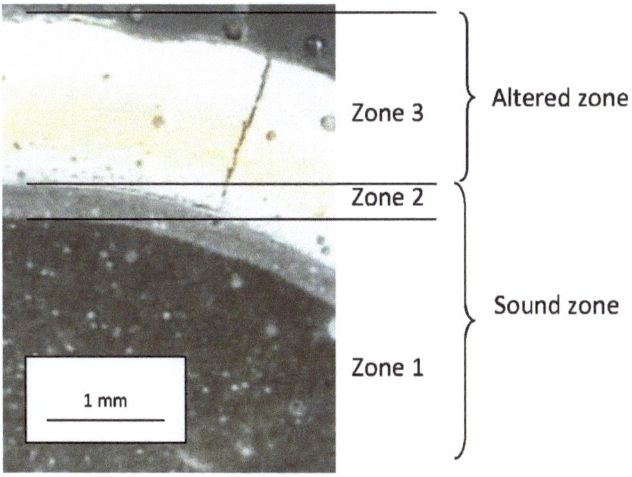

Fig. 6.3 Macroscopic zonation of hardened CEM I pastes immersed in a mix of liquid manure organic acids at a pH of 4 for 4 weeks, observation of a cross-section with video-microscope (Bertron et al. 2005b). Cement paste specimens, made with a water/cement ratio of 0.27, were cylinders 25 mm diameter and 75 mm high. The aggressive solution was a mix of organic acids present in liquid manure: acetic (12.8 g/l), propionic (2.8g/l) butyric (1.6g/l), iso-butyric (0.6g/l) and valeric acids (0.3g/l) for a total acid concentration of 0.278 mol/l. NaOH (2 g/l or 0.05 mol/l) was added to the solution to bring the pH up to 4. The aggressive solution was renewed regularly

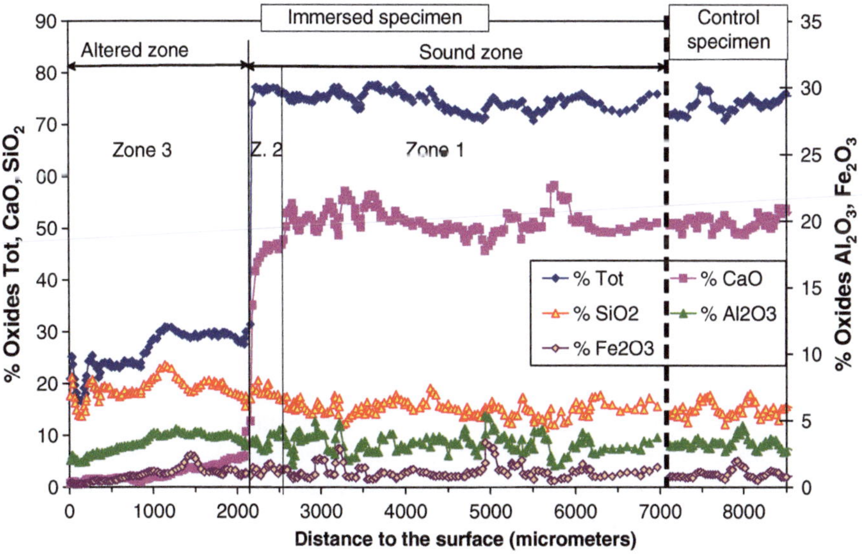

Fig. 6.4 Chemical composition analysed by electron microprobe according to the distance to the surface of a CEM I paste immersed in liquid manure organic acids at pH 4 for 4 weeks (Bertron et al. 2007) (Raw percentages of oxides in sound zone and corrected percentages in altered zone). Elementary analysis with microprobe gives relative mass concentration in the matter in the sounded volume. Thus, because of the decalcification of the altered zone, the absolute variation of other elements (Si, Al, Fe, etc.) is masked by the decrease in the amount of Ca. However, TiO_2 being stable for pH between 4 and 9 (Knauss et al. 2001), the quantity of this oxide was used as a control element to correct the microprobe analysis results in the altered zone so that the absolute variation of each element was represented. The correction ratio was the ratio of the average amount of TiO_2 in the sound zone to that in the altered zone (Bertron et al. 2009b)

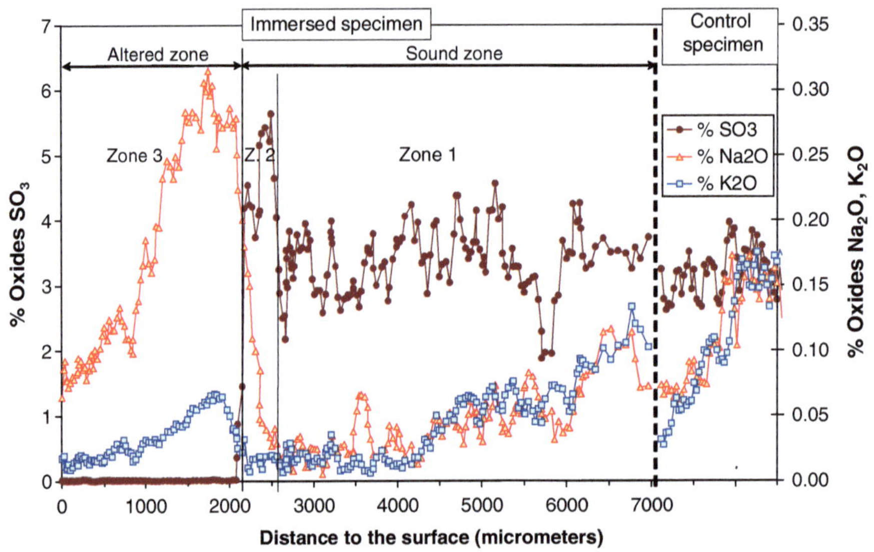

Fig. 6.5 Chemical composition analysed by electron microprobe according to the distance to the surface of CEM I paste immersed in liquid manure organic acids at pH 4 for 4 weeks (Raw percentages of oxides in the sound zone and corrected percentages in the altered zone) (Bertron et al. 2007)

The altered zone presented shade tones between brown-orange (next to sound zone) and whitish next to the solution (Fig. 6.3). Inside the sound zone, two distinct zones were observed: the heart, with the initial colour of the paste (zone 1), and a lighter grey transition zone (zone 2).

The same observations on specimens immersed in acetic acid were made by Pavlìk (1994b) at pH 3.7 but he considered zone 2 as part of the altered zone. The macroscopic zonation matches important chemical and mineralogical changes for both authors; these are shown in Figs. 6.4, 6.5 and 6.6. The altered zone was progressively and almost totally decalcified, and was made up almost entirely of silicon, aluminium and iron (Fig. 6.4). Bertron et al. (2005b, 2007, 2009b) and Oueslati (2011) found that SiO_2 was well preserved in the altered zone, the absolute content of this oxide remaining constant. Al_2O_3 and Fe_2O_3 were also well preserved in the inner part of the altered zone (over about half the thickness) but these oxides were slightly dissolved in the outer part of the degraded zone. Magnesium, found in higher quantities in slag cement, was totally dissolved in the outer part of the altered zone (Bertron et al. 2005b).

The crystalline hydrate phases, notably calcium hydroxide and ettringite, were totally dissolved in the zone in contact with the aggressive solution (Fig. 6.6) at pH 4. Considering the amount of calcium in the degraded zone (Fig. 6.4), C-S-H was almost totally leached except perhaps in the inner part of zone 3, where some low Ca/Si C-S-H may have been preserved. Moreover, the external part of the specimen, which was totally amorphous, was composed of a presumed silica gel (as analysed

Table 6.15 Porosity, specific and apparent densities measured through water intrusion porosity tests of altered and sound zones for CEM I and CEM III paste specimens immersed for 4 weeks in a mix of several organic acids in manure at a pH of 4

	CEM I		CEM III	
	Sound zone	Altered zone	Sound zone	Altered zone
Specific density (kg/m^3)	2625	2433	2672	2436
Apparent density (kg/m^3)	1859	838	1782	1156
Porosity (%)	**30.2**	**65.4**	**35.9**	**52.5**

At the time of the sampling (9–14 weeks of immersion in acetic acid), the degraded layers of the cement paste cylinders were at least 3 or 4 mm thick. After drying at 105°C, a slight withdrawal of the degraded layer had occurred and this layer could be detached easily from the unaltered (or so) core of the paste specimens. Several pieces of the degraded layer were then sampled until a mass of specimen of approximately 3 g was reached and the water intrusion porosimetry test was carried out. The results given in Table 6.15 are thus average values of porosities and densities of the whole altered zone (zones 3 of Figs. 6.3, 6.4 and 6.5).

The porosity created by the organic acid attack is noticeable for both cements. The porosity of the sound zone is higher for CEM III than for CEM I whereas the altered zone of CEM III presents a clearly lower porosity than CEM I. Bertron et al. (2007)

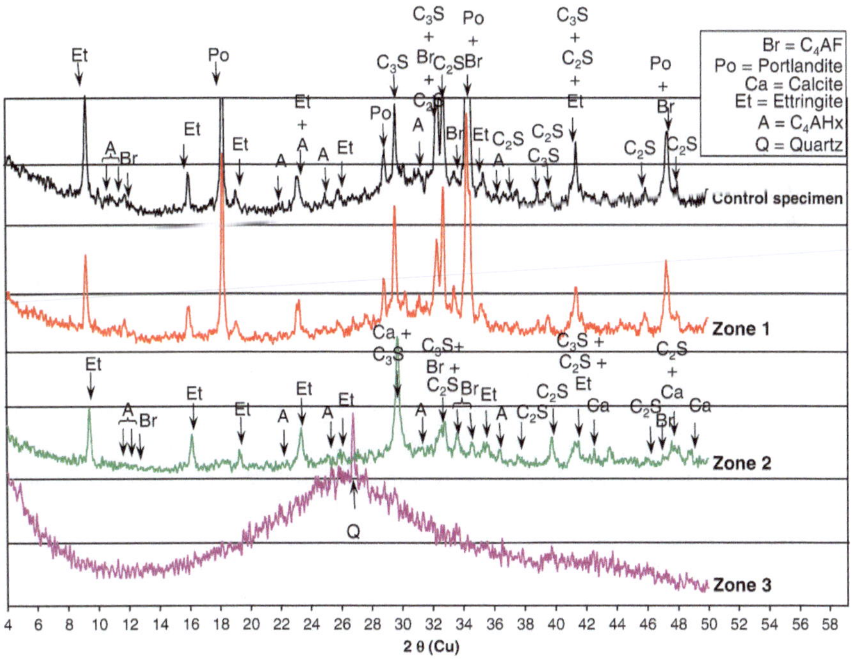

Fig. 6.6 X-ray traces of the control and CEM I specimens immersed in liquid manure organic acids at pH 4 for 4 weeks (Bertron et al. 2007)

with ^{29}Si NMR by Bertron (2004)) containing aluminium and iron, and was enriched in K and Na from the sound zone and from the aggressive solution which contained NaOH (Fig. 6.5).

Bertron et al. (2004b) assumed that the enrichment in alkaline ions in the inner part of the degraded zone was linked with the adsorption of these ions on the remaining low Ca/Si C-S-H. Alkali retention was shown to increase when Ca/Si of the C-S-H decreased (Hong and Glasser 1999; Viallis et al. 1999), because of the negative surface charge of low Ca/Si C-S-H. The amounts of alkaline ions decreased in the outer part of zone 3, which might be linked with the dissolution of these remaining C-S-H. The residual alkaline ions in the outer part of zone 3 may be adsorbed on the surface of the silica gel.

The transition zone between the altered and sound zones (zone 2 on Figs. 6.3, 6.4, 6.5 and 6.6) was characterized by the dissolution of calcium hydroxide and the precipitation of secondary ettringite. This was determined through XRD analysis and confirmed by microprobe profiles showing the slight decrease in CaO content and an increase in SO_3 content in this zone compared to zone 1 (Figs. 6.4, 6.5, and 6.6). The same observations were made by Pavlìk (1994b) using thermal analyses and EDS.

6.4.1.2 Influence of pH and Concentration of Acids on the Degradation Mechanisms

Several authors have studied the influence of pH or acid concentration on the degradation mechanisms by acetic acid and other volatile fatty acids.

One of the purposes of comparing the degradation mechanisms for different pH or concentrations concerns the validation of the experimental conditions used for accelerated tests. Tests performed at pH 4, as presented above, match the aggression by the acidic part of acid whey or silage effluent. However, to simulate liquid manure, for example, the aggressive solution should be designed to have a pH of 6 (lowest pH of this medium, see Sect. 6.2.1.1). In this context, it has been shown that the accelerated test could be conducted at a pH of 4 since alteration mechanisms were almost identical (Bertron et al. 2005b). These tests were performed using the same concentration of acids (0.278 mol/l) and by adding different concentrations of NaOH (2 g/l or 0.05 mol/l and 9.6 g/l or 0.24 mol/l) in order to bring the pH up to 4 and 5.6. The intensity of the degradation, in terms of decalcification and loss of magnesium, was identical at pH of 4 and 6. Bertron noted that the differences in alteration mechanisms were minor and mainly concerned the stability of the anhydrous phases. C_4AF was preserved in the periphery of CEM I paste specimens immersed at pH 6 but not at pH 4. The slag anhydrous phases, globally stable in both solutions, showed a slight decalcification when CEM III pastes were immersed at pH 4.

Table 6.16 Mass ratio Al_2O_3/SiO_2 and Fe_2O_3/SiO_2 in the outer half of the degraded zone of specimens immersed in aggressive solution made of acetic acid or equally aggressive volatile fatty acids

Characteristics of the aggressive solutions		$\frac{Al_2O_3}{SiO_2}$	$\frac{Fe_2O_3}{SiO_2}$	
Acid concentration (mol/l)	pH			Ref.
0.025	3.7	0.288	0.144	a
0.278	4	0.153	0.035	b
0.5	2.5	0.183	0.046	a

Calculated using data from:
[a]Pavlik (1994b)
[b]Bertron (2004; 2007)
These experiments were performed on cement pastes made with ordinary Portland cements for both authors. However, the cement compositions were slightly different: CaO + MgO: 63.79%, SiO_2: 19.10, Al_2O_3: 5.87, Fe_2O_3: 3.18, SO_3: 2.49 for Pavlìk; and CaO: 64.87%, SiO_2: 21.19, Al_2O_3: 3.94, Fe_2O_3: 2.36, MgO: 2.37, SO_3: 3.16 for Bertron. Water/cement mass ratios were 0.4 for Pavlìk and 0.27 for Bertron. The chemical analyses of the pastes were made with EDS for the first and with a microprobe for the second

It was also observed that the transition zone (zone 2 on Fig. 6.3), characterized by the dissolution of calcium hydroxide, was twice as thick at pH 6 (about 400 μm) as at pH 4 (about 200 μm). This increase in the thickness of the transition zone, linked with a lower gradient of pH or acid concentration between the sound zone and the aggressive solution, was also observed by Pavlìk (1994b). This author used different concentrations of acetic acid (0.5 and 0.025 mol/l) for the aggressive solutions with pH of 2.5 and 3.7 unlike Bertron, who used the same acid concentration.

Comparing the results of the two authors for very similar pH (3.7 and 4), it appears that the thickness of the transition zone for 0.025 mol/l acid concentration is far higher than for 0.287 mol/l concentration with the addition of soda. Thus it seems that the acid concentration ranks first in the significant parameters of the attack aggressiveness, before pH.

Pavlìk (1994b) also observed that the colour of the degraded zone varied with the concentration of the aggressive solution, probably expressing different behaviours of aluminium and iron. He reported much more Al and Fe relative to Si in the degraded zone, and notably in its outer part, for the solution with acid concentration of 0.025 mol/l and pH 3.7 than for the one with 0.5 mol/l and pH 2.5 (Table 6.16). This indicated better stability of Al and Fe during the attack. Equally, the relative proportion of Al and Fe to Si was markedly higher for Pavlik at pH 3.7 and c = 0.025 mol/l than for Bertron at pH 4 and c = 0.278 mol/l.

It should be noted that the solubilisation of Al in the outer part of cementitious specimens attacked by organic acids may be cause for concern as Al in potable water is strictly regulated.

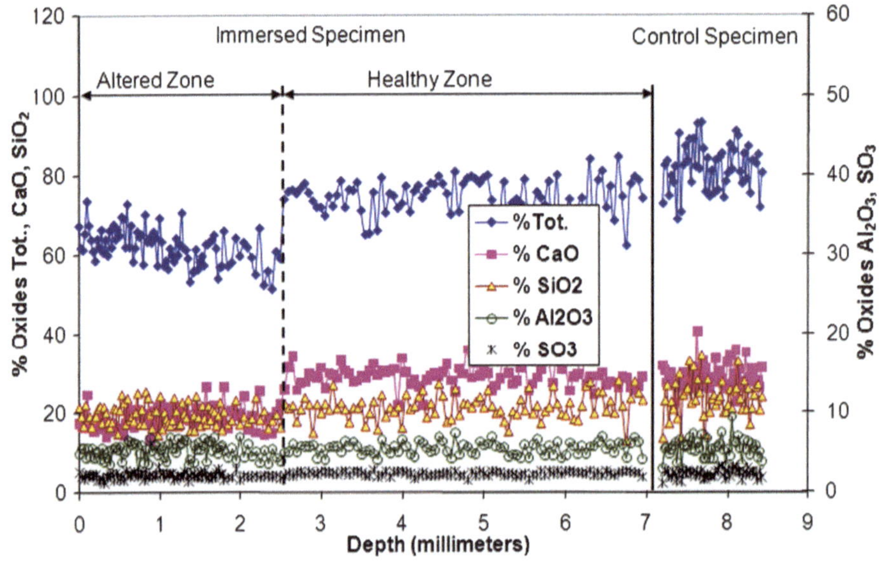

Fig. 6.7 Chemical composition analysed by electron microprobe according to the distance to the surface of (CEM I + slag) paste specimen cured for 3 months and then immersed in acetic acid solution (0.5 M, pH 4) for 3 months (Oueslati 2011) (Raw percentages of oxides in the sound zone and corrected percentages in the altered zone, see Bertron et al. 2009b)

6.4.1.3 Influence of Supplementary Cementitious Materials and Curing Time on the Degradation Mechanisms

Some authors have investigated the mechanisms of alteration of cementitious matrices with supplementary cementitious materials (SCM) added (slag, metakaolin and silica fume) notably (Oueslati 2011; Oueslati and Duchesne 2011a, b, 2012; Larreur-Cayol et al. 2011c; Bertron et al. 2007). With the usual curing time (28 days), the degradation mechanisms are found to be fairly similar to those of OPC pastes (see Sect. 6.4.1.1): almost complete decalcification of the degraded zone, preservation of Si, Al and Fe in this zone and disappearance of S. When the performances of the SCM were enhanced by a longer curing period, Oueslati (2011) noted that the degradation mechanisms were significantly modified (Fig. 6.7) as CaO and SO_3 were almost preserved in the altered zone of OPC + 80% slag pastes cured for 3 months prior to their immersion in the acetic acid solution. For OPC paste, Oueslati did not note a better preservation of calcium with longer curing time. Nevertheless, with 1-month curing time, the outer part of OPC specimens was partially dissolved (acetic acid solution 0.5 M, pH 4), which did not occur when the specimens were previously cured for 3 months.

The beneficial effect of longer curing periods was also observed for OPC paste with 20% metakaolin, but to a lesser extent, as calcium was better preserved in the inner part of the degraded zone only. The increase of the curing period was beneficial for all mixtures (with or without SCMs), significantly decreasing the mass losses and the altered depth while increasing the compressive strength.

Oueslati has also analysed the influence of the curing time on the porosity induced in the altered zones of SCM pastes using mercury intrusion porosimetry. The beneficial effect of a longer curing period was particularly marked for OPC pastes with 80% slag.

6.4.1.4 Mechanisms of Organic-Acid Versus Strong-Acid Attack (Nitric and Hydrochloric Acids)

When calcium salts are dissolved during the attack of the cementitious matrix, organic acid anions have no specific effect in the degradation, which consists of leaching enhanced by the acid/base reaction. Moreover, as shown by De Windt and Devilliers (2010), complexation has little, if any, influence on the cement paste degradation.

The attack by these organic acids may be compared with the attack of strong acids whose salts are soluble in water. This is, for instance, the case for nitric (HNO_3) and hydrochloric (HCl) acids whose calcium salt ($Ca(NO_3)_2.H_2O$ and $CaCl_2.6H_2O$) solubility in water is high (266 g/100 g and 279 g/100 g, resp.).

Pavlìk (1994b) notably compared the degradation mechanisms for cement pastes exposed to nitric acid and acetic acid solutions with the same acid concentrations (0.025 and 0.5 mol/l). The degradation mechanisms were quite similar, notably in terms of widespread decalcification of the matrix and preservation of silicon.

However, as the aggressive solutions had different pH, and as acetic acid was not completely dissociated, unlike nitric acid, some differences, notably linked with the stability of Al_2O_3 and Fe_2O_3 oxides, were found in the degradation mechanisms. The attack by nitric acid was more aggressive.

For 0.025 mol/l nitric acid (pH 1.6), the degraded zone presented notably lower quantities of iron and aluminium than for 0.025 mol/l acetic acid solutions (pH 3.7). Although the concentration of aluminium was quite stable in the whole degraded zone, the relative iron content decreased in the outer part of this zone.

These differences were seen in the degradation aspects of the specimens. Pavlik distinguished two zones inside the altered zone: the outer zone, slightly thicker than the inner one, appeared white whereas the inner zone was brown.

Moreover, the alteration kinetics was found to be markedly lower for acetic acid than for nitric acid (Pavlík 1994a, b, 1996).

Israel et al. (1997) also observed that the effect of acetic acid was comparable to that of hydrochloric acid. The author notably observed a brown-yellow degraded zone identified as an impure silica-gel which was not dissolved and was mechanically stable, though porous. The degradation was observed as a widespread decalcification of the cement paste. However, the anion of this acid had an interaction with the matrix since it formed Friedel's salt in small quantities (Chandra 1988; De Ceukelaire 1992). This salt was dissolved after longer immersion (Israel et al. 1997).

Although the comparison between acetic and nitric or hydrochloric acid attacks shows considerable similarity, greater differences exist between acetic acid and sulphuric acid, notably linked with lower solubility of the corresponding calcium salt: gypsum (Israel et al. 1997).

6.4.2 Attack by Acids with Slightly Soluble and Insoluble Calcium Salts

6.4.2.1 General Considerations

To our knowledge, among organic acids, the calcium salts of which are slightly soluble or insoluble, only the degradation mechanisms by citric, succinic, malic, tartaric and oxalic acids, have been explored in the literature so far (Larreur-Cayol et al. 2011a, b, c; Larreur-Cayol 2012; Bertron et al. 2009a). Among these five acids, two groups can be distinguished:

1. The acids forming calcium salts that are protective to the matrix. This is the case of malic, tartaric and oxalic acids, at least when the attack of ordinary Portland cement matrices is considered.
2. The acids forming calcium salts that are not protective to the matrix: citric and succinic acids are the acids investigated.

This notion of protective/non-protective calcium salt is used by Larreur-Cayol et al. (2011a) and by Bertron et al. (2009a, 2011) with reference to the case of acetic acid: the formation of the calcium salt is said to be protective when the kinetics of degradation with these acids are lower than that observed for acetic acid.

6.4.2.2 Degradation Mechanisms of Oxalic, Malic and Tartaric Acids

The common points between these three acids, as highlighted by Larreur-Cayol et al. (2011a) and by Bertron et al. (2009a, 2011), are:

- the low molar volume of the salt (Table 6.14) compared to that of the hydrated phases $Ca(OH)_2$ and C-S-H, at the expense of which the calcium salts are formed. This low molar volume seems to enable the capillary porosity and the porosity created by the hydrolysis of the matrix to be at least partially and temporarily filled, without the crystallization pressures provoking any visible crack in the matrix.
- the affinity between the salt and the matrix: the salts adhere to the cement paste. However, neither the influence of this factor nor the way to quantify or qualify it have been well apprehended yet, at least by these authors.

It should be recalled that this list is not exhaustive; not all the parameters influencing the degradation mechanisms and their intensity have been identified in the listed literature yet.

Oxalic Acid

As a preliminary remark, it should be stated that some precautions need to be taken in the making of oxalic acid solutions because of the strong chelating effect of the

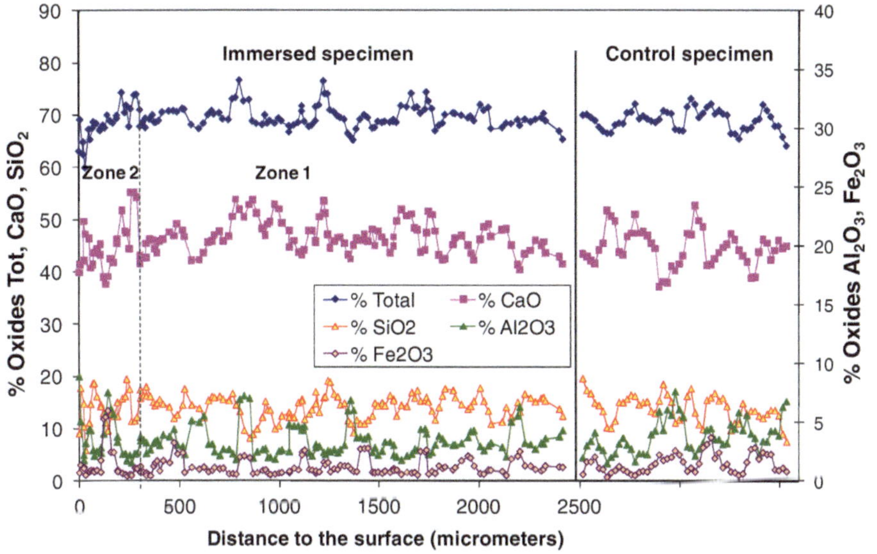

Fig. 6.8 EPMA chemical composition profile of an OPC paste immersed in oxalic acid (0.28 M, pH 0.85) for 1 month and of a control specimen (Larreur-Cayol et al. 2011a)

oxalate anion, which combines with many cations. For this reason, many strong bases (NaOH, KOH, etc.), used to adjust the pH of the solutions, lead to the formation of the corresponding oxalate salt. This necessarily disturbs the analysis of the degradation mechanisms (Bertron et al. 2009a). The results given below were therefore obtained using oxalic acid solutions with no addition of base, which means with a lower pH than many other results described in the rest of the chapter.

The attack by oxalic acid leads to the formation of calcium oxalate mono-hydrate salts $CaC_2O_4.H_2O$, or whewellite, on the surface of immersed specimens, which fully protects the cement matrix as shown by Larreur-Cayol et al. (2011a). At the periphery of the matrix, calcium oxalate is formed without any loss of calcium from the matrix (Figs. 6.8 and 6.9). As a first approach, it may be said that Ca-oxalate salt forms from $Ca(OH)_2$ and, presumably, at least a part of the C-S-H. This salt seems to prevent the penetration of the aggressive species deeper into the sample.

Other experiments performed by these authors using CEM III/C cement pastes immersed in oxalic acid solutions, in the same experimental conditions, showed that the specimens suffered severe degradation (Larreur-Cayol et al. 2011c; Larreur-Cayol 2012).

This result underlines the key role of $Ca(OH)_2$ in the attack by oxalic acid. In cementitious materials incorporating slag, especially at high replacement levels, $Ca(OH)_2$ is consumed to form C-S-H during hydration. Slag cement pastes are known to be more chemically resistant than OPC paste, notably in acidic conditions obtained with acetic acid for example (Oueslati 2011; Oueslati and Duchesne 2011a, b, 2012). However, in the case of the attack by oxalic acid, the absence of $Ca(OH)_2$ is prejudicial to the durability of the matrix.

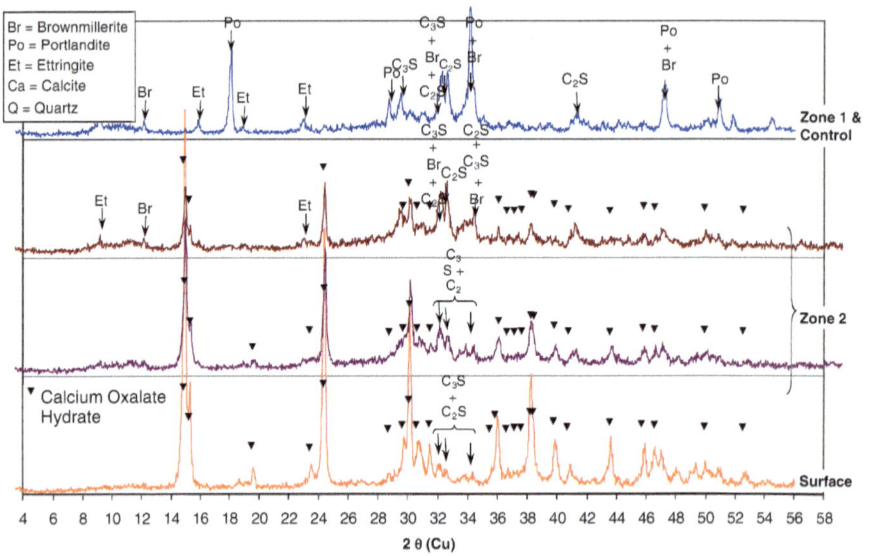

Fig. 6.9 X-ray traces of an OPC paste specimen immersed in oxalic acid for 1 month (Larreur-Cayol et al. 2011a)

Actually, the chemical stability of the Ca-bearing phase of the solid is of prime importance in the determination of the impact of the oxalic acid attack on the matrix (and not only the chemical nature of the cation combining with the oxalate anion to form an insoluble salt). For the oxalate salt to have the most beneficial effect, its formation should occur at as high a pH as possible, so that the further hydrolysis of the material caused by hydronium ion released by oxalic acid can be prevented. In cement paste, $Ca(OH)_2$ is the least stable hydrated phase of the cement paste, its dissolution occurring from pH 12.5 in the interstitial solution (and $[Ca^{2+}] = 20{*}10^{-3}$ M) according to Moranville et al. (2004).

In the case of OPC paste specimens, the highly beneficial effect of calcium oxalate salt is probably related to (i) the sacrificial behaviour of $Ca(OH)_2$, and (ii) the preservation of C-S-H, the main hydrated phase of the cement paste, enabling the mechanical properties and the physical integrity of the matrix to be maintained.

Moreover, as no evolution of the thickness of the outer part of the specimen where Ca-oxalate precipitated was observed over time, it may be assumed that the formation of this salt prevented the penetration of the aggressive species deeper into the sample and that the salt acted as a sealant of the capillary porosity of the outer layer of the matrix. In a CEM I paste with w/c = 0.27, it can be calculated, using Bejaoui and Bary's work (2007), that $Ca(OH)_2$ occupies about 15% of the total volume of the paste. The total porosity of the paste, measured by water intrusion, is about 28–30% (Bertron et al. 2007). The capillary porosity (pore diameter >10 nm) measured by mercury intrusion porosimetry, is 10% (original data). The volume of calcium oxalate being twice that of $Ca(OH)_2$ (66 cm^3/mol vs. 33 cm^3/mol), it may be assumed that the formation of the salt enabled (i) the initial volume of the dissolved $Ca(OH)_2$, (ii) the capillary porosity and, perhaps, (iii) a small part of the hydrate porosity, to be filled without damage to the matrix.

The action of oxalic acid on the cementitious matrix can be compared to that on calcareous rocks. Many natural rocks are affected by oxalic acid produced by lichens: sandstone, granitic rock, basalt, calcareous rocks etc., in quite different ways that combine the physical effect of the penetration of lichens' hyphae inside the rock and the chemical effect of the acid (Chen et al. 2000; Adamo and Violante 2000). Depending on the case, Ca, Mg, Al and Fe are released and/or combined into secondary products, these elements being sensitive to the strong chelating effect of the oxalate anion (Chen et al. 2000; Eick et al. 1996a, b). The action of oxalic acid on calcareous sandstones is expressed by the dissolution of calcite and the formation of Ca-oxalate, which seems to protect the rock against further ageing (Ariño et al. 1995, 1997).

Larreur-Cayol et al. (2011a) could identify Ca-oxalate on cementitious specimens under the form of the monoclinic mono-hydrate whewellite. However, it should be noted that Ca-oxalate also exists in a less stable form: the tetragonal dihydrate weddellite. For example, weddellite serves for lichens as a water absorbing and accumulating substrate which transforms to whewellite when humidity drops (Frost et al. 2003). Weddellite and whewellite very often occur together on the surface of calcareous or non-calcareous historical buildings subjected to biodeterioration notably in the Mediterranean urban environment (Jorge Villar et al. 2004; Monte 2003; Doherty et al. 2007).

Tartaric Acid

The attack by tartaric acid occurs in two stages. It leads to the precipitation of calcium tartrate tetrahydrate $CaC_4H_4O_6.4H_2O$ in the outer part of cement paste specimens, which seems to efficiently protect the sample during the first 2 months of the experiment (Fig. 6.2), limiting the attack to microscopic changes. Then, macroscopic alterations begin. The attack leads to (i) the dissolution of the cement matrix periphery and (ii) the precipitation of a thin layer of calcium tartrate salts adhering to the matrix (outer layer on Figs. 6.10 and 6.11) (Larreur-Cayol et al. 2011a; Bertron et al. 2011). Within the matrix, two zones are observed with similar features to those observed with other acids: a non-altered core (zone 1) and a transition zone (zone 2) (Figs. 6.10 and 6.11). These zones are typical of the hydrolysis of the cement paste caused by the acid-base reactions.

The calcium tartrate layer at the periphery of the samples seems to protect the matrix, at least partially, as the kinetics of alteration remained moderate during the whole experiment performed by Larreur-Cayol et al. (2011a). This phenomenon is comparable to a process used in the wine industry. This process, consisting of brushing the inside walls of tanks with a solution of tartaric acid, helps protect concrete against attack by the acids in wine. However, its protective role is of limited duration since, despite the precipitation of calcium tartrate at the periphery of the matrix, the alteration front progresses to greater depths over time.

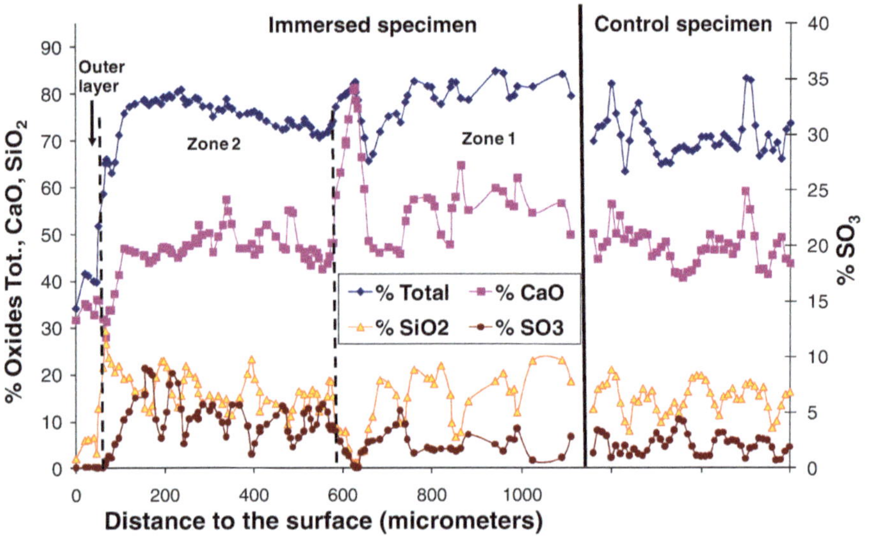

Fig. 6.10 EPMA chemical composition profile of an OPC paste immersed in tartaric acid (0.28 M, pH 4) for 10 months and of a control specimen (Larreur-Cayol et al. 2011a)

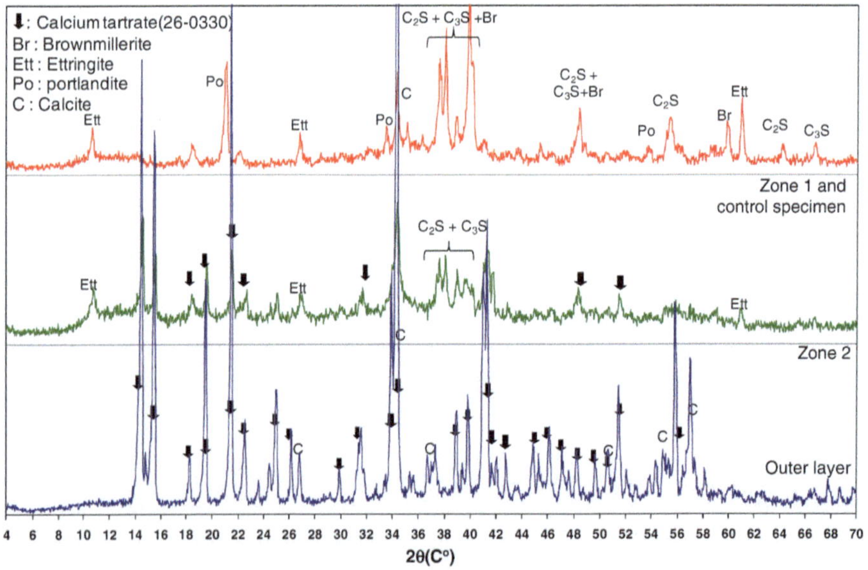

Fig. 6.11 X-ray traces of an OPC paste specimen immersed in tartaric acid for 10 months (Larreur-Cayol et al. 2011a)

The formation of Ca-tartrate is, however, less protective than that of Ca-oxalate, probably because of the higher molar volume of the salt (143 cm^3/mol vs. 66 cm^3/mol), which provokes the break-up of the outer part of the specimen.

Malic Acid

The mechanisms of cementitious matrix degradation by malic acid have been found to be very similar to those by tartaric acid, with the precipitation of calcium malate in the outer part of the specimens (Larreur-Cayol 2012).

6.4.2.3 Mechanisms of Degradation by Citric and Succinic Acids

Citric Acid

Among the acids that have been studied, citric acid is the most aggressive to the cementitious matrix. The attack by citric acid leads to the formation of calcium citrate tetra-hydrate salt, $Ca_3(C_6H_5O_7)_2.4H_2O$ (Bertron et al. 2009a; Larreur-Cayol et al. 2011a). The precipitation of the salt, the molar volume of which is very high (Table 6.14), leads to the break-up of the outer part of the cementitious matrix. This, in turn, explains, at least partially, the very high alteration kinetics of the specimens (Fig. 6.2) as the progression of the alteration front is not slowed by the diffusion of the aggressive species through the altered zone as is usually observed for other acids. Moreover, the calcium citrate salt does not adhere to the matrix; it detaches from the specimen and collects in great quantities at the bottom of the tank. This is why no trace of salt is found inside or at the surface of the altered specimens (Fig. 6.13). Inside the remaining matrix, the chemical and mineralogical patterns (Figs. 6.12 and 6.13) are very similar to those of specimens immersed in acetic acid (Figs. 6.3 and 6.4). The very high aggressiveness of citric acid could also be linked to the poly-acidity of the acid (for one mole of acid, at a pH of 4, two moles of hydronium ions are likely to be released, against one for mono-acids such as acetic or propionic acid), as mentioned by Larreur-Cayol et al. (2011a).

The observations given above relate to conditions of attack such that a sharp pH gradient always exists between the surface and the core of the cementitious specimen. Several authors have studied the influence of organic acids (in particular citric and oxalic acids) on the hydration of cement (Smillie and Glasser 1999; Möschner et al. 2009). Citric acid was supposed to act as set retarder by the sorption of citrate onto the clinker surface which formed a protective layer around the clinker grains retarding their dissolution (Möschner et al. 2009). In a continuously alkaline environment, Smillie and Glasser (1999) found that citrate increased the amount of $CaCO_3$ present in the cement, suggesting that citrate rearranged to form carbonate. It should be noted that such reactions could also occur in external attack by citric acid in zones still at high pH if gradients developed in some situations are not as high as the experiments performed by Bertron (2009b) and Larreur-Cayol et al. (2011a).

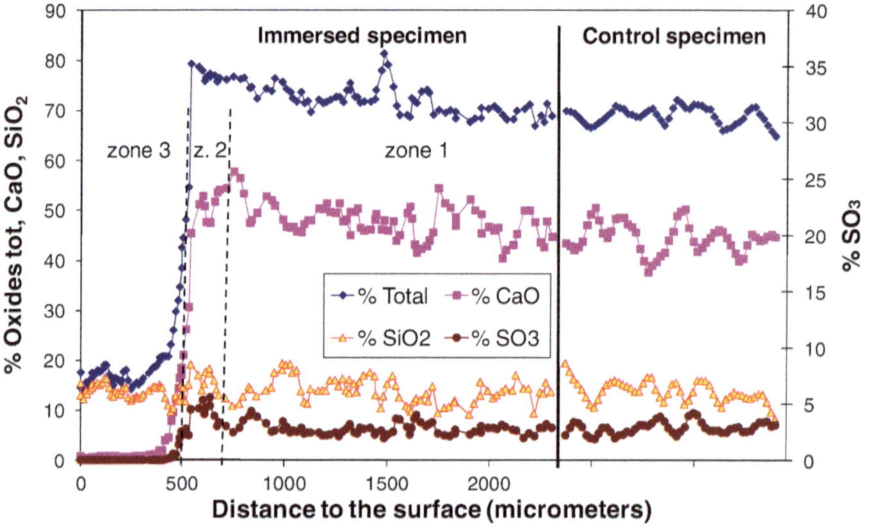

Fig. 6.12 EPMA chemical composition profile of an OPC paste immersed in citric acid (0.28 M, pH 4) for 1 month and of a control specimen (Bertron et al. 2011a)

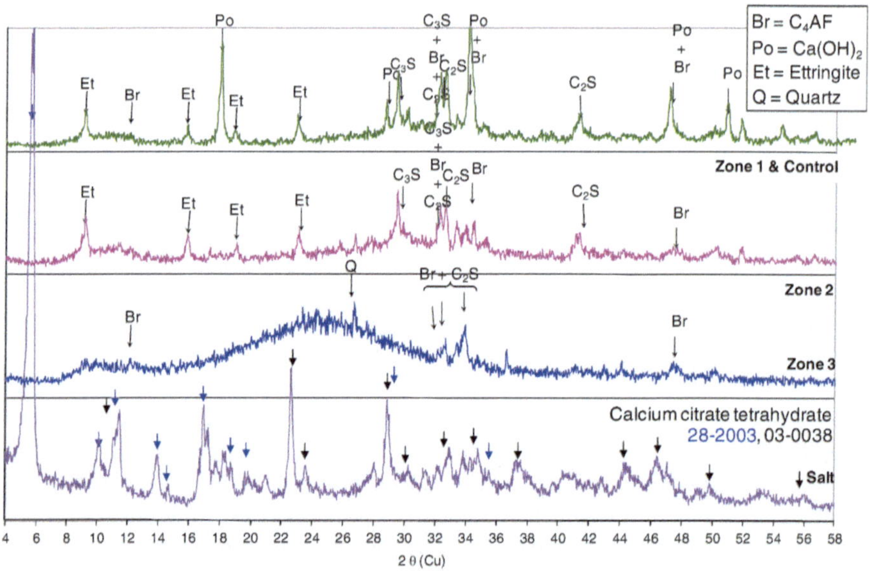

Fig. 6.13 X-ray traces of an OPC paste specimen immersed in citric acid for 1 month (Bertron et al. 2011a)

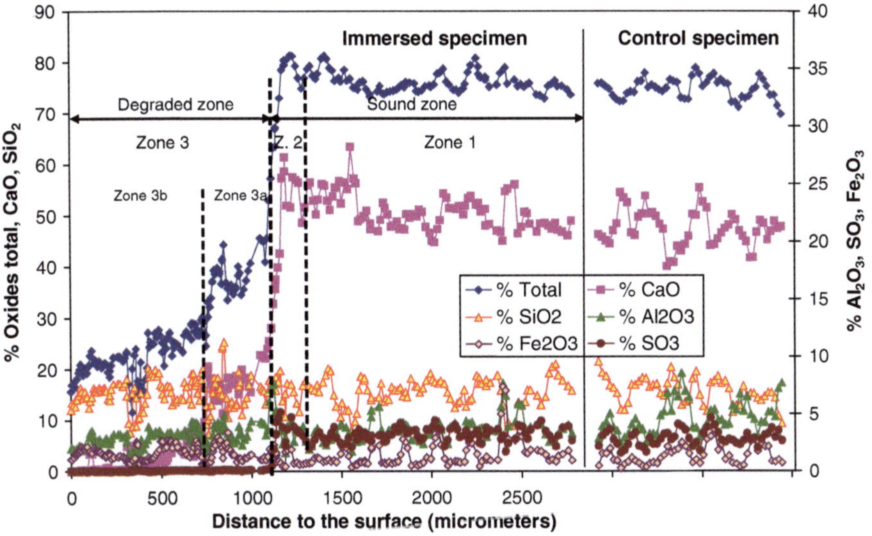

Fig. 6.14 EPMA chemical composition profile of an OPC paste immersed in succinic acid (0.28 M, pH 4) for 1 month and of a control specimen (Bertron et al. 2009a)

Succinic Acid

Regarding succinic acid, the chemical profiles of the specimen immersed in succinic acid are quite similar to those obtained for specimens immersed in acetic acid and other volatile fatty acids except for the inner part of the degraded zone (zone 3a, Fig. 6.14), which is not completely decalcified here. In zone 3a, the precipitation of calcium succinate mono-or tri-hydrate salt (Mathew et al. 1994; Karipides and Reed 1980; Larreur-Cayol 2012) occurs but it does not seem to be protective (or at least as protective as tartaric or malic acid, for example), probably because of the small quantities of salt formed, which may be related to the solubility value (sensibly higher than that of the other protective salts mentioned above, Table 6.12). Nevertheless, calcium succinate does not seem to be destructive either (only moderate cracking is observed on the outer part of the degraded zones) probably thanks to the moderate molar volume of this salt (Table 6.14) (Bertron et al. 2009a; Larreur-Cayol 2012).

It should also be noted that, unlike acetic acid, succinic acid causes the dissolution of the outer part of the degraded zone of the specimens. Bertron et al. (2009a) assume that the higher aggressiveness of succinic acid could be linked to the poly-acidity of this salt.

6.4.3 *Mechanisms with Mixes of Acids*

Very few authors have investigated degradation mechanisms using mixes of acids in order to simulate real effluents.

Larreur-Cayol et al. (2011b) considered mixes of acetic and oxalic acids (0.19 and 0.09 M, respectively; total concentration of acids: 0.28 M, pH 1.57). As seen in Sects. 6.4.1 and 6.4.2, these acids have very different aggressiveness to the cementitious matrix. On the one hand, acetic acid causes leaching of the degraded zone of the cement paste and the formation of a silica gel containing aluminium and iron; on the other, the formation of insoluble calcium oxalate mono-hydrate efficiently protects the matrix against attack by oxalic acid. In the work of Larreur-Cayol et al., cement paste specimens were immersed for 2 months in the aggressive solution, which was regularly renewed (every 2 weeks). It was observed that the pH remained constant throughout the experiment, indicating that there was no release of hydroxyl ion from the cement paste. Monitoring of relative mass losses and degraded depths of the specimens showed that there was no apparent deterioration of the matrix. Chemical and mineralogical analyses with EPMA and XRD showed that the amount of calcium and other elements remained constant in the specimen and that calcium oxalate had precipitated in the outer layer of the specimens. These features are characteristic of the attack by oxalic acid (see Sect. 6.4.2.2.1). The action of oxalic acid is, in this case, dominant over that of acetic acid, in accordance with thermodynamic data listed by De Windt and Devillers (2010) and Larreur-Cayol (2012). Nevertheless, Larreur-Cayol observed that the salts precipitated in the mixture of acids were calcium oxalate mono- and di-hydrate, whewellite and weddellite. Whewellite is the monoclinic form of Ca-oxalate, while the tetragonal weddellite is a less stable form (Lepage and Tawashi 1982). Weddellite and whewellite often occur together on the surface of calcareous or non-calcareous historical buildings that have undergone biodeteriation, notably in the Mediterranean urban environment (Jorge Villar et al. 2004; Monte 2003). This difference in the precipitation form of Ca-oxalate between oxalic acid alone and the acetic-oxalic mixture could be linked to the oxalate- or calcium-dependence nature of the two salts (Daudon et al. 2004, Larreur-Cayol 2012). However, it should be noted that both forms seemed to be protective for the cementitious matrix.

6.5 Conclusion/Summary

In this review chapter, the attack of cementitious matrices by the organic acids found in the agricultural and agrofood industries has been explored. First, in spite of their very variable compositions, some common chemical features have been identified for the acidic part of the waste waters considered. Among agricultural effluents, liquid manure, with pH between 6 and 8, contains about 10 g/l of volatile fatty acids, of which acetic and propionic are the main acids. Silage effluents contain mainly lactic acid (between 5 and 45 g/l) and acetic acid but also butyric acids in much smaller quantities, the resulting pH being 4. Among agrofood waste waters, whey produced by the dairy industry may have pH between 4.5 and 6.5. This effluent contains lactic, acetic and citric acid in various quantities depending notably on the type of dairy product. The concentration in acid is several grams per litre. The sugar industry produces molasses which contains lactic, malic, acetic, oxalic, citric and succinic acids. The fermenting industry produces tartaric, succinic and acetic acid.

The aggressiveness of the attack by these weak acids notably depends on the solubility of the calcium salts. During their action on cement pastes, volatile fatty acids (acetic, propionic and butyric acids) and lactic acid produce calcium salts whose solubility is high to very high in water; although malic, succinic, citric and oxalic acids form slightly soluble to insoluble calcium salts which may then have either a protective or a deleterious effect on the matrix. For organic acids, the calcium salts of which are soluble in water, the major parameter for expressing the aggressiveness of the acids is the dissociation constant, although the polyacidity of the acid can also be mentioned. For organic acids with slightly soluble or insoluble salts, several parameters influence the aggressiveness of the acid: the solubility of the salt, and its physical properties (notably its molar volume). However, further studies are necessary to identify the whole range of parameters that influence the degradation mechanisms: the mesoscopic shape of the salt, its affinity towards the cement matrix, and the kinetics of salt formation, for example, should be investigated.

The attack by the acids in liquid manure, silage juice and whey (except citric acid) is comparable to the attack of strong acids (notably nitric and hydrochloric acid) since the anions of the acids have no specific effect in the degradation, the calcium salts being soluble and the complexes having little or no influence on the alteration. The degradation at pH between 2.5 and 6 results in a well-marked mineralogical and chemical zonation of the cement paste. The degraded zone, which is very porous and has low mechanical strength, is progressively and almost totally decalcified. It is composed of a silica gel containing aluminium and iron. The hydrated and anhydrous crystallized phases are totally dissolved in the zone in contact with the aggressive solution. For the parameters influencing the aggressiveness of the attack, it has been pointed out that the concentration of acid is a more significant parameter than the pH of the solution. The comparison with the attack of strong acids showed significantly higher kinetics for nitric or hydrochloric acid and little difference concerning the mechanisms.

The attack of concrete by malic, tartaric, oxalic, citric and succinic acids leads to the formation of the corresponding calcium salts, at least in the conditions of the experiments found in the literature. The calcium salts are found to be protective for oxalic, tartaric and malic acids, in so far as the kinetics of the matrix degradation with these acids is lower than that with the reference acid: acetic acid. In the case of citric and succinic acid, citrate and succinate calcium salts precipitate and they are not protective. In particular, the formation of citrate salt, the molar volume of which is very high compared to that of $Ca(OH)_2$ or C-S-H, at the expense of which this salt forms, seems to be especially deleterious to the matrix. The attack by this acid is the most aggressive among those investigated in the literature.

Finally, very few studies have focused on mixes of acids, which should, nevertheless, be more representative of real effluents. Further studies are necessary to investigate and clarify the mechanisms and kinetics in these conditions.

References

Adamo, P., Violante, P.: Weathering of rocks and neogenesis of minerals associated with lichen activity. Appl. Clay Sci. **16**(5–6), 229–256 (2000)

Albet, J., Pislor, E., and Pontalier, P.Y.: Process for recovering carboxylic acids from sugar cane industry by-products, In: Proceedings of the 2nd International Congress on Green Process Engineering, Venice, Italy (2009)

Allison, J.D., Brown, D.S., Novo-Gradac, K.J.: MINTEQA2/PRODEF2, A Geochemical Assessment Model for Environmental Systems: Version 3.0 User's Manual. EPA/600/3-91/021. U.S. EPA, Athens, GA, 30605 (1991)

Amyot, A.: Valorisation des sous-produits de maïs sucré sous forme d'ensilage à 30% de matière sèche, Rapport Final IRDA, 131 p, Quebec, Canada (2005)

Amyot, A., Denoncourt, P., Ouattara, B., Lacroix, M.: Qualité de conservation de l'ensilage de maïs protégé avec un bioenrobage, In. Comité des plantes fourragères du Conseil des productions végétales du Québec inc. (ed.) Compte-rendu des Demi-journée d'information scientifique sur les fourrages, pp. 30–34, Victoriaville, Quebec (2003)

Ariño, X., Ortega-Calvo, J.J., Gomez-Bolea, A., Saiz-Jimenez, C.: Lichen colonization of the Roman pavement at Baelo Claudia (Cadiz, Spain): Biodeterioration vs. bioprotection. Sci. Total Environ. **167**(1–3), 353–363 (1995)

Ariño, X., Gomez-Bolea, A., Saiz-Jimenez, C.: Lichens on ancient mortars. Int. Biodeter. Biodegr. **40**(2–4), 217–224 (1997)

Bejaoui, S., Bary, B.: Modeling of the link between microstructure and effective diffusivity of cement pastes using a simplified composite model. Cem. Concr. Res. **37**(3), 469–480 (2007)

Beline, F., Martinez, J., Burton, C.H.: Traitement aérobie du lisier de porcs. Bases théoriques et techniques. Vecteur Environnement **31**, 2 (1998)

Bensted, J.: Effect of silage upon hardened cement and concrete. Il Cemento **1**, 3–10 (1993)

Bernard, C.R., Heduit, M.: Evolution du lisier de porc au cours du stockage. Journées de la recherche porcine en France, 177–184 (1979)

Bertron, A.: Durabilité des matériaux cimentaires soumis aux acides organiques, cas particulier des effluents d'élevage. PhD thesis, INSA Toulouse, 256 p (2004)

Bertron, A., Duchesne, J.: Attack of cementitious materials by organic acids in agricultural and agrofood effluents – A review. In: De Belie, N. (ed.) Proceedings of the RILEM TC 211-PAE Workshop, pp. 11–24, Ghent (2007)

Bertron, A., Escadeillas, G., Duchesne, J.: Cement pastes alteration by liquid manure organic acids: chemical and mineralogical characterization. Cem. Concr. Res. **34**(10), 1823–1835 (2004)

Bertron, A., Duchesne, J., Escadeillas, G.: Attack of cement pastes exposed to organic acids in manure. Cem. Concr. Compos. **27**(9–10), 898–909 (2005a)

Bertron, A., Duchesne, J., Escadeillas, G.: Accelerated tests of hardened cement pastes alteration by organic acids: analysis of the pH effect. Cem. Concr. Res. **35**(1), 155–166 (2005b)

Bertron, A., Coutand, M., Cameleyre, X., Escadeillas, G., Duchesne, J.: L'attaque chimique et biologique des effluents agricoles et agroalimentaires sur les matériaux cimentaires. Matériaux et Techniques **93**, 111–121 (2005c)

Bertron, A., Escadeillas, G., Duchesne, J.: Degradation of cement pastes by organic acids. Mater. Struct. **40**(3), 341–354 (2007)

Bertron, A., Larreur-Cayol, S., Le, T., Escadeillas, G.: Degradation of cementitious materials by some organic acids found in agroindustrial effluents. In: Alexander MG, Bertron A (ed.) Proceedings of the RILEM TC 211-PAE Final Conference, vol. 1, pp. 96–107, Toulouse (2009a)

Bertron, A., Escadeillas, G., De Parseval, P., Duchesne, J.: Processing of electron microprobe data from the analysis of altered cementitious materials. Cem. Concr. Res. **39**(10), 929–935 (2009b)

Bertron, A., Larreur-Cayol, S., Escadeillas, G.: Organic acids' attack on cementitious materials: aggressiveness and degradation mechanisms. In: Instituto de Ciencias de la Construcción

"Eduardo Torroja". CSIC. Palomo, A., Zaragoza, A., Lopez Aguii, J.C. (eds.) Proceedings of the XIIIth International Conference on the Chemistry of Cement, Madrid (2011)

Bories, A., Conesa, F., Boutolleau, A., Peureux, J.L., Tharrault, P.: New treatment of the winery wastewaters by a combined process of distillation-concentration. In: Cemagref (ed.) Proceedings of the 2nd International Specialized Conference on Winery Wastewaters, pp. 127–135, Bordeaux (1998)

Celestine-Myrtil, D.A., Parfait, A.: HPLC determination of organic acids in sugar cane and its industrial by-products. Int. Sugar J. **90**(1070), 28–32 (1988)

Chandra, S.: Hydrochloric acid attack on cement mortar, an analytical study. Cem. Concr. Res. **18** (2), 193–203 (1988)

Chen, J., Blume, H.P., Beyer, L.: Weathering of rocks induced by lichen colonization—a review. Catena **39**(2), 121–146 (2000)

Coca, M., Pena, M., Gonzalez, G.: Variables affecting efficiency of molasses fermentation wastewater ozonation. Chemosphere **60**, 1408–1415 (2005)

Colin, T., Bories, A., Sire, Y., Perrin, R.: Treatment and valorisation of winery wastewater by a new biophysical process (ECCF). Water Sci. Technol. **51**(1), 99–106 (2005)

Daudon, M., Jungers, P., Lacour, B.: Clinical interest of crystalluria study. Ann. Biol. Clin. **62**, 379 –393 (2004)

De Belie, N.: A survey on concrete floors in pig houses and their degradation. J. Agric. Eng. Res. **66**(3), 151–156 (1997)

De Belie, N., Verselder, H.J., De Blaere, B., Van Nieuwenburg, D., Verschoore, R.: Influence of the cement type on the resistance of concrete to feed acids. Cem. Concr. Res. **26**(11), 1717–1725 (1996)

De Belie, N., Debruyckere, M., Van Nieuwenburg, D., De Blaere, B.: Attack of concrete floors in pig houses by feed acids: influence of fly ash addition and cement-bound surface layers. J. Agric. Eng. Res. **68**(2), 101–108 (1997a)

De Belie, N., De Coster, V., Van Nieuwenburg, D.: Use of fly ash or silica fume to increase the resistance of concrete to feed acids. Mag. Concr. Res. **49**(181), 337–344 (1997b)

De Belie, N., Richardson, M., Braam, C.R., Svennerstedt, B., Lenehan, J.J., Sonck, B.: Durability of building materials and components in the agricultural environment: part I, the agricultural environment and timber structures. J. Agric. Eng. Res. **75**(3), 225–241 (2000a)

De Belie, N., Lenehan, J.J., Braam, C.R., Svennerstedt, B., Richardson, M., Sonck, B.: Durability of building materials and components in the agricultural environment, Part III: Concrete structures. J. Agric. Eng. Res. **76**(1), 3–16 (2000b)

De Ceukelaire, L.: The effects of hydrochloric acid on mortar. Cem. Concr. Res. **22**(5), 903–914 (1992)

De La Farge, B., Heduit, M., Brondeau, P., Mongin, J.P., Saugere, D., Cambus, L.: La méthanisationdes lisiers de porcs. In: ITP INRA (ed.) Proceedings of 15ème Journées de la Recherche Porcine, pp. 11–22, Paris (1983)

De Windt, L., Devilliers, P.: Modeling the degradation of Portland cement pastes by biogenic organic acids. Cem. Concr. Res. **40**(8), 1165–1174 (2010)

Doherty, B., Pamplona, M., Miliani, C., Matteini, M., Sgamellotti, A., Brunetti, B.: Durability of the artificial calcium oxalate protective on two Florentine monuments. J. Cult. Heritage **8**(2), 186–192 (2007)

Eick, M.J., Grossl, P.R., Golden, D.C., Sparks, D.L., Ming, D.W.: Dissolution of a lunar simulant as affected by pH and organic anions. Geoderma **74**(1–2), 139–160 (1996a)

Eick, M.J., Grossl, P.R., Golden, D.C., Sparks, D.L., Ming, D.W.: Dissolution kinetics of a lunar glass simulant at 298 K. The effect of pH and organic acids. Geochimica. Cosmochimica. Acta **60**(1), 157–170 (1996b)

Frost, R.L., Yang, J., Ding, Z.: Raman and FTIR spectroscopy of natural oxalates: implications for the evidence of life on mars. Chin. Sci. Bull. **48**(17), 1844–1852 (2003)

Galanos, E., Gray, K.R., Biddelstone, A.J., Thayanithy, K.: The aerobic treatment of silage effluent: effluent characterization and fermentation. J. Agric. Res. **62**(4), 271–279 (1995)

Gelegenisa, J., Georgakakisb, D., Angelidakic, I., Mavrisa, V.: Optimization of biogas production by co-digesting whey with diluted poultry manure. Renew. Energy **32**, 2147–2160 (2007)
Glaus, M.A., van Loon, L.R., Achatz, S., Chodura, A., Fischer, K.: Degradation of cellulosic materials under the alkaline conditions of a cementitious repository for low and intermediate level radioactive waste: Part I: Identification of degradation products. Anal. Chim. Acta **398**(1), 111–122 (1999)
Hong, S.Y., Glasser, F.P.: Alkali binding in cement pastes: part I. The C-S-H phase. Cem. Concr. Res. **29**(12), 1893–1903 (1999)
Israel, D., Macphee, D.E., Lachowski, E.E.: Acid attack on pore-reduced cements. J. Mater. Sci. **32**(15), 4109–4116 (1997)
Johansen, A.G., Vegarud, G.E., Skeie, S.: Seasonal and regional variation in the composition of whey from Norwegian Cheddar-type and Dutch-type cheeses. Int. Dairy J. **12**, 621–629 (2002)
Jorge Villar, S.E., Edwards, H.G., Seaward, M.R.: Lichen biodeterioration of ecclesiastical monuments in northern Spain. Spectrochim. Acta A Mol. Biomol. Spectrosc. **60**(5), 1229–1237 (2004)
Karipides, A., Reed, T.: The structure of diaquasuccinatocalcium (II) monohydrate. Acta Crystallographica. B Struct. Crystallogr. Cryst. Chem. **36**(6), 1377–1381 (1980)
Knauss, K.G., Dibley, M.J., Bourcier, W.L., Shaw, H.F.: Ti(IV) hydrolysis constants derived from rutile solubility measurements made from 100 to 300°C. Appl. Geochem. **16**(9–10), 1115–1128 (2001)
Kok, D.J., Khan, S.R.: Calcium oxalate nephrolithiasis, a free or fixed particle disease. Kidney Int. **46**, 847–854 (1994)
Lange, N.A., Dean, J.A. (eds.): Lange's Handbook of Chemistry, 30th edn. Mc Graw Hill Book Company, New York (1985)
Larreur-Cayol, S.: Attaques des matériaux cimentaires par les acides organiques des effluents agricoles et agro-alimentaires. Ph.D. thesis, Université Paul Sabatier, Toulouse (2012)
Larreur-Cayol, S., Bertron, A., Escadeillas, G.: Alteration of cement-based materials by various organic acids in agro-industrial waste-waters. Cem. Concr. Res. **41**(8), 882–892 (2011a)
Larreur-Cayol, S., De Windt, L., Bertron, A., Escadeillas, G.: Biodeterioration of cementitious matrices by organic acid in agricultural effluents: experiments and modeling. In: Godbout, S. Potvin, L. (eds.) Proceedings of 7th International Symposium on Cement Based Materials for a Sustainable Agriculture, pp. 38–45, Quebec (2011b)
Larreur-Cayol, S., Bertron, A., San Nicolas, R., Escadeillas, G.: Durability of different binders in synthetic agricultural effluents. In: Godbout, S., Potvin, L. (eds.) Proceedings of 7th International Symposium on Cement Based Materials for a Sustainable Agriculture, pp. 56–66, Quebec (2011c)
Lepage, L., Tawashi, R.: Growth and characterization of calcium oxalate dihydrate crystals (weddellite). J. Pharm. Sci. **71**(9), 1059–1062 (1982)
Levasseur, P.: Composition des lisiers de porc, facteurs de variation et méthodes d'évaluation. Techni-Porc **21**(3), 19–25 (1998)
Lide, D.R.: Handbook of Chemistry and Physics, 72nd edn. CRC Press Inc., Boca Raton (1991)
Lodi, S., Rossin, G.: Determination of some organic acids in sugar factory products. J. Chromatogr. A **706**, 375–383 (1995)
Magniont, C., Coutand, M., Bertron, A., Cameleyre, X., Lafforgue, C., Beaufort, S., Escadeillas, G.: A new test method to assess the bacterial deterioration of cementitious materials. Cem. Concr. Res. **41**(4), 429–438 (2011)
Martinez, J., Le Bozec, G.: Déjections porcines et problèmes environnementaux en Europe. Cahiers Agric. **9**, 181–190 (2000)
Mathew, M., Takagi, S., Fowler, B.D., Markovic, M.: The crystal structure of calcium succinate monohydrate. J. Chem. Crystallogr. **24**(7), 437–440 (1994)
McDonald, P., Henderson, A.R., Heron, S.J.E.: The Biochemistry of Silage. Second Edition. Chalcombe Publications, Marlow, Buckinghamshire SL7 3PU, 340 p (quoted by Savoie et al. 2000) (1991)

Medgyesi: Comparaison de l'effet corrosif de certaines acides organiques. In: Valenta O. (ed.) Final Report of the International Symposium on Durability of Concretes, pp. C-431–445. Prague (1969)
Meinck, F., Stooff, H., Kohlschütter, H. (eds.): Les eaux résiduaires industrielles. Masson, Paris (1977). 863 p
Monte, M.: Oxalate film formation on marble specimens caused by fungus. J. Cult. Heritage **4**(3), 255–258 (2003)
Moosbrugger, R.E., Wentzel, M.C., Ekama, G.A., Marais, G.V.R.: Treatment of wine distillery waste in UASB systems – feasibility, alkalinity requirements and pH control. Water Sci. Technol. **28**(2), 45–54 (1993)
Moranville, M., Kamali, S., Guillon, E.: Physicochemical equilibria of cement-based materials in aggressive environments—experiment and modelling. Cem. Concr. Res. **34**(9), 1569–1578 (2004)
Möschner, G., Lothenbach, B., Figi, R., Kretzschmar, R.: Influence of citric acid on the hydration of Portland cement. Cem. Concr. Res. **39**(4), 275–282 (2009)
Müller, D., Heil, M.: Distribution of winery wastewater on land; investigations on ecotoxicology. In: Cemagref (ed.) Proceedings of the 2nd International Specialized Conference on Winery Wastewaters, pp. 115–122, Bordeaux (1998)
O'Donnell, C., Dodd, V.A., O'Kiely, P., Richardson, M.: A study of the effects of silage effluent on concrete: Part 1, significance of concrete characteristics. J. Agric. Eng. Res. **60**(2), 83–92 (1995a)
O'Donnell, C., Dodd, V.A., O'Kiely, P., Richardson, M.: A study of the effects of silage effluent on concrete: Part 2, significance of environmental factors. J. Agric. Eng. Res. **60**(2), 93–97 (1995b)
Orditz, D., Lakel, A., Cronier, J.N.: Wine-making effluents finishing treatment by geo-epuration on siliceous filter. In: Cemagref (ed.) Proceedings of the 2nd International Specialized Conference on Winery Wastewaters, pp. 215–222, Bordeaux (1998)
Ouellet, D.R., Lafreniere, C., Roy, L.R., Berthiaume, R.: Valeur nutritive des ensilages de balles rondes pour les bovins. In Comité des plantes fourragères du Conseil des productions végétales du Québec inc. (ed.) Compte-Rendu des 12 Conférences, Demi-journée d'information scientifique sur les fourrages, Victoriaville, pp. 11–21, Quebec (2001)
Oueslati, O.: Durabilité des matériaux cimentaires soumis aux acides organiques – Résistance chimique, mécanique et de corrosion, Ph.D. thesis, Université Laval, Quebec (2011)
Oueslati, O., Duchesne, J.: Acetic attack of blended cement paste In: Alexander M.G, Bertron A. (eds.) Proceedings of the RILEM TC 211-PAE Final Conference, vol. 1, pp. 108–115, Toulouse (2009)
Oueslati, O., Duchesne, J.: Acetic acid attack of cement matrix: evaluation of durability parameters. In: (Instituto de Ciencias de la Construcción "Eduardo Torroja". CSIC. Palomo A., Zaragoza A., Lopez Aguii J.C. (eds.) Proceedings of the XIIIth International Conference on the Chemistry of Cement, Madrid (2011a)
Oueslati, O., Duchesne, J.: Impact of acetic acid attack on the chemical, physical and mineralogical evolution of cement pastes. In: Godbout S. Potvin L. (eds.) Proceedings of 7th International Symposium on Cement Based Materials for a Sustainable Agriculture, pp. 28–37, Quebec (2011b)
Oueslati, O., Duchesne, J.: The effect of SCMs and curing time on resistance of mortars subjected to organic acids. Cem. Concr. Res. **42**(1), 205–214 (2012)
Patterson, D.C., Walker, N.: The use of effluent from grass silage in the diet of finishing pigs. I. Variation in composition of effluents. Anim. Feed Sci. Technol. **4**, 263–274 (1979)
Paul, J.W., Beauchamp, E.G.: Relationship between volatile fatty acids, total ammonia, and pH in manure slurries. Biol. Wastes **29**(4), 313–318 (1989)
Pavlík, V.: Corrosion of hardened cement paste by acetic and nitric acids. Part I: Calculation of corrosion depth. Cem. Concr. Res. **24**(3), 551–562 (1994a)

Pavlík, V.: Corrosion of hardened cement paste by acetic and nitric acids. Part II: Formation and chemical composition of the corrosion products layer. Ceme. Concr. Res. **24**(8), 1495–1508 (1994b)

Pavlík, V.: Corrosion of hardened cement paste by acetic and nitric acids. Part III: Influence of water/cement ratio. Cem. Concr. Res. **26**(3), 475–490 (1996)

Pavlik, V., Uncik, S.: The rate of corrosion of hardened cement pastes and mortars with additive of silica fume in acids. Cem. Concr. Res. **27**(11), 1731–1745 (1997)

Ribéreau-Gayon, P., Glories, Y., Maujean, A., Dubourdieu, D.: Organic acids in wine. In: Handbook of Enology, Vol 2: The Chemistry of Wine and Stabilization and Treatments, pp. 3–50. Wiley (2006)

Saddoud, A., Hassaïri, I., Sayadi, S.: Anaerobic membrane reactor with phase separation for the treatment of cheese whey. Bioresour. Technol. **98**, 2102–2108 (2007)

Savoie, P., Amyot, A., Lacroix, M., Thierault, R.: Ensilage et environnement. In: Comité des plantes fourragères du Conseil des productions végétales du Québec inc. Compte-rendu des Demi-journée d'information scientifique sur les fourrages, pp. 11–21, Victoriaville, QC (2000)

Seidell, A. (ed.): Solubilities of Inorganic and Organic Compounds – A Compilation of Quantitative Solubility Data from the Periodical Literature, 2nd edn. Van Nostrand D Company, New York (1919). 912 p

Shepderd, H.L.: Performance evaluation of a pilot scale constructed wetland used for treatment of winery process wastewater. In: Cemagref (ed.) Proceedings of the 2nd International Specialized Conference on Winery Wastewaters, pp. 159–167, Bordeaux (1998)

Smillie, S., Glasser, F.P.: Reaction of EDTA, oxalic acid and citric acid with Portland cement. Adv. Cem. Res. **11**(2), 97–101 (1999)

Streit, J., Tran-Ho, L.C., Koenigsberger, E.: Solubility of calcium oxalate hydrates in sodium chloride solutions and urine-like liquors. Monatshefte fur Chemie/Chemical Monthly **129**(12), 1225–1236 (1998)

Taylor, H.F.W.: Cement Chemistry. Academic, London (1990)

Tennis, P.D., Jennings, H.M.: A model for two types of calcium silicate hydrate in the microstructure of Portland cement pastes. Cem. Concr. Res. **30**(6), 855–863 (2000)

Viallis, H., Faucon, P., Nonat, A., Petit, J.C.: Interaction between salts (NaCl, CsCl) and calcium silicate hydrates (C − S − H). J. Phys. Chem. B **103**(25), 5212–5219 (1999)

Viaud, M.N., Briard, D., Wisser, A., Rochard, J., Montcourant, E.: Study of follow-up parameters of biological treatment of winery wastewater by aerated storage. In: Cemagref (ed.) Proceedings of the 2nd International Specialized Conference on Winery Wastewaters, pp. 279–286, Bordeaux (1998)

Walsdorf, N.B., Alexandrides, G., Pak, C.: Ultradense and more soluble and bioavailable preparations of calcium citrate, US Patent 4,814,177:12 p (1989)

Wansbrough, H.: Chemistry in winemaking. In: New Zealand Institute of Chemistry (ed.) Chemical Processes in New Zealand, 2nd edn. New Zealand Institute of Chemistry, Auckland (1998)

Weast, R.C., Astel, M.J., Beyer, W.H.: CRC Handbook of Chemistry and Physics, 67th edn. CRC Press Inc., Boca Raton (1986)

Weydert, M., Van Caenegem, L., Dehon, P.: Resistance of concrete, asphaltic concrete and mastic asphalt to silage juice. J. Agric. Eng. Res. **79**(2), 151–157 (2001)

Williams, A.G.: Organic acids, biochemical oxygen demand and chemical oxygen demand in the soluble fraction of piggery slurry. J. Sci. Food Agric. **34**(3), 212–220 (1983)

Yang, K., Yu, Y., Hwang, S.: Selective optimization in thermophilic acidogenesis of cheese-whey wastewater to acetic and butyric acids: partial acidification and methanation. Water Res. **37**(10), 2467–2477 (2003)

Zabozlaev, A.A., Oganesyan, É.T., Pogorelov, V.I.: Solubilization of poorly soluble calcium salts of organic acids with sorbitol. Pharm. Chem. J. **41**(8), 430–433 (2007)

Zhang, Z., Zhu, J.: Characteristics of solids, BOD_5 and VFAs in liquid swine manure treated by short-term low-intensity aeration for long-term storage. Bioresour. Technol. **97**(1), 140–149 (2006)

Zhang, Z., Zhu, J., Park, K.J.: A bench-scale aeration study using batch reactors on swine manure stabilization to control odour in post treatment storage. Water Res. **40**(1), 162–174 (2006)

Zhu, J., Ndegwa, P., Luo, A.: Temperature effects on pig manure under low level batch aeration. Int. J. Environ. Stud. **60**(5), 523–533 (2003)

Ziegler, D., Heduit, M. Engrais de ferme. Valeur fertilisante. Gestion, Environnement. Institut Technique du Porc, 35 p (1991)

Part II
Modeling Degradation of Cementitious Materials in Aggressive Aqueous Environments

Authors of chapters in Part II

Chapter 7:
P. Le Bescop
B. Lothenbach
E. Samson
K.A. Snyder

Reviewers for Part II

Mark G. Alexander, Manu Santhanam

Chapter 7
Modeling Degradation of Cementitious Materials in Aggressive Aqueous Environments

Patrick Le Bescop, Barbara Lothenbach, Eric Samson, and Kenneth A. Snyder

Contents

P. Le Bescop
CEA, DEN, DPC, SECR, Laboratoire d'Etude du Comportement des Bétons et des Argiles, F-91191 Gif-sur-Yvette, France
e-mail: patrick.le-bescop@cea.fr

B. Lothenbach
Empa, Ueberlandstrasse 129, CH-8600 Duebendorf, Switzerland
e-mail: barbara.lothenbach@empa.ch

E. Samson (✉)
SIMCO Technologies Inc, 1400 Boul. du Parc Technologique, Suite 203, Québec City, QC, Canada G1P 4R7
e-mail: esamson@simcotechnologies.com

K.A. Snyder
NIST, 100 Bureau Drive, Bldg 226, Gaithersburg, MD 20899-8615, USA
e-mail: kenneth.snyder@nist.gov

M. Alexander et al. (eds.), *Performance of Cement-Based Materials in Aggressive Aqueous Environments*, RILEM State-of-the-Art Reports 10,
DOI 10.1007/978-94-007-5413-3_7, © RILEM 2013

Keywords Coupling • Degradation • Electrochemical potential • Multi-ionic • Transport modelling • Thermo-hydro-mechanical models

7.1 Introduction

The vast majority of the world's concrete infrastructure was built around the middle of the twentieth century, and for the most part is now approaching an age of half a century or more. In many cases, the structures are showing signs of degradation mechanisms such as corrosion of the steel reinforcement, alkali-silica reaction (ASR), and freeze-thaw.

The growing stock of degrading concrete structures has led to an increased interest in the durability of cementitious materials. In-depth analyses of the various deterioration mechanisms at the microstructure/chemical level have contributed to our global understanding of the underlying chemical and physical phenomena. At the same time, efforts have been made to model the various degradation mechanisms over the service-life of a structure, taking into account the particular environmental exposure. Long-term modeling can be used to assist engineers in assessing the relative effectiveness of different repair/maintenance solutions to extend the service-life of existing structures, or help predict the long-term behaviour of new structures as a function of their environment.

Most of the service-life modeling efforts have focused on the chloride ingress problem, due to the high cost associated with the corrosion of reinforced concrete structures. But in recent years, models have been developed specifically to address sulfate attack, acid exposure, and calcium leaching. In contrast to chloride ingress, these degradation phenomena involve significant alterations to the microstructure of hydrated cement pastes. Therefore, whereas a model for chloride corrosion relies largely on a single transport parameter (that may be constant in time), reliable models for these other degradation mechanisms must also account for the effects of changes within the microstructure: new mineral phases forming due to chemical reactions and the effect of changing porosity on the transport parameters.

This chapter presents an overview of recent modeling efforts to predict degradation of cementitious materials exposed to aqueous environments. The first part gives the general mathematical framework for ionic transport models. This is followed by a review of different phenomena: decalcification, sulfate attack, and acid exposure.

7.2 Transport Mechanisms

Within the concrete materials research community, the modeling of transport mechanisms began with concern over chloride ingress in concrete. At the time, it was understood that the onset of corrosion was related to the concentration of chlorides at the depth of the steel. Modeling of transport has since expanded to address most of the known degradation mechanisms. The early models were limited to a single-ion transport equation where diffusion was the only mechanism considered. Over the past decade, the need for more refined descriptions of the simultaneous phenomena involved during transport processes prompted the development of multi-ionic models that consider not only diffusion but other transport mechanisms such as water movement under the effect of humidity gradients.

Transport phenomena are usually described by writing the mass conservation equations at the pore level. The equations are then averaged over a Representative Elementary Volume (REV). The averaged mass transport equations are solved, allowing for simulations at the structure scale. The constitutive equation at the pore level for each ionic species is written under the assumption that the two main driving forces behind ionic transport are a sum of the gradient of electrochemical potential and the advection caused by the aqueous solution flow (Bockris and Reddy 1970; Helfferich 1961) (Eq. 7.1):

$$\mathbf{j}_i = \underbrace{-\frac{D_i^o}{RT} c_i \, \mathrm{grad}(\mu_i)}_{\text{electrochemical}} + \underbrace{c_i \mathbf{v}}_{\text{advection}} \tag{7.1}$$

where c_i is the concentration of ionic species i, D_i^o is the self-diffusion coefficient, R is the ideal gas constant, T is the thermodynamic temperature and $\mathbf{v}$ is the velocity of the liquid phase. The electrochemical potential μ_i is defined as:

$$\mu_i = \mu_i^o + RT \ln(\gamma_i c_i) + z_i F \psi \tag{7.2}$$

where μ_i^o is a reference state, γ_i is the chemical activity coefficient, z_i is the valence number of the ionic species, F is the Faraday constant and ψ is the electrical potential. Substituting Eq. 7.2 into Eq. 7.1 yields (Samson and Marchand 2007a):

$$\mathbf{j}_i = -D_i^o \,\mathrm{grad}(c_i) - \frac{D_i^o z_i F}{RT} c_i \,\mathrm{grad}(\psi) - D_i^o c_i \,\mathrm{grad}(\ln \gamma_i) - \frac{D_i^o c_i \,\ln(\gamma_i c_i)}{T} \mathrm{grad}(T) + c_i \mathbf{v} \tag{7.3}$$

Each term on the right-hand side of Eq. 7.3 corresponds to a different mechanism:

- The first term models the movement of ionic species under the effect of concentration gradients (i.e. diffusion).
- The second term involving the electrical potential is responsible for maintaining the macroscopic electroneutrality in the pore solution by altering the velocity of individual species. The electrical potential thus couples each separate flux equation (Sten-Knudsen 2002).
- The chemical activity term (third term) corresponds to a correction to the flux when the ionic strength of the pore solution is high. The chemical activity coefficient in Eqs. 7.2 and 7.3 can be estimated using a model such as the Debye-Hückel or extended Debye-Hückel equations that are valid for ionic strengths around 100 mmol/L, or the Davies correction that is valid for ionic strengths up to 300 mmol/L (Pankow 1994). Pore solutions extracted from cementitious materials have ionic strengths that vary from 300 mmol/L (Hidalgo et al. 2001) to 900 mmol/L (Reardon 1992). Pitzer's model was used by Reardon (1990) to model the paste/solution chemical equilibrium. Samson et al. proposed a modification to the Davies' model (Samson et al. 1999) that extends the applicability to estimating the chemical activity coefficients for 1 mol/L ionic strength solutions.
- The fourth term in Eq. 7.3, which involves temperature, is called the Soret effect. It is the contribution to the flux caused by temperature gradients in the material. In those cases where the system is isotherm, or nearly so, this term can be neglected.
- The fifth term in Eq. 7.3 is the advection caused by the aqueous solution flow.

The constitutive Eq. 7.3 can be simplified in specific cases. If the ionic transport is isothermal, the fourth term containing the temperature gradient can be neglected. Most of the early transport models focused on diffusion and advection. The electrical coupling and chemical activity effects were thus neglected.

To obtain the time-dependent response of the system, the constitutive Eq. 7.3 is substituted in the mass conservation relationship (continuity equation) (Bear and Bachmat 1991):

$$\frac{\partial c_i}{\partial t} + \mathrm{div}(j_i) + r_i = 0 \tag{7.4}$$

where r_i is the reaction rate term accounting for chemical reactions and complexation in the solution. The complexation reactions take place solely within the aqueous phase. The formation of $CaOH^+$ is an example of complexation reaction: $Ca^{2+} + OH^- \leftrightarrow CaOH^+$. At the pore scale, the other types of chemical reactions, i.e. dissolution/precipitation and surface adsorption (including binding (Tang and Nilsson 1993), can be modeled by exchange terms at the aqueous/solid interface).

Combining Eqs. 7.3 and 7.4 gives the complete ionic transport equation in the aqueous phase at the pore scale:

$$\frac{\partial c_i}{\partial t} - \operatorname{div}\left(D_i^o \operatorname{grad}(c_i) + \frac{D_i^o z_i F}{RT} c_i \operatorname{grad}(\psi) + D_i^o c_i \operatorname{grad}(\ln \gamma_i) + \frac{D_i^o c_i \ln(\gamma_i c_i)}{T} \operatorname{grad}(T) - c_i \mathbf{v} \right) + r_i = 0 \quad (7.5)$$

7.2.1 Homogenisation Technique

For practical computations, it is usually not possible to apply the above equations at the pore scale. For example, modeling chloride transport at the pore scale is currently an impossible task. One would have to know the exact geometry of the total porous network. Furthermore, the computational resources required to achieve this task at the structure level are currently unavailable. To circumvent this difficulty, simulations have to be made at the structure scale. The transport equations must be transformed from the pore scale to the structure scale. This can be achieved through the homogenisation, or averaging, technique. The general application of the method can be found in Bear and Bachmat (1991). The technique was specifically applied to cementitious materials in Samson et al. (2005). It should be noted that Johannesson (2003) developed an ionic transport model on the basis of the mixture theory and obtained similar results. In the averaging technique, the equations are integrated over the Representative Elementary Volume (REV), resulting in the equations at the structure scale. The averaged form of Eq. 7.5 is Samson and Marchand (2007a):

$$\frac{\partial(\theta_s C_i^s)}{\partial t} + \frac{\partial(w C_i)}{\partial t} - \operatorname{div}\left(D_i w \operatorname{grad}(C_i) + \frac{D_i z_i F}{RT} w C_i \operatorname{grad}(\Psi) + D_i w C_i \operatorname{grad}(\ln \gamma_i) + \frac{D_i C_i \ln(\gamma_i c_i)}{T} w \operatorname{grad}(T) - C_i \mathbf{V} \right) + w R_i = 0 \quad (7.6)$$

where the uppercase parameters represent the average of the corresponding quantity in Eq. 7.5.

The averaging process introduces the volumetric water content w into the mass transport equation. Also, a term involving the solid phase fraction θ_s and the content of the ionic species i bound to the solid matrix, C_i^s is now included. This term is often used to model the chemical reactions between the pore solution and the hydrated cement paste based on an experimental interaction curve. On the other hand, the term R_i dedicated to homogeneous chemical reactions is neglected[1] in the papers reviewed. More details on the modeling of chemical reactions for specific transport processes will be given in the following sections.

7.2.2 Diffusion Coefficient

The parameter D_i in Eq. 7.6 is the diffusion coefficient at the macroscopic level. It can be related to D_i^o by the expression:

$$D_i = \tau D_i^o \tag{7.7}$$

where τ is the tortuosity of the aqueous phase, a purely geometrical factor accounting for the complexity of the porous network.

Several factors can affect the diffusion coefficient such as the saturation level in the pore space, the temperature, the hydration of the material and the modification to the microstructure caused by chemical reactions. Saetta et al. (1993) proposed that the net effect is a combination of independent factors and can be expressed as separate functions such as:

$$D_i = \tau D_i^o \times S(w) \times G(T) \times H(t) \times M(\phi) \tag{7.8}$$

The function $S(w)$ models the effect of saturation in the pores on the diffusion process. The function $G(T)$ accounts for the temperature effect on ionic diffusion. Because of cement hydration, transport in cementitious materials tends to decrease with time – accounted for by function $H(t)$. This effect is especially important when supplementary cementitious materials such as fly ash are incorporated in the binder. Finally, changes in the porosity, due to hydration or degradation reactions, can also affect the tortuosity. The pore volume can be lowered, such as when gypsum is formed. Conversely, phase dissolution increases the pore volume. The modifications similarly affect the transport properties. This is modeled through the function $M(\phi)$. Various expressions for these functions can be found elsewhere (MacQuarrie and Mayer 2005; Saetta et al. 1993).

[1] The situation is different in groundwater transport, where the homogeneous reactions are an important part of the pollutant movement process (see for instance MacQuarrie and Mayer 2005).

7.2.3 Moisture Transport, Electrochemical Potential and Temperature Effects

To solve the ionic transport equation, other relationships are needed to evaluate water content (w), temperature (T) and electrochemical potential (ψ) fields. Two main methodologies are used to model moisture movement in cementitious materials. The first one is based on a thorough description of all the phases involved in the process: liquid (aqueous solution), water vapour and dry air. Multiple mass conservation equations are involved in obtaining a description of the global moisture fields. Such a model can be found in reference (Mainguy et al. 2001). These models are also reviewed later in this chapter, in the section covering thermo-hydro-mechanical models (Sect. 7.6).

However, most moisture transport models found in the literature are based on a simplified approach, which can be derived from simplifying assumptions. It leads to the single Richards' equation (Samson et al. 2005; Whitaker 1998), which allows evaluating the water content field w:

$$\frac{\partial w}{\partial t} - \mathrm{div}(D_w\,\mathrm{grad}(w)) = 0 \tag{7.9}$$

where D_w is the water content diffusivity. Based on this equation, the fluid velocity in Eq. 7.6 can be calculated according to: $\mathbf{V} = -D_w\, grad(w)$. Equation 7.9 can also be expressed on the basis of the relative humidity (Xi et al. 1994):

$$\frac{\partial w}{\partial h}\frac{\partial h}{\partial t} - \mathrm{div}(D_h\,\mathrm{grad}(h)) = 0 \tag{7.10}$$

Another variable that must be evaluated to solve the transport Eq. 7.6 is the electrochemical potential. Most ionic transport models neglect the electrochemical potential and the electrical coupling between ions. This is the case in groundwater modeling, where the ionic concentrations are in most cases relatively low, at least compared to the pore solution of cementitious materials. Up until recently, this was also the case for ionic diffusion in concrete. However, some recent models now consider this coupling effect, assuming that the high concentrations in the pores may cause strong concentration gradients, in which case the electrochemical potential term in the mass conservation equation is no longer negligible. Some of these models will be reviewed when specific degradation mechanisms are addressed later in the chapter.

Two different approaches have been used to solve the electrochemical potential issue. The first one uses the null current density hypothesis $\sum z_i j_i = 0$ to eliminate the electrochemical potential from the transport equation. This approach was used in Masi et al. (1997) and Truc et al. (2000). The electrochemical potential can also be taken into account using the Poisson's equation that directly relates the potential to the concentration in solution. It is given here in its averaged form (Samson et al. 2005):

$$\mathrm{div}(w\tau\ \mathrm{grad}(\Psi)) + w\frac{F}{\varepsilon}\left(\sum_{i=1}^{N} z_i C_i\right) = 0 \tag{7.11}$$

where ε is the permittivity of the solution (usually assumed to be the same as water) and N is the total number of ionic species in the aqueous phase. The coupling of Poisson's equation with the ionic transport relationship was used in Samson and Marchand (2007a) and Johannesson (2003) to model ionic transport.

Finally, when a structure is exposed to temperature variations, energy conservation equations must be added to the model to predict temperature in cementitious materials. The most comprehensive one consists of resolving the energy balance equation for each phase in the porous medium. This approach was used by Schrefler (2004) to model the temperature and humidity fields in concrete structures exposed to fire. The equations are coupled through balance equations at the interfaces between each phase. However, these terms prove difficult to evaluate and are often neglected.

For most practical cases involving long-term durability analysis, the energy conservation equation can be simplified to the well-known heat conduction relationship:

$$\rho C_p \frac{\partial T}{\partial t} - \mathrm{div}(k\ \mathrm{grad}T) = 0 \tag{7.12}$$

where ρ is the density of the material, C_p is the heat capacity, and k is the thermal conductivity. As emphasized in Samson and Marchand (2007a), this relationship is valid assuming that the heat of hydration effect are negligible after a few days, the heat flow through convection caused by the fluid and gas movement in the material is negligible, and the heat capacity and conductivity parameter can be expressed as an average value of all the individual contribution of the various phases that compose the material. Equation 7.12 has been used in Martín-Pérez et al. (2001), Saetta et al. (1993), and Samson and Marchand (2007a) to evaluate the temperature field in concrete structures.

Temperature affects not only the transport properties, as shown in Eq. 7.8, but also the chemical reactions. Given the slow rate of ionic transport in cementitious materials, all the models reviewed assume that the reaction rates are fast enough to assume thermodynamic equilibrium. In that case only the solubilities are affected by a change in temperature, which can be taken into account using the Van't Hoff relationship (Samson and Marchand 2007a):

$$\ln\left(K_{sp}\right) = \ln\left(K_{sp}^{o}\right) + \frac{\Delta H^{o}}{R}\left(\frac{1}{T^{o}} - \frac{1}{T}\right) \tag{7.13}$$

where K_{sp} is the solubility constant and ΔH^o is the reaction enthalpy. The superscript *o* refers to reference states.

The Van't Hoff equation is approximation assuming that the heat capacity of the system is constant; it is only valid for homogeneous reactions. The Van't Hoff equation might produce significant deviations if applied to heterogeneous reactions over a large range of temperature (Matschei et al. 2007).

7.3 Chemical Reactions

Chemical reactions are included in the mass transport equation (Eq. 7.6) through the terms $\partial(\theta_s C_i^s)/\partial t$ and WR_i which account for heterogeneous and homogeneous chemical reactions respectively. The heterogeneous chemical reactions involve the aqueous phase and interfaces while homogeneous reactions only occur in the bulk (liquid or gas) phase (Rubin 1983). These families of chemical reactions can in turn be divided into two categories: denoted here as equilibrium and kinetic. If the rate of reaction is large with respect to the ionic and fluid transport processes, it is classified as an equilibrium reaction. In this case, the local equilibrium assumption (LEA) is assumed to be preserved throughout the porous system. As emphasized by Steefel and Lasaga (1994), LEA has long been one of the most fundamental tenets of hydrothermal and metamorphic geochemistry and petrology.

The same can be said about cementitious materials (Buil et al. 1992), where the LEA is assumed in most, if not all, reactive transport models. A few dimensional analyses support this assumption. Barbarulo et al. (2000) performed a dimensional analysis that validated LEA in fluid saturated cementitious materials. In Samson and Marchand (2007b), a similar analysis was made based on the Damköhler number for diffusion and advection. (This number characterizes the characteristic time of chemical processes). In both cases, the local equilibrium assumption was justified.

All reactions falling under the sufficiently fast category are modeled through chemical equilibrium equations (Rubin 1983), which are algebraic, as opposed to the partial differential equations used in transport models. Since the equilibrium is, in most cases, expressed through chemical activity, algebraic relations giving the chemical activity coefficients must also be considered while modeling the chemical reactions. The resolution of a transport problem involving chemical reactions is thus called a mixed problem, because it involves algebraic and partial differential equations.

7.3.1 Homogeneous and Heterogeneous Reactions

As indicated previously, the chemical reactions can be divided (Rubin 1983) between two sub-categories: homogeneous and heterogeneous. Homogeneous reactions are those occurring throughout a single phase. This sub-category includes all the complexation reactions, i.e. the formation of products occurring in the aqueous phase. For example, the reaction:

$$Ca^{2+}{}_{(aq)} + OH^{-}{}_{(aq)} \Leftrightarrow CaOH^{+}{}_{(aq)} \tag{7.14}$$

is a homogeneous reaction since it only occurs in the aqueous phase. In this reaction, Ca^{2+} and OH^- are called the primary species and $CaOH^+$ is called the secondary species (Steefel and Lasaga 1994). The general equilibrium relationship for this type of reaction is written as:

$$K_i = \frac{1}{\gamma_i x_i} \prod_{j=1}^{N_c} \left(\gamma_j C_j\right)^{v_{ij}} \tag{7.15}$$

where K_i is the equilibrium constant, x_i is the concentration of the secondary species, γ is the chemical activity coefficient, N_c is the number of components forming the secondary species, c_j is the concentration of the primary species and v_{ij} is the number of moles of primary species j in one mole of secondary species i.

In contrast to the previous category, heterogeneous reactions involve at least two phases (Rubin 1983), typically an aqueous phase an a solid phase (interface). Rubin distinguishes two types of heterogeneous reactions: surface and classical. Surface reactions are either adsorption, in which ions are attracted to the surface of the pore network under the influence of electrostatic forces (Bockris and Reddy 1970), or ion exchange, in which two or more ionic species are exchanged between the surface of the solid and the aqueous phase (Appelo 1996). The classical reactions are precipitation, dissolution, oxidation and reduction. The dissolution of portlandite belongs to this category since it involves the aqueous and the solid phase:

$$Ca^{2+}{}_{(aq)} + 2OH^{-}{}_{(aq)} \Leftrightarrow Ca(OH)_{2(s)} \tag{7.16}$$

Similar to the complexation reaction (Eq. 7.15), the equilibrium of a mineral m can be expressed as a function of a constant K_m (Xu et al. 1999):

$$K_m = \prod_{i=1}^{N_c} C_i^{v^p_{mi}} \gamma_i^{v^p_{mi}} \tag{7.17}$$

where Nc is the number of species entering the composition of the mineral m, c_i is the concentration of species i in solution and v_{mi} is the number of moles of species i in one mole of mineral m.

It should be noted that contrary to Eq. 7.15, the reaction product of dissolution/precipitation reactions is not part of the equilibrium Eq. 7.17, because the solid has an activity coefficient equal to one (Stumm and Morgan 1996). This particularity of the dissolution/precipitation mechanism induces the formation of mineral fronts inside the porous material, as discussed in Rubin (1983). Such fronts can be observed in calcium leaching (e.g. Adenot and Buil 1992; Mainguy et al. 2000) and sulfate attack cases (e.g. Maltais et al. 2004; Planel et al. 2006).

7.3.2 *Thermodynamic Modeling and Databases*

Thermodynamic modeling (Eqs. 7.15 and 7.17) is based on the knowledge of the thermodynamic data, i.e. solubility products, complex formation constants, of all the solid, aqueous and gaseous species that can form. The quality of the results of thermodynamic modeling depends directly on the quality and the completeness of the underlying thermodynamic database.

Thermodynamic data for complexes and solids generally present in geochemical systems, such as gypsum or calcite, have been critically reviewed and can be found in a number of compilations, e.g. (Babushkin et al. 1985; Robie and Hemingway 1995; Shock et al. 1997; Hummel et al. 2002). Specific thermodynamic data for cementitious systems, such as the solubility products of ettringite or hydrogarnet, are generally not included in such generic general databases but are compiled separately in specific databases. It should be noted that such specific databases have always been developed based on a particular general database and thus can be used only together with the respective general database.

Several thermodynamic databases for cement minerals have been compiled and published. The first compilation was published as early as 1965 by Babushkin et al. (1965). Subsequently, a number of other databases for cementitious systems have been published, e.g. (Barret et al. 1983; Reardon 1992; Damidot et al. 1994; Bourbon 2003). The cement-specific-cement database "cemdata2007" covers hydrates commonly encountered in Portland cement systems and is summarised below in Table 7.1. It contains thermodynamic data (solubility product, Gibbs free energy, enthalpy, entropy, heat capacity approximation coefficients and molecular volume) for solids and is applicable in the temperature range 0–100°C; the heat capacity plays an important role in estimating the temperature dependence of solubility product, and it is approximated as follows:

$$C_p = a_0 + a_1 T + a_2 T^{-2} + a_3 T^{-0.5} \tag{7.18}$$

Solubility data have been generally calculated based on a critical review of the available experimental data and on additional experiments to derive missing data or to verify the existing data (Matschei et al. 2007; Möschner et al. 2008). In addition, some data were estimated based on structural analogues. The data collection and selection process is documented in different papers (Lothenbach and Winnefeld 2006; Matschei et al. 2007; Lothenbach et al. 2008; Schmidt et al. 2008).

Even though thermodynamic data have increasingly become available in the last few years, several important gaps still exist, e.g. solubility of hydrotalcite phases, Fe-containing hydrates, uptake of ions in C–S–H.

Table 7.1 Standard thermodynamic properties at 25°C: thermodynamic data for cements "cemdata2007" are taken from Lothenbach and Winnefeld (2006), Matschei et al. (2007), Lothenbach et al. (2008) and Schmidt (2008), data for gypsum, anhydrite, portlandite and calcite from Thoenen and Kulik (2003)

	log K_{S0}	$\Delta_f G°$ (kJ/mol)	$\Delta_f H°$ (kJ/mol)	S° (J/mol.K)	a_0 (J/mol.K)	a_1 (J/mol.K^2)	a_2 (J.K/mol)	a_3 (J/mol. $K^{0.5}$)	V° (cm^3/mol)
(Al-)ettringite[a]	−44.9	−15205.94	−17535	1,900	1,939	0.789			707
Tricarboaluminate[a]	−46.5	−14565.64	−16792	1,858	2,042	0.559	−7.78e6		650
Fe-ettringite[a]	−44.0	−14282.36	−16600	1,937	1,922	0.855	2.02e6		717
Thaumasite	−49.4	−15128.46	−17373	1,883	1,860	0.703	−3.94e6	1,600	663
C_3AH_6[b]	−20.84	−5010.09	−5540	419	292	0.561			150
$C_3AS_{0.8}H_{4.4}$[c]	−29.87	−5368.01	−5855	369	109	0.631	−1.95e6	2,560	143
C_3FH_6[d,b]	−25.16	−4116.29	−4640	439	275	0.627	2.02e6		155
C_4AH_{13}[e,f]	−25.40	−7326.56	−8302	700	711	1.047		−1600	274
C_2AH_8[g]	−13.56	−4812.76	−5433	440	392	0.714		−800	184
$C_4A\bar{S}H_{12}$[f,h]	−29.26	−7778.50	−8750	821	594	1.168			309
C_4ACH_{11}[i]	−31.47	−7337.46	−8250	657	618	0.982	−2.59e6		262
$C_4AC_{0.5}H_{12}$[j]	−29.13	−7335.97	−8270	713	664	1.014	−1.30e6	−800	285
C_2ASH_8[k]	−19.70	−5705.15	−6360	546	438	0.749	−1.13e6	−800	216
C_4FH_{13}[d,e]	−29.4	−6430.94	−7395	737	694	1.113	2.02e6	−1600	286
C_2FH_8[d,g]	−17.6	−3917.38	−4526	476	375	0.780	2.02e6	−800	194
$C_4F\bar{S}H_{12}$[h]	−33.2	−6882.55	−7843	858	577	1.234	2.02e6		322
C_4FCH_{12}[d,i]	−35.5	−6679.20	−7637	737	612	1.157	−5.73e5		290
$C_4FC_{0.5}H_{12}$[d,j]	−33.1	−6440.19	−7363	749	648	1.080	7.24e5	−800	296
C_2FSH_8[d, k]	−23.7	−4809.53	−5453	583	422	0.815	8.91e5	−800	227
CAH_{10}	−7.50	−4622.39	−5320	501	151	1.113		3,200	194
M_4AH_{10}[d,m]	−56.02	−6394.56	−7196	549	−364	4.21	3.75e6	629	220
M_4ACH_9[d]	−51.14	−6580.15	−7374	551	−382	4.24	4.32e6	629	220
M_4FH_{10}[d,m]	−60.0	−5498.84	−6289	586	−381	4.27	5.78e6	629	232
$C_{1.67}SH_{2.1}$(jen.)[n]	−13.17	−2480.81	−2723	140	210	0.120	−3.07e6		78
$C_{0.83}SH_{1.3}$(tob.)[n,o]	−8.0	−1744.36	−1916	80	85	0.160			59
$SiO_{2,am}$[o]	1.476	−848.90	−903	41	47	0.034	−1.13e6		29

$Ca(OH)_2$	−5.2	−897.01	−985	83	187	−0.022	0	−1600	33
$CaSO_4\ 2H_2O$	4.58	−1797.76	−2023	194	91	0.318			75
$CaSO_4$	4.36	−1322.12	−1435	107	70	0.099			46
$CaCO_3$	8.48	−1129.18	−1207	93	105	0.022	−2.6E6		37
syngenite[p]	−7.20	−2884.91	−3172	326	201	0.208	−1.78e6		128
$Al(OH)_3$(am)	0.24	−1143.21	−1281	70	36	0.191			32
$Fe(OH)_3$(mic)	−4.60	−711.61	−844	88	28	0.052			34
C_3S		−2784.33	−2931	169	209	0.036	−4.25e6		73
C_2S		−2193.21	−2308	128	152	0.037	−3.03e6		52
C_3A		−3382.35	−3561	205	261	0.019	−5.06e6		89
C_4AF		−4786.50	−5080	326	374	0.073			130

a_0, a_1, a_2, a_3 are the empirical coefficients of the heat capacity equation: $C^\circ_p = a_0 + a_1T + a_2T^{-2} + a_3T^{-0.5}$; no value = 0.
All solubility products refer to the solubility with respect to the species $Al(OH)_4^-$, $Fe(OH)_4^-$, $SiO(OH)_3^-$, OH^-, H_2O, Ca^{2+}, K^+, Mg^{2+}, CO_3^{2-}, or SO_4^{2-};
Cement shorthand notation is used: A = Al_2O_3; C = CaO; F = Fe_2O_3; H = H_2O; M = MgO; S = SiO_2; $\overline{C}$= CO_2; $\overline{S}$ = SO_3
[a,f]Non-ideal solid solutions
[b,e,g,h,i,j,k,m,n,o]Ideal solid solutions c.f. (Lothenbach et al. 2008)
[c]Precipitates very slowly at 20°C, generally not included in calculations
[d]Tentative values
[p]Density data from Corazza and Sabelli (1967)

By applying thermodynamic modeling to cementitious systems, one implicitly assumes that a quasi-thermodynamic equilibrium exists, i.e. that the liquid (pore solution) and the solid phase (hydrates) are in equilibrium. Often the dissolution of the clinker phases are not considered in such calculations (the pore solutions are generally highly undersaturated with respect to these phases as they dissolve very slowly). Most other precipitation and dissolution processes are sufficiently fast so that the assumption of thermodynamic equilibrium is justified.

In some cases, however, thermodynamically stable phases may not form spontaneously, or they may only form very slowly. Hydrogarnets (C_3AH_6) have been observed to form only after weeks to months in solutions containing Al and Ca (Peppler and Wells 1954). Similarly, siliceous hydrogarnets ($C_3AS_xH_{6-2x}$; $x \leq 1$) have been synthesized at 105°C (Matschei et al. 2007), but do not form at ambient temperatures. Thus, the formation of siliceous hydrogarnets is often suppressed in thermodynamic calculations, to agree with the experimental observations.

In spite of the apparent simplicity of the equilibrium equations, the resolution of the chemical system that represents hydrated cement can become complex when one tries to account for the complexity: the large number of minerals present (see Table 7.1), the speciation in solution, and the interactions of the species (activity corrections). Due to the complexity of the system, computer codes are generally used to obtain the stable phase composition.

Two approaches to solving the geochemical equilibrium equations are used in geochemical modeling codes: (i) solving the law of mass action equations, or (ii) minimizing the Gibbs free energy of the system. The first approach is common and available in many speciation codes such as MINEQL (Westall et al. 1986), EQ3/6 (Wolery 1992), PHREEQE and its derivatives (Brown et al. 1991), and CHESS (van der Lee 1997) and extensively described in textbooks (Appelo and Postma 1996). It requires log K values (see e.g. Table 7.1, Eqs 7.14 and 7.17) for the aqueous complexes, solids or gases at the temperature and pressure of interest. To calculate the composition of a system, mass and charge balances equations are solved iteratively using Newton–Raphson iterations.

The Gibbs free energy minimization (GEM) approach is represented e.g. by GEMS (Kulik et al. 2004). Thermodynamic equilibrium in a system is obtained when there is no longer any spontaneous tendency for change, i.e. at equilibrium the Gibbs free energy of the system is at a minimum. The GEM approach is based on mass and charge balance of the whole system and the equilibrium composition is calculated automatically from the stochiometrically possible phases. Activity coefficients are calculated for each phase separately and several variable liquid and/or solid solutions might be considered. (An example of this approach is given in Part I, Chap. 2 (Sect. 2.4.3) of this Report)

Although in principle both approaches should yield the same equilibrium composition (assuming the same reaction constants), there may exist local minima when using either method. "Settling in a local minimum" is more common when using the mass action approach, so steps must be taken to ensure finding the global minimum.

7.4 Coupling Transport and Chemistry

7.4.1 *Linear Transport Equations Coupled with Chemistry*

The early models for reactive transport in porous materials attempted to substitute the chemical equilibrium relationships such as Eqs. 7.15 and 7.17 into a simplified version of the conservation equation (7.6):

$$\frac{\partial(\theta_s C_i^s)}{\partial t} + \frac{\partial(wC_i)}{\partial t} - \mathrm{div}(D_i w\ \mathrm{grad}(C_i) - C_i\mathbf{V}) + wR_i = 0 \tag{7.19}$$

where the electrical coupling between the ions and chemical activity effects are neglected. This simplified transport equation has been used extensively in groundwater transport where it is argued that the high velocity fluid flow overwhelms other transport mechanisms. In that case, the diffusion coefficient is replaced by the dispersion tensor **D*** (Bear and Bachmat 1991; Fetter 1999), which is the same for all species.

This is called the Direct Substitution Approach (DSA), where the transport equations can be expressed according to different formulations (Kirkner and Reeves 1988). The next paragraphs summarize three different formulations of the DSA, as described by Kirkner and Reeves. The different formulations were devised assuming saturated materials and a linear transport operator, i.e. electrical coupling and chemical activity effects were neglected. While the saturated assumption could be extended to unsaturated cases, it will be shown that DSA works only if a linear transport model with a uniform dispersion coefficient is used.

7.4.1.1 Direct Substitution Approach

The N reactive species in solution are divided into formulations of the N_c components and N_x secondary (or complex) species. The equations governing the transport of the N_c components and N_x secondary species are:

$$\frac{\partial \phi c_j}{\partial t} + \mathrm{div}\left(c_j\mathbf{v} - \phi\mathbf{D} * \mathrm{grad}\ c_j\right) = r_j^c \quad j = 1, ..., N_c \tag{7.20}$$

$$\frac{\partial \phi x_i}{\partial t} + \mathrm{div}(x_i\mathbf{v} - \phi\mathbf{D} * \mathrm{grad}\ x_i) = r_i^x \quad i = 1, ..., N_x \tag{7.21}$$

where ϕ is the porosity and the reaction terms in Eq. 7.19 have been lumped in the r_j^c and r_i^x terms. The subscript i of the dispersion tensor has been omitted since it is assumed that the mechanical dispersion dominates over diffusion, which leads to all species having the same dispersion coefficient. In Eqs. 7.20 and 7.21, the porosity replaces the water content of Eq. 7.19 because of the saturated pores assumption.

The total soluble concentration u_j of each species is then introduced as:

$$u_j = c_j + \sum_{i=1}^{N_x} A_{ij} x_i \tag{7.22}$$

where A_{ij} is a stoichiometry coefficient matrix linking the main species to the complexes.

Substituting Eq. 7.15 in Eq. 7.22 gives:

$$u_j = c_j + \sum_{i=1}^{N_x} A_{ij} K_i \prod_{k=1}^{N_c} c_k^{A_{ik}} \tag{7.23}$$

The reaction rates r_j^c are defined as:

$$r_j^c = r_j^{c(aq)} + r_j^{c(s)} + r_j^{c(p)} \tag{7.24}$$

where:

$$r_j^{c(aq)} = -\sum_{i=1}^{N_x} A_{ij} r_i^x \quad \text{reaction rate of } j \text{ due to} \tag{7.25}$$

$$r_j^{c(s)} = -\frac{\partial \phi s_j}{\partial t} \quad \text{reaction rate of } j \text{ due to sorption or IX} \tag{7.26}$$

$$r_j^{c(p)} = -\sum_{i=1}^{N_p} B_{ij} \frac{\partial \phi p_i}{\partial t} \quad \text{reaction rate of } j \text{ due to diss./prec.} \tag{7.27}$$

In Eqs. 7.25, 7.26, and 7.27, s_j corresponds to the concentration of the main species j sorbed on the solid phase, p_i is the concentration of mineral i, N_p is the number of immobile species, and B_{ij} is a stoichiometry coefficient matrix. Kirkner and Reeves showed that it is possible to write the sorbed concentrations as:

$$s_j = f_j^{sc}(c_1, c_2, \ldots, c_{N_c}) \tag{7.28}$$

Using Eq. 7.23, it is possible to write Eq. 7.28 as a function of the total concentration u_j:

$$s_j = f_j^{su}(u_1, u_2, \ldots, u_{N_c}) \tag{7.29}$$

Summing the mass conservation Eqs. 7.20 and 7.21 and using the total concentration u_j (Eq. 7.23) eliminates the reaction rates $r_j^{c(aq)}$. This operation is possible because all species have the same dispersion coefficient. Substituting the rate expressions (7.26) and (7.27), and the sorption function (7.29) in the summed conservation equations yield:

Formulation A:

$$\frac{\partial(\phi u_j)}{\partial t} + \frac{\partial\left(\phi f_j^{su}\right)}{\partial t} + \sum_{k=1}^{N_p} B_{kj} \frac{\partial(\phi p_k)}{\partial t} + L(u_j) = 0 \quad j = 1, \ldots, N_c \tag{7.30}$$

where *L(.)* is the linear transport operator:

$$L(.) = \text{div}[(.)\mathbf{v} - \phi\, \mathbf{D}^* \text{grad}(.)] \tag{7.31}$$

The complexation reactions are implicitly included in the mass conservation equation as part of the definition of u_i (Eq. 7.23).

It is possible to further simplify the transport equations by introducing the total concentration w_j as the main variable:

$$w_j = u_j + s_j + \sum_{k=1}^{N_p} B_{kj} p_k k \quad j = 1, \ldots, N_c \tag{7.32}$$

From this definition, it is possible to write all the concentration variables as a function of the total concentration variable (Kirkner and Reeves 1988):

$$c_j = f_j^{cw}\left(w_{1,\ldots,}w_{N_c}\right) \quad j = 1, \ldots, N_c \tag{7.33}$$

$$p_j = f_j^{pw}\left(w_{1,\ldots,}w_{N_c}\right) \quad j = 1, \ldots, N_p \tag{7.34}$$

$$u_j = f_j^{uw}\left(w_{1,\ldots,}w_{N_c}\right) \quad j = 1, \ldots, N_c \tag{7.35}$$

$$s_j = f_j^{sw}\left(w_{1,\ldots,}w_{N_c}\right) \quad j = 1, \ldots, N_c \tag{7.36}$$

Based on these functions, it is possible to write Eq. 7.30 as:

Formulation B:

$$\frac{\partial(\phi w_j)}{\partial t} + L\left[f_j^{uw}\left(w_{1,\ldots,}w_{N_c}\right)\right] = 0 \tag{7.37}$$

or alternatively:

Formulation C:

$$\frac{\partial(\phi w_j)}{\partial t} + L(w_j) = L\left[f_j^{sw}\left(w_{1,\ldots,}w_{N_c}\right)\right] + \sum_{k=1}^{N_p} B_{kj} L\left[f_k^{pw}(w_1, \ldots, w_{N_c})\right] \tag{7.38}$$

These different formulations have been used to solve reactive transport problems. Valocchi et al. (1981) used formulation A (Eq. 7.30) in their model, which considered ion-exchange reactions but did not include complexation nor dissolution/precipitation. Miller and Benson (1983) and Jennings et al. (1982) also used a formulation similar to A while considering complexation and ion exchange.

Although the chemical reactions are included in formulations B and C, the total concentration they provide cannot be used directly and must be converted back to c_i and x_i variables. To do so requires solving the algebraic chemical expressions separately. This is discussed, for instance, in (Cederberg et al. 1985). These formulations hint at the operator splitting approaches presented in the following section. However, they are considered under the DSA category because the transport equations still need to be solved simultaneously (Yeh and Tripathi 1989).

It should be noted that all papers cited previously only consider complexation and sorption reactions. As mentioned in (Kirkner and Reeves 1988), the discontinuities (fronts) induced by the presence of dissolution/precipitation reactions require a special treatment. Some aspects are discussed in (Kirkner et al. 1984). Most notably, the discontinuities associated with the precipitate content require the use of Dirac-type interpolation in a finite element formulation, which differs from the classical C^o shape functions (Zienkiewicz and Taylor 1989). Lichtner (1985) presented a model dealing with dissolution/precipitation reactions as a moving front problem. The approach includes specific mass conservation equations at the front locations, which must be added to the global mass conservation equations.

7.4.1.2 Operator Splitting Approaches

One of the first models that separated transport and chemistry was developed by Grove and Wood (1979). The model was used to predict transport of calcium, magnesium, and sulfate in groundwater. The transport equations (linear advection diffusion equation (ADE), see Eq. 7.31) are first solved without considering any reaction terms. Next, the chemical reactions are solved using the concentration estimated by the transport step as an initial estimate. The algorithm then proceeds to the next time step and solves the next transport/chemistry sequence.

What really triggered interest for splitting transport and chemistry in reactive transport models was the very influential paper published by Yeh and Tripathi (1989). Their main argument for splitting transport and chemistry was to reduce CPU time. It also avoids most of the problems associated with the dissolution/precipitation reactions. Although their arguments were criticized (Saaltink et al. 2001) because they were theoretical and did not reflect actual calculation results, the paper was so highly regarded that DSA almost completely disappeared from that point on. Since then, splitting approaches have been adopted as the preferred method to model reactive transport problems despite the implicit numerical error that they introduce in the calculations (Kanney et al. 2003; Jacques et al. 2006).

The possibility of coupling existing transport and chemical models together proved too attractive compared to the numerical problems and computational burden associated with the fully coupled Direct Substitution Approach.

The operator splitting approaches are divided into two main categories:

- Sequential Iterative Approaches (SIA) and
- Sequential Non-Iterative Approaches (SNIA).

In the iterative approach, iterations are performed between the transport and chemistry until convergence is reached. In the SNIA, transport and chemistry are solved sequentially without iterations. The model described at the beginning of this section (Grove and Wood 1979) belongs to the SNIA category.

The SIA method consists mainly in expressing the formulations given in the previous section so that the transport equations can be solved individually before chemistry is considered on a node per node basis (Yeh and Tripathi 1989). For instance, Walsh et al. (1984) used transport equations corresponding to formulation C (Eq. 7.38) solved using an explicit time stepping scheme. This linearizes the transport equations, which can be solved individually. The total concentrations are then separated into the solute, sorbed, and mineral components upon solving the chemical equilibrium equations. In Engesgaard and Kipp (1992), a model dealing with complexation, precipitation/dissolution, and oxidation/reduction was presented. The transport equations are expressed according to formulation A (see Eq. 7.30), where the reaction terms are considered as explicit source/sink terms. During the first iteration, the transport equations are first solved with the reaction terms set to zero. After chemistry, the source/sink terms are estimated based on the variations in the amount of precipitate and included in the transport equations. The iterations continue until a stable solution is reached. A similar algorithm was used in Simunek and Suarez (1994).

At the other end of the operator-splitting spectrum is the SNIA, exemplified by the paper from Grove and Wood (1979) mentioned at the beginning of this section. SNIA can be selected over SIA for some reasons. Since it completely decouples transport and chemistry, it can prove easier to implement different transport and chemistry schemes in a single simulation platform. Removing the iterations between transport and chemistry can potentially reduce calculation times. However, as mentioned in Steefel and MacQuarrie (1994), SNIA requires a smaller time step than SIA to achieve the same level of accuracy. Strategies to reduce the operator-splitting error in SNIA, by controlling the grid spacing and time step, are discussed in Jacques et al. (2006). Comparisons between SNIA and SIA are provided in Steefel and MacQuarrie (1994), Walter et al. (1994), and Xu et al. (1999). In most cases, SNIA tends to use less CPU time to get the same results. But the performances of SNIA are dependent on the type of chemical reactions. In Xu et al. (1999), the results showed numerical dispersion when ionic exchange simulations were performed with this algorithm. The dispersion resulted in less sharp concentration fronts than with the SIA algorithm.

Although all references cited previously were concerned with contaminant transport in groundwater, some papers have been specifically devoted to concrete barriers. In recent papers (De Windt et al. 2004; Trotignon et al. 2007), the authors used a SIA algorithm to simulate the long term durability of a concrete slab in contact with soil. In both papers, the simulations were performed on the layered concrete/soil system. In Marty et al. (2009), a similar problem was considered using a SNIA algorithm.

7.4.2 Nonlinear Transport Equations Coupled with Chemistry

All the papers cited in Sects. 7.4.1 and 7.4.2 describe models based on linear transport operator, Eq. 7.31. This allows introducing the total soluble concentration given by Eq. 7.23 into the mass transport equation. It is then possible to eliminate the source/sink terms associated with the complexation reactions, which gives Eq. 7.30. This is possible when the fluid phase flow in the pore network induces a dispersion factor that is strong enough to overcome the individual self-diffusion coefficient of the soluble species.

In the case of concrete, the permeability of the material is so low that high velocity flows are never encountered, except in macro-cracks. In that case, the coupling terms in the flux expression (7.6) may have a significant influence on the transport of ions. But the difficulties involved with solving these terms have led to the development of models dedicated to concrete that were still using advection diffusion-type equations similar to Eq. 7.20 for long-term durability predictions. For example, most of the models dedicated to chloride ingress are still based on a single ADE equation, as reviewed in the Part I of this report, dealing with chemical degradation (see for instance Saetta et al. 1993; Nagesh and Bhattacharjee 1998; Martín-Pérez et al. 2001). As mentioned previously, multi-ionic models based on the splitting-operator approach have been used to assess the long term durability of concrete barriers in contact with groundwater (De Windt et al. 2004; Trotignon et al. 2007; Marty et al. 2009).

Some multi-ionic models dedicated to species ingress in concrete have been developed considering the electrical coupling term in Eq. 7.6. The model presented by Truc et al. (2000) accounts for electrical coupling in modelling the chloride ingress in concrete structures. Chloride binding in the hydrated cement paste is approximated by a relationship similar to Eq. 7.28. Samson and Marchand presented a multi-ionic model based on Eq. 7.6 that was used to model chloride (Samson and Marchand 2007a) and sulfate ingress (Samson and Marchand 2007b). Their model uses a SNIA algorithm to couple transport with the chemical equilibrium relationships. Samson and Marchand (2007b) showed that neglecting the electrical coupling may lead to erroneous predictions of sulfate ingress rate, emphasized also in Galindez and Molinero (2010). That type of model is able to consider the different self-diffusion coefficients of the considered species while maintaining the electroneutrality of the solution. However, the extensive computational time to solve the transport equations, coupled with the absence of self-diffusion data for secondary species (Li and Gregory 1974), forced the authors to neglect homogeneous complexation reactions. No study could be found to assess the impact of this on long-term durability simulations.

7.5 Modeling Degradation Mechanisms

7.5.1 Decalcification

The decalcification process consists in the leaching of hydroxide, alkali ions, and calcium from the system when it is in contact with an aqueous solution with a pH much lower than the pore solution. The following sections deal with modeling of specific degradation mechanisms in cementitious materials, due to aggressive aqueous solutions. These can be related to the corresponding mechanisms covered in Part I, and draw on the material presented earlier in this chapter.

Calcium leaching in cement-based materials is a coupled chemical equilibrium/diffusion phenomenon. Most models found in literature are based on a simplified version of Eq. 7.6 (Buil et al. 1992; Haga et al. 2005; Kuhl et al. 2004a; Mainguy et al. 2000; Yokozeki et al. 2004), developed for saturated and isothermal material cases. Instead of considering multiple ionic species, the model only considers the mass conservation equation of calcium in the pore solution:

$$\phi(x,t)\frac{\partial C(x,t)}{\partial t} = D(x,t)\frac{\partial^2 C(x,t)}{\partial x^2} - \frac{\partial C_S(x,t)}{\partial t} \tag{7.39}$$

where $C(x,t)$ is the Ca^{2+} concentration in the liquid phase, $C_S(x,t)$ is the Ca content in the solid phase, $\phi(x,t)$ is the porosity and $D(x,t)$ is the effective diffusion coefficient of Ca^{2+} ions. Equation 7.39 neglects phenomena such as chemical activity, convection and electrical coupling. The calcium content in solid phase $C_S(x,t)$, and its derivative, is calculated from its relationship with calcium concentration in solution (Fig. 7.1) (Haga 2005; Kuhl et al. 2004a; Mainguy 2000).

Another approach, based on a multi-ionic transport description, consists in determining the calcium content in solid by solving the chemical equilibrium between the minerals and the ionic species in the pore solution. The modeling of calcium leaching of hardened cement pastes in deionized water, by coupling equation (7.6) and the dissolution/precipitation equilibrium of $Ca(OH)_2$ and C–S–H, was presented by Maltais et al. (2004). The dissolution of portlandite and the decalcification of C–S–H were defined by their solubility constants $K_{Ca(OH)_2}$ and K_{CSH} respectively (Table 7.2).

The different leaching models reviewed take into account the evolution of the porosity of the material as solid phases dissolve. This increase of porosity as calcium is leached is given by Haga et al. (2005), Kuhl et al. (2004a), Maltais et al. (2004), and Yokozeki et al. (2004):

$$\phi_{leaching} = \frac{M_{Ca(OH)_2}}{d_{Ca(OH)_2}}\left(C_{S,Ca(OH)_2{}^0} - C_{S,Ca(OH)_2}\right) \tag{7.40}$$

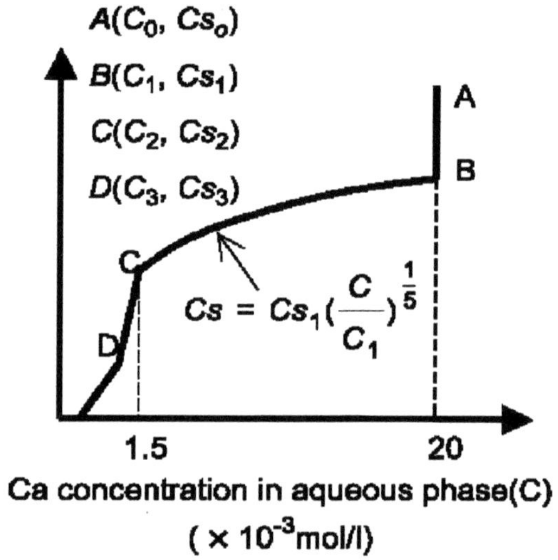

Fig. 7.1 Relationship between Ca concentrations in the solution and the solid (Daimon et al. 1977; Haga et al. 2005)

Table 7.2 Solubility constants of portlandite and C–S–H

Name	Chemical composition	Expression for equilibrium (K_{sp})	$-\log K_{sp}$ [a]
Portlandite	$Ca(OH)_2$	$\{Ca^{2+}\}\{OH^-\}^2$	5.2
C–S–H (Maltais et al. 2004)	0.65 Ca $(OH)_2$ + CaH_2SiO_4	$\{Ca^{2+}\}\{OH^-\}^2$	6.2
C–S–H (Berner 1992; Henocq et al. 2006)	x $Ca(OH)_2$ + CaH_2SiO_4 or $5CaO.5SiO_2.10.5H_2O$	$\{Ca^{2+}\}\{H_2SiO_4^{2-}\}$ or $\{Ca^{2+}\}^5.\{H_3SiO_4^-\}^6.\{OH\}^4/\{H_2O\}^{0.5}$	*f*(C/S)[a]

Berner (1992), Henocq et al. (2006), and Maltais et al. (2004)
[a]$-\log K_{sp}$ is a function of C/S ratio according to the empirical function *f*

where $M_{Ca(OH)2}$ is the molar mass of portlandite and $d_{Ca(OH)2}$ is the phase density. The superscript *o* designates the initial state. This pore volume increase modifies the calcium diffusion coefficient. Relationships between porosity changes and diffusion coefficients (the function $M(\phi)$ in Eq. 7.8) are given Table 7.3.

Figure 7.2 presents simulations results from the simplified and multi-ionic approaches compared to experimental data (Maltais et al. 2004; Yokozeki et al. 2004). The results in Fig. 7.2a were obtained by solving Eq. 7.6 coupled with chemical equilibrium while results in Fig. 7.2b were obtained by solving Eq. 7.39 coupled with the calcium in solid phase relationship given by Fig. 7.1.

7.5.2 External Sulfate Attack

External sulfate attack can be summarized as the penetration of sulfate from the environment into the concrete pore structure. The chemical reactions between the ions in the pore solution and the hydrated phases of the cement paste lead to

Table 7.3 $D = M(\phi)$ relationships found in literature

Relationship	Reference
$D = e^{(9.95\phi - 29.08)}$	Mainguy et al. (2000)
$D = (\phi(x,t)/\phi_0)^n D_0$	Haga et al. (2005)
$D = (0.001 + 0.07\phi^2 + H(\phi - \phi_c)1.8(\phi - \phi_c)^2)D_0$	Garboczi and Benz (1992)
$D = \frac{1-cG_{vol}}{1-dS_{vol}} P_{vol} f(\phi - \phi_{gel}) D_0$	Yokozeki et al. (2004)

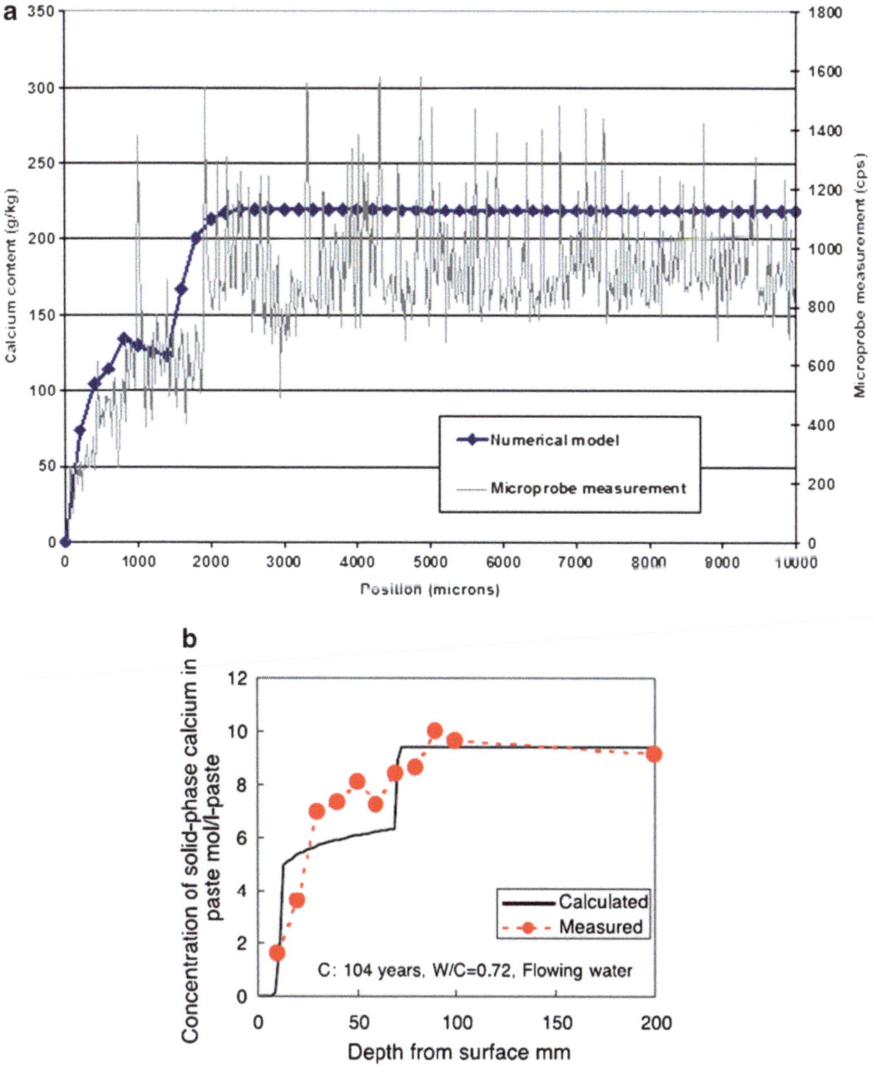

Fig. 7.2 Simulations of calcium leaching: (**a**) on cement paste (Maltais et al. 2004) and (**b**) on mortar (Yokozeki et al. 2004)

Table 7.4 Solubility constants of solid phases involved during sulfate ingress in hydrated cement systems

Name	Chemical formula	Expression for equilibrium (K_{sp})	$-\log K_{sp}$
Ettringite	$3CaO.Al_2O_3.3CaSO_4.32H_2O$	$\{Ca^{2+}\}^6\{OH^-\}^4\{SO_4^{2}\}^3\{Al(OH)_4^-\}^2$	44.0
Monosulfate	$3CaO.Al_2O_3.CaSO_4.12H_2O$	$\{Ca^{2+}\}^4\{OH^-\}^4\{SO_4^{2-}\}\{Al(OH)_4^-\}^2$	29.4
Gypsum	$CaSO_4.2H_2O$	$\{Ca^{2+}\}\{SO_4^{2-}\}$	4.6

Maltais et al. (2004) and Samson and Marchand (2007b)

the formation of ettringite and gypsum, which may cause expansion and cracking. Similar to decalcification, models were developed under multi-ionic and single-ion approaches. Because of the mechanical effect of sulfate ingress, transport models were also coupled with mechanical calculations in order to predict material expansion.

Single-ion transport models are presented in Tixier and Mobasher (2003) and Gospodinov et al. (1999). They are based on the mass conservation of sulfate for isothermal and saturated conditions. Both approaches neglect ionic coupling and activity effects. Assuming that chemical reactions involving sulfate can be modeled according to $R = K(s)c$, Eq. 7.6 can in this case be simplified as:

$$\frac{\partial c}{\partial t} - \mathrm{div}(D\ \mathrm{grad}(c)) + K(s)c = 0 \tag{7.41}$$

where c is the sulfate concentration, D is the diffusion coefficient and $K(s)$ controls the rate of chemical reaction.

The parameter K depends on the content of sulfate-bearing solid phases s. Knowing the relationship between the solid phases s and the sulfate concentration c, it is possible to estimate the amount of sulfate that forms through time. In Tixier (2003), the solid phase content is calculated on the basis of the following type of relationship:

$$\frac{\partial s}{\partial t} = f(K, c, s) \tag{7.42}$$

The estimation of the amount of sulfate-bearing solid formed upon contaminant ingress is then used in Tixier and Mobasher (2003) to calculate volume variations that are associated with strain. The model was used to calculate linear expansion of materials immersed in sodium sulfate solutions.

Similarly, some multi-ionic models found in the literature focus solely on estimating the alteration of the paste's microstructure upon sulfate exposure, while other models go one step further and perform mechanical calculations. The first case, without mechanical calculations, is illustrated in Maltais et al. (2004) and Samson and Marchand (2007b). The multi-ionic model for unsaturated cementitious materials presented by the authors is based on Eqs. 7.6, 7.9, and 7.11. The chemical reaction terms are eliminated from Eq. 7.6. The chemical equilibrium relationship of the sulfate-bearing solid phases, shown in Table 7.4, is solved in

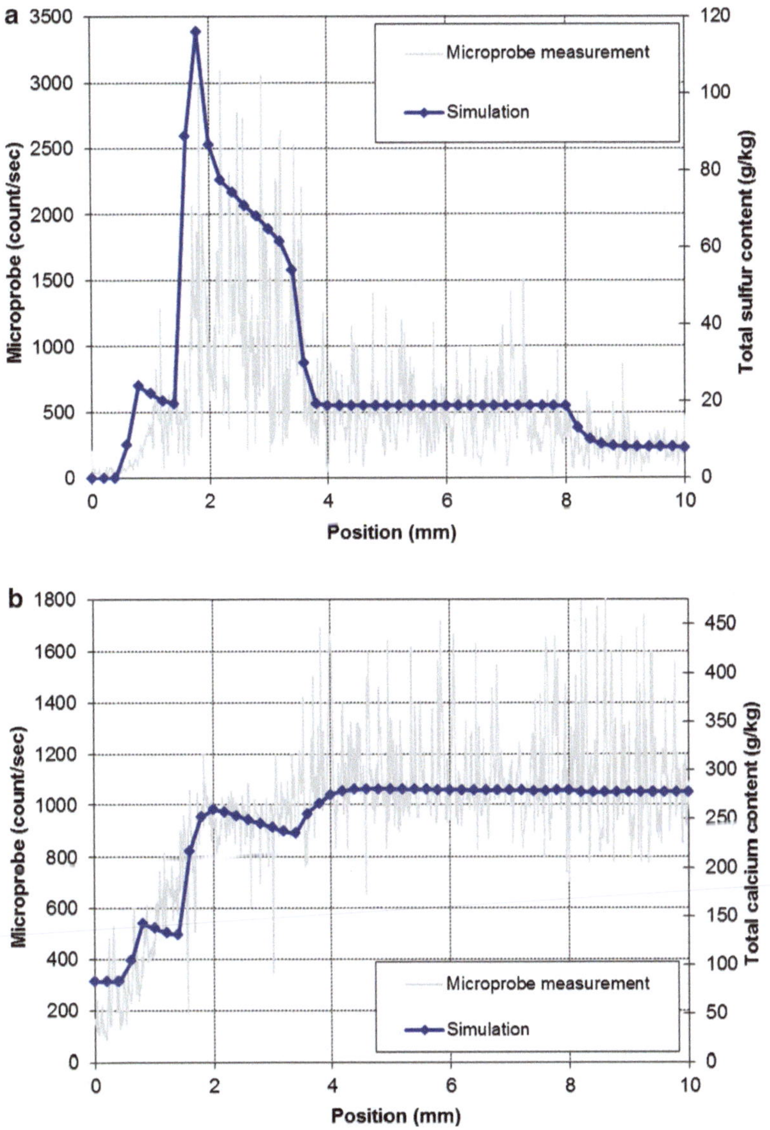

Fig. 7.3 Prediction of the (**a**) sulfur profile and (**b**) calcium content of a hydrated cement paste exposed to sodium sulfate for 1 year (Samson and Marchand 2007b)

a separate module. Using this approach, the authors were able to predict the formation of an ettringite front in the material as well as the presence of a gypsum peak near the surface. Both predictions match well with the measured sulfur content (Fig. 7.3a). Since other ionic species are considered in the calculations, this type of approach allows for modeling other phenomena occurring during sulfate ingress.

In this particular case, the authors were able to predict the simultaneous loss of calcium due to the leaching of the pore solution (Fig. 7.3b). This constitutes a certain advantage of the multi-ionic approach over single-ion models. A similar multi-ionic approach (Lothenbach et al. 2010) was presented in a model that does not consider electrical coupling between species.

Multi-ionic models that incorporate expansion calculations were presented in Schmidt-Döhl and Rostasy (1999), Rigo et al. (2005), and Bary et al. (2010). The model presented in Rigo et al. (2005) also incorporates a probabilistic component to perform Monte-Carlo calculations to predict the expansion of mortar bars exposed to sodium sulfate solutions.

It should be noted that all models reviewed for sulfate attack were only validated for Na_2SO_4 exposure. Technically, multi-ionic models could easily be tested for exposure to $CaSO_4$ and $MgSO_4$ solutions by considering the proper cations and related solid phases. But no comparison with experimental data could be found in the literature.

7.5.3 *Acid Attack*

In contrast to decalcification and sulfate attack, the modeling of concrete structures exposed to acid solutions has received very little attention. Many experimental studies showed that materials exposed to these low pH environments exhibit a degradation front where the rate of mass loss is proportional to the square root of time (Pavlik 1994):

$$d = K\sqrt{t} \tag{7.43}$$

where d is the degraded thickness, K is the proportionality constant which depends on the type of material and exposure conditions, and t is the time. This empirical relationship summarizes most of the work dedicated to the study of acid attack on cementitious materials.

Recently, a model was developed by Beddoe and Hillbig (2007), which provides a multi-ionic framework for this particular case of exposure. The model considers the electrical coupling between the fluxes of ionic species Eq. 7.6. The rate of dissolution of the binder matrix is controlled by a kinetic expression:

$$\frac{\partial c_{ca}}{\partial t} = k_{ca} s_{ca} c_H \tag{7.44}$$

where c_{ca} is the concentration of calcium in the pore solution, k_{ca} is a rate constant, s_{ca} is the amount of potentially soluble calcium in concrete and c_H is the proton (H^+) concentration in the pore solution.

To maintain the equilibrium of the various components of the acid reaction, the equilibrium relationship of the $HX \leftrightarrow H^+ + X^-$ reaction is enforced:

$$\frac{c_H c_X}{c_{HX}} = K_a \tag{7.45}$$

where K_a is the acid constant. The modeling results showed that the degraded concrete thickness follows a square root of time-type of relationship (Beddoe and Schmidt 2009).

7.6 Thermo-Hydro-Mechanical (THM) Models

Many degradation phenomena observed on concrete structures are directly related to water content and temperature gradients in the material. Some of these degradation phenomena occur at early ages, such as drying shrinkage and heat of hydration-induced crack formation. In other cases, the damage may occur later in the life of the structure, even though the material has gained strength from the hydration process. This is the case, for instance, for structures exposed to freezing and thawing cycles and to long-term deformations induced by creep. Damage also occurs as a result of long exposure to a chemically aggressive environment, as reviewed in the previous section.

Modeling chemical damage in this context is performed using thermo-hydro-mechanical (THM) approaches. While many different THM models have been proposed, they are in most cases based on solving the conservation equations for the gas and fluid phases, coupled to mechanical and thermal relationships. The global THM framework is reviewed in the following section. THM models based on simplified fluid transport equations are also reviewed.

7.6.1 General Overview

The general framework outlined in this section closely follows the description found in Gawin and Schrefler (1996). Other THM models can be found in Baggio et al. (1995), Bary et al. (2000), Khalili and Loret (2001), Obeid et al. (2001), Gawin et al. (2006), and Bary et al. (2008).

To simulate moisture transfer, most THM models found in the literature are based on separate mass conservation equations for the gas, water vapour and liquid vapour phases. The mass conservation equations for the fluid phases, under the assumption of a rigid solid matrix, can be expressed as Mainguy et al. (2001):

Liquid phase:

$$\frac{\partial(\phi S \rho_l)}{\partial t} + \mathrm{div}\left(\phi S_{\rho_l v_l}\right) + \mu_{l \to gv} = 0 \tag{7.46}$$

Vapour phase:

$$\frac{\partial(\phi\rho_v(1-S))}{\partial t} + \mathrm{div}(\phi(1-S)\rho_v\mathbf{v}_v) + \mu_{l\to v} = 0 \tag{7.47}$$

Dry air phase:

$$\frac{\partial(\phi\rho_a(1-S))}{\partial t} + \mathrm{div}(\phi(1-S)\rho_a\mathbf{v}_a) = 0 \tag{7.48}$$

where the subscript l, v and a stand for the liquid, vapour and dry air phases respectively, ϕ is the porosity, S is the saturation of the liquid phase, ρ is the density, $\mathbf{v}$ is the velocity (flow) vector, and $\mu_{l\to v}$ is a source term accounting for the condensation/vaporization at the liquid/gas interface.

In most THM models, the coupling with mechanics allows consideration of cases where the solid matrix assumption is not valid. In this case, terms involving the deformation of the matrix are added to the mass transport Eqs. 7.46 and 7.47. To keep the model to a reasonable size, it is possible to combine the liquid and vapour phase conservation equations to eliminate the condensation/vaporization term. Also, relationships between vapour pressure, water content, saturation and capillary pressure can be used so that the resulting equation is based on a single state variable. This approach was used in Baggio et al. (1995), Gawin and Schrefler (1996), and Gawin et al. (2006). The following paragraphs give the basic conservation equations considered in most THM models (Gawin and Schrefler 1996).

7.6.1.1 Water Species (Liquid + Vapour)

The mass conservation equation for the combined liquid and gaseous water is given by:

$$\phi\frac{\partial((1-S)\rho_v)}{\partial t} + \phi\rho_l\frac{\partial S}{\partial t} + \alpha(1-S)\rho_v\frac{\partial(\mathrm{div}\,\mathbf{u})}{\partial t} + bS\rho_l\frac{\partial(\mathrm{div}\,\mathbf{u})}{\partial t}$$
$$+\mathrm{div}\left(\rho_v v_g\right) + \mathrm{div}(\rho_l\mathbf{v}_l) + \mathrm{div}\left(p_g\mathbf{v}_v^d\right) = 0 \tag{7.49}$$

where $\mathbf{u}$ is the displacement vector of the solid matrix, the subscript g refers to the gas phase, α is Biot's constant, $\mathbf{v}_g$ is the bulk velocity of the gas phase and $\mathbf{v}_v{}^d$ is the flow of vapour due to diffusion. Constitutive relationships provide the flow equations for the different phases. For the liquid phase, the flow equation corresponds to Darcy's law:

$$v_l = -\frac{k_l}{\mu_l}(\mathrm{grad}(\rho_l) - \rho_l\mathbf{g}) \tag{7.50}$$

where k_l is the liquid permeability of the material, μ_l is the dynamic viscosity of the liquid phase, p_l is the liquid pressure, and **g** is the gravity acceleration vector. A similar expression is used for the gas phase flow:

$$\mathbf{v}_g = -\frac{k_g}{\mu_g}\left(\text{grad}\,(\rho_g) - \rho_g \mathbf{g}\right) \tag{7.51}$$

Finally, the flow of vapour due to diffusion can be expressed as (Whitaker 1977):

$$\mathbf{v}_v^d = -D_v \text{grad}\left(\frac{\rho_v}{\rho_g}\right) \tag{7.52}$$

where D_v is the diffusion coefficient of vapour.

To express Eq. 7.49 as a function of a single state variable for water, most models rely on the definition of capillary pressure p_c, which relates liquid and gas pressures (Bear 1988):

$$\rho_c = \rho_g - \rho_l \tag{7.53}$$

and Kelvin's relationship (Kaviany 1995):

$$\rho_c = -\frac{p_l RT}{M_w} \ln\left(\frac{\rho_v}{\rho_v^s}\right) = \frac{\rho_l RT}{M_w} \ln(H) \tag{7.54}$$

where R is the ideal gas constant, T is the temperature, M_w is the molar mass of water, $p_v{}^s$ is the saturated vapour pressure, and H is the relative humidity. Also, the ideal gas law can be used to relate the vapour density of Eq. 7.52 to the vapour pressure:

$$\frac{\rho_v}{M_w} = \frac{n_v}{V_g} = \frac{\rho_v}{RT} \tag{7.55}$$

where n_v is the number of vapour molecules in the gas phase volume V^g. Substituting relationships (7.53), (7.54), and (7.55) in the flow Eqs. 7.50, 7.51, and 7.52 allow expression of the water species conservation equation as either liquid pressure, vapour pressure, or capillary pressure. If the relationship between saturation or water content and capillary pressure is known, such as Van Genuchten's relationship (Van Genuchten 1980), it is possible to use these state variables instead.

A comprehensive discussion on the choice of the state variable can be found in Gawin et al. (2006). The authors suggest using the pressure variables instead of saturation or water content because these variables are not continuous at the interface between two materials having different porosities and moisture retention characteristics. It is thus not possible with these variables to solve problems where different material layers (material interfaces) are present.

Dry air:

The mass conservation equation for dry air is given by Gawin and Schrefler (1996):

$$\phi \frac{\partial((1-S)\rho_a)}{\partial t} + \alpha(1-S)\rho_a \frac{\partial(\text{div } \mathbf{u})}{\partial t} + \text{div}(\rho_a \mathbf{v}_g) - \text{div}\left(\rho_g \mathbf{v}_v^d\right) = 0 \quad (7.56)$$

Similar to the water species relationship, flow Eqs. 7.51 and 7.52 are substituted in (7.56) to yield the full dry air conservation relationship.

7.6.1.2 Energy Conservation

The evolution of temperature in the porous material is calculated from the energy balance relationship. Most authors assume that the different phases (solid matrix, liquid and gas) of the material are in thermal equilibrium. In these conditions, the energy balance equation is given by Gawin and Schrefler (1996):

$$\rho\, C_p \frac{\partial T}{\partial t} + \left(C_{pl}\rho_l \mathbf{v}_l + C_{pg}\rho_g \mathbf{v}_g\right) \text{ grad (T)} - \text{div}\left(\lambda_{eff} \text{ grad }(T)\right) + \sum f_k = 0 \quad (7.57)$$

where ρ is the average density of the material, C_p is the specific heat, λ_{eff} is the effective thermal conductivity, and f_k represents the various sink or source terms that affect temperature.

Many different phenomena may contribute to the term f_k in Eq. 7.57. Gawin et al. (2002a, b) reported terms associated with phase change (evaporation) and dehydration of water. In Ulm and Coussy (1995, 1998), Gawin et al. (2006), the modeling of the heat generated during the hydration of concrete and its coupling with mechanics is discussed.

In Khalili and Loret (2001), the authors devised a THM model without assuming thermal equilibrium between the solid, liquid, and gas phases. In that case, three separate energy balance equations similar to (7.57) were considered for each phase, with thermal exchange terms between the phases being added.

7.6.1.3 Mechanical Coupling

In many THM models, the time-dependent mass and energy conservation equations are coupled to the static linear momentum balance equation, assuming that the deformations induced by the fluids and energy transport are slow enough to be considered as a series of static mechanical problems. In that case, the linear momentum balance equation is given by Gawin and Schrefler (1996):

$$div(\sigma) + \rho \mathbf{b} = 0 \quad (7.58)$$

where $\boldsymbol{\sigma}$ is the total stress tensor, **b** is a body force vector, usually corresponding to gravity, and ρ is the average density of the material, expressed as a weighted average of the separate phase density:

$$\rho = (1 - \phi)\rho_s + \phi S \rho_l + \phi(1 - S)\rho_g \tag{7.59}$$

where ρ_s is the density of the solid matrix. The constitutive equation is expressed as a function of the effective stress $\boldsymbol{\sigma}'$, which was first introduced in soil mechanics (Gray and Schrefler 2001). The effective stress expression for an isotropic material is given by Bary et al. (2000):

$$\sigma' = \sigma + bp\mathrm{I} \tag{7.60}$$

where b is the Biot scalar parameter, p is the average pressure and **I** is the identity tensor. The average pressure is given by Gawin and Schrefler (1996):

$$p = S p_l + (1 - S) p_g \tag{7.61}$$

The effective stress relationship links the pressure variations in the material resulting from moisture content variations with mechanical deformations. As an alternative to the relative stress expression, Coussy derived an incremental form of Eq. 7.60 (Coussy 1995):

$$d\sigma' = d\sigma + (dp^g - S dp^c)\mathrm{I} \tag{7.62}$$

which in most cases yields results similar to Eq. 7.60 according to Gray and Schrefler (2001).

The simplest mechanical coupling case corresponds to a linear elastic model, where the total stress tensor is related to the elastic strain by:

$$\sigma = \mathbf{E_o} : \varepsilon^e \tag{7.63}$$

where ε^e is the elastic strain contribution and $\mathbf{E_o}$ is the elasticity matrix. The total strain $\boldsymbol{\varepsilon}$ can be expressed as (Gawin and Schrefler 1996):

$$\varepsilon = \varepsilon^e + \varepsilon^T + \varepsilon^o \tag{7.64}$$

where ε^T is the strain caused by thermo-elastic expansion and ε^o is the autogeneous strain. The thermal strain is written as (Cervera et al. 1999):

$$\varepsilon^T = \alpha_\mathrm{T}\left(\mathrm{T} - \mathrm{T}^{\mathrm{ref}}\right) \mathrm{I} \tag{7.65}$$

where α_T is the thermal dilatation coefficient and T^{ref} is a reference temperature for which the thermal strain is null.

In Gawin et al. (2002a, b), the linear elastic model was expanded to a nonlinear damage model to simulate concrete exposed to temperature above the critical point of water. In that case, the stress–strain relationship is written as:

$$\sigma = (1 - \mathrm{d})\mathbf{E}_0 : \varepsilon^e \tag{7.66}$$

where d is the scalar damage parameter (Mazars and Pijaudier-Cabot 1989). Kahalili and Loret (2001) presented a THM model where the mechanical formulation is based on an elasto-plastic constitutive law to simulate moisture induced shrinkage. In Baggio et al. (1995), the authors devised a THM approach where the mechanical model is based on viscoelasticity and damage. Similar to the SNIA algorithm of reactive transport modeling, the calculations are performed in two sequential steps. The hydro-thermal simulations are performed first, and then the mechanical analysis is performed using a separate code.

The THM framework described in the previous section has been expanded theoretically in Schrefler and Pesavento (2004). Averaging the microscopic mass balance and constitutive equations over a Representative Elementary Volume (REV), they obtained a set of conservation equations that includes interface terms that can be used to model different exchange phenomena between the various phases. The final system of equations is made of 30 equations with 30 corresponding independent variables. Although the global model could hardly be solved when all equations are considered, it provides a framework for devising specific applications.

7.6.2 Simplified THM Models

Given the complexity of the equations involved, solving THM models can be a daunting task. Not surprisingly, simplifications have been made in some cases to keep the numerical problem manageable. In other cases, the model can be simplified due to the nature of the problem to be solved. This is the case for instance in Bary et al. (2000). The authors present a model to simulate the damage caused by high fluid pressure on concrete structures such as dams. In this case, the material can be considered saturated and the temperature uniform, which simplifies the set of equations.

Many authors assume that the gas pressure is constant in the porous material. This allows removing dry air from the set of equations and also simplifies the remaining water species conservation equation as shown in Chijimatsu et al. (2000), Obeid et al. (2001), and Bary et al. (2008). In Meschke and Grasberger (2003), the authors used a nonlinear diffusion approach similar to the approach proposed by Bazant and Najjar (1971) but based on the capillary pressure state variable instead of the relative humidity.

7.7 Coupled Ionic Transport and Mechanical Models

The concern over the long-term stability of concrete in contact with chemically aggressive environments has initiated the development of models that attempt at coupling THM models with reactive transport approaches. A number of phenomena that are important in degradation of concrete structures are governed by a combination of chemical, physical, and mechanical processes: effects of locally formed precipitates on porosity and tortuosity, stress and material damage caused by precipitation of expanding minerals, increased porosity of material by leaching of matrix components, and decreased material strength by leaching of matrix components

Understanding these effects and quantitative estimation of long-terms effects requires models that can accommodate the different process descriptions. Integration of these models needs to be tight, because chemical and physical properties change dynamically in time and place. However, coupling all the mechanisms is very challenging from a numerical standpoint. It also implies knowing a large number of parameters for which no testing methods yet exist. Accordingly, most efforts at coupling transport with mechanical damage have relied so far on simplified approaches.

7.7.1 *Simplified Transport Coupling*

Most attempts to link reactive transport in cementitious materials with THM models have been concerned with the decalcification of the hydrated cement paste and the impact it has on the mechanical resistance of the material and the possible damage induced to the matrix. In all the papers reviewed for this chapter (Gérard et al. 1998; Ulm and Coussy 1998; Kuhl et al. 2004a, b), the concentration of calcium was used as the driving mechanism behind the matrix dissolution. Accordingly, the transport of ions was limited to a single equation modeling the diffusion of calcium, similar to Eq. 7.39:

$$\frac{\partial s}{\partial t} + \frac{\partial wc}{\partial t} - \mathrm{div}(D_{ca}\mathrm{grad}\,(c)) = 0 \qquad (7.67)$$

where c is the concentration of calcium in the pore solution of the material, s is the amount of calcium in the solid phase (as portlandite and C–S–H), w is the water content, and D_{ca} is the diffusion coefficient of calcium. The first term on the left hand side of Eq. 7.67 accounts for the calcium released in the pore solution upon portlandite and C–S–H dissolution. In all models cited previously, the calcium concentration is estimated from the C/S ratio as shown in Fig. 7.4 (the experimental data in Fig. 7.4b were taken from Berner 1992). Going from Fig. 7.4a, b is hypothetical at best since no direct relationship exists between the amount of

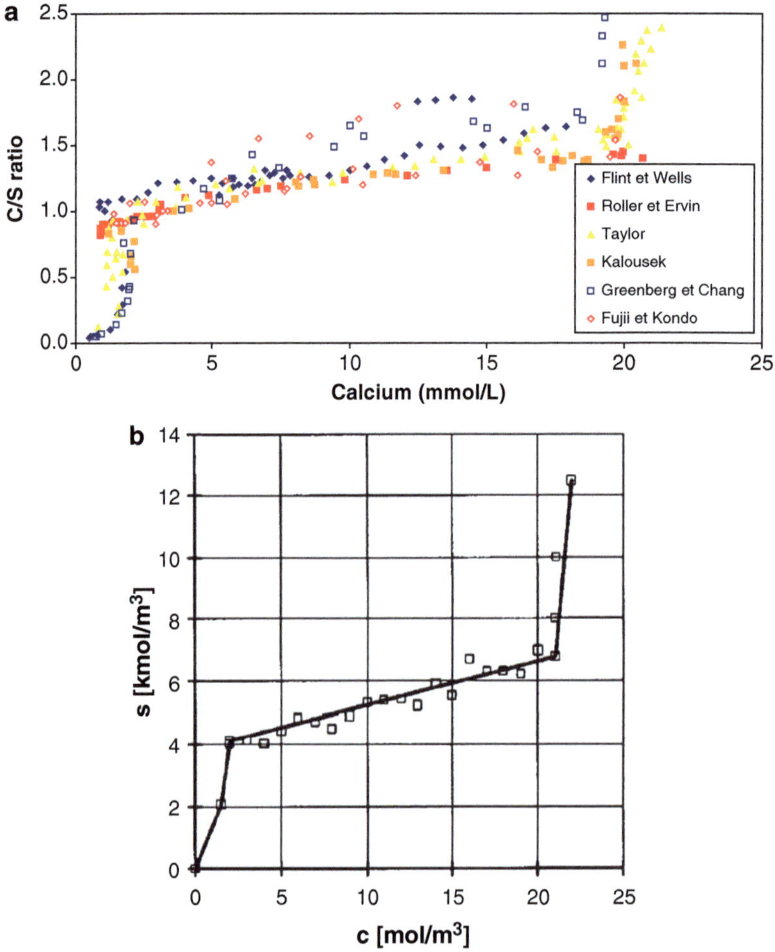

Fig. 7.4 Relationship between (**a**) CaO/SiO_2 vs. calcium in solution (Berner 1992) (**b**) solid content vs. calcium in solution (Ulm 1999)

portlandite and C–S–H and the calcium in pore solution, as emphasized by the equilibrium relationship of Eq. 7.17.

To couple transport and matrix dissolution to the mechanical model, (Gérard et al. 1998) introduced an ageing parameter V_d (0:1) that depends on the amount of calcium dissolved. According to Fig. 7.4, it can also be related to the calcium in solution. The ageing parameter is then incorporated in the damage model (Eq. 7.66) as:

$$\sigma = (1 - V_d)(1 - d)\mathbf{E_0} : \varepsilon^e \tag{7.68}$$

According to this formulation, the ageing parameter reduces the mechanical properties of the material as the matrix is dissolved. In Gérard et al. (1998), the authors also consider a feedback effect that modifies the transport parameters as a function of the damage level.

The modeling approach just described has receive special attention from the nuclear waste storage community, where the extended service-life required by the concrete structures (1,000+years) makes modeling essential to safety assessment. Modeling specifically for this context is described in Sellier et al. (2009). The extended service-life requirement of nuclear waste storage facilities also emphasizes the role of other phenomena such as creep and hydration. This was considered in the model presented by Buffo-Lacarrière and Sellier (2009) where the mechanical constitutive model incorporates terms for creep. Leaching was modeled on the basis of calcium, similar to Eq. 7.67, and is coupled with an hydration model that predicts the evolution of temperature and water content in the material as a function of time (Buffo-Lacarrière et al. 2007).

7.7.2 Coupling Damage with Multi-Ionic Transport

As shown in the previous section, most models that attempt to couple reactive transport and THM are limited to simplified ionic diffusion equations. Very few models that couple multi-ionic reactive transport and damage mechanics could be found. The model developed by Planel (2002) uses a reactive transport model based on the SIA algorithm and linear transport equations (see Eq. 7.31) to simulate the precipitation of ettringite and gypsum on exposure to groundwater containing sulfate. The mechanical portion of this model is based on a damage model similar to Eq. 7.66, where the damage parameter is a function of the volume of ettringite precipitated compared to the initial porosity of the material. The dissolution of C–S–H and its impact on the pore volume is also considered.

Stora et al. (2009, 2010) adopted an analogous approach to model leaching, with multi-ionic reactive transport equations fully coupled with mechanical behaviour and an anisotropic damage model. The mechanical and diffusive properties are upgraded from the newly calculated mineral concentrations at each time step by applying analytical upscaling techniques. This model was extended to sulfate attack in Bary et al. (2011) by accounting for both crystallization pressures depending on the ionic activities in pore solution, and volume of precipitated ettringite (Tixier and Mobasher 2003).

Sarkar et al. (2010) modeled sulfate attack by coupling multi-ionic transport equations and mechanical damage. The transport equations are a simplified version of Eq. 7.6 where diffusion and chemical activity effects are considered without the electrodiffusion term. The mass transport equations are coupled to the chemical equilibrium model ORCHESTRA (Meeussen 2003). Upon minerals formation or dissolution, stresses in the material are calculated and translated into modifications to the transport properties.

7.8 Conclusion

The review of models for decalcification, sulfate attack and acid exposure show that, for these cases, the multi-ionic approach is receiving much attention, compared to chloride ingress predictions. By considering more than one ionic species, the models can predict several simultaneous alteration mechanisms in hydrated cement pastes. While the decalcification and sulfate attack cases have clearly been the subject of more research efforts, the recent publication of a multi-ionic model for acid exposure (Beddoe and Hillbig 2007) shows the interest for this type of approach.

On the other hand, THM models offer very fine descriptions of moisture, temperature and mechanics, but most of these models are coupled to simplified ionic transport models. These models mostly track a single species, and the chemistry is summarized as a function that relates the amount bound in the material to the concentration in the pore solution.

Both approaches suffer from serious flaws. However, a fully coupled THM, multi-ionic chemical model involves numerical and computational efforts, and for now seems out of reach. Nevertheless, papers published in recent years have shown an increase in the complexity of models, especially compared to the early years of modeling chloride ingress using Fick's law. With the constant improvement of computational capacities, the goal of considering complex transport, chemistry and mechanics in a single model is not impossible.

Notwithstanding recent progress, the field of cementitious materials modeling is still many years behind the modeling work being done for transport of contaminants in soil, where models have evolved from single-ion models many years ago, to now incorporate tens of species, secondary species and minerals. The efforts in the geochemical modeling field now focus on solving computational problems to improve calculation time using massively parallel computers, multi-thread processes and so on. These efforts are essential because the geochemistry community has already concluded that the relative value of multi-ionic models compared to simplified ones is no longer questioned. It is appropriate and timeous that the same mindset be exhibited by the cementitious materials community.

References

Adenot, F., Buil, M.: Modeling of the corrosion of the cement paste by deionized water. Cem. Concr. Res. **22**, 489–496 (1992)

Appelo, C.A.J.: Multicomponent ion exchange and chromatography in natural systems. Rev. Mineral. Reactive Transp. Porous Media **34**, 193–227 (1996)

Appelo, C.A.J., Postma, D.: Geochemistry, Groundwater and Pollution. A.A. Balkema, Rotterdam (1996)

Babushkin, V.I., Matveev, O.P., Mčedlov-Petrosjan, O.P.: Thermodinamika Silikatov. Strojisdat, Moskau (1965)

Babushkin, V.I., Matveyev, G.M., Mchedlov-Petrossyan, O.P.: Thermodynamics of Silicates. Springer, Berlin (1985)

Baggio, P., Majorana, C.E., Schrefler, B.A.: Thermo-hygro-mechanical analysis of concrete. Int. J. Numer. Methods Fluids **20**, 573–595 (1995)

Barbarulo, R., Marchand, J., Snyder, K.A., Prené, S.: Dimensional analysis of ionic transport problems in hydrated cement systems. Cem. Concr. Res. **30**, 1955–1960 (2000)

Barret, P., Bertrandie, D., Beau, D.: Calcium hydrocarboaluminate, carbonate, alumina gel and hydrated aluminates solubility diagram calculated in equilibrium with CO_{2g} and with Na_{aq}^{+} ions. Cem. Concr. Res. **13**, 789–800 (1983)

Bary, B., Bournazel, J.P., Bourdarot, E.: Poro-damage approach applied to hydro-fracture analysis of concrete. J. Eng. Mech. **126**, 937–943 (2000)

Bary, B., Ranc, G., Durand, S., Carpentier, O.: A coupled thermo-hydro-mechanical-damage model for concrete subjected to moderate temperatures. Int. J. Heat Mass Transf. **51**, 2847–2862 (2008)

Bary, B., Leterrier, N., Deville, E., Le Bescop, P.: Coupled chemo-transport-mechanical modeling and simulation of mortar external sulfate attack. In: Proceedings – CONMOD 2010, pp. 65–78. Lausanne, Switzerland (2010)

Bary, B., Leterrier, N., Deville, E., Le Bescop, P.: Coupled mechanical and chemo-transport model for the simulation of cementitious materials subjected to external sulfate attack. In: Proceedings – NUWCEM 2011, Avignon, France (2011)

Bazant, Z.P., Najjar, L.J.: Drying of concrete as a nonlinear diffusion problem. Cem. Concr. Res. **1**, 461–473 (1971)

Bear, J.: Dynamics of Fluid in Porous Media. Dover, New-York (1988)

Bear, J., Bachmat, Y.: Introduction to Modeling of Transport Phenomena in Porous Media. Kluwer Academic, Dordrecht (1991)

Beddoe, R.E., Hillbig, H.: Acid attack on concrete – a new computer model. In: Setzer, M.J. (ed.) Proceedings of the 5th International Essen Workshop, pp. 387–398. Aedification Publishers, Essen (2007)

Beddoe, R.E., Schmidt, K.: Effect of concrete composition on resistance of concrete to acid attack. In: Concrete in Aggressive Aqueous Environments, vol. 1: Performance, Testing and Modeling, RILEM Proceedings 63, Toulouse, France, pp. 187–195 (2009)

Berner, U.R.: Evolution of pore water chemistry during degradation of cement in a radioactive waste repository environment. Waste Manag. **12**, 201–219 (1992)

Bockris, J.O.'.M., Reddy, A.K.N.: Modern Electrochemistry – An Introduction to an Interdisciplinary Area. Plenum Press, New York (1970)

Bourbon, X.: Chemical conceptual model for cement based materials ~ mineral phases and thermodynamic data. ANDRA Technical Report C.NT.ASCM.03.026.A, ANDRA, Chatenay-Malabry, France (2003)

Brown, P., Haworth, A., Sharland, S., Tweed, C.: HARPHRQ: a geochemical speciation program based on PHREEQE – users guide. Report NSS/R188, Harwell Laboratory, Oxfordshire, United Kingdom (1991)

Buffo-Lacarrière, L., Sellier, A.: Numerical modeling of effects of chemical evolution on mechanical behaviour of concrete. In: Concrete in Aggressive Aqueous Environments, vol. 1: Performance, Testing and Modeling, RILEM Proceedings 63, Toulouse, France, pp. 167–178 (2009)

Buffo-Lacarrière, L., Sellier, A., Escadeillas, G., Turatsinze, A.: Multiphasic finite element modeling of concrete hydration. Cem. Concr. Res. **37**, 131–138 (2007)

Buil, M., Revertegat, E., Oliver, J.: A model of the attacks of pure water or undersaturated lime solutions on cement. In: Stabilization and Solidification of Hazardous Radioactive and Mixed Wastes, 2nd vol. ASTM STP 1123. pp. 227–241. Philadelphia, USA (1992)

Cederberg, G.A., Street, R.L., Leckie, J.O.: A groundwater mass transport and equilibrium chemistry model for multicomponent systems. Water Resour. Res. **21**, 1095–1104 (1985)

Cervera, M., Olivier, J., Prato, T.: Thermo-chemo-mechanical model for concrete. II: Damage and creep. J. Eng. Mech. **125**, 1018–1027 (1999)

Chijimatsu, M., Fujita, T., Kobayashi, A., Nakano, M.: Experiment and validation of numerical simulation of coupled thermal, hydraulic and mechanical behavior in the engineered buffer materials. Int. J. Numer. Anal. Methods Geomech. **24**, 403–424 (2000)

Corazza, E., Sabelli, C.: The crystal structure of syngenite, $K_2Ca(SO_4)_2$*(H_2O). Zeitschrift für Kristallographie **124**, 398–408 (1967)

Coussy, O.: Mechanics of Porous Continua. Wiley, Chichester (1995)

Daimon, M., Abo-El-Enein, S.A., Hosaka, G., Goto, S., Kondo, R.: Pore structure of calcium silicate hydrate in hydrated tricalcium silicate. J. Am. Ceram. Soc. **60**, 110–114 (1977)

Damidot, D., Stronach, S., Kindness, A., Atkins, M., Glasser, F.P.: Thermodynamic investigation of the CaO-Al_2O_3-$CaCO_3$-H_2O closed system at 25° C and the influence of Na_2O. Cem. Concr. Res. **24**, 563–572 (1994)

De Windt, L., Pellegrini, D., van der Lee, J.: Coupled modeling of cement/claystone interactions and radionuclide migration. J. Contam. Hydrol. **68**, 165–182 (2004)

Engesgaard, P., Kipp, K.L.: A geochemical transport model for redox-controlled movement of mineral fronts in groundwater flow systems: a case of nitrate removal by oxidation of pyrite. Water Resour. Res. **28**, 2829–2843 (1992)

Fetter, C.W.: Contaminant Hydrogeology. Prentice-Hall, Upper Saddle River (1999)

Galindez, J.M., Molinero, J.: On the relevance of electrochemical diffusion for the modeling of degradation of cementitious materials. Cem. Concr. Compos. **32**, 351–359 (2010)

Garboczi, E.J., Bentz, D.P.: Computer simulation of the diffusivity of cement-based materials. J. Mater. Sci. **27**, 2083–2092 (1992)

Gawin, D., Schrefler, B.A.: Thermo-hydro-mechanical analysis of partially saturated porous materials. Eng. Comput. **13**, 113–143 (1996)

Gawin, D., Pesavento, F., Schrefler, B.A.: Modeling of hygro-thermal behaviour and damage of concrete at temperature above the critical point of water. Int. J. Numer. Anal. Methods. Geomech. **26**, 537–562 (2002a)

Gawin, D., Pesavento, F., Schrefler, B.A.: Simulation of damage-permeability coupling in hygro-thermo-mechanical analysis of concrete at high temperature. Commun. Numer. Methods. Eng. **18**, 113–119 (2002b)

Gawin, D., Pesavento, F., Schrefler, B.A.: Hygro-chemo-mechanical modeling of concrete at early ages and beyond. Part I: Hydration and hygro-thermal phenomena. Int. J. Numer. Methods. Eng. **67**, 299–331 (2006)

Gérard, B., Pijaudier-Cabot, G., Laborderie, C.: Coupled diffusion-damage modeling and the implications on failure due to strain localidsation. Int. J. Solids Struct. **35**, 4107–4120 (1998)

Gospodinov, P.N., Kazandjiev, R.F., Partalin, T.A., Mironova, M.K.: Diffusion of sulfate ions into cement stone regarding simultaneous chemical reactions and resulting effects. Cem. Concr. Res. **29**, 1591–1596 (1999)

Gray, W.G., Schrefler, B.A.: Thermodynamic approach to effective stress in partially saturated porous media. Eur. J. Mech. A/Solids **20**, 521–538 (2001)

Grove, D.B., Wood, W.W.: Prediction and field verification of subsurface-water quality changes during artificial recharge, Lubbock, Texas. Groundwater **17**, 250–257 (1979)

Haga, K., Sutou, S., Hironaga, M., Tanaka, S., Nagasaki, S.: Effects of porosity on leaching of Ca from hardened ordinary Portland cement paste. Cem. Concr. Res. **35**, 1764–1775 (2005)

Helfferich, F.: Ion Exchange. McGraw-Hill, New York (1961)

Henocq, P., Marchand, J., Samson, E., Lavoie, J.A.: Modeling of ionic interactions at the C-S-H surface – application to CsCl and LiCl solutions in comparison with NaCl solutions. In: Marchand, J. et al. (eds.) 2nd International Symposium on Advances in Concrete through Science and Engineering, RILEM Proceedings PRO, vol. 51. RILEM Publications, Quebec City (2006)

Hidalgo, A., De Vera, G., Climent, M.A., Andrade, C., Alonso, C.: Measurements of chloride activity coefficients in real Portland cement paste pore solutions. J. Am. Ceram. Soc. **84**, 3008–3012 (2001)

Hummel, W., Berner, U., Curti, E., Pearson, F.J., Thoenen, T.: Nagra/PSI chemical thermodynamic data base 01/01. USA, also published as Nagra Technical Report NTB 02–16. Universal Publishers, Wettingen (2002)

Jacques, D., Simunek, J., Mallants, D., van Genuchten, M.T.: Operator splitting errors in coupled reactive transport codes for transient variably saturated flow and contaminant transport in layered soil profiles. J. Contam. Hydrol. **88**, 197–218 (2006)

Jennings, A.A., Kirkner, D.J., Theis, T.L.: Multicomponent equilibrium chemistry in groundwater quality models. Water Resour. Res. **18**, 1089–1096 (1982)

Johannesson, B.F.: A theoretical model describing diffusion of a mixture of different types of ions in pore solution of concrete coupled to moisture transport. Cem. Concr. Res. **33**, 481–488 (2003)

Kanney, J.F., Miller, C.T., Kelley, C.T.: Convergence of iterative split operator approaches for approximating nonlinear reactive transport problems. Adv. Water Resour. **26**, 247–261 (2003)

Kaviany, M.: Principles of Heat Transfer in Porous Media. Springer, New York (1995)

Khalili, N., Loret, B.: An elasto-plastic model for non-isothermal analysis of flow and deformation in unsaturated porous media: formulation. Int. J. Solids Struct. **38**, 8305–8330 (2001)

Kirkner, D.J., Reeves, H.W.: Multicomponent mass transport with homogeneous and heterogeneous chemical reactions: Effect of the chemistry on the choice of numerical algorithm, 1. Theory. Water Resour. Res. **24**, 1719–1729 (1988)

Kirkner, D.J., Reeves, H.W., Jennings, A.A., et al.: Finite element analysis of multicomponent contaminant transport including precipitation-dissolution reactions. In: Laible, J.L. (ed.) Finite Elements in Water Resources, pp. 309–318. Springer, New York (1984)

Kuhl, D., Bangert, F., Meschke, G.: Coupled chemo-mechanical deterioration of cementitious materials. Part I: Modeling. Int. J. Solids Struct. **41**, 15–40 (2004a)

Kuhl, D., Bangert, F., Meschke, G.: Coupled chemo-mechanical deterioration of cementitious materials. Part II: numerical methods and simulations. Int. J. Solids Struct. **41**, 41–67 (2004b)

Kulik, D., Berner, U., Curti, E.: Modeling geochemical equilibrium partitioning with the GEMS-PSI Code, in PSI Scientific Report vol IV. In: Smith, B., Gschwend, B. (eds.) Nuclear Energy and Safety, pp. 109–122 (ISSN 1423-7334). Paul Scherrer Institute, Villigen, Switzerland (2003)

Li, Y.H., Gregory, S.: Diffusion of ions in sea water and in deep-sea sediments. Geochim. Cosmochim. Acta **38**, 703–714 (1974)

Lichtner, P.C.: Continuum model for simultaneous chemical reactions and mass transport in hydrothermal systems. Geochim. Cosmochim. Acta **49**, 779–800 (1985)

Lothenbach, B., Winnefeld, F.: Thermodynamic modeling of the hydration of Portland cement. Cem. Concr. Res. **36**, 209–226 (2006)

Lothenbach, B., Matschei, T., Möschner, G., Glasser, F.P.: Thermodynamic modeling of the effect of temperature on the hydration and porosity of Portland cement. Cem. Concr. Res. **38**, 1–18 (2008)

Lothenbach, B., Bary, B., Le Bescop, P., Schmidt, T., Leterrier, N.: Sulfate ingress in Portland cement. Cem. Concr. Res. **40**, 1211–1225 (2010)

MacQuarrie, K.T.B., Mayer, K.U.: Reactive transport modeling in fractured rock: a state-of-the-science review. Earth Sci. Rev. **72**, 189–227 (2005)

Mainguy, M., Tognazzi, C., Torrenti, J.-M., Adenot, F.: Modeling of leaching in pure cement paste and mortar. Cem. Concr. Res. **30**, 83–90 (2000)

Mainguy, M., Coussy, O., Baroghel-Bouny, V.: Role of air pressure in drying of weakly permeable materials. J. Eng. Mech. **127**, 582–592 (2001)

Maltais, Y., Samson, E., Marchand, J.: Predicting the durability of Portland cement systems in aggressive environments – laboratory validation. Cem. Concr. Res. **34**, 1579–1589 (2004)

Martín-Pérez, B., Pantazopoulou, S.J., Thomas, M.D.A.: Numerical solution of mass transport equations in concrete structures. Comput. Struct. **79**, 1251–1264 (2001)

Marty, N.C.M., Tournassat, C., Burnol, A., Giffaut, E., Gaucher, E.C.: Influence of reaction kinetics and mesh refinement on the numerical modeling of concrete clay interactions. J. Hydrol. **364**, 58–72 (2009)

Masi, M., Colella, D., Radaelli, G., Bertolini, L.: Simulation of chloride penetration in cement-based materials. Cem. Concr. Res. **27**, 1591–1601 (1997)

Matschei, T., Lothenbach, B., Glasser, F.P.: Thermodynamic properties of Portland cement hydrates in the system CaO-Al_2O_3-SiO_2-$CaSO_4$-$CaCO_3$-H_2O. Cem. Concr. Res. **37**, 1379–1410 (2007)

Mazars, J., Pijaudier-Cabot, J.P.: Continuum damage theory – application to concrete. J. Eng. Mech. **115**, 345–365 (1989)

Meeussen, J.C.L.: ORCHESTRA: an object-oriented framework for implementing chemical equilibrium models. Environ. Sci. Technol. **37**, 1175–1182 (2003)

Meschke, G., Grasberger, S.: Numerical modeling of coupled hygromechanical degradation of cementitious materials. J. Eng. Mech. **129**, 383–392 (2003)

Miller, C.W., Benson, L.V.: Simulation of solute transport in a chemically reactive heterogeneous system: model development and application. Water Resour. Res. **19**, 381–391 (1983)

Möschner, G., Lothenbach, B., Rose, J., Ulrich, A., Figi, R., Kretschmar, R.: Solubility of Fe-ettringite ($Ca_6(Fe(OH)_6)_2(SO_4)_3 \cdot 26H_2O$). Geochim. Cosmochim. Acta **72**, 1–18 (2008)

Nagesh, M., Bhattacharjee, B.: Modeling of chloride diffusion in concrete and determination of diffusion coefficients. ACI Mater. J. **95**, 113–120 (1998)

Obeid, W., Alliche, A., Mounajed, G.: Identification of the physical parameters used in the thermo-hygro-mechanical model (application to the case of cement mortar). Transp. Porous Media **45**, 215–239 (2001)

Pankow, J.F.: Aquatic Chemistry Concepts. Lewis Publishers, Chelsea (1994)

Pavlík, V.: Corrosion of hardened cement paste by acetic and nitric acids Part I: Calculation of corrosion depth. Cem. Concr. Res. **24**, 551–562 (1994)

Peppler, R.B., Wells, L.S.: The system of lime, alumina, and water from 50 to 250°C. J. Res. Natl. Bur. Stand. **52**, 75–92 (1954)

Planel, D.: Les effets couples de la precipitation d'espèces secondaires sur le comportement mecanique et la degradation chimique des bétons, Ph.D. thesis (in French), Marne-La-Vallée, France (2002)

Planel, D., Sercombe, J., Le Bescop, P., Adenot, F., Torrenti, J.M.: Long-term performance of cement paste during combined calcium leaching-sulfate attack: kinetics and size effect. Cem. Concr. Res. **36**, 137–143 (2006)

Reardon, E.J.: An ion interaction model for the determination of chemical equilibria in cement/water systems. Cem. Concr. Res. **20**, 175–192 (1990)

Reardon, E.J.: Problems and approaches to the prediction of the prediction of the chemical composition in cement/water systems. Waste Manag. **12**, 221–239 (1992)

Rigo, E., Schmidt-Döhl, F., Krauss, M., Budelmann, H.: Transreac: a model for the calculation of combined chemical reactions and transport processes and its extension to a probabilistic approach. Cem. Concr. Res. **35**, 1734–1740 (2005)

Robie, R.A., Hemingway, B.S.: Thermodynamic properties of minerals and related substances at 298.15 K and 1 bar (10^5 Pascals) pressure and at higher temperatures. U. S. Geological Survey Bulletin 2131, Virginia, USA (1995)

Rubin, J.: Transport of reacting solutes in porous media: relation between mathematical nature of problem formulation and chemical nature of reactions. Water Resour. Res. **19**, 1231–1252 (1983)

Saaltink, M.W., Carrera, J., Ayora, C.: On the behavior of approaches to simulate reactive transport. J. Contam. Hydrol. **48**, 213–235 (2001)

Saetta, A., Scotta, R., Vitaliani, R.: Analysis of chloride diffusion into partially saturated concrete. ACI Mater. J. **90**, 441–451 (1993)

Samson, E., Marchand, J.: Modeling the effect of temperature on ionic transport in cementitious materials. Cem. Concr. Res. **37**, 455–468 (2007a)

Samson, E., Marchand, J.: Modeling the transport of ions in unsaturated cement-based materials. Comput. Struct. **85**, 1740–1756 (2007b)

Samson, E., Lemaire, G., Marchand, J., Beaudoin, J.J.: Modeling chemical activity effects in strong ionic solutions. Comput. Mater. Sci. **15**, 285–294 (1999)

Samson, E., Marchand, J., Snyder, K.A., Beaudoin, J.J.: Modeling ion and fluid transport in unsaturated cement systems in isothermal conditions. Cem. Concr. Res. **35**, 141–153 (2005)

Sarkar, S., Mahadevan, S., Meeussen, J.C.L., van der Sloot, H., Kosson, D.S.: Modeling and sensitivity analysis of cementitious materials degradation under combined calcium leaching and external sulfate attack. Cem. Concr. Comp. **32**, 241–252 (2010)

Schmidt, T., Lothenbach, B., Romer, M., Scrivener, K.L., Rentsch, D., Figi, R.: A thermodynamic and experimental study of the conditions of thaumasite formation. Cem. Concr. Res. **38**, 337–349 (2008)

Schmidt-Döhl, F., Rostásy, F.S.: A model for the calculation of combined chemical reactions and transport processes and its application to the corrosion of mineral-building materials. Cem. Concr. Res. **29**, 1039–1045 (1999)

Schrefler, B.A.: Multiphase flow in deforming porous material. Int. J. Numer. Methods Eng. **60**, 27–50 (2004)

Schrefler, B.A., Pesavento, F.: Multiphase flow in deforming porous material. Comput. Geotech. **31**, 237–250 (2004)

Sellier, A., Buffo-Lacarrière, L., El Gonouni, M., Bourbon, X.: Behaviour of HPC nuclear waste storage structures in leaching environment. In: Concrete in Aggressive Aqueous Environments, vol. 1: Performance, Testing and Modeling, RILEM Proceedings 63, Toulouse, France , pp. 142–166 (2009)

Shock, E.L., Sassani, D.C., Willis, M., Sverjensky, D.A.: Inorganic species in geologic fluids: correlations among standard molal thermodynamic properties of aqueous ions and hydroxide complexes. Geochim. Cosmochim. Acta **61**, 907–950 (1997)

Simunek, J., Suarez, D.L.: Two-dimensional transport model for variably saturated porous media with major ion chemistry. Water Resour. Res. **30**, 1115–1133 (1994)

Steefel, C.I., Lasaga, A.C.: A coupled model for transport of multiple chemical species and kinetic precipitation/dissolution reactions with application to reactive flow in single phase hydrothermal systems. Am. J. Sci. **294**, 529–592 (1994)

Steefel, C.I., MacQuarrie, K.T.B.: Approaches to modeling of reactive transport in porous media. Rev. Mineral. Reactive Transp. Porous Media **34**, 83–129 (1994)

Sten-Knudsen, O.: Biological Membranes – Theory of Transport, Potentials and Electric Impulses. Cambridge University Press, Cambridge (2002)

Stora, E., Bary, B., He, Q.-C., Deville, E., Montarnal, P.: Modelling and simulations of the chemo–mechanical behaviour of leached cement-based materials: leaching process and induced loss of stiffness. Cem. Concr. Res. **39**, 763–772 (2009)

Stora, E., Bary, B., He, Q.-C., Deville, E., Montarnal, P.: Modelling and simulations of the chemo-mechanical behaviour of leached cement-based materials: interactions between damage and leaching. Cem. Concr. Res. **40**, 1226–1236 (2010)

Stumm, W., Morgan, J.J.: Aquatic Chemistry – Chemical Equilibria and Rates in Natural Waters, 3rd edn. Wiley Interscience, New York (1996)

Tang, L., Nilsson, L.-O.: Chloride binding capacity and binding isotherms of OPC pastes and mortars. Cem. Concr. Res. **23**, 247–253 (1993)

Thoenen, T., Kulik, D.: Nagra/PSI chemical thermodynamic database 01/01 for the GEM-Selektor (V.2-PSI) geochemical modeling code. Villigen. http://gems.web.psi.ch/doc/pdf/TM-44-03-04-web.pdf, PSI (2003). Accessed 2010

Tixier, R., Mobasher, B.: Modeling of damage in cement-based materials subjected to external sulfate attack. 1: Formulation. J. Mater Civil Eng **15**, 305–313 (2003)

Trotignon, L., Devallois, V., Peycelon, H., Tiffreau, C., Bourbon, X.: Predicting the long term durability of concrete engineered barriers in a geological repository for radioactive waste. Phys. Chem. Earth **32**, 259–274 (2007)

Truc, O., Ollivier, J.P., Nilsson, L.O.: Numerical simulation of multi-species diffusion. Mater. Struct. **33**, 566–573 (2000)

Ulm, J.F., Coussy, O.: Modeling of thermochemomechanical couplings of concrete at early ages. J. Eng. Mech. **121**, 785–794 (1995)

Ulm, J.F., Coussy, O.: Couplings in early-age concrete: from material modeling to structural design. Int. J. Solids Struct. **35**, 4295–4311 (1998)

Ulm, J.F., Torrenti, J.M., Adenot, F.: Chemoporoplasticity of calcium leaching in concrete. J. Eng. Mech. **125**, 1200–1211 (1999)

Valocchi, A.J., Street, R.L., Roberts, P.V.: Transport of ion-exchanging solutes in groundwater: chromatographic theory and field simulations. Water Resour. Res. **17**, 1517–1527 (1981)

Van der Lee, J.: Modélisation du comportement géochimique et du transport des radionucléides en présence de colloïdes. Thèse de l'Ecole Nationale Supérieure des Mines de Paris, 200 p (1997)

Van Genuchten, M.T.: A closed form equation for predicting the hydraulic conductivity of unsaturated soils. Soil Sci. Soc. Am. **44**, 892–898 (1980)

Walsh, M.P., Bryant, S.L., Shechter, R.S., Lake, L.W.: Precipitation and dissolution of solids attending flow through porous media. AIChE J. **30**, 317–328 (1984)

Walter, A.L., Frind, E.O., Blowes, D.W., Ptacek, C.J., Molson, J.W.: Modeling of multicomponent reactive transport in groundwater 1. Model development and evaluation. Water Resour. Res. **30**, 3137–3148 (1994)

Westall, J., Zachary, J., Morel, F.M.M.: MINEQL- a computer program for the calculation of the chemical equilibrium composition of aqueous systems. Technical Report 86–01, Department of Chemistry, Oregon State University (1986)

Whitaker, S.: Simultaneous heat, mass and momentum transfer in porous media: a theory of drying. Adv. Heat Transf. **13**, 119–203 (1977)

Whitaker, S., et al.: Coupled transport in multiphase systems: a theory of drying. In: Hartnett, J.P. (ed.) Advance in Heat Transfer, vol. 31, pp. 1–104. Academic Press, San Diego, USA (1998)

Wolery, T.: EQ3/6: A software package for geochemical modeling of aqueous systems: package overview and installation guide (version 7.0). UCRL-MA-1106662 PTI ed., Lawrence Livermore National Laboratory, Livermore (1992)

Xi, Y., Bazant, Z.P., Molina, L., Jennings, H.M.: Moisture diffusion in cementitious materials – moisture capacity and diffusivity. Adv. Cem. Based Mater. **1**, 258–266 (1994)

Xu, T., Samper, J., Ayora, C., Manzano, M., Custodio, E.: Modeling of non-isothermal multi-component reactive transport in field scale porous media flow systems. J. Hydrol. **214**, 144–164 (1999)

Yeh, G.T., Tripathi, V.S.: A critical evaluation of recent developments in hydro-geochemical transport models of reactive multichemical components. Water Resour. Res. **25**, 93–108 (1989)

Yokozeki, K., Watanabe, K., Sakata, N., Otsuki, N.: Modeling of leaching from cementitious materials used in underground environment. Appl. Clay Sci. **26**, 293–308 (2004)

Zienkiewicz, O.C., Taylor, R.L.: The Finite Element Method, 4th edn. McGraw-Hill, London (1989)

Part III
Methods for Testing Concrete Degradation in Aggressive Aqueous Environments

Authors of chapters in Part III

Chapter 8:
N. De Belie

Chapter 9:
M. Castellote

Chapter 10:
K. Van Tittelboom
N. De Belie
R.D. Hooton

Chapter 11:
M.G. Alexander
N. De Belie

Chapter 12:
K.L. Scrivener
N. De Belie

Chapter 13:
Q. Yuan
M. Santhanam

Chapter 14:
M. Santhanam

Chapter 15:
A. Bertron

Reviewers for Part III

M.G. Alexander, A. Bertron, J. Duchesne, B. Lothenbach, M. Santhanam, K.L. Scrivener

Chapter 8
General Considerations

Nele De Belie

Contents

Keywords Concrete • Performance • Durability • Test methods • Classification • Simulation

N. De Belie (✉)
Magnel Laboratory for Concrete Research, Department of Structural Engineering,
Ghent University, Ghent, Belgium
e-mail: nele.debelie@ugent.be

M. Alexander et al. (eds.), *Performance of Cement-Based Materials in Aggressive Aqueous Environments*, RILEM State-of-the-Art Reports 10,
DOI 10.1007/978-94-007-5413-3_8, © RILEM 2013

8.1 Introduction

In recent decades, numerous concrete structures have shown severe durability problems. Degradation mechanisms such as alkali silica reaction, chloride penetration, carbonation, acid attack, biogenic sulfuric acid attack, etc. have necessitated the complete renewal of some structures. Apart from the inconvenience and even occasional loss of life caused by the failure of structures, these degradation problems have caused huge financial losses. These include not only the cost of repair and restoration of structures, but also the economic losses caused for example by traffic problems during the temporary closure of bridges, roads and tunnels. It is therefore important to tackle these problems by formulating the most appropriate concrete compositions and by taking protective measures.

On the other hand new concrete types have been developed in recent years, such as high and ultra high strength concrete, polymer cement concrete, self compacting concrete, high volume fly ash concrete, etc. Durability specifications are not always available for such materials. Furthermore, concrete performance may be highly dependent on the exact mix composition and the compatibility between cement and constituents such as admixtures. Therefore, the need for suitable test methods and performance-based indicators increases.

The main aim of PART III is to critically review existing test methods to assess performance of cementitious materials in aggressive aqueous environments, as well as test methods which can be used to characterise and rate relative performance, with a view to informing long term predictions.

While this introduction deals with some general considerations, the chapters that follow will give further information on test methods for specific aggressive environments: soft or carbonated waters (leaching), organic acids, sulfates, ammonium and magnesium salts, inorganic acids, chlorides and biogenic corrosion.

8.2 Classification of Test Methods for Performance of Cementitious Materials in Aggressive Aqueous Environments

There are different ways in which test methods can be classified when it is desired to assess the performance of cementitious materials in aggressive aqueous solutions. This section considers five possible classifications depending on what is considered to be the important governing variable.

8.2.1 According to the Type of Aggressive Species

Differences between chemical and microbiological degradation mechanisms can be perceived. Chemical degradation includes attack by soft water, acids (organic acids such as lactic and acetic acid and inorganic acids such as H_2SO_4), and salts which might include ions such as CO_3^{2-}, SO_4^{2-}, Mg^{2+}, NH_4^+ and Cl^-. These different ions may occur alone or in combination; in the latter case interaction may be important, the combined effect being more or less severe than the sum of individual effects. Regarding this issue NBN EN 206-1 (2001) "Concrete – Part 1: Specification, performance, production and conformity" states that "where two or more aggressive characteristics lead to the same class, the environment shall be classified into the next higher class, unless a special study for this specific case proves that it is not necessary".

Microbiological degradation mechanisms feature aggressive substances produced by micro-organisms, such as in the sulfur cycle causing biogenic sulfuric acid attack in sewers. The substances causing the degradation may be the same as for chemical degradation and act in the same way on the cement matrix, but the fact that micro-organisms are involved has an additional effect. For example, certain organisms may favour growing on certain substrates more than on others. This may therefore demand the use of special test methods. For instance Live/Dead® analysis during microbiological tests by Vincke (2001) showed a more rapid colonisation by micro-organisms of the surface of slag cement samples compared to ordinary Portland cement samples.

8.2.2 According to the Aggressive Environment

Aggressive aqueous environments may include marine environments, sewers, underground structures, hydraulic structures, chemical plants, industrial structures, liquid-containing structures, and even aggressive (acid) rainfall. Different tests may be required according to the nature of the environment, or at the least the test results should be interpretable in terms of the relevant aggressive environment.

8.2.3 According to the Scale of the Test Method

A difference can be made between small scale laboratory tests, large scale simulation tests, and in situ tests. In the first two types of test method, additional parameters may be changed to accelerate the degradation (see Sect. 8.3). For instance, in the work of De Belie and Verstraete, when configuring a test method for biogenic sulfuric acid attack, first a test procedure for small concrete samples (20 × 20 × 50 mm) was developed (Vincke et al. 1999). Later, this test procedure was scaled up to obtain more representative results for concrete with large aggregates, for which samples were cylinders of 80 mm diameter and 15 mm thickness, taken some mm below the inner surface of the concrete pipes (De Belie et al. 2004). Investigating the degradation of slatted floors in animal houses by organic acids formed in acidified meal-water mixtures, first small concrete prisms

(40 × 40 × 80 mm) were used as test specimens (De Belie et al. 1996, 1997a, b). Later, an automated test procedure on larger concrete specimens was developed (De Belie et al. 1998, 2002). To validate the results, in situ degradation experiments were also carried out (De Belie 1997). The group of Mori, when simulating biogenic sulfuric acid attack, carried out laboratory scale tests in a simulation chamber on mortar specimens of 40 × 40 × 160 mm, as well as in a demonstration plant consisting of a pipe with diameter 150 mm and a length of 20 m (Mori et al. 1992).

The scale of the test method can have a significant influence on the test results, since it may affect factors such as the specimen surface area/liquid ratio, rate of replenishment of aggressive substances, presence of an interfacial transition zone (concrete vs. mortar specimens), use of real-life or simulated aggressive liquids, choice of an accelerated or close-to-reality test, etc. Research carried out on mortar or cement paste specimens cannot always be extrapolated to concrete (Cohen and Mather 1991). Because of the presence of aggregates, a transition zone may appear in concrete between the aggregates and the paste. Due to the transition zone effect, the microstructures of pure paste and paste within concrete are different. The effects of transition zones extend far beyond the aggregate surface and individual effects of surfaces in concrete overlap. Therefore paste in concrete generally has a different microstructure than pure paste.

8.2.4 According to Combination with Mechanical Action or Not

When only chemical action is present, a slowly growing layer of degraded material is often formed on the surface of the specimen and this can slow down further reactions. When the chemical action is combined with mechanical action, particularly abrasion, this will remove the degraded layer and leave a new surface for chemical attack; this may accelerate the degradation process. Common ways of exerting an abrasive action in laboratory tests are manual or automated brushing, and immersing in water that is shaken or stirred. The effect of mechanical action is also very clear in the case of sulfuric acid attack. When carrying out accelerated degradation tests on cylindrical test specimens in a 0.5% sulfuric acid solution, an alternating increase and decrease of the radius could be noticed, corresponding to alternating expansion of the concrete due to immersion and formation of reaction products and subsequent material loss due to brushing (Monteny et al. 2001).

8.2.5 According to the Method of Measuring Degradation

Different measures for quantifying degradation may be implemented in test procedures. Regarding experimental methods to determine sulfate resistance for instance, the Koch/Steinegger method (Steinegger 1970) implies measurement of the residual flexural strength, while Wittekindt (1960; ASTM C1012 2004) implement a criterion based on length changes.

In general, degradation measurements may include thickness change, mass loss, linear or volumetric expansion, residual strength, pH change of liquid, calcium released in liquid, hydrogen ion consumption, loss of elastic modulus, etc. These measurements may be supplemented by SEM or XRD analysis to examine the microstructure. Bertron et al. (2004) measured the altered depth visualised with phenolphthalein, in relation to the time of exposure, to investigate the kinetics of progression of the dissolution front for cement pastes immersed in a mix of five organic acids. Chemical modifications in different zones were characterised with an electron microprobe, analysing Ca, Si, Al, Fe, Mg, S, K, Na. Schneider and Chen (2005) also used phenolphthalein indicator to determine penetration depths of ammonium nitrate in concrete specimens subjected simultaneously to flexural loads. Maltais et al. (2004) used microprobe analysis together with SEM and XRD analyses to establish chemical composition of phases after degradation of Portland cement systems in deionised water and sodium sulfate solutions. Neuenschwander et al. (2006) and Schmidt et al. (2006) examined the use of a non-destructive method using ultrasonic surface waves (leaky Rayleigh waves) to quantify degradation of mortar samples in sulfuric acid solutions.

The choice of the degradation measure may lead to different conclusions regarding the relative performance of concrete types. For instance in microbiological tests, De Belie et al. (2004) used both mass loss and thickness change as parameters to judge degradation. Samples with limestone aggregates showed a somewhat higher mass loss than samples with inert aggregates, while their thickness change was lower. This effect could be partly explained by taking the density of the concrete constituents into account. For concrete with limestone aggregates, the removed material consisted of "concrete" (= aggregates + cement mortar) with a density of about 2,400 kg/m^3. For the concrete with inert aggregates, by contrast, the removed material consisted in this stage only of cement mortar with a density of around 2,000 kg/m^3. This partly explains why limestone concrete with a lower average attack depth, can still register a higher mass loss. This effect would probably disappear when degradation proceeds and large aggregates are removed from the matrix. The large differences between thickness and mass results could also suggest that concrete with limestone aggregates underwent a somewhat larger expansion than concrete with inert aggregates.

Therefore, one single measure may not suffice to characterise the degradation sufficiently. Mass loss, for instance, may well be used when no secondary products precipitate, otherwise it may be the result of a combination of several phenomena. In particular, mass gain may actually occur at the start of testing with degradation mechanisms such as acids and sulfates, due to the formation of secondary reaction products. It is therefore recommended to use multiple relevant indicators to investigate the resistance of concrete to a particular degradation mechanism (Cohen and Mather 1991; Bassuoni et al. 2006): for instance for sulfate attack, mass changes, expansion, dynamic modulus of elasticity and flexural strength can be monitored.

De Belie et al. (2004) combined measurements of thickness with an automated laser profilometer, mass loss, and calcium and sulfate concentrations in the microbial suspension, to quantify biogenic sulfuric acid degradation.

8.3 Acceleration of Degradation in Simulation Tests

The resistance of concrete to aggressive aqueous environments can be tested in different ways:

- Realistic concentrations of the aggressive agents can be used in combination with a sensitive method to detect deterioration and an extrapolation method used to predict the degradation in the future.
- Accelerated tests can be performed. The degradation rate can be increased by means of higher concentrations of the aggressive medium, higher temperature, greater contact surfaces, and so on.

The advantage of carrying out the experiments in a realistic way and keeping the factors responsible for the deterioration unchanged is the certainty that the deterioration process itself has not been changed. To be able to detect differences in degradation with methods close to reality within a reasonable time, sensitive methods are needed. Rombén (1978) determined a relationship between the acid consumption and the attack depth of concrete caused by hydrochloric acid attack. He measured the amount of acid needed to keep the pH of the solution constant and the amount of Ca^{2+} released in the solution as a function of the time. He also determined the Ca^{2+} content of the concrete. Combination of these data allowed him to estimate the depth of attack as a function of time.

A disadvantage of close-to-reality investigations is that only the very early stages of the attack can be measured. To estimate the attack of the concrete at a later stage an extrapolation of the obtained relationship has to be used. This implies the risk that new mechanisms of a type not noticeable during the measuring time, may arise in a later period of the deterioration. The rate of attack in the case of sulfuric acid probably follows relationships which are difficult to extrapolate, due to the special mechanism of expansion followed by material loss, which is characteristic for this acid (Rombén 1978).

One of the common ways to investigate the chemical resistance of concrete is to carry out accelerated tests in the laboratory. The advantage of these methods is that the entire life of the specimen in question can be simulated. An acceleration of the process can be achieved in different ways. The concentration of the aggressive solution can be increased and/or the reaction surface can be increased by using specimens with a large surface area-to-volume ratio. Also, physical effects such as temperature or wetting/drying cycles can be used. Several such factors affecting the rate and extent of attack are summarised below.

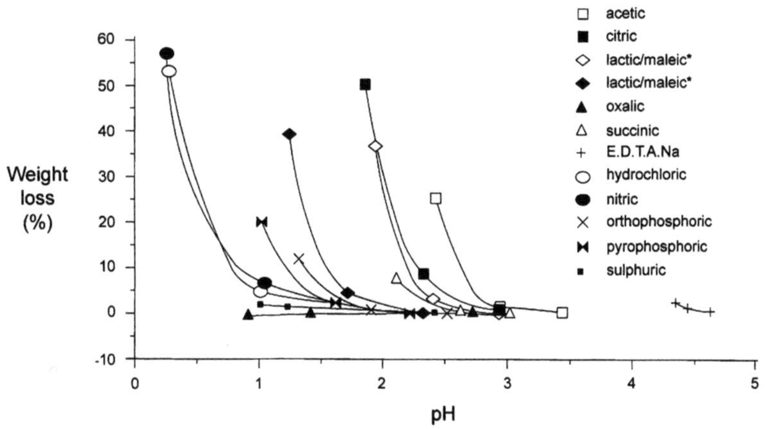

Fig. 8.1 Weight loss (%) of concrete samples as a function of the pH of the solution (Redrawn after Bayoux et al. 1990; *the difference between lactic and maleic acid is not clear in the original publication)

8.3.1 Strength of the Acid in the Attacking Solution

Strong acids are normally very aggressive (for a given concentration), because they decompose the cement hydrates easily. Nevertheless, some strong acids, such as phosphoric acid, are less aggressive than weak organic acids like citric, lactic or malic acid, since they do not dissociate readily.

8.3.2 pH and Concentration of the Attacking Solution

The pH value is very often used to define the concentration of an acid solution, because it is easier to measure pH than concentration. When the pH decreases from 7, the solution becomes more acidic and it will become more aggressive to concrete. Therefore it is of practical interest to define the acceptable limits of the pH without exaggerated corrosion. Figure 8.1 (Bayoux et al. 1990) shows that acceptable limits for a certain concrete depend on the acid, because of the differences in acid strength.

It is interesting to observe that the pH limits are lower for strong acids such as hydrochloric or nitric, than for weak acids such as lactic or acetic. This can be explained because weak acid solutions have a higher pH than strong acid solutions for a given concentration. For example, 0.1 N acetic acid has a pH of 2.9, while the pH of 0.1 N hydrochloric acid equals 1. Thus, for a given pH, strong acids have a lower concentration than weak acids and may be less aggressive.

De Belie et al. (2002) also found in experiments with the TAP (test equipment for accelerated degradation) that lactic/acetic acid attack on concrete was more

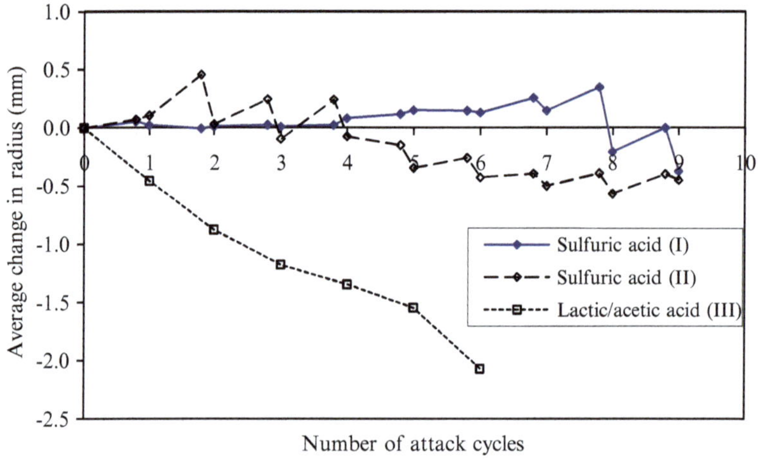

Fig. 8.2 Average change in radius vs. number of attack cycles for a TAP degradation test with a sulfuric or lactic/acetic acid solution, respectively (De Belie et al. 2002)

severe than sulfuric acid attack (see also Chap. 11). Figure 8.2 shows the average change in radius for concrete samples subjected to sulfuric or lactic/acetic acid attack. For sulfuric acid attack, the simulation liquid consisted of 0.5% sulfuric acid in water (pH = 0.8–1.0). The concrete mix proportions of reference specimen I were based on a mix design commonly used for sewer pipes, featuring a high sulfate resistant (HSR) Portland cement (CEM I 42.5 HSR/LA) and a w/c ratio of 0.40. For specimen II, 30 kg/m^3 silica fume was added. The lactic/acetic acid degradation was carried out with a simulation liquid consisting of lactic and acetic acid in water, both with concentrations of 30 g/L (pH = 2.0–2.2). The reference concrete specimen III had a mix design commonly used for production of prefabricated concrete slats, comprising an ordinary Portland cement (CEM I 42.5R) and a w/c ratio of 0.39.

For the same number of attack cycles (keeping in mind furthermore that the cycles for the sulfuric acid attack were twice as long), the average decrease in radius in the lactic/acetic acid solution was much more pronounced than for the sulfuric acid solution (−2.1 mm for concrete III, compared to +0.13 and −0.42 mm for concretes I and II, respectively, after six cycles). This was despite the pH of the sulfuric acid solution (pH = 0.8–1.0) being significantly lower than the pH of the organic acid solution (pH = 2.0–2.2).

Changing the concentration of the aggressive liquid in order to obtain acceleration of the degradation process in laboratory tests may result in unwanted effects. For instance, in the case of sulfate corrosion, changing the concentration could change the attack mechanism. Cohen and Mather (1991) warn that for sulfate attack, at low SO_4^{2-} concentrations (less than 1,000 mg SO_4^{2-}/L), deterioration of the concrete is mainly due to the formation of ettringite. At high concentrations the formation of gypsum is the main cause of deterioration. Cohen and Mather

(1991) conclude that in the case of sulfate attack, results obtained in the laboratory under accelerated test conditions by increasing sulfate concentration cannot be used to predict actual behaviour in the field.

Bertron et al. (this Report, Part I) advise to consider five parameters to compare the aggressiveness of different acids: (i) the solubility of the acid salts in water (mainly calcium salts), (ii) the acids' dissociation constants, (iii) the mono- or poly-acidity of the salts (iv) the constants of the organometallic complexes formed in solution, and (v) the physical properties of the salts.

A distinction between different laboratory tests can also be based on the pH being kept constant during the test (e.g. by automatic titration) or not. In the latter case the pH will usually increase during the test because of leaching of free lime and alkalis from the concrete; simulation liquids can then be changed at certain intervals. Fourie and Alexander (2007) found that allowing the pH to increase from 1.0 to 1.5 during a test with hydrochloric acid, resulted in a significantly lower mass loss and hydrogen ion consumption compared to the situation when the pH was controlled between 1.00 and 1.05 or between 0.95 and 1.00.

Bertron et al. (2005) investigated the effect of pH on mineralogical and chemical modifications occurring in the cementitious matrix of ordinary Portland cement and slag cement pastes submitted to organic acid mixes. They wanted to assess if an aggressive solution of pH 4 used to accelerate alteration kinetics would not change the degradation mechanism in comparison to a solution with a pH of 6, which is more realistic in the case of liquid manure attack. They found that the kinetics of alteration (mass loss and altered depths) was nine fold higher in the solution with a pH of 4, but that the alteration mechanisms were sensibly identical. The modifications in both cases included decalcification, disappearance of crystallized and amorphous hydrated phases, and probable formation of silica gel containing aluminium and iron. Minor alteration mechanism differences mainly concerned the stability of anhydrous phases such as C_4AF which is stable at pH of 6, but not at a pH of 4. Also a slight decalcification of slag anhydrous phases was seen at a pH of 4.

8.3.3 *Anion/Cation Combined with the Aggressive Ion in a Salt Solution*

The anion or cation combined with the aggressive ion in a salt solution will partly determine the aggressiveness. Some examples are:

- Magnesium ions are particularly detrimental in combination with sulfate ions. Apart from the effect of Mg^{2+} itself, the reaction produces calcium sulfate, which can cause further damage by formation of expansive ettringite.
- Magnesium sulfate has a much more damaging effect than either calcium or sodium sulfate.

- The anion combined with NH_4^+ is also important, as it determines the solubility of the calcium salt produced. Ammonium oxalate, ammonium acetate and ammonium carbonate have a limited aggressiveness. Ammonium bicarbonate and nitrate are more harmful and the anion of ammonium sulfate and chloride results in an extra detrimental effect.
- The cation combined with NO_3^- determines the aggressivity of the nitrate. A 10% calcium or alkali nitrate solution would not attack concrete, while aluminium nitrate will. Ammonium nitrate would be most harmful.
- Chlorides which produce unstable and water soluble compounds when reacting with $Ca(OH)_2$, for example $MgCl_2$, $CaCl_2$, NH_4Cl, $FeCl_3$, $CuCl_2$ and $AlCl_3$, could damage concrete. Other chlorides, such as NaCl and KCl, which do not react with lime would only degrade concrete in very high concentrations.

8.3.4 Physical State of the Attacking Medium

Dry, non-hygroscopic solids do not attack dry concrete, but some will attack moist concrete. A moist, reactive solid can attack concrete, as can aggressive liquids and solutions. Dry gases, if aggressive, may come into contact with sufficient moisture within the concrete to make attack possible. Moist, aggressive gases would tend to be more destructive.

8.3.5 Temperature

Temperature may affect the rate of attack in two different ways. The common effect is that chemical activity usually increases exponentially, approximately doubling with each 10°C rise in temperature. Temperature may also affect the rate of attack indirectly. As temperature rises, the moisture content of the concrete is reduced, making it drier, but more permeable to additional fluid. As temperature falls, it may sometimes cause sufficient thermal shrinkage to open small cracks and allow greater penetration of liquid into the concrete.

8.3.6 Replenishment of the Aggressive Medium

For example, a concrete structure may be placed in an acidic soil, but if the acidity is not replenished, the available acid may be quickly neutralized with little or no damage to the concrete.

8.3.7 Alternate Wetting and Drying

In some cases the process can be accelerated by means of wetting and drying cycles which allow uptake of aggressive agents through convective processes, being an order of magnitude faster than diffusion. De Belie et al. (2002) applied an alternate wetting and drying procedure in their TAP equipment, by mounting concrete cylinders on rotating horizontal axes (about 1 revolution per hour) and rotating them through containers with simulation liquids. Cycles of chemical attack were followed by abrasion using rotary brushes. The change in dimension of the concrete specimens and the surface roughness were determined through a non-contact distance measurement with laser sensors. Alternate wetting and drying may also increase the crystallisation pressures. Dissolved substances may migrate through the concrete and deposit at or near the surface from which evaporation occurs. The deposit may be the original substance or a reaction product formed in the concrete. This effect can be seen in the familiar 'efflorescence' on walls of concrete, brick or stone. Progressive deposition can be disruptive.

8.3.8 Alternate Heating and Cooling

Alternate heating and cooling may be harmful. Freezing and thawing can damage concrete through physical effects.

8.3.9 Pressure

Pressure is seldom of consequence in its effect on chemical reaction rates, but can greatly increase the degree to which attacking substances penetrate the concrete. Partial pressures of aggressive gases may be of importance, because they are a measure of concentration.

8.4 Other Important Points

Other important points to be considered include the effect of the age of test specimens on the results of accelerated tests. For instance, considering concrete with fly ash compared to ordinary Portland cement concrete, it may be that the resistance in an accelerated test is higher for OPC concrete at the age of 28 days, while it is higher for fly ash concrete after some years. Under real conditions the aggressive species may exert their effect over a certain time period. Concentrating the effect at a certain age of the specimens may favour some concrete types over

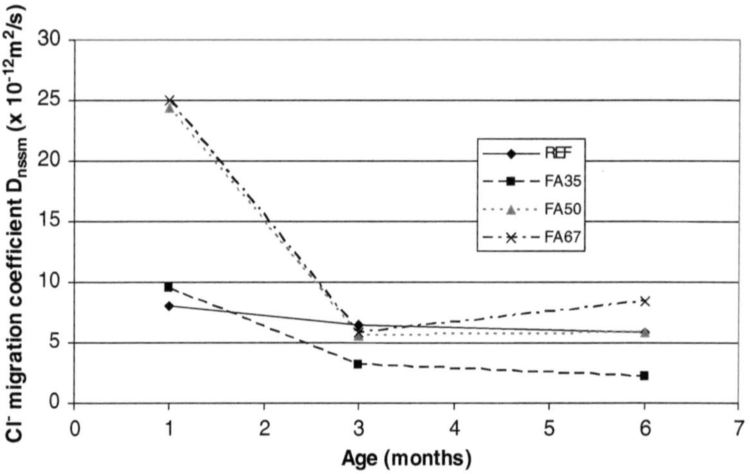

Fig. 8.3 Chloride migration coefficients of concrete specimens in which 0, 35, 50 and 67% of the cement was replaced with type F fly ash, versus age at the start of the test (Lammertijn 2007)

others. Figure 8.3 for instance (Lammertijn 2007) shows the non steady-state chloride migration coefficient D_{nssm} in a CTH test (NT Build 492 1999) for four concrete types with 400 kg/m^3 cementitious material (CM) and a w/cm ratio of 0.4 prepared according to NBN B15-001 (2004). The reference concrete (REF) was made with ordinary Portland cement CEM I 52.5 N; in FA35, FA50 and FA67, respectively 35, 50 and 67% of the cement was replaced with type F fly ash (1.96% CaO). The concrete was stored at 20°C and more than 90% relative humidity until the age of testing. CTH tests were carried out at the age of 1, 3 and 6 months. At 1 month, chloride migration coefficients are much higher for FA50 and FA67 than for REF and FA35, whereas at 3 months they become similar to the reference concrete.

Also the choice of the reference should be considered. Often test results are relative results, in comparison with a reference concrete. The choice of this reference is therefore of great importance. For example, the relative sulfate resistance of a certain concrete type may be completely different when compared to a reference with ordinary Portland cement or with high sulfate resistant (Portland) cement. Factors such as mix procedure, curing, specimen size, etc. should always be detailed.

8.5 Conclusions

The choice of a test method to rate relative performance of different cementitious materials in aggressive aqueous environments will have a major effect on the nature of the degradation mechanisms and the usefulness of the test results. Parameters such as the scale of the test method, physical state of the attacking medium, the pH and concentration of the solution, temperature, rate of replenishment, mechanical action, alternate wetting and drying, alternate heating and cooling, pressure, etc. should be carefully selected and should always be reported together with test results. Also the choice of degradation measure may lead to different conclusions regarding relative performance of concrete types. Often a combination of multiple relevant indicators will be necessary. Sample preparation procedures and the concrete age at the time of testing are also of utmost importance. The choice of the reference should be considered, since test results are often in comparison with a reference concrete.

In the following chapters, an overview is presented of test methods related to some of the most important degradation mechanisms in cementitious materials. After a discussion of test methods for degradation in soft or carbonated waters, the methodology for testing sulfate resistance is critically reviewed. Tests in which inorganic acid and biogenic sulphuric acid are the aggressive agents are the topic of the next two chapters. Chlorides can also alter the products of cement hydration and cause other forms of concrete deterioration, therefore test methods for chloride transport are discussed in the next chapter. The last two chapters focus on test methods for attack by magnesium salts, and by organic acids.

References

ASTM C1012-04: Standard test method for length change of hydraulic-cement mortars exposed to a sulfate solution (2004)

Bassuoni, M.T., Nehdi, M.L., Greenough, T.R.: Enhancing the reliability of evaluating chloride ingress in concrete using the RCPT ASTM C 1202 rapid chloride penetrability test. J. ASTM Int. **3**(3), 1–13 (2006)

Bayoux, J.P., Letourneux, J.P., Marcdargent, S., Verschaeve, M.: Acidic Corrosion of High Alumina Cement. Seminar High Alumina Cement, London (1990)

Bertron, A., Escadeillas, G., Duchesne, J.: Cement pastes alteration by liquid manure organic acids: chemical and mineralogical characterization. Cem. Concr. Res. **34**(10), 1823–1835 (2004)

Bertron, A., Duchesne, J., Escadeillas, G.: Accelerated tests of hardened cement pastes alteration by organic acids: analysis of the pH effect. Cem. Concr. Res. **35**(1), 155–166 (2005)

Cohen, M.D., Mather, B.: Sulfate attack on concrete – research needs. ACI Mater. J. **88**(1), 62–69 (1991)

De Belie, N.: On-farm trial to determine the resistance of different concrete slats for fattening pigs. J. Agric. Eng. Res. **68**(4), 311–316 (1997)

De Belie, N., Verselder, H.J., De Blaere, B., Van Nieuwenburg, D., Verschoore, R.: Influence of the cement type on the resistance of concrete to feed acids. Cem. Concr. Res. **26**(11), 1717–1725 (1996)

De Belie, N., Debruyckere, M., Van Nieuwenburg, D., De Blaere, B.: Attack of concrete floors in pig houses by feed acids. Influence of fly ash addition and cement-bound surface layers. J. Agric. Eng. Res. **68**(2), 101–108 (1997a)

De Belie, N., De Coster, V., Van Nieuwenburg, D.: The use of fly ash or silica fume to increase the resistance of concrete to feed acids. Mag. Concr. Res. **49**(181), 337–344 (1997b)

De Belie, N., Verschoore, R., Van Nieuwenburg, D.: Resistance of concrete with limestone sand or polymer additions to feed acids. Trans. ASAE **41**(1), 227–233 (1998)

De Belie, N., Monteny, J., Taerwe, L.: Apparatus for accelerated degradation testing of concrete specimens. Mater. Struct. **35**(251), 427–433 (2002)

De Belie, N., Monteny, J., Beeldens, A., Vincke, E., Van Gemert, D., Verstraete, W.: Experimental research and modelling of the effect of chemical and biogenic sulfuric acid on different types of commercially produced concrete sewer pipes. Cem. Concr. Res. **34**(12), 2223–2236 (2004)

Fourie, C., Alexander, M.G.: A dynamic hydrochloric acid test for evaluating acid resistance of sewer pipe concrete. In: Workshop Performance of Cement-based Materials in Aggressive Aqueous Environments, Ghent (2007)

Lammertijn, S.: Carbonation and chloride penetration in high-volume fly ash concrete (in Dutch). M.Sc. thesis, Faculty of Engineering, Ghent University, 186 p (2007)

Maltais, Y., Samson, E., Marchand, J.: Predicting the durability of Portland cement systems in aggressive environments – laboratory validation. Cem. Concr. Res. **34**(9), 1579–1589 (2004)

Monteny, J., De Belie, N., Vincke, E., Verstraete, W., Taerwe, L.: Chemical and microbiological tests to simulate sulfuric acid corrosion of polymer-modified concrete. Cem. Concr. Res. **31**(9), 1359–1365 (2001)

Mori, T., Nonaka, T., Tazaki, K., Koga, M., Hikosaka, Y., Noda, S.: Interactions of nutrients, moisture and pH on microbial corrosion of concrete sewer pipes. Water Res. **26**(1), 29–37 (1992)

NBN B15-001: Supplement to NBN EN 206-1 – Concrete – Specification, Performance, Production and Conformity. NBN, Brussels (2004)

NBN EN 206-1: Concrete – Part 1: Specification, Performance, Production and Conformity. NBN, Brussels (2001)

Neuenschwander, J., Schmidt, T., Lüthi, T., Romer, M.: Leaky Rayleigh wave investigation on mortar samples. Ultrasonics **45**(1–4), 50–55 (2006)

NT Build 492: Concrete, Mortar and Cement Based Repair Materials. Chloride Migration Coefficient from Non-Steady-State Migration Experiments. Nordtest, Espoo (1999)

Rombèn, L.: Aspects on testing methods for acid attack on concrete. CBI Forsk. **78**(1), 1–61 (1978)

Schmidt, T., Neuenschwander, J., Romer, M., Lüthi, T.: Investigation of physical and microstructural changes in the surface of mortar samples exposed to sulphuric acid. In: Malhotra, V.M. (ed.) Proceedings of the 7th CANMET/ACI International Conference on Durability of Concrete, Montreal, SP 234 (2006)

Schneider, U., Chen, S.W.: Deterioration of high-performance concrete subjected to attack by the combination of ammonium nitrate solution and flexure stress. Cem. Concr. Res. **35**(9), 1705–1713 (2005)

Steinegger, H.: Prüfung und beurteilung der sulfatwiderstandsfähigkeit van zementen. Zement-Kalk-Gips **23**(2), 67–71 (1970)

Vincke, E., Verstichel, S., Monteny, J., Verstraete, W.: A new test procedure for biogenic sulfuric acid corrosion of concrete. Biodegradation **10**(6), 421–428 (1999)

Vincke, E.: Biogenic sulfuric acid corrosion of concrete: microbial interaction, simulation en prevention. Ph.D. thesis, Faculty of Agricultural and Applied Biological Sciences, Ghent University, 208 p (2001)

Wittekindt, W.: Sulfatbeständige zemente und ihre prüfung. Zement-Kalk-Gips **13**(2), 565–572 (1960)

Chapter 9
Tests for Leaching and Degradation in Soft or Carbonated Waters

Marta Castellote

Contents

Keywords Leaching • Soft water • Carbonated water • Extractions tests • Dynamic tests • Batch tests • Flow tests • Accelerated methods

9.1 Introduction

Leaching can be defined in a general manner as the extraction of certain materials from a solid by a liquid. In a porous solid, leaching is a combined diffusion-dissolution/precipitation process where the concentration gradients between the pore solution and the pure water cause diffusion of the extracted materials from the pore solution to the surrounding water. The reduction in concentration of the targeted species in the pore solution forces the dissolution of more material.

Specifically, in the case of concrete, there are two fields in which leaching is important:

M. Castellote (✉)
Instituto de Ciencias de la Construcción "Eduardo Torroja", Madrid, Spain
e-mail: martaca@ietcc.csic.es

M. Alexander et al. (eds.), *Performance of Cement-Based Materials in Aggressive Aqueous Environments*, RILEM State-of-the-Art Reports 10,
DOI 10.1007/978-94-007-5413-3_9, © RILEM 2013

- On the one hand, leaching refers to the degradation process caused by the decalcification of concrete due to the dissolution of phases of the hydrated cement paste, mainly portlandite and C–S–H exposed to water, with a composition varying from pure water to salty water such as seawater. The consequences of leaching are an increase of the porosity and permeability, and a loss of mechanical strength. This phenomenon affects structures in contact with water for long periods of time: dams, water pipes, etc. In recent decades, leaching has been identified as very relevant in the durability and service life prediction of nuclear waste containers.
- On the other hand, leaching is the main process for contaminants, stabilised and/or solidified within a solid matrix, to enter the environment. This phenomenon affects the confinement of toxic and radioactive elements in cementitious or pozzolanic binder materials, and on the potential emission of dangerous substances from the material to the environment. Information on leaching referring to this field can be found in USEPA (2003a, b and c) and Van der Sloot (2007).

This report considers the former point on degradation of the concrete itself; however, it is very important to be aware of the fact that most research on leaching and most of the commonly used standard tests have been developed within the framework of the second field. The reader is also referred to the corresponding Chap. 4 on this subject in Part I of this Report.

9.2 General Considerations Regarding Leaching of Concrete

When the cement paste comes in contact with water, the solid phases which are stable at the alkaline pH of the concrete are no longer in thermodynamic equilibrium and get dissolved depending on their solubility properties (Berner 1992; Reardon 1992; Matschei et al. 2007). The dissolution-induced alteration is considered to be one of the major factors that alter the physical properties of the cement paste.

Traditionally, calcium is used as an indicator of chemical deterioration because it is the main element of hydrates and plays an essential role in the chemistry of the cement. Leaching by deionized water induces calcium and hydroxide concentration gradients which continuously decrease from the sound zone to the exposed surface of the material. This causes the diffusion of calcium and hydroxide ions from the pore solution to the aggressive solution, and thus lowers the amount of calcium concentration in the pore solution (Adenot and Buil 1992; Faucon et al. 1997; Haga et al. 2005; Alonso et al. 2006).

The altered material can be seen as a layered system composed of an unaltered core and different zones separated by dissolution or precipitation fronts and progressive decalcification of C–S–H (Adenot and Buil 1992; Andac and Glasser 1999; Glasser et al. 2008). The degraded zone induced by water exposure is characterized by a decalcification of C–S–H inducing silicate polymerization.

Decalcification changes the bulk density and the pore structure of the hydrated cement paste leading to an increase of the porosity, permeability and a loss of mechanical strength. The increase of pore volume is larger when there is a higher initial amount of $Ca(OH)_2$ (Haga et al. 2005), which can be attributed to the dissolution of $Ca(OH)_2$, while the porosity created by C–S–H decalcification is almost negligible (Carde et al. 1996; Mainguy et al. 2000).

Loss of calcium leads to the dissolution of portlandite and secondary precipitation of AF_m, ettringite and calcite (Faucon et al. 1997, 1998a, b). The precipitation of these minerals takes place in the innermost part of the degraded zone while they are dissolved in the outermost part of the altered zone (Faucon et al. 1997, 1998a, b). Some authors have pointed out the important role that other elements such as Fe, Al, sulfates and Mg could play in this process (Faucon 1997; Porteneuve 2001; Llorente et al. 2009a, b). It has been postulated that one or several species could act as remediator of the damaged C–S–H gel stabilising the structure generated by the loss of calcium.

9.3 Variables in the Leaching Tests

In general, it can be said that leaching tests are conducted to examine mass transfer of "leachate" from the solid to the liquid, termed the "leachant". As the different tests are defined by different experimental variables, the choice of the method has an effect on the results (Alonso et al. 2006; Llorente et al. 2009b). Therefore, it is recommended to perform a range of leaching tests in order to completely understand the leaching behaviour of a material (Stegemann and Coté 1990; DD ENV 12920 1998) and/or try to reproduce the mechanisms of the expected specific scenarios of the concrete degradation (Alonso et al. 2006).

Classification of leaching tests is a complex issue due to the high number of variables involved. There are several key variables that distinguish leaching tests (WTC, Wastewater Technology Center 1990; La Grega et al. 1994; van der Sloot et al. 1997, among others), and that are summarised in (Perera et al. 2004) as the following:

- Sample preparation
- Leachant characteristics (pH and composition)
- Mode and method of contact
- Liquid-to-solid ratio (L/S)
- Leachant renewal
- Contact time
- Temperature
- Leachate-solid separation
- Analysis of the leachate

Each combination of these variables in a more or less sophisticated experimental set-up makes a different leaching method. Therefore, a wide variety of different tests has been developed and found in bibliography, some of them as standard methods (in a regional, national or continental coverage, and mainly developed for wastes) and others varying the parameters according to the author's ideas in order to adapt to a specific scenario as closely as possible.

9.4 Classification and Description of Leaching Methods

All these methods can be classified according to different criteria; the most commonly used classifications are those based on renewal of the leachant, testing hierarchy, and the difference between "natural" and "accelerated" methods.

9.4.1 Classification According to Renewal of the Leachant

According to the renewal of the leachant, leaching tests can be classified as (WTC 1990; DOE 2003):

- Extraction tests
- Dynamic tests

This classification is based on whether the leachant is renewed (in the case of the latter) or not (the former).

9.4.1.1 Extraction Tests

Extraction tests include all tests that bring a specific amount of leachant in contact with a specific amount of material for a specific amount of time (WTC 1990), usually agitated and using a powdered or crushed sample.

The assumption made when conducting a single extraction test is that a steady-state condition (equilibrium) is achieved by the end of the testing period, though this may not necessarily be the case in practice. Reaching equilibrium in single extraction leach testing is critical to predicting leaching behaviour over long periods of time (EQM 1998). Therefore, the objective of this kind of test is to obtain the maximum amount of extracted species without any rate-limiting mechanisms, thus eliminating or decreasing as much as possible the physical barriers to mass transport (Perera et al. 2004).

Single extraction leaching tests, among others, include the following standard methods (DOE 2003) that have been used for testing degradation in cementitious matrices:

- ASTM D 3987, Standard Test Method for Shake Extraction of Solid Waste with Water
  This method employs water as the extraction fluid. A water/solid waste mixture (L/S = 20:1, volume to mass {v/m}) is mechanically agitated for 18 h using a device that rotates sample containers on a central axis at a rate of 29 rotations/min. After rotation, the aqueous phase is separated from the solid material (decantation followed by filtration) for analysis. Particle size reduction is not required.
- NEN 7341, Dutch Total Availability Leaching Test
  The availability test (NEN 7341) is an agitated extraction test performed at two controlled pH values, pH = 4 and pH = 7. In this test, a finely ground sample (<125 μm) is leached at a 50:1 L/S (v/m) ratio for 3 h.
- CEN/TS 14429:2005 Characterization of waste - Leaching behaviour test – Influence of pH on leaching with initial acid/base addition
  Separate test portions are leached at L/S equal to 10 for 48 h with leachants containing pre-selected amounts of acid or base in order to reach stationary pH values at the end of the extraction period. Each leachant is added in three steps in the beginning of the test. At least eight final pH values are required, covering at minimum the range from pH 4 to 12 (both included i.e. the lowest value = 4 and the highest value = 12). The tests are carried out at a fixed contact time at the end of which equilibrium condition can be assumed to be reached for most constituents in most waste materials to be characterized. The equilibrium condition as defined in the standard is verified at the end of the extraction period.
  The standard prEN 14429 has been developed from the NEN 7341 and the so called Acid Neutralization Capacity (ANC) tests (Stegemann and Coté 1991). In the ANC method, 11 sub-samples of ground waste are extracted for 48 h, each with an increasing amount of nitric acid. The pH values of the extracts are measured and plotted as a titration curve. In Hidalgo et al. (2001), nitric acid is continuously added in an inert atmosphere, evolution of pH is recorded by titration and liquid and solid phases are taken at different pHs corresponding to progressive degradation stages. At the time of writing this report, there is a technical specification: CEN/TS 14997 – pH dependent leaching test (continuous pH control), in publication (www.leaching.net).
- EN 12457/1-4, Compliance Test for Granular Waste Materials and Sludges
  This test consists of four procedures:

  - Part 1 – One-stage batch test at a L/S ratio of 2 L/kg (v/m) for materials with high solid content and with particle size <4 mm (with or without size reduction)
  - Part 2 – One-stage batch test at a L/S ratio of 10 L/kg (v/m) for materials with particle size <4 mm (with or without size reduction)

- Part 3 – Two-stage batch test at a L/S ratio of 2 L/kg (v/m) and 8 L/kg (v/m) for materials with high solid content and with particle size <4 mm (with or without size reduction)
- Part 4 – One-stage batch test at a L/S ratio of 10 L/kg (v/m) for materials with particle size <10 mm (with or without size reduction)

9.4.1.2 Dynamic Tests

Dynamic tests include all tests that continuously or intermittently renew the leachant to maintain a driving force for leaching and generate information as a function of time providing information about the kinetics of contaminant mobilization (WTC 1990).

There are three primary types of dynamic tests (DOE 2003):

- Serial batch tests
- Flow-around tests
- Flow-through tests

(a) Serial batch tests

In a serial batch test, a portion of a granular sample is mixed with the leaching fluid and agitated at a set L/S ratio for a specified period of time. The leachate is then separated from the solids and replaced with a fresh leaching fluid until the desired number of leaching periods has been completed. Data from these tests can be used to infer temporal release of leachable constituents. Examples of standardised serial batch tests are:

- ASTM D 4793, Standard Test Method for Sequential Batch Extraction of Waste with Water

This method employs water as the extraction fluid. The procedure calls for ten sequential extractions of the same sample of waste with water followed each time by separation of the aqueous phase for analysis. The final pH of the leachate is to reflect the interaction of the leaching fluid with the buffering capacity of the waste. 100-g sample (on a dry weight basis) is extracted at a L/S ratio of 20–1 (v/m). The sample/water mixture is constantly agitated for approximately 18 h and then filtered using a pressure filtration device equipped with a 0.45 or 0.8-μm filter. The damp solid material is transferred back to the extraction vessel and re-extracted with a fresh volume of water. The procedure is performed ten times in sequence on the same sample to generate ten aqueous solutions. At least the first four extraction sequences must be conducted without interruption.

(b) Flow-around tests

Flow-around tests, as their name implies, are usually applied to monolithic samples. The sample is placed in the test vessel and leaching fluid is added. Leachant is renewed at specified intervals and the driving force to maintain leaching is by diffusion. A draft for a new EN flow-around test method, is presently in development. Examples of existing standard methods of flow-around test are:

- NEN 7345, Tank Leach Test

A monolithic specimen is subjected to leaching in a closed tank to evaluate surface area-related release. The leaching fluid (demineralised water) is renewed after 8 h and 1, 2, 4, 9, 16, 36, 64 days using a leaching fluid with a volume of five times the volume of the material to be tested (i.e., a L/S of 5:1, v/v).

- ANSI/ANS-16.1, Measurement of the leachability of solidified low level radioactive wastes by short-term test procedure (1986).

A monolithic specimen is subjected to leaching in a closed tank to evaluate surface area-related release. The leaching fluid (demineralised water) is renewed after 2, 7, 24, 48, 72, 96, 120, 456, 1,128 h and 90 days using a leaching fluid with a volume of ten times the volume of the material to be tested.

(c) Flow-through tests
In a flow-through test, leaching fluid is passed, either intermittently or continuously, through an open container packed with a porous solid sample of either monolithic or granular material; leaching is produced under advective conditions. The leachate is periodically sampled and analyzed for the parameters of interest. The results are used to examine contaminant release over time and as a function of L/S ratio. Examples of standard flow-through tests given in (DOE 2003) are:

- ASTM D 4874, Standard Test Method for Leaching Solid Waste in a Column Apparatus

This generates aqueous leachate from a solid using a column apparatus. The maximum particle size for the column procedure is 10 mm. This procedure uses reagent water in a continuous up-flow mode to leach a sample of a solid material.

- NEN 7343, Column Test

In the NEN 7343 column test, demineralised water, adjusted to pH 4, is passed upward through a column of ground waste (<4 mm). Seven consecutive leachate fractions are collected, each representing an L/S ratio within the range of 0.1–10 L/kg (v/m). Total test duration is approximately 21 days.

- prEN 14405, Upflow percolation test

This test provides for sequential flushing of a column packed with granular (<4 mm) material with acidified water at increasing liquid to solid ratios (L/S = 0.1–10, v/m). The column is eluted by pumping the leaching fluid from the bottom of the column to the top, minimizing the creation of channels and column plugging.

9.4.2 Classification Based on a Testing Hierarchy

Another classification is that made by CEN TC 292 "Characterisation of waste". Here a classification is made based on the hierarchy in testing, coupling a generic approach at characterisation level with compliance or verification methods at material specific level, with the following categories (Van der Sloot and Dijkstra 2004):

- *Characterization tests*: Used for basic characterization of the release behaviour of the material.
- *Compliance tests*: with the purpose to "check" whether a material (still) complies with the behaviour of a reference material (tested with a characterization test) and/or complies with regulations.
- *On-site verification/Quality control tests* have the purpose to determine quickly (within a short time) if a material (or conditions) complies with earlier determined or expected behaviour in its practical application.

From this classification, developed for dangerous substances, almost the only category used for degradation of concrete is the characterization tests, the first step in the CEN TC 292 test hierarchy. The methods include one representative of each group of the previous classification criteria with the exception of the serial batch tests: Upflow percolation test, (PrEN 14405); Tank test (Method of the type of NEN 7345); and pH dependence test (PrEN 14429).

A summary of the main characteristics of the cited standard methods is given in Table 9.1, again pointing out that these methods are only a sample of all the existing leaching methods, standardised and non-standardised.

As mentioned, apart from the standard tests, other non-standardised leaching tests can be found in the literature, which have been developed in order to study some specific aspect of the degradation of concrete; some of these introduce small variations or are an adaptation of the standards to a particular scenario for concrete. The most typical modification is based on the attempt to accelerate the process.

9.4.3 Classification Based on the Difference Between "Natural" and "Accelerated" Methods

Leaching of concrete can be considered a slow process. Lagerblad (1999) observed 5–10 mm leached depth in concretes exposed for 100 years to stagnant natural water with anhydrous grains that had hydrated during leaching, inducing a densification of the material with new Portlandite formed from this re-hydration. Therefore, the interest in developing and evaluating the changes that take place when concrete is submitted to accelerated leaching tests has increased, especially in relation to long-term radioactive waste storage.

Table 9.1 Summary of the main characteristics of the cited standard leaching test methods

Method	Title	L/S Vol/mass	Leachant (pH)	Duration	Sample
Extraction tests					
ASTM D 3987	Standard Test Method for shake extraction of solid waste with water	20:1	Water	18 h	Granulated
NEN 7341	Availability Test	50:1	Water acidified with HNO_3 to pH 4 and 7; pH at constant value	3 h per step	<125 μm
prEN 14429: 2002	Characterization of waste – leaching behaviour tests – Influence of pH on leaching with initial acid/base addition	10 (l/kg)	at least eight different final values of pH covering the minimum range of 4–12	48 h	
EN 12457/ 1-4	Compliance Test for Granular waste materials and sludges	2 L/kg and 10 L/kg	Water		<4 and <10 mm
Dynamic tests: Serial Batch Tests					
ASTM D 4793	Standard Test Method for sequential batch extraction of waste with water	20:1	Water	18 h	No reduction required
Dynamic tests: Flow-around Test					
NEN 7345	Tank Leach Test	5:1 (v/v)	Reagent water acidified to pH 4 with HNO_3	8 h and 1, 2, 4, 9, 16, 36, 64 days	Monolith >40 mm
ANSI/ ANS-16.1-1986	Measurement of the leachability of solidified low-level radioactive wastes by a short-term procedure	10:1	Water	2, 7, 24, 48, 72, 96, 120, 456, 1,128 h and 90 days	Monolith
Dynamic tests: Flow-through Tests					
ASTM D 4874	Standard Test Method for Leaching Solid Waste in a Column Apparatus		Water	Days to months	10 mm

(continued)

Table 9.1 (continued)

Method	Title	L/S Vol/mass	Leachant (pH)	Duration	Sample
NEN 7343	Column Test	Seven fractions: L/S range of 0.1–10 L/kg	Reagent water acidified to pH 4 with HNO_3	21 days	95% <4 mm
prEN 14405	Upflow percolation test	0.1–10	Acidified water		<4 mm

Therefore, another typical classification is that of "natural" versus "accelerated" methods. Even though the dividing limit is not clear, as in specific natural environments some of the conditions can be quite severe. In general, in the category of accelerated methods, the following tests are included:

- The extraction tests, as defined previously
- Methods that stress one or several of the variables involved (pH, composition, time of contact, temperature etc.)
- Methods changing the driving force for leaching.

The accelerated methods are being used increasingly, in spite of their limitations, to study the mechanisms that take place in the cement paste and their correlation with natural degradation processes.

9.4.3.1 Acceleration by Use of Acid Solutions

The majority of the accelerated procedures are carried out with strongly acidified solutions instead of deionised water. Ammonium nitrate solution, at different concentrations (6 M is the most commonly used one), is currently used to cause the degradation of cementitious materials (Lea 1965) since this aggressive salt is supposed to only accelerate the rate of calcium leaching by intensifying the chemical gradient, without interacting with the specific leaching-based phenomena, i.e., portlandite $Ca(OH)_2$ dissolution and C–S–H decalcification (Carde et al. 1997). The associated ions, nitrates, do not interact with the cementitious matrix, and the pK_a value of the acidic-basic couple NH_4^+/NH_3 (9.25) is under the pH of portlandite dissolution (12.5). A concentrated NH_4NO_3 solution maintains its pH around its pK_a value (Perlot et al. 2009). Nowadays, an increasing number of researchers are using ammonium nitrate solution in order to assess leaching ability of concrete mixes, among others, Schneider and Chen (2003), Perlot (2005), Kamali et al. (2008), Perlot et al. (2009). Moudilou et al. (2002) used a combination of an acidic solution (HNO_3 solution with pH equal to 5) and a possibility to vary the temperature in the 20–40°C range in a dynamic leaching system. Furthermore, they integrated a boiler in their system in order to concentrate the leachate so that concentrations of leached trace metals would be sufficient for precise analytical determination. Nitric acid is also used in the leaching test proposed by the GEF8 working group of the French concrete standardization committee. They select the pH value in the test, based on the required exposure class according to EN 206–1: pH = 5.5 for XA1, pH = 4.5 for XA2 and pH = 4.0 for XA3. The performance of the concrete is judged based on the amount of leached calcium ions per unit area of concrete and the mean depth of the coloration front obtained by spraying phenolphthalein indicator solution (Jacquemot 2009) (see also Chap. 5 in Part I of this Report).

9.4.3.2 Acceleration by Use of Electrical Fields

Among the accelerated tests that change the driving force for leaching, one very promising technique is the use of electrical fields applied to hardened concrete. Concrete has an internal pore network partially or completely filled with an aqueous phase having different ions. When it is subjected to an external electrical field, it behaves as an electrolyte and ions are able to move driven by the electrical field, causing the disruption of the equilibrium of the solid phases with the pore solution. This leads, among other effects, to an acceleration of the dissolution of cement hydrates. The preliminary work concerning this acceleration method is described by Kuroi and Sueyoshi (1987). Later, Saito et al. (1992) applied a voltage drop of 25 V to mortar specimens and obtained dissolution of Ca^{2+} with a rate approximately 60 times higher than that of a system in which water was renewed from time to time under no potential gradient. They also stated, through XRD of the specimens, that the deterioration mode of the hydrates was the same with the external potential applied as without it. Since then, several authors have applied this electrical acceleration processes (Gerard 1996; Faucon et al. 1998a, b, Le Marechal et al. 1998; Saito and Nakane 1999; Gerard et al. 1999; Saito and Deguchi 2000; Castellote et al. 2002, 2003, 2004a, b) which improved the understanding of the effects of these treatments.

Concerning the comparison of the migration method with the other "non-accelerated" methods, Gerard (1996) found similar mechanisms as in some acid scenarios with acceleration factors between 50 and 100 with deionised water and higher than 500 with a buffer solution and electrical field of 1,300 V/m strength over the ends of the specimen. Castellote et al. (2003), testing ultra high performance concrete, found that some of the mixes behaved in a similar way when submitted to an electrical field and when they were leached according to the standard ANSI/ANS-16.1-1986. The acceleration factor was such that 5.5 Coulombs/cm^2 of charge passed was equivalent to 24 ± 2 years in an ANSI test. However, other mixes seemed to degrade through a different mechanism, and it was not possible to establish the same correlation between charge passed and time for every sample.

9.4.3.3 Acceleration by Use of Hydraulic Pressure

A further type of acceleration which changes the driving force for leaching is the use of hydraulic pressure. In other words, a flow of water is forced under pressure through the sample. It corresponds to the method for measuring the water permeability of concrete, collecting the fluid passing through the sample as leachate and analysing the residual microstructure of the solid (Ballivy et al. 1992; Colin 1990; Palardy 1996; Palardy et al. 1998; Hidalgo et al. 2004; Perlot 2005; Perlot et al. 2009). From the basic system there can be different variations: with this kind of method, Perlot (2005) simulated the environment in a deep repository with pressure generated by the lithostatic layer and an increase in temperature due to the radioactive wastes.

As a final consideration regarding accelerated tests, it is necessary to continue with the research on accelerated leaching methods, linking the different tests to the different scenarios of degradation and characteristics of the materials, and developing and calibrating them properly.

9.5 Final Considerations

There are a large number of standard and non-standard leaching tests, which results in a limited comparability of the results, the problem being even bigger when using accelerated tests.

As far as the large number of leaching methods is concerned, it is important to point out that in April 2006 a CEN/TC, TC 351 "Construction products: Assessment of release of dangerous substances", was established with the scope to develop horizontal standardised assessment methods for harmonised approaches relating to the release (and/or the content when this is the only practicable or legally required solution) of regulated dangerous substances under the Construction Products Directive (CPD) taking into account the intended conditions of use of the product. It addresses emission to indoor air, and release to soil, surface water and groundwater. Even though these harmonised methods of leaching are not directly developed for investigating the degradation process caused by the decalcification of concrete exposed to water, as said, most of the methods used for concrete in this field are based on the existing waste standards or their modifications. Therefore, it is expected that the leaching methods finally approved by the CEN/TC 351 (working at the time of writing this report), and those that are based on the already existing leaching methods, will be mostly used in the near future, also for leaching of hydrated phases of cement.

In March 2011, a Technical report was published entitled CEN/TR 16142:2011 "Concrete. A study of the characteristic leaching behaviour of hardened concrete for use in the natural environment", prepared by the TC CEN/TC 51. This report deals with developments by a consortium of Dutch/German institutes, as officially reported to the European Commission in EUR 17869-EN, leading to a performance test method for characterizing the leaching behaviour of hardened concrete for use in contact with the natural environment. The pre-normative research underpinning this TR included a literature survey and three progressively staged inter-laboratory studies, finding a poor reproducibility using a characterization method consisting of two different leaching procedures: an availability test and a tank diffusion test.

In any event, the choice of the method to be used will depend on the aim of the tests, mainly concerning ranking of concretes or prediction of the service life. Provided that the different methods themselves have an important effect on the obtained results, even for ranking of concretes, it is recommended to perform a range of leaching tests in order to completely understand the leaching behaviour of a material under different conditions and/or use a leaching method that reproduces the mechanisms of the expected specific scenarios of the concrete degradation.

For wastes, in addition to the specific leaching methods, there is a more general framework; the EN 12920:2006 + A1:2008, "Characterization of waste – Methodology for the determination of the leaching behaviour of waste under specified conditions". According to this standard, the leaching tests are only one part of the total methodology, the results of the tests being the so called "parameter specific test". The same philosophy can be applied to the leaching degradation of concrete. Thus, the application of the leaching test method alone would not be sufficient for the determination of the detailed leaching behaviour under specified conditions. A complete methodology should include the description of the scenario, performance of the accordingly selected leaching method, modelling of the process, and validation of the results.

References

Adenot, F., Buil, M.: Modelling of the corrosion of the cement paste by deionized water. Cem. Concr. Res. **22**(2–3), 489–496 (1992)

Alonso, C., Castellote, M., Llorente, I., Andrade, C.: Ground water leaching resistance of high and ultra high performance concretes in relation to the testing convection regime. Cem. Concr. Res. **36**(9), 1583–1594 (2006)

Andac, M., Glasser, F.P.: Long-term leaching mechanism of Portland cement-stabilized municipal solid waste fly ash in carbonated water. Cem. Concr. Res. **129**(8), 179–186 (1999)

Ballivy, G., Colin, J.C., Mnif, T.: Effectiveness of injected cement grouts mortar. Harsh environmental condition. In: ASCE Proceedings, GT Division, pp. 588–600, New Orleans (1992)

Berner, U.R.: Evolution of pore water chemistry during degradation of cement in a radioactive waste repository environment. Waste Manag. **12**(2–3), 201–219 (1992)

Carde, C., François, R., Torrenti, J.M.: Leaching of both calcium hydroxide and C–S–H from cement paste: modeling the mechanical behaviour. Cem. Concr. Res. **26**(8), 1257–1268 (1996)

Carde, C., Escadeillas, G., François, R.: Use of ammonium nitrate solution to stimulate and accelerate the leaching of cement pastes due to deionised water. Mag. Concr. Res. **49**(181), 295–301 (1997)

Castellote, M., Andrade, C., Alonso, C., Turrillas, X., Kvick, Å., Terry, A., Vaughan, G., Campo, J.: Synchrotron radiation diffraction study of the microstructure changes in cement paste due to accelerated leaching by application of electrical fields. J. Am. Ceram. Soc. **85**(3), 631–635 (2002)

Castellote, M., Llorente, I., Andrade, C., Alonso, C.: Accelerated leaching of ultra high performance concretes by application of electrical fields to simulate their natural degradation. Mater. Struct. **36**(2), 81–90 (2003)

Castellote, M., Turrillas, X., Andrade, C., Alonso, C., Rodríguez, M.A., Kvick, A.: In-situ accelerated leaching of cement pastes by application of electrical fields monitored by Synchrotron X-Ray diffraction. Appl. Phys. Mater. Sci. Process. **79**(3), 661–669 (2004a)

Castellote, M., Turrillas, X., Llorente, I., Alonso, C., Andrade, C., Campo, J.: A neutron diffraction study of changes induced in aluminous cement paste by the application of external electrical fields. Phys. B Conden. Matter **350**(1–3), E561–E564 (2004b)

Colin, J.C.: Développement d'un perméamètre pour stimuler les conditions entourant les roches à moyenne profondeur. Mémoire de maîtrise, Université de Sherbrooke, 122 p, Canada (1990)

DOE, Department of Ecology, Washington state: An Assessment of Laboratory Leaching Tests for Predicting the Impacts of Fill Material on Ground Water and Surface Water Quality. Publication No. 03-09-107 (2003)

British Standards Institution: Characterisation of waste. Methodology for the determination of leaching behaviour of waste under specified conditions, DD ENV 12920. British Standards Institution, London (1998)

EQM, Environmental Quality Management: Survey of Oily Waste Leaching Procedures: Technology Overview and Suggested Protocols, Final Report. EPA Contract 68-W6-0068, WA#4 (1998)

Faucon, P.: Concrete durability: physico-chemistry of the alteration by water (in French). Ph.D. thesis, Génie Civil, L'Universite de Cergy Pontoise, 260 p, France (1997)

Faucon, P., Adenot, F., Jorda, M., Cabrillac, R.: Behaviour of crystallised phases of Portland cement upon water attack. Mater. Struct. **30**(8), 480–485 (1997)

Faucon, P., Adenot, F., Jacquinot, J.F., Petit, J.C., Cabrillac, R., Jorda, M.: Long-term behaviour of cement pastes used for nuclear waste disposal: review of physico-chemical mechanisms of water degradation. Cem. Concr. Res. **28**(6), 847–857 (1998a)

Faucon, P., Gerard, B., Jacquinot, J.F., Marchand, J.: Water attack of a cement paste: towards an improved accelerated test? Adv. Cem. Res. **10**(2), 67–73 (1998b)

Gerard, B.: Contribution des couplages mécanique-chimique-transfert dans la tenue à long terme des ouvrages de stockage de déchets radioactifs. Ph.D. thesis, Laboratoire de Mécanique et Technologie, Génie Civil, Université Laval, 380 p, Canada (1996)

Gerard, B., Pijaudier-Cabot, G., Le Ballego, C.: Calcium leaching of cement based materials: a chemo-mechanics application. In: Eligehausen, R. (ed.) Construction Materials Theory and Application, pp. 313–329. Hans-Wolf Reinhart. Zum 60 Geburtstag. ibiden-Verlag, Stuttgart (1999)

Glasser, F.P., Marchand, J., Samson, E.: Durability of concrete – degradation phenomena involving detrimental chemical reactions. Cem. Concr. Res. **38**(2), 226–246 (2008)

Haga, K., Sutou, S., Hironaga, M., Tanaka, S., Nagasaki, S.: Effects of porosity on leaching of Ca from hardened ordinary Portland cement paste. Cem Concr. Res. **35**(9), 1764–1775 (2005)

Hidalgo, A., Andrade, C., Alonso, C.: An accelerated leaching test to evaluate the long term behaviour of concrete in waste disposal. L'Industria Italiana del Cemento **766**, 498–507 (2001)

Hidalgo, A., Llorente, I., Alonso, C., Andrade, C.: Evaluation of concretes for high level nuclear waste repositories. In: Vázquez, E., Hendriks, Ch.F, Janssen, G.M.T. (eds.) International RILEM Conference on the Use of Recycled Materials in Buildings and Structures, pp. 945–952. RILEM Publications SARL, Barcelona (2004)

Jacquemot, F.: Leaching performance test assessing durability of concrete exposed to chemical attack. In: Alexander, M.G., Bertron, A. (eds.) Proceedings of the RILEM TC 211-PAE Conference Concrete in Aggressive Aqueous Environments: Performance, Testing and Modeling, pp. 372–379. RILEM Publications SARL, Toulouse (2009)

Kamali, S., Moranville, M., Leclercq, S.: Material and environmental parameter effects on the leaching of cement pastes: experiments and modeling. Cem. Concr. Res. **38**(4), 575–585 (2008)

Kuroi, T., Sueyoshi, T.: Basic study on softening phenomenon of cement paste due to action of electric current. Gen. Meet. Cem. Assoc. p164. Tokyo, Japan (1987)

La Grega, M.D., Buckingham, P.L., Evans, J.C.: Stabilisation and solidification. In: Hazardous Waste Management, pp. 641–704. McGraw-Hill, New-York (1994)

Lagerblad, B.: Texture and chemistry of historic concrete subjected to prolonged hydratation. In: The Nordic Concrete Federation (ed.) Proceedings of the Workshop on Water in Cement Paste & Concrete Hydration and Pore Structure, pp. 148–154. Skagen, Denmark (1999)

Le Maréchal, J., Gerard, B., Marchand, J., Gagnon, J.P., Didry, O.: New accelerated leaching experiment: the LIFT procedure. In: Malhotra, V.M. (ed.) Proceedings Fourth CANMET/ACI/ JCI International Conference, Tokushima, Japan, SP 179–55 (1998)

Lea, F.M.: The action of ammonium salts on concrete. Mag. Concr. Res. **17**(52), 115–116 (1965)

Llorente, I., Castellote, M., Gonzalez-Arrabal, R., Ynsa, M.D., Muñoz-Martin, A., de Viedma, P. G., Castillo, A., Martínez, I., Andrade, C., Zuloaga, P., Ordoñez, M.: PIXE/RBS as a tool to study cementitious materials: application to the dynamic leaching of concrete. Nucl. Instrum. Methods Phys. Res. B **267**(23–24), 3670–3674 (2009a)

Llorente, M., Castellote, M., Andrade, C.: Comparison between several methods for determining the resistance of concrete to leaching. In: Alexander, M.G., Bertron, A. (eds.) Proceedings of the International RILEM TC 211-PAE conference Concrete in Aggressive Aqueous Environments, Performance, Testing and Modeling, pp. 356–362. Rilem Publications, Toulouse, France (2009b)

Mainguy, M., Tognazzi, C., Torrenti, J.M., Adenot, F.: Modelling of leaching in pure cement paste and mortar. Cem. Concr. Res. **30**(1), 83–90 (2000)

Matschei, T., Lothenbach, B., Glasser, F.P.: The AFm phase in Pórtland cement. Cem. Concr. Res. **37**(2), 118–130 (2007)

Moudilou, E., Bellotto, M., Defosse, C., Serclerat, I., Baillif, P., Touray, J.C.: A dynamic leaching method for the assessment of trace metals released from hydraulic binders. Waste Manag. **22** (2), 153–157 (2002)

Palardy, D.: Etude comparative sur la durabilité de coulis à base de ciment soumis à différentes conditions retrouvées potentiellement dans un site de stockage souterrain de déchets nucléaires. Mémoire de Maîtrise, Université de Sherbrooke, 372 p, Canada (1996)

Palardy, D., Onofrei, M., Ballivy, G.: Microstructural changes due to elevated temperature in cement based grouts. Adv. Cem. Based Mater. **8**(3–4), 132–138 (1998)

Perera, A.S.R., Al-Tabbaa, A., Reid, J.M., Stegemann, J.A.: State of practice report UK stabilisation/solidification treatment and remediation: Part 4: Testing and performance criteria. In: STARNET State of Practice Report. www-starnet.eng.cam.ac.uk (2004)

Perlot, C.: Influence de la décalcification de matériaux cimentaires sur les propriétés de transfert application au stockage profond de déchets radioactifs. Ph.D. thesis, UPS Toulouse/ Sherbrooke University, 283 p (2005)

Perlot, C., Carcassès, M., Ballivy, G.: Experimental protocols set up for leaching simulation under static and dynamic conditions. In: Alexander, M.G., Bertron, A. (eds.) Proceedings of the International RILEM TC 211-PAE conference Concrete in Aggressive Aqueous Environments, Performance, Testing and Modeling, pp. 363–371. Rilem Publications, Toulouse, France (2009)

Porteneuve, C.: Caractérisation des Bétons par Résonance. Magnétique Nucléaire: Application à I'Etude de I'Altération par I'Eau. Ph.D. thesis, Sciences aux Materiaux, L'Universite Paris VI, France (2001)

Reardon, E.J.: Problems and approaches to the prediction of the prediction of the chemical composition in cement/water systems. Waste Manag. **12**(2–3), 221–239 (1992)

Saito, H., Deguchi, A.: Leaching tests on different mortars using accelerated electrochemical method. Cem. Concr. Res. **30**(11), 1815–1825 (2000)

Saito, H., Nakane, S.: Comparison between diffusion test and electrochemical acceleration test for leaching degradation of cement hydration products. ACI Mater. J. **96**(2), 208–211 (1999)

Saito, H., Nakane, S., Ikari, S., Fujiwara, A.: Preliminary experimental study on the deterioration of cementitious materials by an acceleration method. Nucl. Eng. Des. **138**(2), 151–155 (1992)

Schneider, U., Chen, S.W.: Influence of concentration of ammonium nitrate solutions on concrete. In: Malhotra, V.M. (ed.) Symposium Proceedings of 6th CANMET/ACI International Conference on Durability of Concrete, pp. 439–447, Thessaloniki, Greece (2003)

Stegemann, J.A., Coté, P.L.: Summary of an investigation of test methods for solidified waste evaluation. Waste Manag. **10**(1), 41–52 (1990)

Stegemann, J.A., Coté, P.L.: Test methods for solidified Waste evaluation – A cooperative program, report EPS 3/HA/8, Environment Canada, Ottawa, Ontario, Canada (1991)

United States Environmental Protection Agency USEPA Method 1312: Synthetic precipitation leaching procedure. SW846. On-line test methods for evaluation of solid wastes, physical chemical methods. http://www.epa.gov/epaoswer/hazwaste/test/pdfs/1312.pdf (2003a). Accessed Oct 2011

United States Environmental Protection Agency, USEPA Method 1311: Toxicity characteristic leaching procedure. SW846 On-line test methods for evaluation of solid wastes, physical chemical methods. http://www.epa.gov/epaoswer/hazwaste/test/pdfs/1311.pdf (2003b). Accessed Oct 2011

United States Environmental Protection Agency: Method 1320: Multiple extraction procedure. SW846. On-line test methods for evaluation of solid wastes, physical chemical methods. http://www.epa.gov/epaoswer/hazwaste/test/pdfs/1320.pdf (2003c). Accessed Oct 2011

Van der Sloot, H.A.: Harmonisation of leaching tests. http://www.leaching.net (2007) Accessed Oct 2011

Van der Sloot, H.A., Dijkstra, J.J.: Development of horizontally standardized leaching tests for construction materials: a material based or released based approach? ECN-C–04-060 report (2004)

Van der Sloot, H.A., Heasman, L., Quevauviller, P.: Harmonisation of Leaching/Extraction Tests. Studies in Environmental Science 70 Series, Elsevier Science B.V., Amsterdam (1997)

Wastewater Technology Centre: Compendium of waste leaching tests. Report EPS 3/HA/7, Environment Canada (1990)

Chapter 10
Test Methods for Resistance of Concrete to Sulfate Attack – A Critical Review

Kim Van Tittelboom, Nele De Belie, and R. Doug Hooton

Contents

K. Van Tittelboom • N. De Belie (✉)
Magnel Laboratory for Concrete Research, Department of Structural Engineering, Ghent University, Technologiepark – Zwijnaarde 904, B-9052 Ghent, Belgium
e-mail: kim.vantittelboom@ugent.be; nele.debelie@ugent.be

R.D. Hooton
Department of Civil Engineering, University of Toronto, 35 St. George Street, Toronto, M5S 1A4, Canada
e-mail: hooton@civ.utoronto.ca

M. Alexander et al. (eds.), *Performance of Cement-Based Materials in Aggressive Aqueous Environments*, RILEM State-of-the-Art Reports 10,
DOI 10.1007/978-94-007-5413-3_10, © RILEM 2013

Keywords Sulfate • Ettringite • Thaumasite • Salt crystallization • Delayed ettringite formation • Standard methods • Degradation measurements • Expansion

10.1 Introduction

Sulfate attack comprises a series of chemical reactions between sulfate ions and the components of hardened concrete. As these reactions may lead to cracking, spalling or strength loss of concrete structures, appropriate test methods are needed to determine the resistance of concrete under sulfate exposure. Accelerated test methods are most suitable since sulfate attack is typically a long term process. The current ASTM C1012 (2004) test method accelerates the attack mechanism by using a solution with a high sulfate concentration in which mortar specimens are immersed. The SVA procedure (see Table 10.1 for more detail) uses smaller specimens to obtain results earlier (Mielich and Öttl 2004). In the Wittekindt method not only smaller specimens are used but also the w/c-ratio is increased (Wittekindt 1960). However, these tests still require several months. Test methods such as ASTM C452 (2006) and the Chatelier-Anstett test use a mixture of cement and gypsum. Since in this case the sulfate source is located internally, no more time is needed for sulfate ingress. With these test methods results are obtained after 2 weeks, but the attack mechanism no longer represents field conditions in a realistic way.

Another problem relates to the way of quantifying the degree of degradation. SVA (Mielich and Öttl 2004), Wittekindt (Wittekindt 1960), Duggan (Scott and Duggan 1986), ASTM C1012 (2004) and C452 (2006) use expansion measurements. Other researchers have proposed the use of decrease in compressive strength; in the rapid electrochemical test, current is measured to determine the depth of sulfate penetration. Depending on the selected degradation measure, different conclusions can be drawn regarding the performance of concrete under sulfate attack. In this chapter, an overview of the existing test methods is presented and a critical discussion is performed.

Reactions between sulfates and some cement components can lead to the deterioration of concrete structures. Cement contains different clinker minerals such as tricalcium aluminate (C_3A). If ground cement clinker is mixed with water, the C_3A reacts violently leading to rapid stiffening. To obtain a better workability of the cement paste, the reaction between C_3A and water must be delayed. Therefore, a small amount of sulfate is added to the cement powder in the form of gypsum ($CaSO_4{\cdot}2H_2O$) or anhydrite ($CaSO_4$). When mixing with water, the C_3A reacts with the gypsum or anhydrite, leading to the formation of ettringite ($C_3A(C\underline{S})_3H_{30-32}$). The ettringite forms a coating on the surface of the C_3A and delays the hydration

Table 10.1 Overview of test methods

Test method	Sample size	Type of sample	W/C	Curing + treatment	Sulfate location	Sulfate concentration	Solution exchange	Measurements	Measuring time	Criterion
ASTM C1012-04 (2004)	25.4 × 25.4 × 285 mm	Mortar specimens with stainless steel studs	0.46/0.485	23.5 ± 0.5 h in mould at 35 ± 3°C, demoulding, in $Ca(OH)_2$ saturated water at 23 ± 2.0°C to 20 ± 1 MPa (f_{ccub}) when 20 MPa is reached initial length is measured and specimens are exposed to the test solution at 23°C	External	352 mol of Na_2SO_4/m^3		Length change	**6 months to 1 year** 1, 2, 3, 4, 8, 13 and 15 weeks 4, 6, 9 and 12 months	Blended cements 0.034%(3*) 0.037% (4*) 0.039%(5*) 0.041%(6*) Portland cements 0.01% (3*) 0.011% (4*) 0.012% (5–6*) *Number of samples
ASTM C1012-04' (**Ferraris et al.** 2005)	10 × 10 × 40 mm	Specimens of mortar or cement paste with pins embedded in both ends		Moulds in closed plastic bag (100%RH) in curing cabinet at 22 ± 2°C, after 24 h demoulded and placed in lime water in curing cabinet, stud screwed in end pins (3 or 4 days after mixing), epoxy around studs and 5 mm on top side of specimen, after drying of epoxy again in limewater until 7 days after casting, exposure to sulfate solution at 23 ± 0.5°C	External	50 g/l Na_2SO_4		Length change	Measured every day for first 2 weeks, then 1x every week until deterioration commences	
ASTM C1012-04" (**Guirguis and Cusack** 2001)	15 × 40 × 160 mm	Mortar specimens, stainless steel studs		Cured for 7 days (2 days in mould and 5 days in water)	External	352 mol of Na_2SO_4/m^3		Length change		
CEN (Mauris 2005)	20 × 20 × 160 mm	Mortar specimens	0.50	1 day storage in mould at 20°C and 95%RH, demoulding, 27 days pre-storage in H_2O, after which storage is continued in sodium sulfate solution	External	16 000 mg SO_4/l (Na_2SO_4)		Length change	**140 days**	
Flat prism Procedure (Wittekindt 1960)	10 × 40 × 160 mm	Mortar specimens, stainless steel inserts	0.60	20–24 h in mould (20°C, 95% RH), 13 days in distilled water at 20°C and normal RH or 5 days in fog room at 40°C and 100%RH, stored in distilled water or Na_2SO_4 solution of two different concentrations	External	29 800 mg SO_4/l (Na_2SO_4)	Weekly	Length change	**56 days**	$\Delta L/L_0 \leq 0.5$ mm/m at 56 days

(continued)

Table 10.1 (continued)

Test method	Sample size	Type of sample	W/C	Curing + treatment	Sulfate location	Sulfate concentration	Solution exchange	Measurements	Measuring time	Criterion
SVA (Mielich and Öttl 2004)	10 × 40 × 160 mm	Mortar specimens	0.50	2 days in mould at 20°C and >90%RH, 12 days in $Ca(OH)_2$ saturated water, following storage in 20°C sodium sulfate solution	External	29 800 mg SO_4/l (Na_2SO_4)	Monthly	Length change	**91 days** measurements at 0, 28, 56, 70, 91 days	$\Delta L/L_0 \leq 0.5$ mm/m at the age of 91 days
AS 2350.14 (1996)	40 × 15 × 150 mm	Mortar specimens with gage studs		Between 44 h and 48 h after mixing → demoulding, cured for 7 days (2 days in mould and 5 days in lime saturated water)	External	0.352 mol/l Na_2SO_4	Every 2 weeks	Length change	**16 weeks** measurements at age of 7 days (after curing and before immersion), every 2–16 weeks	
Small prism procedure (Steinegger 1970)	10 × 10 × 60 mm	Mortar specimens		24 h in mould in H_2O at 35 ± 3°C, cured for 24 or 48 h at 22°C and 95%RH, demoulding, suspended in distilled water during 21 days afterwards ½ specimens stored in distilled water and ½ stored in test solution (for 56 days)	External	25 000 mg SO_4/l (Na_2SO_4)		Flexural strength – in H_2O: B_w, in solution: B_s	**77 days,** 56 days exposure to solution after 21 days curing	$B_s/B_w \geq 0.7$ after 56 days of exposure
MNS (Mulenga et al. 1999)	40 × 40 × 160 mm	Concrete samples		After 28 days preliminary storage at 20°C the test specimens are rapidly saturated with sulfate solution by applying a negative pressure of 150 mbar, test specimens are then stored at 8°C in a sodium sulfate solution	External	33 400 mg SO_4/l (Na_2SO_4)		Uniaxial tensile strength	**112–148 days**	Ratio: uniaxial tensile strength of specimens stored in sulfate to those stored in water, at least 0.80 after 84 days or 0.70 after 120 days of exposure
Mehta and Gjorv (1974)	12.5 × 12.5 × 12.5 mm	Cement paste	0.50	Specimens cured in moist environment at 50°C for 7 days, after curing period, specimens divided in three groups, one group placed in calcium sulfate solution, second group placed in sodium sulfate solution and third group subjected to a compressive strength test	External	Saturated $CaSO_4$ solution (0.12% SO_3) and a 4% Na_2SO_4 solution (2.1% SO_3)	Solutions titrated with 1 N H_2SO_4 every 24 h till colour of Bromothymol Blue indicator changes to yellow (pH 6.0–7.6)	Compressive strength	**35 days**	

Notched beam test (Hughes and Grounds 1985)	250 × 25 × 20 mm	OPC and OPC/pfa paste	0.47	Mould covered and rotated for 18 h; after demoulding, spec. placed in saturated $Ca(OH)_2$ solution to cure at 35°C for 1, 4 or 12 weeks; notch sawn in centre of one 25 mm wide face of beam with 3 mm blade, to a depth of 10 mm, then extended further 3 mm with 150 μm blade, before beam immersed in sulfate solution; beam span = 200 mm, crosshead speed 0.05 mm/min; load recorded and central deflection monitored; tests conducted in saturated $Ca(OH)_2$ solution, measurements done before and during immersion of beams in sulfate solution	External	0.7 M ± 0.01 M Na_2SO_4 solution		Bending strength	Curing period of 1, 4 or 12 weeks; when OPC/pfa spec. cured for 1 week, T_c occurs at 21 days; when cured for 12 weeks, T_c occurs after 114 days.
Soaking and drying test (Almeida 1991)	150 mm cubes → 50 mm cubes	Concrete cubes		Samples cured by immersion in water at 20°C for 28 days; after immersion for 2 h in sulfate solution at 20°C, samples dried at 105°C for 10–15 h. Samples kept in oven for at least 16 h, cooled at room temperature and again placed in solution. Cycle repeated 120 times	External	14 g Na_2SO_4 mixed with 86 g water	After 15-cycle series, the solution is renewed.	Weight change and compressive strength	28 days curing followed by 120 test cycles
Test for high strength concrete in sulfate environment (Park et al. 1999)	Ø100 × 200 mm 50 × 50 × 50 mm	Concrete specimens	0.28 0.36 0.54		External	(a) Pure water (b) 10% Na_2SO_4 (c) 10% $MgSO_4$ (d) Mix of 10% Na_2SO_4 and $MgSO_4$	Monthly renewed (pH maintained at 7)	Change in comp. str.; Length change, change in weight	**270 days** Measurements at 60, 90, 120, 150, 210 and 270 days

(continued)

Table 10.1 (continued)

Test method	Sample size	Type of sample	W/C	Curing + treatment	Sulfate location	Sulfate concentration	Solution exchange	Measurements	Measuring time	Criterion
Rapid electro-chemical test (Tumidajski and Turc 1995)	Middle 51 mm of 200 × 102 mm cylinders	Concrete specimens	0.42 0.50 0.60		External in combination with 60 V direct current delivered to specimen	88 g/l Na_2SO_4 anolyte; 12 g/l NaOH catholyte	Hourly	Initial current and total charge vs. sulfate front	**6 h**	
Half-cell potential measurement (Assaad et al. 2002)	Ø100mm × 100 mm	Conc cyl, reinf. bar in centre Ø10 mm × 90 mm, top coated with epoxy	0.40 0.50	Copper-copper sulfate electrode and a high impedance voltmeter used to read E_{corr}, 6 potential readings taken every 20 mm along 3 meridians at 0°,120°, 240°, for total of 18 readings per specimen	External	Partially immersed in Na_2SO_4 (20000 ppm SO_4^{2-}) subjected to H_2S (1,000 ppm H_2S) Subjected only to H_2S gas, N_2, air (0.1% H_2S, 9.9% N_2, 90% air)	Control circuit keeps gas in test chamber at required concentration	Electrochem. pot. of steel; reflects ionic condition between bars and pore fluid of concrete	**36 months** 23 test cycles of 5 weeks each	> − 200 mV: risk < 10%; −200 mV to −350 mV: risk ≈ 50% <−350 mV: risk > 90%
ASTM C452-06 (2006)	25 × 25 × 285 mm	Mortar specimens	0.46/0.485	Cured 22–23 h at 23 ± 2°C, 95%RH, moulds removed, specimens placed for at least 30 min in water before initial measurement. During test, specimens stored in water with volume water/sample ≤ 5 at 23 ± 2°C	Internal (mixed in sample)	7.0% SO_3 content by mass	Water replenished every 7 days for first 28 days, every 28 days thereafter	Length change	**14 days;** first measurement after 24 h, second at age of 14 days	0.01% (3*) 0.011% (4*) 0.012% (5–6*) *Number of samples
ASTM C1038-04 (2004)	25 × 25 × 285 mm	Mortar specimens with gauge studs		22.5 ± 0.5 h in moist room; after demoulding, in water or saturated lime water at 23 ± 2°C (30 min), initial measurement, specimens afterwards again stored in saturated lime water	Internal (cement with added sulfate)			Length change	**14 days;** measurements at 24 h ± 15 min and 14 days	
Duggan test (Scott and Duggan 1986)	Ø22 mm × 50 ± 5 mm	Concrete cores	0.40	Specimens soaked in distilled water (21°C) for 72 h, then heated to 82°C for 24 h, air cooled to room temperature for 1 h and soaked in dist. water (21°C) for 24 h; cycle repeated twice; subsequently specimens	Internal (mixed in sample)			Length change	**31 days**	Expansion ≤ 0.05%

				heated to 82°C for 72 h, air cooled for 1 h; specimens then placed in plastic jar and submerged in dist. water (21°C) while length change measurements taken for 20 days					
Le Chatelier-Anstett (Blondiau 1961)	Ø80 mm × 30 mm	Cement paste	0.50	Cured 24 h at 23°C, 100%RH, demoulding, stored in water for 15–20 days, pressed to 20 MPa during 1 min; specimens stored in contact with wet paper for 90 days.	Internal (mixed in sample)	50 g gypsum added to 100 g cement		Diameter increase	**20 + 90 days;** measurements at 24 h, 7, 28 and 90 days after curing	1.25% diameter increase at 28 days
TAP-test (De Belie et al. 2002)	Ø270 mm × 70 mm	Concrete cylinders	0.40	Samples rotated in acid solution at speed of 1.04 rev/h (each cycle = 12 days)	External	0.5% H_2SO_4	Renewed after each cycle	Change of cylinder radius and surface roughness	**108 days;** 9 cycles of 12 days → 108 days	
Leaky Rayleigh wave (Neuensch-wander et al. 2006; **Schmidt et al.** 2006)	40 × 160 × 10 mm	Mortar specimens		40 × 160 × 240 mm specimens stored in limewater for 28 days, reduced to 40 × 160 × 12 mm specimens, ground on both sides to obtain 40 × 160 × 10 mm specimens, exposed to 4 different test solutions, ultrasonic measurements in demineralised water as coupling medium	External	(a) Lime water (b) 0.1 mol H_2SO_4 (c) 0.5 mol H_2SO_4 (d) Sandblasting of wet sample		Reduction in sound velocity		
BSA test (Vincke et al. 1999)	20 × 20 × 50 mm	Concrete prisms glued on glass plate		Samples incubated in 20-fold diluted aqueous solution of biological sulfur suspension; for control samples, inactivated suspension used; 4 cycles containing three steps were performed: (a) Specimen placed in H_2S atmosphere for 3 days; (b) Specimen suspended in recipient containing	External	(a) 250 ppmvl H_2S (b) 950 ml tap water/ 100 mg/l $(NH_4)_2SO_4$/ 10 mg/l K_2HPO_4/ 50 ml biological sulfur suspension		During step (a) and (b): total Ca^{2+} content released; During step (b): change in pH and sulfate concentration; After every cycle: change in height (10 different points)	**51 days** 3 cycles of 17 days → 51 days	

(continued)

Table 10.1 (continued)

Test method	Sample size	Type of sample	W/C	Curing + treatment	Sulfate location	Sulfate concentration	Solution exchange	Measurements	Measuring time	Criterion
				600 ml of simulation solution for 10 days while specimen stirred; (c) stirring specimens for 2 days in recipient containing milli-Q water; (d) specimens dried for 2 days + slightly brushed				After procedure: change in weight		
BSA-test (De Belie et al. 2004)	Ø80 × 15 mm	Concrete cylinders glued on plastic plates of 120 × 90 mm		Samples incubated in 20-fold diluted aqueous solution of biological sulfur suspension; for control samples, inactivated suspension used; four test cycles containing four steps performed at 28°C: (a) H_2S incubation chambers of 10 L for 3 days; (b) specimen hung in 600 ml of solution for 10 days while placed on rotary shaker (90 rpm); (c) specimens in milli-Q water while on rotary shaker (for 2 days); (d) drying for 2 days at 28°C	External	100 mg/l $(NH_4)_2SO_4$ and 10 mg/l K_2HPO_4 + bacteria		Ditto above, except that: After every cycle: thickness of blocks measured (automatically)	**68 days;** 4 cycles of 17 days → 68 days	
Sand et al. (1987)	600 × 110 × 70 mm	Concrete samples		Samples standing in 10 cm of water at 30°C and a pH of 7.0; whole test chamber kept at 30°C and saturated air humidity, test blocks periodically sprayed with *Thiobacilli thiooxidans* cultures, H_2S acted as a substrate for the bacteria	External	10 mg/l H_2S		Weight loss	**1 year**	
Mori et al. (1992)	40 × 40 × 160 mm	Mortar specimens	0.65	Specimens cured in water at 20–22°C for 4 weeks: H_2S concentration of simulation chamber kept	External	400 ppm H_2S		Reduction in cross section	**6 months**	

				at 400 ppm by regulating flow rates of 4% Na_2S and 1.5 N HCl, specimens placed in sewage, AT basal medium or distilled water; during first 2 months, tests specimens inoculated with *Thiobacilli thiooxidans* every 2 weeks					
Mori et al. (1992)	40 × 40 × 160 mm	Mortar specimens	0.65	Specimens cured in water at 20–22°C for 4 weeks; samples placed vertically in a pipe with a diam. 15 cm, length 20 m so that bottom 4–8 cm submerged in sewage; temp. ranged from 12 to 30°C; H_2S concentration in atmosphere of pipe ranged from 25–300 ppm	External	25–300 ppm H_2S		Reduction in cross section	**6 months**

of the latter. However, the formation of ettringite will also lead to expansion and an increase in volume of the cement paste. At this stage the expansion is innocuous because the volume increase can occur unobstructed in the fresh cement paste. This formation of ettringite is called primary ettringite formation.

The deterioration of concrete due to sulfate attack is caused by the formation of ettringite in the hardened cement paste, called secondary ettringite formation. When sulfate ions, coming from an internal or external source, react with C_3A in the presence of moisture, ettringite is formed, causing considerable expansion and cracking of the concrete structure.

Another reaction, occurring when concrete comes into contact with, for example, a sodium sulfate solution, is the conversion of calcium hydroxide (CH) into secondary gypsum, which can also cause expansion and cracking. With magnesium sulfate, the magnesium reacts with hydroxyl ions from calcium hydroxide to form brucite, in addition to gypsum. Magnesium can also substitute calcium in C-S-H, resulting in softening and strength loss. During sulfate attack, calcium hydroxide (CH) and calcium silicate hydrate (C-S-H) phases may also be dissolved and leached out. At low temperature and in the presence of soluble carbonate and reactive silicate, sulfate attack can lead to the formation of thaumasite. In contrast to ettringite, thaumasite does not necessarily cause expansion, but it leads to loss in strength and decomposition of the microstructure (Schneider et al. 2002). (See also Chap. 2 in Part I of this Report).

Therefore, damage can occur due to chemical attack on the aluminate phases in the cementing materials, due to chemical attack on the calcium-silicate matrix, due to acidic dissolution of the matrix, or due to physical attack due to sulfate salt crystallization in pores near surfaces subjected to drying. This complexity, combined with the slow rate of many forms of sulfate attack, conspires against adoption of a single performance test method, and in some cases requires adoption of prescriptive approaches (such as w/cm limits, or limits on allowable cement types). For many decades, many national building codes and specifications have adopted a combination of performance and prescriptive options to address sulfate attack, and in general this has greatly reduced the potential for failures due to sulfate attack. However, a number of sulfate-related problems such as ettringite formation, sulfate salt crystallization, and thaumasite sulfate attack, are not adequately addressed in current standards (Hooton and Brown 2009).

In the European Standard EN197-1 (2000), there is currently no test method for evaluating the sulfate resistance of a Portland or blended cement, due to lack of agreement on a common test method, in spite of several European countries previously having had test methods (Hooton and Brown 2009). As a result, in 2006 the CEN committee drafted an amendment A2 to EN197-1 (2000) to prescriptively allow a family of seven types of cement for use in sulfate resistant applications (not yet adopted). These types include three CEM I cements with either 0, 3, or 5% Bogue C_3A in the clinker, as well as slag cements CEM III/B (66–80% slag) and CEM III/C (81–95% slag), and pozzolan cements CEM IV/A (20–35% pozzolan) and CEM IV/B (36–55% pozzolan).

10.2 Types of Sulfate Attack on Concrete

In the present context, concrete may be deteriorated due to sulfate attack, sulfuric acid attack or biogenic sulfuric acid attack.

Sulfate attack on concrete can be subdivided into external and internal attack. External sulfate attack occurs when water, containing dissolved sulfates, penetrates into the concrete. It comprises calcium aluminate reactions, thaumasite formation and sulfate salt crystallization. Internal sulfate attack occurs when a sulfate source is incorporated in the concrete when mixed. It encompasses the use of over-sulfated cements and the appearance of delayed ettringite formation.

When sulfate solutions enter the concrete matrix through pores, sulfate ions may react with sources of alumina to form ettringite and with sources of calcium to form gypsum, resulting in cracking or strength loss. The thaumasite form of sulfate attack (TSA) requires a source of sulfate as well as carbonate. Sulfate salt crystallization encompasses precipitation of sulfate salts in concrete pores near drying surfaces (Hooton and Brown 2009).

Deleterious expansion in concrete may also occur when excessive amounts of sulfates are present in the cement or when contaminated aggregates are used. As the sulfate source is incorporated in the concrete when mixed, this form of sulfate attack is called internal sulfate attack. Delayed ettringite formation (DEF) is a special form of internal sulfate attack, often occurring in heat-cured precast concrete (Hooton and Brown 2009).

When concrete is attacked by sulfuric acid, there are two important aspects. On the one hand, there is the reaction with the sulfate ion; on the other hand, the action of the hydrogen ion has to be considered. It can be stated that sulfate attack comprises only one aspect of sulfuric acid attack (Monteny et al. 2000). The sulfuric acid first reacts with the calcium hydroxide in the concrete to form gypsum. This reaction is associated with an increase in volume which will cause deterioration. However damage is also caused by the reaction between the formed gypsum crystals and the C_3A minerals, which leads again to the formation of expansive ettringite (Monteny et al. 2001). Sulfuric acid attack is covered in Chap. 11 of the Report (Part III).

Another important phenomenon is biogenic sulfuric acid corrosion, which occurs often in sewer systems (Monteny et al. 2001). The main cause of degradation is the corrosion of concrete due to the in-situ production of sulfuric acid by bacteria. This so-called bacteriogenic corrosion is discussed in more detail in Chap. 12 (Part III) of the Report.

As the reactions described above may lead to cracking, spalling or strength loss of concrete structures, it is important that the materials selection and mix design of concrete, in susceptible regions, be carefully considered to resist sulfate attack. Because of this, reliable and reproducible test methods are needed. The advantages and disadvantages of existing test methods will be discussed in this chapter. As it is difficult to test thaumasite formation, delayed ettringite formation and sulfate salt crystallization, not many methods are available to test these types of sulfate attack. As a consequence these forms of sulfate attack are discussed separately in Sect. 10.6 of this chapter.

10.3 Performance of Concrete under Sulfate Attack

To know whether certain types of concrete can be utilized in a sulfate rich environment, standards are needed. Two different types of standards could be used: prescriptive or performance standards. Prescriptive standards describe a product by its composition, while performance standards prescribe how well the product must perform. Prescriptive standards, such as ASTM C150 (2009) limit the tricalcium aluminate (C_3A) level for different cement types in order to control the resistance against sulfate attack, but there is also an optional performance test for sulfate resistant cements, ASTM C452 (Hooton and Brown 2009). Since sulfate resistance of concrete depends on many factors, and not only on the C_3A content, Ferraris et al. (2005) concluded that prescriptive standards are not sufficient. According to these researchers it would be better to utilize performance-based standards, using physical tests to evaluate sulfate resistance. For example the ASTM C1012 (2004) standard test evaluates the expansion of mortar prisms, immersed in a sodium sulfate solution, and requires them to have expansion below a certain limit.

The resistance of concrete against sulfate attack can be tested in two different ways. Realistic attack mechanisms can be used in combination with very sensitive methods to detect deterioration (see Sect. 10.3). Subsequently an extrapolation method can be used to calculate degradation in the future. A second way to perform tests is by using accelerated test methods in which higher sulfate concentrations, higher water to cementitious materials ratios (w/cm), smaller test specimens, etc. are utilized to gain results early (De Belie 2007).

The advantage of carrying out the experiments in a realistic way, and keeping the factors responsible for deterioration unchanged, is the certainty that the deterioration process itself has not been changed. However, sensitive equipment is needed to detect the deterioration within a reasonable time. A disadvantage of these close-to-reality test methods is that only the first stages of the attack mechanism can be measured. To estimate the behaviour of the concrete at later stages, an extrapolation of the obtained relationship has to be used. This implicates the risk that new attack mechanisms, which were not noticeable during the measuring time, arise in a later period of the deterioration and affect the accuracy of the test results. However, in the study performed by Monteiro and Kurtis (2008), where concrete samples remained submersed in a 2.1% Na_2SO_4 solution for 40 years, it was seen that after a certain initiation time, the expansion of the samples followed a definite scaling law. Nevertheless, this initiation time was found to be anywhere between 1 and 10 years.

Because of this, most common methods to investigate the resistance of concrete against sulfate attack incorporate accelerated tests. The advantage of this procedure is that the whole life of the specimen can be simulated while results are obtained rapidly. An acceleration of the attack mechanism can be achieved in different ways. The concentration of the sulfate solution can be increased or the reaction surface can be increased by using specimens with large surface area/volume ratio. In some cases, the process can be accelerated by means of continuous wetting and drying cycles, increasing the crystal pressure. Raising the temperature can also increase the process speed. Sometimes high w/c-ratios are used to obtain a higher permeability of the concrete specimens and also the addition of gypsum to the cement paste is employed, so no extra time is needed for penetration of the sulfate into the concrete. Because of the rapid change of materials and working parameters in the field, accelerated performance tests are required, to evaluate the resistance of concrete against sulfate attack, as an on-time response to the changing conditions.

10.4 Overview of Test Methods

In the extensive (but non-exhaustive) Table 10.1, an overview of test methods, for determining the resistance against sulfate attack, is given. Different tests are available to simulate the three types of concrete attack: sulfate attack, sulfuric acid attack and biogenic sulfuric acid attack.

Some of the methods used for testing sulfate attack simulate external attack by means of immersion of the samples in a test solution or by performing continuous wetting and drying cycles. Others, such as the ASTM C452 (2006) and Chatelier-Anstett test method (Talero 2002), simulate internal sulfate attack by mixing the cement with gypsum. Different sulfate solutions and concentrations are used. Sometimes the same test solution is used during the whole testing period, and at other times the solution is replenished periodically. In the SVA test method (Mielich and Öttl 2004) and the Wittekindt test method (Wittekindt 1960), the test solution is replenished every week. In the Soaking and Drying test, introduced by Almeida (1991), the solution is renewed after every 15-cycle series. Mehta and Gjorv (1974) used a circulating solution in which the pH is kept constant by manual titration with H_2SO_4. Some of the test methods, for example the MNS (Mulenga, Nobst and Stark) test method (Mulenga et al. 1999) and the Duggan method (Scott and Duggan 1986), use concrete samples, others test mortar or cement paste with different w/c-ratios. Different samples sizes and shapes are used, trying to accelerate the tests. Wittekindt (Wittekindt 1960), CEN (Mauris 2005) SVA (Mielich and Öttl 2004), ASTM C1012 (2004) and C452 (2006) use length change measurements as degradation parameter while in the Chatelier-Anstett method (Blondiau 1961) diameter increase is used. Decrease in flexural strength is used in the Koch-Steinegger (Steinegger 1970) and Notched Beam test (Hughes and Grounds 1985). All the different parameters, mentioned above, influence the duration of each test method, which varies from a few days to a few months.

Test methods, where sulfuric acid is used as an aggressive liquid, are in most cases similar to those for sulfates. Some examples are outlined in Chap. 11. The test specimens, mostly mortars, are submerged in a sulfuric acid solution or are exposed to wetting and drying cycles. The test solution is replenished at regular times or pH is kept at a certain level by titration. The deterioration of specimens is characterised by investigating mass loss, change in radius and/or surface roughness. Change in compressive strength is also used as an indicator for the attack. A key difference between the different test procedures is whether or not the test specimens are brushed before degradation measurements are taken.

Biogenic sulfuric acid corrosion consists of a complex process. Several researchers tried to simulate this type of corrosion as it happens in situ by creating optimum conditions for the bacteria, trying to increase the rate of corrosion. The rate of attack is determined by mass loss of the test specimens, decrease of the cross-section and the change in pH of the water in which the concrete blocks are submerged (see Chap. 12, Part III).

The half-cell potential measurement test (Assaad et al. 2002, 2005) determines the resistance of reinforced concrete against steel corrosion due to sulfate attack. Assaad et al. (2005) concluded that, for structures made of reinforced concrete in a sulfate environment, concrete corrosion is the critical factor and steel corrosion is less of a problem. The results of this test indicate that the quality of the concrete is the key to the best protection of the steel in sulfate exposure conditions.

Besides the above mentioned test methods there are some other approaches used to study sulfate attack. Brown and Hooton (2002) presented the data obtained from a Scanning Electron Microscope with Energy Dispersive X-ray (SEM/EDX) analysis of concrete specimens immersed in a sulfate solution for 21 years. The authors were able to clearly identify the formation of ettringite, gypsum and thaumasite in the pores. Borsoi et al. (2000) preferred to use X-ray Diffraction (XRD) for the detection of degradation products. Also X-ray Fluorescence (XRF) can be used to establish the elemental composition of the samples (Mauris 2005). Besides, there are some other supplementary techniques used by researchers with the aim of obtaining further information about the progress of sulfate attack on cementitious materials. Paglia et al. (2002) used Ion Chromatography (IC) measurements to investigate sulfate adsorption. They also used pH and conductivity measurements to define the total amount of adsorbed sulfates. By using the impulse excitation technique (Mauris 2005) the E-modulus can be determined, giving an indication of the damage due to sulfate attack.

10.5 Critical Review of Test Methods

10.5.1 *Influence of the Type of Sulfate Solution*

The use of the term 'sulfate attack' has led to most research focusing on the effects of the SO_4-ion alone. In reality, attack by different sulfate solutions, such as those containing Ca, Na, Mg, Fe as the cation, proceeds differently with respect to the mechanism of attack and the manifestation of damage. For instance, in the case of sodium sulfate (Na_2SO_4) attack, tricalcium aluminate (C_3A) and calcium hydroxide (CH) are attacked, but the calcium silicate hydrates (C-S-H) remain intact. CH is dissolved, as calcium is required for the formation of gypsum and ettringite, and transformed into sodium hydroxide (NaOH). As the presence of NaOH in solution ensures the continuation of high alkalinity in the system, the C-S-H gel, being the main strength-producing component of the hydrated cement paste, remains intact.

The reaction products, ettringite and gypsum, lead to extensive cracking, caused by the expansive stresses generated. The reaction of calcium sulfate ($CaSO_4$) is similar except that damage occurs more slowly, because the solubility of Ca^{2+} ions is lower and because $CaSO_4$ supplies the calcium needed for the reaction that otherwise would be removed from CH. However, when the attacking solution consists of magnesium sulfate ($MgSO_4$), all cement minerals are attacked, including C-S-H. Formation of the relatively insoluble and poorly alkaline brucite ($Mg(OH)_2$), due to reaction with dissolved CH, reduces the stability of the C-S-H gel. In addition, $MgSO_4$ directly attacks calcium silicate hydrate, transforming it into gypsum and weak magnesium silicate hydrate (M-S-H). In this case surface damage is imperceptible until late stages of attack, because of a dense brucite layer on the surface, while reduction in strength begins quite early. The attack of sulfates in seawater is again different because of the presence of a substantial quantity of Cl-ions. Chlorides penetrate into the hydrated cement paste and bind the C_3A to form chloroaluminate compounds. The lowered availability of C_3A can reduce the damage caused by sulfate attack due to a direct reduction in the quantity of ettringite that can be formed (Santhanam 2007; Hooton and Brown 2009).

Different cations do not only show different attack and damage mechanisms, they have also different solubility. Rebel et al. (2005) stated that the solubility of magnesium and sodium is respectively 35.7 g and 28.1 g in 100 g of water at 25°C, which is much higher than the solubility of calcium, which is 0.205 g in 100 g of water at 25°C. This difference is important because the lower the solubility of the sulfate, the lower its possible concentration in the groundwater and the lower the possibility that the sulfate would penetrate into the concrete structure. However, it must be noted that if even low amounts of other cations such as sodium or magnesium are present together with calcium, its solubility increases dramatically. As mostly more than one cation is involved in the soil, Rebel et al. (2005) stated that it would be good to test the solubility in a water soluble sulfate test in order to estimate how much sulfate in the soil would be dissolved in the ground water, become mobile and migrate into the concrete. From these results the destructiveness of the sulfate source to the concrete may be predicted.

Many of the test methods listed in Table 10.1 use a solution of Na_2SO_4 as the attacking sulfate solution. The resistance against sulfate attack of the cementitious specimens is based upon the results gained from these tests. This can lead to improper design in the case when concrete structures are exposed to a sulfate solution other than the Na_2SO_4 test solution. According to Santhanam (2007), using a low C_3A content might be favourable in the case of exposure to Na_2SO_4 solutions, but may not be effective when the structure is exposed to $MgSO_4$ solutions and in particular, in the case of sulfuric acid attack, where the prevailing low pH conditions lead to a direct attack of the C-S-H. It is important that, before testing, some research is done to see which types of sulfate solutions are present in the environment of the structure and that, to a certain extent, a similar solution is used in the test method.

Since different sulfate solutions cause different types of damage, the use of an appropriate parameter for the description of damage is needed. For example, while expansion may be commonly observed in the case of Na_2SO_4 attack, strength loss is a more appropriate indicator of $MgSO_4$ attack. But as the sulfates found in ground water are often a combination of calcium, sodium and magnesium sulfate, Santhanam (2007) concluded that, using one single parameter, such as expansion, to define damage may not be sufficient.

10.5.2 Size and Shape of Specimens

Most laboratory tests use small test specimens, because handling and testing are easier with small specimens. An even more important reason for using small specimens is to obtain test results early by accelerating the attack process. However, Santhanam (2007) stated that the disadvantage is that patterns and mechanisms of failure are often different between small laboratory specimens and real life structures.

The ASTM C1012 (2004) test method uses mortar prisms with dimensions of $25.4 \times 25.4 \times 285$ mm. Ferraris et al. (2005, 2006) tried to accelerate this test method by using smaller test specimens. They used $10 \times 10 \times 40$ mm test prisms and stated that the same conclusions regarding the potential sulfate resistance could be obtained using either method. However, the new method, whereby expansion measurements are performed on smaller prisms, requires less than one third of the testing time compared to ASTM C1012. As the ASTM C1012 test method takes between 6 months and 1 year to perform, Hooton (2008) also considers that it would be preferable to replace this method by a new one, using smaller specimens, to obtain results faster, provided that the smaller specimens give reliable results.

Apart from the dimensions of the specimens, their shape may also influence the test results. Ferraris et al. (1997) used three different specimen geometries: spheres, prisms and cylinders with various diameters of 25, 50 and 75 mm. The cylinders were sealed at the ends, so penetration of sodium sulfate was transverse to the cylinders' axis. It was observed that the larger diameter cylinders expanded more slowly than the smaller diameter cylinders. This observation suggested that the transport rate of sulfate ions into the material governs the deterioration. Prisms with dimension of $25 \times 25 \times 279$ mm, tested at the same time, expanded like the 25 mm diameter cylinders, confirming that the size and not the shape was the important factor in the time scale of the expansion due to sulfate attack. However, the cracking pattern in both situations was different. The stress concentrations introduced by the sharp edges on the sides and ends of the prisms and the ends of the cylinders led to cracking along the edges first, which tended to dominate the subsequent random surface cracking and spalling. The spheres showed random surface cracking only, followed by spalling.

10.5.3 Measured Degradation Parameters

Table 10.1, listing several test methods, shows that different degradation measurements are used. Wittekindt (Wittekindt 1960), SVA (Mielich and Öttl 2004), CEN (Mauris 2005), ASTM C1012 (2004) and C452 (2006) use length change as an indication of degradation. Koch-Steinegger (Steinegger 1970) uses decrease in flexural strength while the (biogenic) sulfuric acid corrosion tests (De Belie et al. 2004, De Belie 2007; Monteny et al. 2000, 2001; Vincke et al. 1999) mostly use change in mass as degradation measurement. Also change in radius, reduction in sound velocity, electrochemical potential, change in pH, etc. are used as parameters to quantify degradation.

Expansion measurement appears to be the most common method used to predict or evaluate the sulfate resistance because these non-destructive tests are easy to perform. Expansion is mainly caused by the formation of gypsum and ettringite. As mentioned by Almeida (1991) gypsum occupies a volume 2.24 times larger than that of the original compounds and the formation of ettringite leads to a volume increase of 2.5–3.27 times. However, it should be mentioned that not all researchers agree with the fact that gypsum formation causes expansion (Tian and Cohen 2000). Besides, Atkinson and Hearne (1990) observed that only about 1/20 of the solid volume increase due to ettringite formation appears as bulk expansion, because reaction products can form in the pore structure. Therefore, they stated that the rate of degradation is not linearly related to the rate of expansion.

Mehta et al. (1979) described the inadequacy of length change measurements. Alite cements (C_3S) tested using the ASTM C452 (2006) test method exhibited minimal expansion because the cements contained no C_3A, and therefore would not produce expansive ettringite upon exposure to a sulfate solution. However, after 6 years of curing, spalling occurred, and it was discovered the C-S-H had converted into gypsum and aragonite. Mehta et al. (1979) were of the opinion that sulfate deterioration, most often reported in the field, is not caused by ettringite formation, rather it is due to the decomposition of CH and C-S-H to gypsum, and by conversion of these hydration products to aragonite. Neither of the currently accepted ASTM test methods predicts this form of damage. Because both ASTM C452 (2006) and C1012 (2004) use mortar bar expansion as a measure of sulfate resistance, only the ettringite form of sulfate attack is considered.

Another disadvantage of the C1012 (2004) test method is described by Ferraris et al. (2005). In the ASTM C1012 test method the sulfate solution penetrates from both the end and side surfaces of the test prisms. This penetration from multiple directions results in the prism ends being more completely affected than the mid-length portions of the prism. The highly affected area extends about half the length of the measuring pin, which is then lifted by the expansion of the outer mortar. This would indicate that the early expansion measured is not the result of the entire cross-section reacting to the sulfate infiltration and that the core, up to that point, has served to restrain expansion. Measurements of expansion according to ASTM C1012 are based on the assumption that the whole specimen is expanding uniformly at the same time, while it was observed in one case by Ferraris et al. (2005, 2006) that only a small fraction around the pin of the test specimen was responsible for most of the measured expansion. As expansion is measured as a percentage of the total gauge length affected (in ASTM standards, the gauge length is taken as the free distance between the embedded ends of the length change studs), the real expansion near the pin could be 50 times larger than reported. ("Real expansion" means the change in length due to the chemical reaction divided by the affected length of the specimen). It has not been confirmed that this is a common occurrence, but two solutions were proposed by Ferraris et al. (2005, 2006) to circumvent this problem. The first solution consists of protecting the ends of the specimens with epoxy (the end faces and 5 mm down the sides of the prisms), so sulfate penetration cannot occur from these surfaces. The second solution reduces the cross-section of the specimens to shorten the time necessary to permeate the specimen with the sulfate. The combination of these modifications would protect the end surfaces and pin region from sulfate penetration while decreasing the time for the solution to permeate the test bar cross-section, producing a more effective configuration to measure expansion of the exposed sides of the specimen.

Cao et al. (1997) argued that the amount of expansion might not necessarily indicate specimen failure and that in some cases low expansion could be associated with signs of crumbling, indicating early failures. This suggested that reduction in strength would be a better parameter to assess sulfate resistance.

The accelerated test method described by Mehta and Gjorv (1974) uses loss in strength as parameter for sulfate resistance. Cracking caused by ettringite or gypsum formation and ettringite expansion as well as loss of C-S-H will adversely affect compressive strength. Thus, both forms of sulfate attack are considered when loss in strength is used as measure of damage experienced during sulfate exposure.

However, using change in strength as indicator would require a large number of test specimens, especially for prolonged tests. Strength loss measurements are destructive test methods and therefore Ferraris et al. (2006) stated that the number of specimens needed would be too large to be feasible. Another disadvantage, quoted by Paglia et al. (2002), is that, under particular conditions, the hydrates formed during the attack may fill the pores resulting in an unexpected strength increase during initial stages of attack.

Mass loss can also be used to determine the degradation rate. Al-Amoudi (1995) stated that for different cement blends, stored in mixed magnesium sulfate or sodium sulfate solutions, the deterioration is best assessed by the mass loss rather than length change. Mass loss is often used in biogenic sulfuric acid corrosion tests. However, when using this parameter we have to be aware that the formation of reaction products can also lead to an increase in mass. Only after a given time of test, the mass of the material that disintegrates from the specimen exceeds that of the material retained inside, making mass variation negative.

Modelling criteria for sulfate attack could be successfully developed if one unified degradation parameter was used but, due to the complicated mechanism of sulfate attack, it is not possible to use only one parameter to predict sulfate related damage in all environmental conditions. Expansion measurements are often proposed as a representative parameter for damage but Santhanam (2007) stated that, in many field studies, surface softening and loss of structural stability are the key damage indicators. Since the choice of the degradation measure may lead to different conclusions, regarding the relative performance of concrete types, one single measure may not suffice to characterize the degradation sufficiently. Therefore it is recommended to use multiple relevant indicators to investigate the resistance of concrete against sulfate attack. For instance a combination of mass loss, expansion measurements, decrease in flexural strength and E-modulus might be used.

10.5.4 Parameters Influencing the Performance of Cementitious Materials in Sulfate Solutions

External sulfate attack will occur when a sulfate solution penetrates the concrete and chemically reacts with its constituents, mainly the cement matrix. Thus, factors affecting the sulfate resistance of concrete are not only those influencing the chemical reaction with the cement matrix, but also those influencing the possible sulfate transport through the pore system. When a higher w/c-ratio is used, a more porous concrete is obtained and transport of sulfates happens faster.

Therefore, to predict the resistance of concrete against sulfate attack, it is necessary to develop a protocol that separates the various mechanisms, such as absorption of sulfate solution, diffusion of sulfates into the pore structure, and chemical reaction between the sulfates and the hydration products. According to Ferraris et al. (2006) these mechanisms cannot all be examined in one single test method. They proposed to determine the sorptivity and diffusion coefficient of the specimen before testing the chemical reaction between cement paste and sulfate ions. The first two properties to investigate depend on the mixture design of the concrete and, to a lesser extent, on the specific cement used, while the last property depends uniquely on the composition of the cement used. Therefore, it can be conceived that a "non-sulfate resistant" cement in a low sorptivity (or low diffusion coefficient) concrete will have a longer service life than the same cement in a high sorptivity (or high diffusion coefficient) concrete. Ferraris et al. (2006) concluded that at least three tests are needed to determine the service life of a concrete structure in regard to sulfate attack: a water absorption test on concrete, a diffusion test on concrete and a test for the sulfate attack resistance of the cement. As most of these procedures are shorter than the current standard tests, it is expected that the determination of the cement and concrete performance under sulfate attack, using these tests, would require less testing time.

10.5.5 Cement Versus Concrete

Most of the test methods on sulfate attack, listed in Table 10.1, are carried out on cement paste or mortar specimens. The question may arise if these samples behave in the same way as concrete under sulfate attack.

Cohen and Mather (1991) state that the behaviour of cement paste and concrete under sulfate attack cannot be compared. Because of the presence of aggregates, a transition zone appears in concrete between the aggregates and the paste. Due to the effect of the transition zone, the microstructure of pure paste and paste within concrete are different. Since the effects of transition zones extend far beyond the aggregate surface and the individual effects of surfaces in concrete overlap, Cohen and Mather (1991) stated that little or no paste in concrete has the same microstructure as pure paste. The authors suggested that when research is done on the resistance of concrete against sulfate attack, experiments both on concrete and on cement paste should be carried out and compared.

Bentur and Cohen (1987) also investigated if pure cement paste could be used as a representative model to extrapolate concrete behaviour. The effects of the transition zone were studied and this research proved that the expansion in the transition zone was limited. Moreover, when additives such as silica fume are added to mortar or concrete, the transition zone is no longer visible in the matrix, resulting in almost no micro-structural difference between concrete and cement paste. Bentur and Cohen (1987) proved that, in this latter case, the use of cement paste, to explain concrete behaviour, is possible. Some other comparisons between the behaviour of concrete and cement paste were carried out by Grabowski et al. (1992) using the Duggan (Scott and Duggan 1986) test method (involving severe wet/dry heat cycling prior to exposure to water as detailed in Table 10.1). It was seen that the rate of concrete expansion followed the rate of cement expansion and did not depend on the aggregate size.

10.5.6 Field-Like Conditions in Accelerated Tests

There is a need for test methods simulating field-like conditions of concrete under sulfate attack. However it is not always possible to create field-like conditions in the laboratory. The primary difficulties are related to the scale of testing, as well as the need for results within a short time period. As discussed before, most test methods use small sized specimens because they are easier to handle and especially because they accelerate the degradation process. There are also other methods to accelerate tests but it is important that they mimic, to a certain degree, the behaviour of concrete under field conditions.

ASTM E632 (1996) describes the conditions which have to be fulfilled by accelerated tests. According to this standard, accelerated aging tests are: "Tests in which the degradation of building materials is intentionally accelerated over that expected in service. One basic assumption in this methodology is that the mechanisms of action in the laboratory and the field have to be identical."

In the following sections, conditions used in accelerated test methods are compared with the field conditions and the possible impact on test results is discussed.

10.5.6.1 Sulfate Concentration

One way to shorten the duration of a test method is by changing the concentration of the sulfate solution in order to obtain acceleration of the degradation process. The highest sulfate concentrations in groundwater, in a so-called highly aggressive environment, where the lower limits of exposure are defined by ACI 201.2R-92 (1992) and EN 206 (2000) standards, are 10,000 and 6,000 mg/L SO_4^{2-} respectively, while the ASTM C1012 (2004) test method proposes the use of a sulfate concentration of 50,000 mg/L Na_2SO_4 which is equivalent to 34,000 mg/L SO_4^{2-}. Although in some areas of North America, sulfate concentrations in excess of 34,000 mg/L SO_4^{2-} are not unusual (e.g. parts of Colorado and South Dakota), the concentrations used in the ASTM C1012 (2004) test method are much higher than the values specified by the ACI and EN standards.

This increase in concentration may result in unwanted effects, since changing the sulfate concentration can change the attack mechanism. Biczok (1967) and Cohen and Mather (1991) mention that, for attack by sodium sulfate solution at a low concentration of sulfates (<1,000 mg/L SO_4^{2-}), the primary product is ettringite, while at high concentrations (>8,000 mg/L SO_4^{2-}) gypsum is the main product. In the intermediate range (1,000–8,000 mg/L SO_4^{2-}), both gypsum and ettringite are observed. In the case of magnesium sulfate attack, ettringite production is observed at low concentrations (<4,000 mg/L SO_4^{2-}), a mixture of ettringite and gypsum forms at intermediate concentrations (between 4,000 and 7,500 mg/L SO_4^{2-}), and magnesium corrosion dominates at high concentrations (>7,500 mg/L SO_4^{2-}). Glasser (2009) also found that gypsum formation depends on both the absolute concentration of the test solution and on the mass ratio (the mass of the sulfate solution in which the specimens are submerged divided by the mass of cement present in the samples). The higher the mass ratio and sulfate concentration, the more gypsum is formed. Cohen and Mather (1991) concluded that in the case of sulfate attack, laboratory results with accelerated test conditions by increasing sulfate concentration could not be used to predict actual behaviour in the field. (Glasser and Matschei reach the same conclusion – see Chap. 2, Sect. 2.4.3, Part I).

On the other hand, Munn et al. (2008) performed several experiments on the resistance of geopolymer concrete against sulfuric acid attack by using different concentrations of the attacking solution. They observed that geopolymer concrete in a 10% acid for 8 weeks had similar strength reduction as specimens of the same concrete soaked in 1% acid for about 14 months.

Hughes and Grounds (1985) developed a sensitive test method, which could be performed under realistic sulfate concentrations, while data could still be obtained within a reasonable time. In the Notched Beam test (a bending test performed on a notched beam), a localized concentration of stresses occurs at the tip of the notch.

By concentrating the sulfate attack and its evaluation in a particular zone (the tip of the notch), the notched beam test has proved to be a sensitive measure of the sulfate resistance, so high sulfate concentrations are not needed to perform this test. In this test the maximum load, P_{max}, increased until a critical immersion time was reached, when it suddenly decreased to below its initial value. Since the increase in P_{max} occurs before the onset of rapid expansion, the notched beam test is not only a sensitive technique, but gives also an earlier indication of the deterioration than does the expansion test, while the final ranking of sulfate resistance is similar for the two techniques. However, because of the relative complexity of this test method, the notched beam test has not yet been accepted as a commercial test.

10.5.6.2 pH of the Sulfate Solution

Investigations by Mehta (1975), Brown (1981) and Glasser (2009) suggest that continuous immersion of test specimens, as suggested by several test methods, is not a valid representation of the field conditions, since the pH of the attacking solution rapidly changes from neutral (~7) to basic (~12) due to a leaching of the alkalis from the concrete into the surrounding sulfate solution. Furthermore, the sulfate concentration in the solution decreases during continuous immersion. This could lead to a discrepancy in using lab experiments to predict field behaviour. In general, lab specimens, subjected to continuous immersion, are able to withstand the attack longer than the corresponding field exposure specimens. This is because the field specimens are exposed to a sulfate solution with an almost constantly maintained sulfate concentration and pH-value.

Mehta (1975), Brown (1981) and Ferraris et al. (1997) are some of the researchers pointing out the importance of having a constant pH-solution throughout the testing period. Since in real field conditions, the sulfate concentration of the groundwater would always remain the same, testing methods should also aim to maintain the same pH throughout the testing. Moreover, in their cited work, Ferraris et al. (1997) showed that the attack progresses much faster in a constant pH environment, favouring the shortening of the time needed to obtain test results.

Merriman (1955) was the first to attempt to carry out a test method in a controlled environment. In 1933 he developed the so-called Merriman Slab test. In this test, specimens were immersed in a 10% Na_2SO_4 solution in which sulfate concentration was maintained constant by titrating the solution to neutrality on a daily basis using sulfuric acid (H_2SO_4). Mehta and Gjorv (1974) developed a test method in which the pH and the sulfate concentration were kept constant at a desired level. In their experiments Mehta and Gjorv (1974) kept the pH of the immersion solution at the original pH by manual titration with H_2SO_4; later Mehta (1975) automated the titration by using a pH-stat. This arrangement permitted pH control in the range of 5.7–6.7. Brown (1981) further developed this test and investigated the effects of pH of the sulfate solution on strength and expansion. He found that a control of the pH at 7 significantly increased the rate of sulfate attack, as measured by either strength loss or expansion, compared to ASTM

C1012 (2004). Brown (1981) also found that fixing the hydroxyl ion concentration in solution, via controlled additions of sulfuric acid, also fixed sulfate ion concentration in solution, thus leading to a more field-like situation. Clifton et al. (1999) used an apparatus similar to that used by Brown (1981).

Glasser (2009) proposed to stabilize the sulfate concentration during the test by keeping the mass ratio ((mass of aqueous phase)/(mass of cement)) as high as practical and certainly > 10. He also proposed to monitor the decline of aqueous sulfate and to renew the solution when the sulfate concentration decreased by more than 10% of its initial value. In order to control the pH of the solution, two methods are quoted by Glasser (2009): bubbling air through the solution or acid titration using nitric acid (HNO_3).

Mehta (1975) pointed out that sulfate solutions, with constant but different pH value, can lead to different reaction products. For instance weak solutions of sodium, calcium or magnesium sulfate can react with the aluminate constituents of the cement paste to form ettringite, which leads to expansion and cracking of the concrete. However stronger solutions are somewhat acidic (6.0–6.5 pH) and are therefore capable of entering into chemical reaction with the calcium hydroxide present in the cement paste. Finally, sulfate solutions of weaker cations, such as iron can be highly acidic (2–3 pH) and can decompose even the calcium silicate hydrate gel, the main cementing constituent. The acidic attack, involving the conversion of calcium hydroxide and calcium silicate hydrate gel to gypsum, is accompanied by surface softening and strength loss. Mehta (1975) stated that the constant pH is a desirable feature of testing, since it allows simulation of field exposure conditions. The pH can be maintained at that of sea water, soil water, and if necessary a variable pH environment can be simulated.

In his experiments to determine the acid resistance of sewer pipe concrete, Fourie (2007) used sulfuric acid solutions at a pH of 1.0. According to Fourie (2007), a pH of 1.0 is an ideal strength for testing as it is similar to that of sulfuric acid environments generated by the *Acidithiobacillus thiooxidans* species, which are responsible for the degradation of sewer pipe concrete.

However, unlike the real macro size concrete elements, small sized samples are thoroughly affected by the leaching process, as a result of a low pH solution. The data obtained during the test cannot quantify the contribution of the leaching process in the loss of strength observed. Another misleading effect, of low pH solutions, is its interference to the stability of the primary and the secondary ettringite that has formed in the samples. Ettringite is stable at pH 11.5, unstable at pH 10 and highly unstable at pH 6. In this case, since the cement paste cubes are so small, the decomposition of the existing ettringite crystals in the entire sample is highly probable. However, Mehta (1975) has observed ettringite in mortar specimens tested in a sulfate solution held at pH 6.2 indicating that the pH gradients in those samples were large enough to allow ettringite formation.

10.5.6.3 Wet/Dry Cycles

In field conditions concrete is also subjected to environmental effects such as wetting and drying, which could accelerate the degradation procedure. Wetting/drying cycling, as used by Almeida (1991), De Belie et al. (2004) and De Belie (2007), or partial immersion in solution, as used for the half cell potential measurements (Assaad et al. 2002, 2005), could lead to extremely high sulfate concentrations and cause rapid deterioration due to sodium sulfate crystallization. Almeida (1991) found that in the soaking and drying test, the porosity and capillary absorption controlled the performance of concrete more than the chemical composition. The highest resistance to sulfate attack was found in concretes with superplasticizer. This chemical product, used for reducing the w/c-ratio in concretes keeping the workability constant, leads to a lower porosity and thus to a better resistance of the concrete during the soaking and drying test.

Studies of wet and dry cycling by the U.S. Bureau of Reclamation indicate that the accompanying degradation is more rapid than that associated with continuous immersion. From a comparison of the times for specimens to reach an expansion of 0.5%, it was estimated that 1 year of accelerated testing (wetting and drying) produced damaging effects that equalled 8 years of continuous immersion in the solution (Clifton et al. 1999; Santhanam 2007).

However none of the above-mentioned studies were concerned with identifying the mechanisms of degradation by wetting and drying cycles, which poses the question: Was the degradation process in these studies purely physical involving crystal pressure or were chemical processes also involved such as the formation of ettringite? Clifton et al. (1999) stated that depending on the time of drying, both physical and chemical processes may be involved, with the chemical processes likely dominating at slower drying rates.

10.5.6.4 Internal Sulfate Sources

As mentioned before, external sulfate attack occurs when sulfates penetrate from the outside environment through capillary pores and cracks into the microstructure of the concrete. Significant symptoms of attack do not occur until sufficient sulfate enters the microstructure and reacts. The internal form of chemical sulfate attack is likely very similar to the mechanism of external sulfate attack, except that the diffusion of sulfate from outside of the concrete is not a necessary precursor, as the sulfate is already within the concrete at the time of hardening. Felekoglu et al. (2006) reported that concrete subjected to internal sulfate attack showed external symptoms of damage very much earlier than the external form of attack, because of the instant availability of sulfates.

The ASTM C452 (2006) test method, which uses a sulfate addition in the mix itself, can potentially provide a basic assessment on the reactiveness of tested cements in a relatively short time, such as 2 weeks. But this test method does not represent field conditions in a realistic way. The initial 7% SO_3 content of the mortar causes acceleration in the progress of all reactions, shortening the time to get the test results. This causes an attack which is much more severe than those that may occur in a real environment. Contrary to recommendations of ASTM E632 (1996) the test conditions of ASTM C452 (2006) do not simulate field exposure of concrete to sulfate environment, which involves the ingress of sulfates into concrete and the subsequent exfoliation of the attacked layers, opening up free surfaces and accelerating the ingress of sulfate into the concrete (Clifton et al. 1999).

While ASTM C150 (2009) sulfate levels allowed in (over-sulfated) cements were increased several times from 1941 to 1971 (Biczok 1967) as cement compositions and finenesses changed to allow better optimization of sulfate contents, more recently it was decided to allow the SO_3 limits to be exceeded provided that it could be demonstrated that the optimum SO_3 content was above the stated limit (typically using the ASTM C563 (2007) optimum sulfate test). In this case a 14-day mortar bar expansion test (ASTM C1038 (2004), which had been previously adopted in Canadian CSA cement standards) was also required to be performed to show that the level of SO_3 in the cement would not result in adverse expansions from internal sulfate attack. This performance test appears to have been satisfactory when a 14-day expansion limit of 0.020% was specified. This test and expansion limit has been extended to blended cements and combinations of cementitious materials in the CSA A3001 (2004) specifications.

10.5.6.5 Concrete Permeability

Since some disadvantages are associated with the internal inclusion of SO_3, other methods were sought to accelerate the diffusion process. According to Moir (1999), increasing the w/c-ratio to 0.60, and thus increasing the porosity of the concrete, may be used to shorten the time scale necessary to discriminate between cement pastes which are inherently sulfate resisting and those which are not. However, not all types of concrete have equal permeability at equal w/c-ratio, which may lead to misinterpretation of the results. For example, concrete made with ordinary Portland cement or blended cement, both at the same w/c-ratio, may have a different permeability and thus the degree of sulfate ingress and deterioration due to sulfate attack may differ. Hooton and Brown (2009) stated that prior to the use of water-reducing admixtures and supplementary cementing materials or blended cements, concrete with equal w/c-ratio had almost equal permeability but now a concrete can be made less permeable at a given w/c-ratio by the use of water reducing or superplasticizing admixtures which act to lower the porosity and permeability. Additionally, pozzolans or slag as part of the cementitious matrix will reduce, in most cases, the permeability. To speed up the duration of the test, it would be better to test concrete samples with the same high permeability instead of testing samples with the same high w/c-ratio. Based on these results, one can be able to analyze the sulfate resistance of different cement compositions. However, rapid tests are needed to determine the permeability.

Using more permeable samples may be useful to accelerate the degradation process, however, one should be aware that when more pores are present in the concrete matrix, the sulfate solution will penetrate more quickly but as there is more place for the reaction products, expansion may occur later.

Besides Monteny et al. (2000), in a study of attack by sulfuric acid, stated that the degradation rate may even be higher when using lower w/c-ratio. Due to lower w/c-ratio, the porosity will decrease, so sulfate ingress and corrosion should be lower but as the relative amount of cement increases, this may implicate a higher rate of acidic attack.

It should be noted that ACI, EN and most other standards limit maximum w/c-ratio for sulfate exposures, with lower w/c-ratio required for more severe exposures, since the first line of defence is to limit the rate of sulfate penetration.

10.5.6.6 Supplementary Cementitious Materials

In spite of an increased potential for ettringite formation due to the supply of supplementary sources of alumina, blended cement concretes, if produced at low w/c-ratio, are generally regarded as more resistant to sulfate attack because the resistance to fluid penetration is increased by the pozzolanic reaction.

In the ASTM C452 (2006) test method, calcium sulfate is added to the mortar during mixing, in order to accelerate the test by eliminating the time for sulfate ingress. As stated before, a disadvantage of this procedure is that sulfate attack starts immediately after mixing. This results in insufficient time for the pozzolanic action of mineral additions in blended cements. Therefore, this test method is only appropriate for Portland cement and not for blended cements and cementitious combinations containing supplementary cementing materials (SCM), as these systems rely on reduced permeability and reduced calcium hydroxide content to provide resistance, and this requires hydration prior to sulfate exposure.

A solution for this problem, suggested by Clifton et al. (1999), would be to combine gypsum with a partially hydrated cement paste and make compressed specimens that are exposed to a sulfate solution. By this approach, ASTM C452 (2006) may become applicable to all types of hydraulic cements.

10.5.6.7 Test Temperature

Concrete structures might be exposed to sulfate attack when they come into contact with groundwater containing sulfate. The temperature at which the attack occurs can differ sharply from the temperature of 20–23°C normally used in laboratory test methods. In Germany and neighbouring countries, for example, the average soil temperature is only about 9°C. Caution is needed since the mechanism of damage may change at temperatures less than 15°C. For example, the thaumasite-form of sulfate attack will occur at low temperatures, leading to loss in strength and, in extreme cases, to complete destruction. In contrast to the classical expansion reactions by ettringite and gypsum, this type of damage is not necessarily accompanied by expansion (Schneider et al. 2002).

Additionally, due to the inverse solubility curves of both calcium hydroxide and calcium sulfate, raising temperatures will not necessarily accelerate sulfate attack, and in fact, may decrease it. Further, temperatures above about 70°C will destabilize ettringite.

In attempting to reduce the duration of test methods, the environmental conditions of the test do not always duplicate the conditions encountered in the field. Therefore, it is likely that the material response and service life might not be the same in the laboratory and in the field.

10.6 Special Types of Sulfate Attack

10.6.1 Thaumasite Attack (Hooton and Brown 2009)

Thaumasite was first identified in deteriorating concrete in 1965 by Erlin and Stark (1965), then much later by Bickley et al. (1994), and has been extensively investigated by Crammond (2002) and the Thaumasite Expert Group (UK Expert Report 1999) in the UK. Ettringite and thaumasite are frequently found together in deteriorated concrete. Thaumasite can potentially form in the absence of external sulfate due to carbonation, which can decompose both ettringite (releasing sulfates) and the C-S-H binder (releasing hydrous silica). External sulfate, or an inadvertent gross excess of internal sulfate, can also contribute to thaumasite formation. Similarly, if a source of readily soluble $CaCO_3$ is present within the concrete, thaumasite formation may occur. Thus, thaumasite may form as a consequence of sulfate ingress, of carbonation, or both. Thaumasite preferentially forms under the cold, wet, alkaline conditions typically experienced by buried concrete structures. While originally thought to occur only in concretes exposed to cool temperatures, thaumasite formation has also been observed in concrete placed in temperate climates (Crammond 2002).

The use of sulfate resisting cement does not prevent the formation of thaumasite, because the calcium silicate hydrates (C-S-H), not the aluminate phases, are attacked by external sulfates. The replacement of C-S-H by thaumasite results in the softening of the cement paste matrix into a white, mushy non-cohesive mass.

Currently there are no standards specifically related to the prevention of thaumasite formation and there is a lack of standardized test methods. However, papers published on thaumasite typically have tended to use mortar bar expansions or mortar cube strengths where the specimens have been stored at 5-10°C and monitored for change over time (Akpinar 2006). Hooton (personal communication) approached this by casting two sets of ASTM C1012 (2004) mortar bars which are immersed in sodium sulfate solution as per the standard, but with one set of bars stored at 5°C. While results are incomplete, it is anticipated that this approach could be used to simultaneously assess the resistance of different cementitious systems to both the ettringite and thaumasite forms of sulfate attack.

10.6.2 Sulfate Salt Crystallization (Hooton and Brown 2009)

Damage due to precipitation of sulfate salts in concrete pores near drying faces can occur in tunnels, culverts, slabs on grade or foundations under wetting/drying cycles, or a simple evaporative front. The sulfates then precipitate as salts as the water evaporates near the drying surface (Verbeck 1967). These salts can accumulate in pores and undergo phase changes due to changes in temperature and relative humidity, resulting in expansive pressures. A common reversible transformation is between thenardite ($NaSO_4$) and mirabilite ($NaSO_4 \cdot 10H_2O$), which, if occurring in capillary pores, can cause progressive surface damage. According to Scherer (2004), during wetting and drying cycles, it is the crystallization pressure from formation of mirabilite during wetting that does most of the damage.

Many of the early studies in North America addressed physical sulfate attack from salt crystallization, without acknowledging this directly. In the western part of Canada and the US, problems of premature deterioration of concrete in contact with the ground were observed in the early twentieth century. These soils contained extremely high sulfate contents and were often called 'alkaline soils' in the literature. This led to several research projects both on the roles of cement composition and concrete mixture proportions.

From the 1920s, even before the chemical reactions were understood, it was found that regardless of cement C_3A content, having low w/cm and low unit water contents of concrete mixtures were essential to obtain resistance to the actions of sulfate soils. Wilson and Cleve (1928) at the Portland Cement Association (PCA) conducted two field site studies starting in 1921 in Montrose, Colorado and at Medicine Lake, South Dakota. Over 2,000 cylinders, 250 × 60 mm, were semi-immersed in sulfate soils, while the air would have been at low relative humidity. After 7 years exposure, they reported that concretes with a w/c-ratio of 0.45 or less were relatively undamaged, while those with higher w/c suffered significant damage, as shown in Fig. 10.1. The deterioration almost certainly included evaporative transport and sulfate salt crystallization.

The PCA later carried out a series of cyclic sodium sulfate immersion and evaporation tests on 150 × 150 × 900 mm beams in Sacramento, California. It was found that both w/c and cement type influenced the results (Verbeck 1967; Stark 2002). While it was concluded that most of the damage occurred in the evaporative zone due to sulfate salt crystallization, little petrographic work was undertaken.

Several large laboratory studies were carried out at the US Bureau of Reclamation (USBR) for over 40 years. Concrete cylinders, 75 × 150 mm, were either immersed in 2.1% sodium sulfate solution or exposed to alternate wetting (in that solution) and drying cycles. An expansion of 0.5% or a loss of 40% in dynamic modulus of elasticity was considered as failure. After 40 years exposure of cements with a range of C_3A and w/c, it was found that all the mixtures with a w/c-ratio of 0.45 or less did not fail, regardless of C_3A content, as shown in Fig. 10.2 taken from a summary of the USBR work published by Monteiro and Kurtis (2003). Some mixtures did not fail with w/c as high as 0.53, but others in that w/c range (0.45–0.53) were damaged.

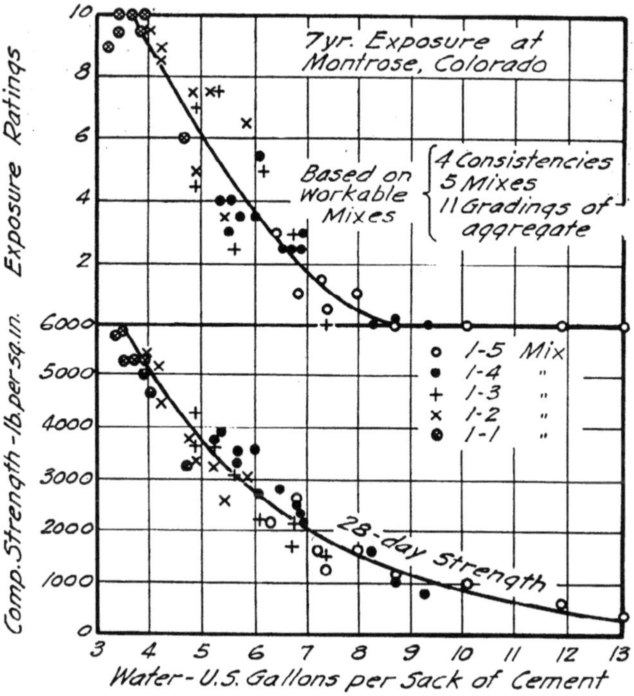

Fig. 10.1 Concrete visual ratings and strength after 7-years exposure to sulfate soils in Montrose, Colorado (Wilson and Cleve 1928). Notes: 4, 6, and 8 gal./sack are equivalent to w/c = 0.36, 0.55, and 0.73 respectively. An exposure rating of 10 is undamaged, while 0 indicates total destruction

Since obtaining a low w/cm concrete is the solution to evaporative transport and subsequent salt crystallization, the maximum w/cm limits in the concrete standards provide protection in these situations.

There is still a need for a performance test method to evaluate concretes with different permeabilities or resistances to diffusion rather than by prescribing maximum w/cm limits. A semi-immersion test was developed by Ferraris et al. (2005) where either 100 × 200 mm concrete cylinders or cores, or 50 × 100 mm mortar cylinders were immersed to half of their height in sodium sulfate solution with a layer of paraffin oil to prevent evaporation from the solutions. The top of the cylinders were exposed to three different conditions in air at 23°C at either 30 and 80% RH, or 80% RH at 40°C to obtain different possibilities of thenardite or mirabilite crystallization, as per Fig. 10.3. Spalling was observed in the air-exposed portions of the cylinders under all 3 conditions, but the fastest deterioration occurred for the 23°C-30% RH condition where thenardite formation would be favoured along with precipitation of supersaturated solution of mirabilite. In addition to visual observations of damage, diameters were measured at pre-determined locations with time.

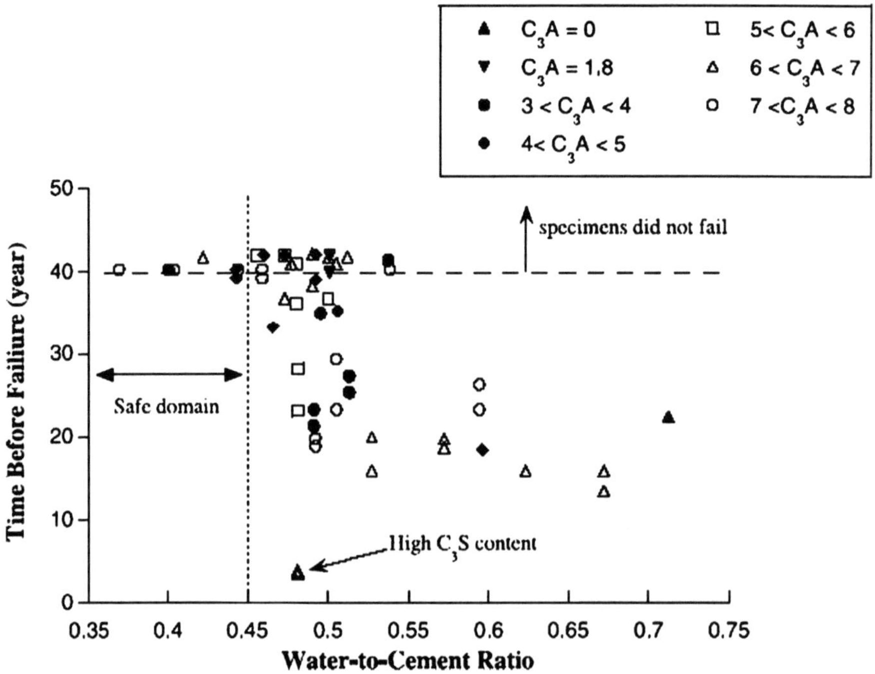

Fig. 10.2 Time to failure as a function of w/c for cements of various C_3A contents (Monteiro and Kurtis 2003).

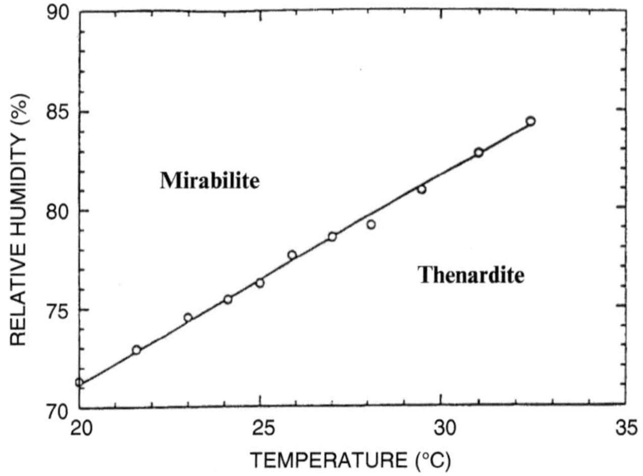

Fig. 10.3 Effect of temperature and humidity on thenardite - mirabilite phase changes (Folliard and Sandberg 1994).

10.6.3 Delayed Ettringite Formation (Hooton and Brown 2009)

The term delayed ettringite formation (DEF) was first used to describe a form of deterioration in heat-cured precast concrete railway ties (Heinz and Ludwig 1987). It was postulated that the normal early formation of ettringite was prevented due to the very high curing temperatures used, and that when exposed to moisture in service, deleterious amounts of ettringite formed resulting in expansion and cracking of the concrete. Since then, DEF has been implicated as a cause of deterioration in numerous other cases, mostly involving precast concrete elements that have been subjected to intentional heat curing during the production process or unintentional high temperature exposures at early ages. These problems have resulted in restrictions being placed on the curing process, including limits on pre-set times, heating and cooling rates, and the maximum allowable curing temperature (e.g. in Germany and in the Canadian CSA A23.4 (2009) standard for precast concrete). German experience with these curing limits is that these and similar practices in Europe have been successful in eliminating damage due to DEF (Thomas and Skalny 2006; Skalny and Locher 1997; Stark et al. 1998).

As an alternative to placing limitations on curing practices, many different methods for assessing the potential for DEF problems have been proposed, but given the lack of clear field cases where DEF is the sole form of damage, assessing their reliability is difficult. Various chemical indicators of Portland cements, such as C_3A, SO_3, alkali contents, and fineness have been proposed (Taylor 1994) but none have shown to relate to mortar or concrete expansions, outside of the set of specimens used to develop them. Based on a large experimental study performed for the PCA (Hooton, unpublished), the most reliable approach appears to be a performance-based test proposed by Kelham (1996), using expansion of mortar bars that have been exposed to elevated temperature curing cycles. As clearly shown by Famy et al. (2001), leaching of alkalis from the bars (or removal of alkalis from pore solution by ASR reactions) is an important issue and will accelerate DEF expansions. Therefore, ASTM C1038 (2004) mortar bars (which have the same mortar composition as ASTM C1012 (2004) bars) can be cast, the moulds sealed in a container over water, then exposed to elevated temperatures of at least 70°C, and monitored for expansion while subsequently stored in water (to leach alkalis). While an expansion limit of 0.10% is typically used for mortar bar tests, the question is what exposure time is adequate? The data of Ramlochan et al. (2003, 2004), indicate that a 1-year exposure period may be required if bars are stored in lime-saturated water, but Taylor et al. (2001) showed that 6 months appears to be adequate if bars are stored in plain water. Further, the beneficial effects of alumina-bearing pozzolans or slag additions can be assessed using this type of test (Ramlochan et al. 2003, 2004).

10.7 Selection of an Appropriate Test Method Based Upon the Anticipated Field Conditions

The common element to improve resistance to any form of sulfate attack is to design the concrete to reduce the rate of sulfate ingress. While most standards limit W/C or w/cm as an indirect way of limiting the permeability of concrete to aggressive solutions, this does not mean that all types of concrete will have equal permeability at equal w/cm. Therefore, it would seem more logical to replace maximum w/cm limits with maximum limits on permeability. Many fluid or ionic transport tests are time consuming and expensive, and so it would be easier to use a rapid index test that determines values related to permeability. In the CSA A23.1 (2000) concrete standard, the ASTM C1202 (2004) rapid resistance to chloride penetration test is used for concretes exposed to aggressive chemicals using a limit of 1,500 Coulombs at 56 days of age for concretes that otherwise have maximum w/cm = 0.40 and 35 MPa strength. ACI is considering a similar approach (Bickley et al. 2006). Other rapid tests such as a bulk resistivity test or the NordTest NTBuild 492 rapid migration test could be equally well considered. Even though all these tests were originally designed as rapid methods to evaluate a concrete's resistance to chloride penetration, they would apply equally well as indicators of sulfate ingress. A modification of ASTM C1202 (2004) was already suggested by Tumidajski and Turc (1995) for evaluation of sulfate resistance. They proposed a rapid test, based on electrochemical polarization, for the ingress of sulfate into concrete. The influence of the composition on the initial current delivered and the total charge accumulated over 6 h was investigated and the results corresponded with the sulfate penetration depth determined from 2 year ponding experiments.

Once a test limit is placed on the permeability of the concrete, then a series of tests could be performed to evaluate the chemical resistance of the cementitious materials to the various types of sulfate attack. It would be advantageous to use a single paste or mortar mix and cast single types of specimens that can then be placed in the required suite of environments. Specimens such as the ASTM C1012 (2004) bars, the Wittekindt (Wittekindt 1960) thin mortar bars or even the NIST mini paste prisms (Schneider et al. 2002) could be used and exposed to solutions of sodium and magnesium sulfate at both 23 and 5°C, with expansion or visual deterioration used as failure criteria. This would generally address both the ettringite and thaumasite forms of attack. As stated previously, physical sulfate attack will be minimized by limiting the w/cm or the permeability. But to prove the concept, a test program is needed where the rapid tests are evaluated against long-term performance of concretes, made with a range of w/cm, permeabilities and cementing materials, and exposed to environments simulating the various forms of sulfate attack, some of which is being currently done (Thomas and Skalny 2006).

10.8 Limitations of Test Methods

The desirable attributes of a good test method include (a) reliability, (b) reproducibility, (c) that it addresses a relevant performance issue (d) it uses realistic/appropriate boundary conditions, and (e) it is rapid and as simple as practical without overly sacrificing the other criteria (Hooton and Brown 2009).

However, it must be realized that while a test may generally address a relevant performance concern, it typically cannot mimic all relevant boundary conditions and in-situ scenarios. For example, most sulfate resistance tests only measure the resistance of a cementitious binder (typically in a mortar sample) when fully immersed in a sulfate solution (e.g. ASTM C1012 (2004) or the German thin mortar bar test). While these tests are relevant to the ettringite form of sulfate distress from attack on the cement aluminates, they do not relate to situations such as (a) where the concrete is also exposed to evaporative transport (e.g. physical sulfate salt crystallization pressures such as from reversible mirabilite-thenardite transformations), or (b) where concretes exposed to cool temperatures and carbonation favour formation of thaumasite by an attack on the C-S-H, or (c) where different types of sulfate salts and concentration levels, than those tested, are experienced.

It is doubtful that a single test could address all of these issues, and testing for all possible exposures would be both cost- and time-prohibitive. Thermodynamic calculations have been used to predict the phase assemblages that would occur for different starting materials placed in different solutions at different temperatures (Matschei et al. 2007; Glasser et al. 2008). These could then be used to determine appropriate test methods, or potentially in lieu of them. Development of predictive models, combined with rapid index tests and making use of relevant performance databases may provide useful solutions eventually. While models may be difficult to include directly in standards and building codes, they could be used to develop a more rational set of requirements than simply using sulfate ion concentrations as the method of exposure classification, and for selection of appropriate materials options for use in standards.

10.9 Conclusions and Recommendations

Many test methods are available to determine the resistance of concrete to sulfate attack, sulfuric acid attack or biogenic sulfuric acid attack. Some test methods take a long time to perform. In the case of ASTM C1012 (2004), the 6 months to 1 year required for testing would almost certainly cause problems to a company striving to maintain its production. This has led to the development of more accelerated tests, based on the methodology outlined in ASTM E632 (1996). Tests are accelerated by using higher sulfate concentrations, higher temperatures, smaller tests specimens, etc. However the question may rise if these accelerated test methods still reflect field conditions. Moreover, comparison studies between these test methods as well as inter-laboratory comparisons indicate wide scatter in results.

The following measures are recommended for improving the testing procedures for sulfate resistance:

- When a concrete structure is exposed to a sulfate rich environment, it is sensible to test the resistance of the concrete composition beforehand. The type of sulfate environment needs to be studied and the sulfate solubility and concentration need to be determined, so as to simulate during the test. For example when sodium and magnesium sulfates are present it is not sufficient to immerse a sample in a Na_2SO_4 solution (ASTM C1012) but the effect of both types of cations must be considered. Degradation due to Na_2SO_4 attack may be estimated by measuring expansion while in the case of $MgSO_4$ attack, strength loss may be used as a degradation parameter.
- If thaumasite attack or sulfate salt crystallization is expected, specific tests need to be performed.
- Using higher sulfate concentrations to accelerate the test is not always appropriate because the attack mechanisms may change. It is better to use smaller samples, replace concrete by mortar, or use specimens with a higher permeability bearing in mind that these interventions may also cause some negative effects.
- As the permeability of the concrete is a very important parameter, it would be of significant interest to develop fast tests to measure this parameter.
- It is also important that the sulfate concentration remains constant during the test, e.g. by using a high volume of sulfate solution in comparison with the volume of the specimens used. However, when the sulfate concentration drops below a certain level, the solution needs to be replenished.
- The pH of the solution also needs to be kept constant. This can be done by means of titration with sulfuric or nitric acid. Nevertheless, in case of an in situ situation of a limited amount of aggressive liquid compared to the concrete surface area, it might be appropriate not to keep the pH constant, in order to assess the buffering capacity of the material.
- The temperature used during the test should be near to the in situ temperature.
- Finally before adopting a new test method, a round robin test should be held to test its reproducibility.

References

Akpinar, P.: Multi-approach studies on the mechanisms of sulfate induced degradation of cementitious materials. Ph.D. thesis, UPC, Barcelona, Spain (2006)

Al-Amoudi, O.S.B.: Performance of 15 reinforced concrete mixtures in magnesium-sodium sulphate environments. Constr. Build. Mater. **9**(3), 149–158 (1995)

Almeida, I.R.: Resistance of high strength concrete to sulfate attack: soaking and drying test. In: Concrete Durability, pp. 1073–1092. American Concrete Institute, ACI SP-100, Farmington Hills (1991)

American Concrete Institute ACI 201.2R-92: Guide to Durable Concrete. American Concrete Institute, Farmington Hills (1992)
AS 2350.14-1996: Methods of Testing Portland and Blended Cements Method 14: Length Change of Portland and Blended Cement Mortars Exposed to a Sulfate Solution. Standards Australia, Homebush (1996)
Assaad, V.F., Jofriet, J.C., Negi, S.C., Hayward, G.L.: Corrosion of concrete in farm buildings. In: De Belie, N., Sonck, B. (eds.) Proceedings of the IVth International Symposium for Concrete for a Sustainable Agriculture – Agro, Aqua and Community Applications, pp. 149–156. Ghent, Belgium (2002)
Assaad, V.F., Jofriet, J.C., Negi, S.C., Hayward, G.L.: Laboratory tests of sulphide and sulphate corrosion of reinforced concrete. In: Sánchez Espinosa, E., Garcimartin, M.A. (eds.) Proceedings of the Vth International Symposium on Concrete for a Sustainable Agriculture, pp. 179–190. El Escorial, Spain (2005)
ASTM International C1012-04: Standard Test Method for Length Change of Hydraulic-Cement Mortars Exposed to a Sulfate Solution. (2004)
ASTM International C1038-04: Standard Test Method for Expansion of Hydraulic Cement Mortar Bars Stored in Water. (2004)
ASTM International C150-09: Standard Specification for Portland Cement. (2009)
ASTM International C452-06: Standard Test Method for Potential Expansion of Portland-Cement Mortars Exposed to Sulphate. (2006)
ASTM International C563-07: Standard Test Method for Approximation of Optimum SO_3 in Hydraulic Cement Using Compressive Strength. (2007)
ASTM International E632-82: Standard Practice for Developing Accelerated Tests to Aid Prediction of the Service Life of Building Components and Materials. (1996)
Atkinson, A., Hearne, J.A.: Mechanistic model for the durability of concrete barriers exposed to sulfate-bearing groundwaters. In: Materials Research Society Symposium Proceeding, vol. 176, pp. 149–156. Boston (1990)
Bentur, A., Cohen, M.D.: Effect of condensed silica fume on the microstructure of the interfacial zone in Portland cement mortars. J. Am. Ceram. Soc. **70**(10), 738–743 (1987)
Bickley, J.A., Hemmings, R.T., Hooton, R.D., Balinski, J.: Thaumasite related deterioration of concrete structures. In: Mehta, P.K. (ed.) Concrete Technology: Past, Present and Future, pp. 159–175. American Concrete Institute, ACI SP-144, Farmington Hills (1994)
Bickley, J.A., Hooton, R.D., Hover, K.C.: Performance specifications for durable concrete: current practice and limitations. Concr. Int. **28**(9), 51–57 (2006)
Biczok, I.: Concrete Corrosion Concrete Protection. Chemical Publishing Company Inc., New York (1967)
Blondiau, L.: Considérations diverses relatives à l'essai de résistance chimique au sulfate de calcium suivant le processus Le Chatelier-Anstett. Revue Matériaux de Construction Travaux Publics III **546**, 189–200 (1961)
Borsoi, A., Collepardi, S., Coppola, L., Troli, R., Collepardi, M.: Sulfate attack on blended Portland cements. In: Malhotra, V.M. (ed.) Proceedings of Fifth CANMET/ACI International Conference on Durability of Concrete, Barcelona, pp. 417–432, (2000)
Brown, P.W.: An evaluation of sulfate resistance of cements in a controlled environment. Cem. Concr. Res. **11**(5–6), 719–727 (1981)
Brown, P.W., Hooton, R.D.: Ettringite and thaumasite formation in laboratory concretes prepared using sulfate-resisting cements. Cem. Concr. Compos. **24**(3–4), 361–370 (2002)
Cao, H.T., Bucea, L., Ferguson, O. Sulfate resistance of cementitious materials – mechanisms, deterioration process, testing and influence of binder. In: Concrete Institute of Australia Proceedings of 18th Biennial Conference, Adelaide, pp. 263–268, (1997)
Clifton, J.R., Frohnsdorff, G., Ferraris, C. Standards for evaluating the susceptibility of cement-based materials to external sulfate attack. In: Materials Science of Concrete – Sulfate Attack Mechanisms, Special Volume, Proceedings, Quebec, 5–6 October 1998, pp. 337–355, (1999)
Cohen, M.D., Mather, B.: Sulfate attack on concrete – research needs? ACI Mater. J. **88**(1), 62–69 (1991)

Crammond, N.: The occurrence of thaumasite in modern construction – a review. Cem. Concr. Compos. **24**(3–4), 393–402 (2002)
CSA A23.1: Concrete Materials and Methods of Concrete Construction/Methods of Testing for Concrete. Canedian Standards Association, Canada (2000)
CSA A23.4-09: Precast Concrete – Materials and Construction. Canedian Standards Association, Canada (2009)
CSA A3001: Cementitious Materials for Use in Concrete. Canedian Standards Association, Canada (2004)
De Belie, N.: Evaluation of methods for testing concrete degradation in aggressive solutions. In: De Belie, N. (ed.) Workshop on Performance of Cement-based Materials in Aggressive Aqueous Environments – Characterization, Modelling, Test Methods and Engineering Aspects, Ghent, pp. 63–74, (2007)
De Belie, N., Monteny, J., Taerwe, L.: Apparatus for accelerated degradation testing of concrete specimens. Mater. Struct. **35**(251), 427–433 (2002)
De Belie, N., Monteny, J., Beeldens, A., Vincke, E., Van Gemert, D., Vertraete, W.: Experimental research and prediction of the effect of chemical and biogenic sulfuric acid of different types of commercially produced concrete sewer pipes. Cem. Concr. Res. **34**(12), 2223–2236 (2004)
EN 197-1: Cement. Composition, Specifications and Conformity Criteria for Low Heat Common Cements. (2000)
EN 206: Concrete – Part 1: Specification, Performance, Production and Conformity. (2000)
Erlin, B., Stark, D.C.: Identification and occurrence of thaumasite in concrete. In: Highway Research Record 113. Highway Research Board, Washington, DC (1965)
Famy, C., Scrivener, K., Atkinson, A., Brough, A.: Influence of the storage conditions on the dimensional changes of heat-cured mortars. Cem. Concr. Res. **31**(5), 795–803 (2001)
Felekoglu, B., Ramyar, K., Tosun, K., Musal, B.: Sulfate resistances of different types of Turkish Portland cements by selecting the appropriate test methods. Constr. Build. Mater. **20**(9), 819–823 (2006)
Ferraris, C.F., Clifton, J.R., Stutzman, P.E., Garboczi, E.J.: Mechanisms of degradation of Portland cement-based systems by sulfate attack. In: Scrivener, K.L., Young, J.F. (eds.) Mechanisms of Chemical Degradation of Cement-Based Systems, pp. 185–192. Spon Press, London (1997)
Ferraris, C., Stutzman, P., Peltz, M., Winpigler, J.: Developing a more rapid test to assess sulfate resistance of hydraulic cements. Res. Natl. Inst. Stan. Technol. **110**(5), 529–540 (2005)
Ferraris, C.F., Stutzman, P., Snyder, K.: Sulfate resistance of concrete: a new approach. PCA R&D Serial Number 2486. (2006)
Folliard, K.J., Sandberg, P.: Mechanism of concrete deterioration by sodium sulfate crystallization. In: Malhotra, V.M. (ed.) Durability of Concrete, pp. 933–945. American Concrete Institute, ACI SP-145, Farmington Hills (1994)
Fourie, C.W.: Acid resistance of sewer pipe concrete. M.Sc thesis, University of Cape Town (2007)
Glasser, F.P.: The thermodynamics of attack on Portland cement with special reference to sulfate, concrete in aggressive aqueous environments, performance, testing and modeling. In: Alexander, M.G., Bertron, A. (eds.) Proceedings Concrete in Aggressive Aqueous Environments, Toulouse, pp. 3–17, (2009)
Glasser, F.P., Marchand, J., Samson, E.: Durability of concrete – degradation phenomena involving detrimental chemical reactions. Cem. Concr. Res. **38**(2), 226–246 (2008)
Grabowski, E., Czarnecki, B., Gillott, J.E., Duggan, C.R., Scott, J.F.: Rapid test of concrete expansivity due to internal sulfate attack. ACI Mater. J. **89**(5), 469–480 (1992)
Guirguis, S., Cusack, D.: Developing performance specification for cement. In: Concrete Institute of Australia Conference, Perth (2001)
Heinz, D., Ludwig, U.: Mechanism of secondary ettringite formation in mortars and concretes subjected to heat treatment. In: Concrete Durability. American Concrete Institute, SP-100 2, Farmington Hills (1987)
Hooton, R.D.: Bridging the gap between research and standards. Cem. Concr. Res. **38**(1), 247–258 (2008)

Hooton, R.D., Brown, P.W.: Development of test methods to address the various mechanisms of sulfate attack. In: Alexander, M.G., Bertron, A. (eds.) Proceedings Concrete in Aggressive Aqueous Environments, Toulouse, pp. 280–297, (2009)

Hughes, D.C., Grounds, T.: The use of beams with a single edge notch to study the sulfate resistance of OPC and OPC/PFA pastes. Mag. Concr. Res. **37**(131), 67–74 (1985)

Kelham, S.: The effect of cement composition and fineness on expansion associated with delayed ettringite formation. Cem. Concr. Compos. **18**(3), 171–179 (1996)

Matschei, T., Lothenbach, B., Glasser, F.P.: The role of calcium carbonate in cement hydration. Cem. Concr. Res. **37**(4), 551–558 (2007)

Mauris, F.: Influence of reactive aluminium oxide on the expansion of sulphate-resisting Portland cements cured at 40°C and stored in sodium sulphate solution. M.Sc thesis (2005)

Mehta, P.K.: Evaluation of sulfate-resisting cements by a new test method. J. Am. Concr. Inst. **72** (10), 573–575 (1975)

Mehta, P.K., Gjorv, O.E.: A new test for sulfate resistance of cements. Test. Eval. **2**(6), 510–515 (1974)

Mehta, P.K., Pirtz, D., Polivka, M.: Properties of alite cements. Cem. Concr. Res. **9**(4), 439–450 (1979)

Merriman, T.: Fort peck specifications. In: Bogue, R.H. (ed.) The Chemistry of Portland Cement, p. 717. Reinhold Publishing Co., New York (1955)

Mielich, O., Öttl, C.: Practical investigation of the sulfate resistance of concrete from construction units. Otto-Graf-J. **15**, 135–152 (2004)

Moir, G.: Development of sulfate resistance performance test for cements. In: Naus, D. (ed.) Workshop design of durability of concrete, DURANET Conference, Berlin, pp. 49–55, (1999)

Monteiro, P.J.M., Kurtis, K.E.: Time to failure for concrete exposed to severe sulfate attack. Cem. Concr. Res. **33**(7), 987–993 (2003)

Monteiro, P.J.M., Kurtis, K.E.: Experimental asymptotic analysis of expansion of concrete exposed to sulfate attack. ACI Mater. J. **105**(1), 62–71 (2008)

Monteny, J., Vincke, E., Beeldens, A., De Belie, N., Taerwe, L., Van Gemert, D., Vertraete, W.: Chemical, microbiological, and in situ test methods for biogenic sulfuric acid corrosion of concrete. Cem. Concr. Res. **30**(4), 623–634 (2000)

Monteny, J., De Belie, N., Vincke, E., Verstraete, W., Taerwe, L.: Chemical and microbiological tests to simulate sulfuric acid corrosion of polymer-modified concrete. Cem. Concr. Res. **31**(9), 1359–1365 (2001)

Mori, T., Nonaka, T., Tazaki, K., Koga, M., Hikosaka, Y., Noda, S.: Interactions of nutrients, moisture and pH on microbial corrosion of concrete sewer pipes. Water Res. **26**(1), 29–37 (1992)

Mulenga, D.M., Stark, J., Nobst, P. Praxisnahes Prüfverfahren zum Sulfatwiderstand von Beton und Mörtel mit und ohne Flugasche. In: Beiträge zum DafStb – Forschungskolloquium 37, pp. 197–207. Bauhaus-Universität Weimar (1999)

Munn, R., Song, X., Marosszeky, M.: Experimental study of geopolymer concrete resistance to sulfuric acid attack. In: Türkeri, N., Şengül, Ö. (eds.) Proceeding of the Eleventh International Conference on Durability of Building Materials and Components, Istanbul, pp. 885–892, (2008)

Neuenschwander, J., Schmidt, T., Lüthi, T., Romer, M.: Leaky Rayleigh wave investigation on mortar samples. Ultrasonics 45(1–4), 50–55 (2006)

Paglia, C., Wonbacher, F., Böhni, H., Sommer, M.: An evaluation of the sulfate resistance of cementitious material accelerated with alkali-free and alkaline admixtures laboratory vs. field. Cem. Concr. Res. **32**(4), 665–671 (2002)

Park, Y.S., Suh, J.K., Lee, J.H., Shin, Y.S.: Strength deterioration of high strength concrete in sulfate environment. Cem. Concr. Res. **29**(9), 1397–1402 (1999)

Ramlochan, T., Zacarias, P., Thomas, M.D.A., Hooton, R.D.: The effect of pozzolans and slag on the expansion of mortars cured at elevated temperature, Part I: expansive behaviour. Cem. Concr. Res. **33**(6), 807–814 (2003)

Ramlochan, T., Hooton, R.D., Thomas, M.D.A.: The effect of pozzolans and slag on the expansion of mortars cured at elevated temperature, Part II: microstructural and microchemical investigations. Cem. Concr. Res. **34**(8), 1341–1356 (2004)

Rebel, B., Detwiler, R.J., Gebler, S.H., Hooton, R.D.: The right sulfate test makes a difference. Concr. Int. **27**(2), 49–52 (2005)

Sand, W., Bock, E., White, D.C.: Biotest system for rapid evaluation of concrete resistance to sulfur-oxidizing bacteria. Mater. Perform. **26**(3), 14–17 (1987)

Santhanam, M.: Sulphate attack from an engineering perspective. In: De Belie, N. (ed.) Workshop on Performance of Cement-based Materials in Aggressive Aqueous Environments – Characterization, Modelling, Test Methods and Engineering Aspects, Ghent, pp. 106–113, (2007)

Scherer, G.W.: Stress from crystallization of salt. Cem. Concr. Res. **34**(9), 1613–1624 (2004)

Schmidt, T., Neuenschwander, J., Romer, M., Lüthi, T.: Investigation of physical and microstructural changes in the surface of mortar samples exposed to sulphuric acid. In: Malhotra, V.M. (ed.) Proceedings of the Seventh CANMET/ACI International Conference on Durability of Concrete, Montreal, pp. 561–570, (2006)

Schneider, M., Puntke, S., Sylla, H.M., Lipus, K.: The influence of cement on the sulfate resistance of mortar and concrete. Cem. Int. **1**, 2–16 (2002)

Scott, J.F., Duggan, C.R.: Potential new test for alkali-aggregate reactivity. In: Grattan-Bellew, P. E. (ed.) Concrete Alkali-Aggregate Reaction, pp. 319–323. Noyes, New Jersey (1986)

Skalny, J., Locher, F.W.: Curing practices and internal sulfate attack – the European experience. Cem. Concr. Aggreg. **21**(1), 59–63 (1997)

Stark, D.C.: Performance of Concrete in Sulfate Environments. Portland Cement Association Report RD129, 28p. (2002)

Stark, J., Bollmann, K., Seyfarth, K.: Ettringite – cause of damage, damage intensifier or uninvolved third party. ZKG Intl. **5**(51), 280–292 (1998)

Steinegger, H.: Prüfung und beurteilung der sulfatwiderstandsfähigkeit van zementen. Zement-Kalk-Gips **23**(2), 67–71 (1970)

Talero, R.: Kinetochemical and morphological differentiation of ettringites by the Le Chatelier-Anstett test. Cem. Concr. Res. **32**(5), 707–717 (2002)

Taylor, H.W.F.: Delayed ettringite formation. In: Grutzeck, N.W., Sarker, S.L. (eds.) Advances in Cement and Concrete, pp. 122–131. American Society of Civil Engineers, New York (1994)

Taylor, H.F.W., Famy, C., Scrivener, K.L.: Review: delayed ettringite formation. Cem. Concr. Res. **31**(5), 683–693 (2001)

Thomas, M.D.A., Skalny, J.: Chemical resistance of concrete. In: Lamond, L.F., Pielert, J.H. (eds.) STP 169D, Significance of Tests and Properties of Concrete and Concrete-Making Materials, pp. 253–273. ASTM International, Philadelphia (2006)

Tian, B., Cohen, M.D.: Does gypsum formation during sulfate attack on concrete lead to expansion? Cem. Concr. Res. **30**(1), 117–123 (2000)

UK Expert Report: The Thaumasite Form of Sulfate Attack: Risks, Diagnosis, Remedial Works and Guidance on New Construction. Report of the Thaumasite Expert Group. DETR, London (1999)

Tumidajski, P.J., Turc, I.: A rapid test for sulfate ingress into concrete. Cem. Concr. Res. **25**(5), 924–928 (1995)

Verbeck, G.J.: Field and laboratory studies of the sulfate resistance of concrete. In: Swenson, E.G. (ed.) Performance of Concrete: Resistance of Concrete to Sulphate and Other Environmental Conditions, pp. 113–124. University of Toronto Press, Toronto (1967)

Vincke, E., Verstichel, S., Monteny, J., Verstraete, W.: A new test procedure for biogenic sulfuric acid corrosion of concrete. Biodegradation **10**(6), 421–428 (1999)

Wilson, R., Cleve, A.: Brief summary of tests of the effect of sulfate soils and waters on concrete. In: McMillan, F.R. (ed.) Report of the Director of Research for the Year Ending November 1928, pp. 70–74. Portland Cement Association (1928)

Wittekindt, W.: Sulfatbeständige zemente und ihre prüfung. Zement-Kalk-Gips **13**(2), 565–572 (1960)

Chapter 11
Testing for Degradation by Inorganic Acids

Mark G. Alexander and Nele De Belie

Contents

Keywords Inorganic acid • Hydrochloric acid • Sulfuric acid • pH • Abrasion • Cyclic submersion

11.1 Introduction

Acid testing of concrete specimens is not simple. Questions to be considered are: the type of acid to be used; the representativeness of the concrete mixtures tested; the nature of the measurements to characterise degradation; and how the results are to be applied to practical design and prediction of service life. It was mentioned in

M.G. Alexander (✉)
Department of Civil Engineering, University of Cape Town, Cape Town, South Africa
e-mail: mark.alexander@uct.ac.za

N. De Belie
Magnel Laboratory for Concrete Research, Department of Structural Engineering, Ghent University, Ghent, Belgium
e-mail: nele.debelie@ugent.be

M. Alexander et al. (eds.), *Performance of Cement-Based Materials in Aggressive Aqueous Environments*, RILEM State-of-the-Art Reports 10,
DOI 10.1007/978-94-007-5413-3_11, © RILEM 2013

Chap. 8 (Sect. 8.3) that a key factor in determining the aggressiveness of acids is the solubility of the corrosion product. In general, where the corrosion product is insoluble or only sparingly soluble, and is deposited on the surface of the concrete, the apparent aggressiveness of the acid is much reduced. The opposite situation occurs where corrosion products are soluble. Therefore, comparison of results between different test methods that use different acids and which treat the corrosion products developed in different ways can be problematic.

This chapter covers two test methods for determining acid attack on concrete specimens. They are both 'dynamic' in that the specimens are rotated in a solution and corrosion products are removed. However, they differ in the acids used and in the measurement of the key parameters. They were also developed for different reasons: that at the University of Cape Town was developed to characterise concrete mixtures for sewer pipes, while that at the Magnel Laboratory for Concrete Research at the University of Ghent was developed primarily for testing concretes for use in agro-industries, but also for sewer pipe concrete on occasions. Therefore, the latter method is often also used for testing of organic acids, although the principles are also applicable for testing of inorganic acids. This should be borne in mind when reading this chapter. It is important to stress that it may not be possible to directly generalise the methods reviewed here, developed for specific conditions. However, the principles may indeed be generalised.

Section 6.4 in Part I on mechanisms of attack by inorganic acids should also be recalled. For inorganic acid attack of concrete, dissolution is the main mechanism; however, insoluble silica residues remain and can adhere to the concrete. Further, depending on the acid, other residues of varying solubility may build up on the concrete surface, affecting the rate of acid attack. The pH range of testing is also important, since, for example alumina-based materials only solubilise at pH of 4 or less.

11.2 Method Developed at the University of Cape Town – Dynamic Hydrochloric Acid Resistance Test

11.2.1 Introduction

The process of biogenic acid formation in sewers and subsequent attack of concrete sewer pipes is highly complex, involving a dynamic interaction between sewer environment (aqueous and gaseous), biological activity, and pipe materials. Consequently, it is not easy to simulate in the laboratory. However, knowledge of the inorganic acid resistance of concrete mixtures is useful in terms of studying the impacts of the materials used. With this aim in mind, a 'dynamic' hydrochloric acid resistance test has been developed at the University of Cape Town.

The acid formed biogenically in sewers is sulfuric acid. The test method detailed here, however, uses hydrochloric acid, which needs explanation. During preliminary testing with sulfuric acid, it was found that the acid solution rapidly became saturated with sulfate ions (within less than an hour), thus reducing the aggressiveness of the acid. An adhering gypsum layer also built up on the specimen surface, further hindering acid attack. It was observed that in a sewer, specifically at the water line where flowing sewage constantly removes the corrosion layer, fresh concrete surfaces are continually available for acid attack. It was thus important that the test simulate this worst-case condition. Considering that, for the same hydrogen ion concentration, a solution of sulfuric or hydrochloric acid has the same aggressivity to concrete, sulfuric acid can be replaced in the test with hydrochloric acid. The products of corrosion by hydrochloric acid are far more soluble than those of sulfuric acid, and thus the solution remains 'aggressive' for longer periods between solution renewals. This simplifies the test method and better simulates the worst-case condition in sewers.

The test involves submerging cylindrical concrete specimens (ϕ 80 mm diam., L = 150 mm) in a pH 1.0 hydrochloric acid solution, under 'dynamic' conditions in which the specimens are continuously rotated and brushed. The procedure described herein ensures repeatable test conditions to consistently measure and reliably compare acid resistance, by measuring mass loss and hydrogen ion consumption rates. Standardising the method involved establishing specific limits for variation of test parameters such as pH, solution changes and specimen cleaning. A constant pH requires strict control to ensure specimens are exposed to equally corrosive hydrochloric acid solutions. Solution renewals limit accumulation of suspended solids and dissolved ions, thus allowing accurate titration measurements of hydrogen ion concentration. A cleaning/brushing technique allows removal of the corrosion debris and ensures the specimens remain in constant contact with the hydrochloric acid solution.

11.2.2 Dynamic Hydrochloric Acid Resistance Test Method

The method determines the acid resistance of concrete containing dolomitic or calcareous aggregates, intended for the manufacture of sewer pipes. Such aggregates are 'sacrificial', i.e. they are consumed in the acid reactions, together with the cementitious binder. A sufficient thickness of concrete is provided on the inside face of the pipe, and the consumption of this internal sacrificial layer ensures that the pipe retains its structural integrity at the end of the service life, thus allowing rehabilitation of the pipes if needed. The method described can, however, also be used for siliceous or non-dissolving aggregates, provided profile measurements can be taken of the degraded surface of the specimens, so as to determine the maximum depth of attack.

Test specimens are cylindrical, 80 mm diameter × 150 mm, giving an initial nominal surface area of approximately 478 cm^2. Specimens are prepared using heavy mechanical compaction of 'dry' mixes, which simulates the conditions during pipe manufacture to ensure a representative concrete. Cores of an adequate

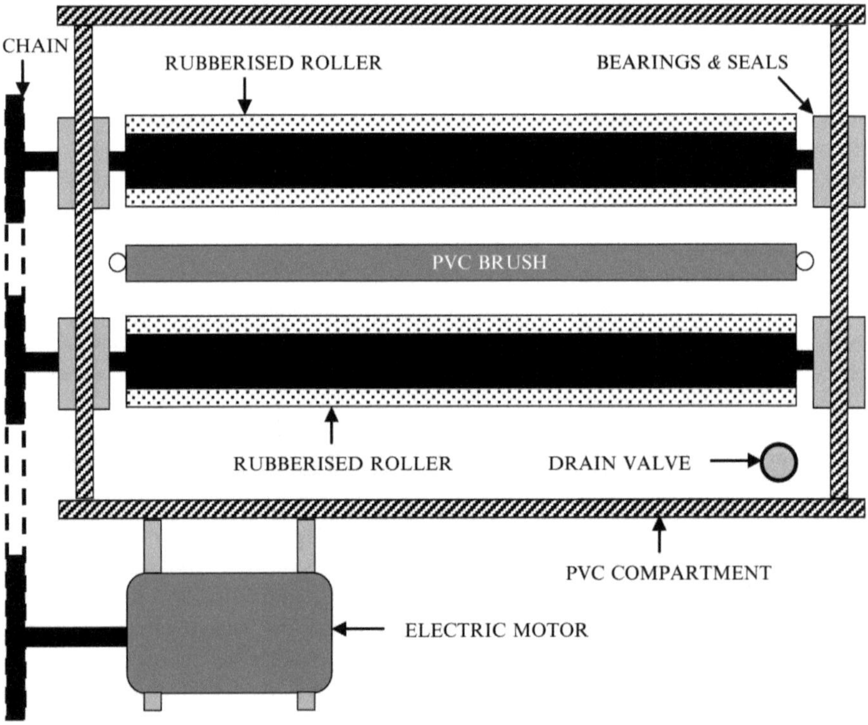

Fig. 11.1 Schematic arrangement of test rig illustrating the plan of one compartment, and showing the layout of components (not to scale)

size and free from reinforcing taken from concrete pipe(s) can also be used, if the quality of actual pipes needs to be evaluated.

11.2.3 Description of Test Rig

The rig consists of one or more solution-containing compartments made of 10 mm poly-vinyl chloride (PVC) sheeting (Fig. 11.1), covered with a PVC lid to prevent evaporation and contamination of the solution. The full-capacity of a compartment is approximately 72 L (800 × 300 × 300 mm). Each compartment contains two rubberised rollers, a PVC brush and a PVC drain valve. The rig should preferably be mounted on a metal frame allowing it to stand approximately 1,150 mm above the floor. All materials in contact with the acidic test solution must be acid resistant.

Rollers made of stainless steel rubberised with neoprene (50 mm O.D.) span the length of the compartment, positioned 50 mm apart at the bottom centre of the compartment. They are rotated via a motor-driven chain. The chain runs over sprockets attached to the rollers on the outside of the compartments. The electric

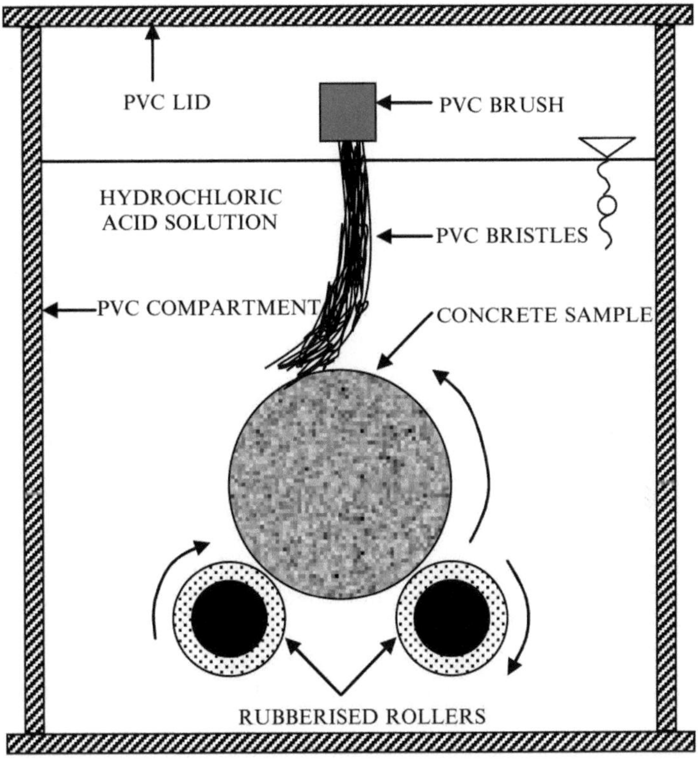

Fig. 11.2 Schematic arrangement of test rig illustrating the cross-section of a single compartment, showing rotation and brushing of a concrete specimen on the rubberised rollers (not to scale)

motor must be able to run for long periods, and rotates the specimens at approximately 16 rpm.

A brush spanning the length of the compartment is positioned at the top centre. The brush sweeps the surface of the specimen to remove loosely adhering debris (Fig. 11.2). A drain valve underneath the compartment allows removal of the corrosion debris.

11.2.4 Test Procedure

A compartment is filled with approximately 50 L of hydrochloric acid solution (pH = 1). A standard cylindrical specimen, approximately 2 kg mass, is pre-saturated with tap water, immersed in the solution, and rotated at approximately 16 rpm. The brush on the top surface of the specimen dislodges corrosion debris. Attack is characterised by measurements of mass loss, specimen dimensions, and hydrogen ion consumption, with time. A pH of between 1.00 and 1.05 is maintained for the duration of the test. Solutions are renewed every 24 h of running time to ensure accurate control of the hydrogen ion concentration. A test is run for a period sufficient to ensure the physical and chemical factors influencing the attack are reflected in the mass loss and hydrogen ion consumption measurements. A recommended test period is approximately 96 h of running time, although 48 h will often suffice to establish clear trends.

Testing four specimens made from the same mix gives reliable results. A rig consisting of four separate compartments is recommended allowing simultaneous testing of the four specimens. Hydrogen ion consumption is important in assessing acid resistance. Mass loss and acid solubility of concrete are not directly related. Mass loss is a function of the dissolution rate of the acid-soluble fraction and fall-out rate of the acid-insoluble fraction.

The hydrochloric acid solutions must be regularly renewed to ensure precise control and measurement of hydrogen ion concentration. Frequency of renewal depends on the rate of attack: with higher rates, the solution is more rapidly saturated with insoluble particles and soluble ions such as calcium and aluminium, which affect the titration method used to determine the hydrogen ion consumption during the test. Cloudiness due to suspension of insoluble particles in the solution makes detection in the colour change of the indicator difficult. Soluble aluminium ions complex with the hydroxide ions and precipitate out of solution masking the end-point of the titration. Solution renewals, however, do not affect the mass loss measurements. Solutions are renewed every time the colour change of the indicator is difficult to detect or on noticing a precipitate during the titration, making the colour change sluggish. Testing as described above allows renewals on average once every 24 h to ensure accurate determination of the hydrogen ion concentration.

11.2.5 Further Information – Hydrochloric Acid Resistance Test

The information given in the following sections provides further information on the development of the hydrochloric acid test method. The question of the use of hydrochloric acid rather than sulfuric acid in the test method has been covered in the Introduction to this section. The sections below deal with aspects of 'standardising' the test method, and give some typical results obtained. Details are given in Fourie (2007), Fourie and Alexander (2007, 2009), and Alexander and Fourie (2010).

11.2.5.1 Standardising the pH of the Hydrochloric Acid Solution

The aim of the investigation described below was to establish the limits in which the pH can vary without significantly affecting the mass loss and hydrogen ion consumption measurements.

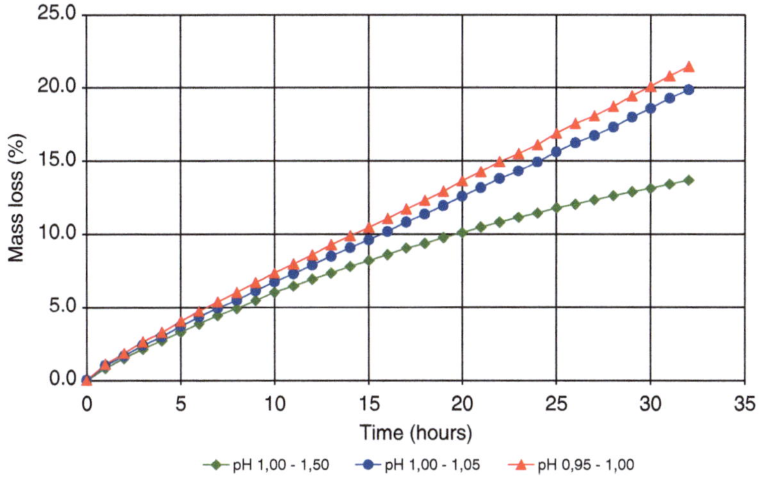

Fig. 11.3 Mass losses with time of Olifantsfontein aggregate specimens in hydrochloric acid solutions of different strengths to establish pH control limits

Dolomitic aggregate concrete specimens (3 No. air-cured for 28 days – good quality aggregate source) were immersed in separate hydrochloric acid solutions (50 L) in which the pH of one solution was allowed to vary between 1.00 and 1.50 and that in the other solutions strictly controlled between the limits of 1.00–1.05, and 0.95–1.00, for 32 h. Specimens were continuously brushed during rotation and removed and cleaned under running water once every hour prior to weighing. Hydrogen ion concentrations of the solutions were determined prior to re-immersion of the specimens. Dimensions of the specimens were measured every 8 h. Solutions in which the pH was controlled were renewed after 24 h.

Mass loss and hydrogen ion consumption rates in the solution in which pH varied from 1.00 to 1.50 continuously decreased with time (Figs. 11.3 and 11.4). Specimen mass loss rates in the solutions in which the pH was controlled, by addition of acid, remained constant with time and hydrogen ion consumption rates increased with time. After 32 h, mass loss and hydrogen ion consumption of the specimen in the solution at pH 0.95–1.00 (21.4% and 12.67 mg/cm^2) was similar to that of the specimen in the solution at pH 1.00–1.05 (19.8% and 11.55 mg/cm^2). Mass loss and hydrogen ion consumption of the specimen in solution of pH varying from 1.00 to 1.50 was significantly lower (13.7% and 7.94 mg/cm^2).

A steady increase in pH from 1.00 to 1.50 relates to a substantial decrease in hydrogen ion concentration. Acid-soluble compounds dissolve at slower rates resulting in a decrease in mass loss and hydrogen ion consumption rates. Conversely, in the pH controlled solutions, acid-soluble compounds dissolve at a steady rate due to the constant supply of hydrogen ions resulting in constant mass loss rates. An increase in the hydrogen ion consumption rates with time was due to the exposure of the dolomite stone with time.

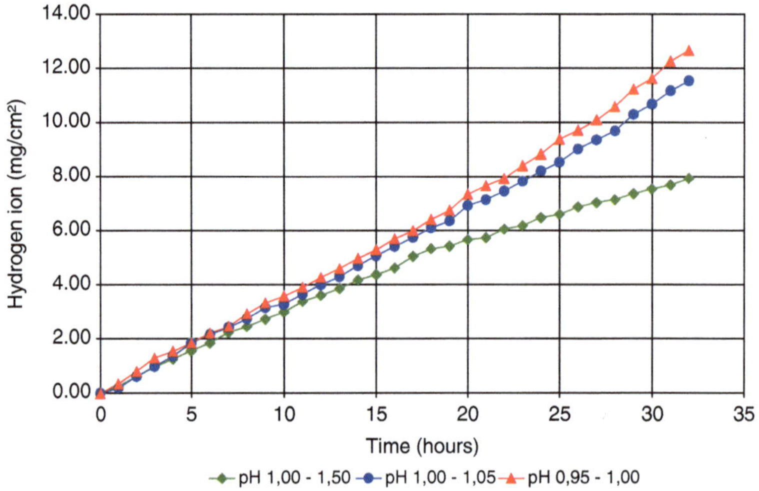

Fig. 11.4 Hydrogen ion consumptions with time of Olifantsfontein aggregate specimens in hydrochloric acid solutions of different strengths to establish pH control limits

Allowing pH of the hydrochloric acid solution to vary from 1.00 to 1.50 significantly decreased the aggressiveness of the solution. An acid-soluble concrete decreases the pH of the solution at a faster rate compared with a concrete of lower acid solubility in the same initial solution. Such variation in the aggressiveness of the solutions results in misleading measurements of mass loss and hydrogen ion consumption with time. Strictly controlling the pH of the solution eliminates this variability. A steady attack rate is maintained allowing accurate comparisons of mass loss and hydrogen ion consumption rates to assess acid resistance. Thus, pH of the solutions must be maintained between 0.95 and 1.05.

Solution renewals every 24 h prevent murky conditions from corrosion debris and allow for easy detection of indicator colour change denoting the titration end-point. Solution renewals do not affect mass loss and hydrogen ion consumption rates, but allow for accurate determination of the hydrogen ion concentrations.

11.2.5.2 Standardising Cleaning of the Specimens

The specimen brushing technique should not cause specimen abrasion but only remove corrosion debris accumulating on the specimens. Removal of the debris ensures that the concrete specimens remain in direct contact with the hydrochloric acid solution, eliminating variability arising from different accumulation rates.

A good quality dolomite stone, with particle sizes between 6.7 and 9.5 mm, and siliceous filler sand were used for the specimens. Acid insolubility of the dolomite (measured in a standard acid digestion test) was 2.95%.

Specimens (4 No., air-cured for 28 days) were each immersed and continuously rotated in separate hydrochloric acid solutions maintained at pH 1.00–1.05 for 32 h.

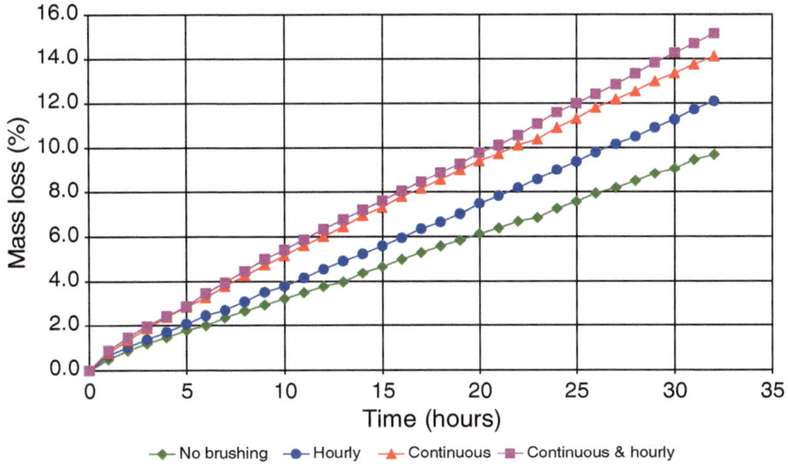

Fig. 11.5 Mass losses with time of concrete specimens in hydrochloric acid solutions maintained at a pH of 1.00–1.05 to monitor effects of brushing

A specimen in one of the solutions was left without brushing so as to assess the effects of the different brushing techniques on the remaining three specimens:

- cleaning specimens hourly under running water with a soft-nylon-bristle brush
- cleaning specimens continuously in solution with a soft-PVC-bristle brush
- cleaning specimens continuously in solution, and hourly under running water, with the respective brushes

Specimen mass losses were determined hourly, and hydrogen ion concentrations once every 4 h. Solutions were renewed after 24 h, and dimensions were determined every 12 h.

Specimen mass loss and hydrogen ion consumption rates remained constant with time (Figs. 11.5 and 11.6). After 32 h, mass loss and hydrogen ion consumption of the specimen brushed continuously and cleaned hourly (15.2% and 10.57 mg/cm^2), and that of the specimen only brushed continuously (14.1% and 9.93 mg/cm^2) were similar. The corresponding measurements on the specimen cleaned hourly (12.1% and 7.29 mg/cm^2) and those of the specimen without brushing (9.7% and 6.50 mg/cm^2) were significantly lower.

Corrosion debris on the specimen hinders mobility of the ions involved in the acid attack. A gel-like layer with acid-insoluble particles is developed on specimens not continuously brushed. Stagnant conditions in the layer slow influx of hydrogen ions from the solution and allow build up of calcium chloride, aluminium hydroxide and iron hydroxide, which reduce dissolution rates of acid-soluble compounds. Continuous brushing removes the debris and gives higher mass loss and hydrogen ion consumption rates. Continuous brushing with hourly cleaning removes any firmly bonded debris, which the more gentle sweeping action of the soft-PVC-

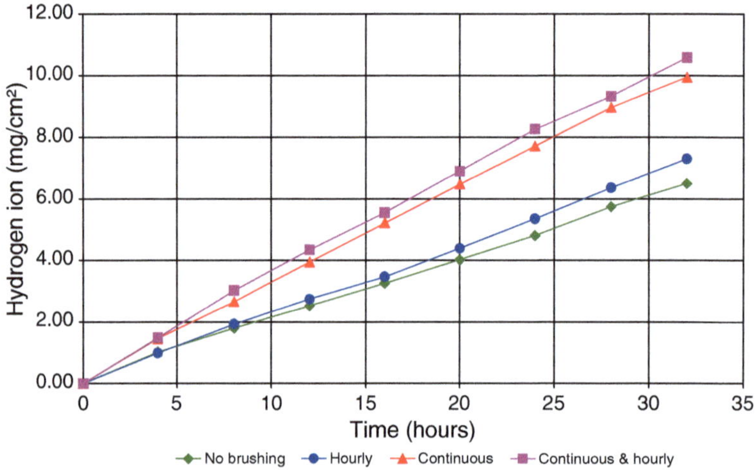

Fig. 11.6 Hydrogen ion consumptions with time of concrete specimens in hydrochloric acid solutions maintained at a pH of 1.00–1.05 to monitor effects of brushing

bristle brush cannot, and results in slightly higher mass loss and hydrogen ion consumption rates.

11.3 Method Developed at the University of Ghent – Accelerated Degradation Tests (TAP)

This method was designed to accelerate the deterioration process through alternate wetting and drying, because this procedure simulates the real-life situations under investigation. It is described in detail by De Belie et al. (2002) and can be used for testing degradation in organic as well as in inorganic acid solutions. Three cylinders (ϕ diam. 270 mm, h = 70 mm) of each concrete mixture are subjected to a cyclic procedure of immersion in an acidic solution and drying in air (Fig. 11.7). The cylinders, fixed on horizontal axes, rotate at a speed of 1 revolution per hour through separate containers. Each point of the outer circumference is submerged during 1/3 of the rotation time. After each cycle, which usually lasts for 6 days for lactic/acetic acid attack and 12 days for sulfuric acid attack, the cylinders are dried in air and brushed with rotary brushes to remove weakly adhering concrete particles (Fig. 11.8). In Table 11.1 the different steps in the test procedure are related to different stages occurring during the degradation in practice.

The corrosion of the specimens is measured using laser sensors, connected to a computer (Fig. 11.8). The sensor amplifiers supply a voltage signal, which is linearly related to the distance between the concrete surface and the sensor. Five measurements per mm are taken along the concrete surface. A software trigger is programmed to start the measurement of a cylinder profile when the raised edge of

Fig. 11.7 TAP, the apparatus for accelerated degradation testing developed at Ghent University

Fig. 11.8 Brushing of the TAP cylinders (*left*) and measurement of the distance with lasers (*right*)

a stainless steel angle, fixed on the concrete cylinder, passes the laser beam. The first 50 measurements of a contour line are performed on the horizontal part of the angle steel, which acts as reference plate. The three sensor heads are mounted on a mechanical device which can always be placed in the same position on the frame of the TAP. This device allows the position of the sensor heads to be adjusted in a direction parallel to the cylinder axes in steps of 0.5 mm with an accuracy of 0.01 mm. In this way it is possible to scan the circumference of a cylinder every time at the same position along the cylinder length. By moving the system several times, for every cylinder several parallel profiles can be measured, equally distributed along the cylinder length. For the experiments discussed here, 5 profiles (sulfuric acid attack) to 7 profiles (lactic/acetic acid attack) per cylinder were measured. The laser measurements can also be used to calculate the surface roughness of the concrete, expressed by means of the R_a-value (BS1134 1972).

In the lactic/acetic acid test procedure, the radius of the cylinders is determined at the beginning of the experiment and after each cycle of alternating immersion,

Table 11.1 Relation between different steps in the TAP test procedure and real-life situations

TAP	Lactic/acetic acid attack	Sulfuric acid attack
	(a) concrete floors in animal houses, (b) silage storage structures	(a) sewers (b) manure storage structures
Cyclic immersion and chemical attack	(a) wetting by feed + water residues and manure, drying in between (b) wetting: effluent release from silage	(a and b) wetting/drying by fluctuation of the wastewater/ manure level
Drying in air	(a and b) drying during periods that the compartment or silo is not in use	(a) dry weather: low wastewater level (b) emptying of manure storage facility
Abrasion by brushing	(a) abrasion by animals, cleaning with brushes or high-pressure hose	(a) high flow rates because of high loading or rain; abrasion by flowing water, turbulence
	(b) abrasion by animals (in self-feeding silo), cleaning and farm machinery (block cutters)	(b) abrasion by cleaning (less common)

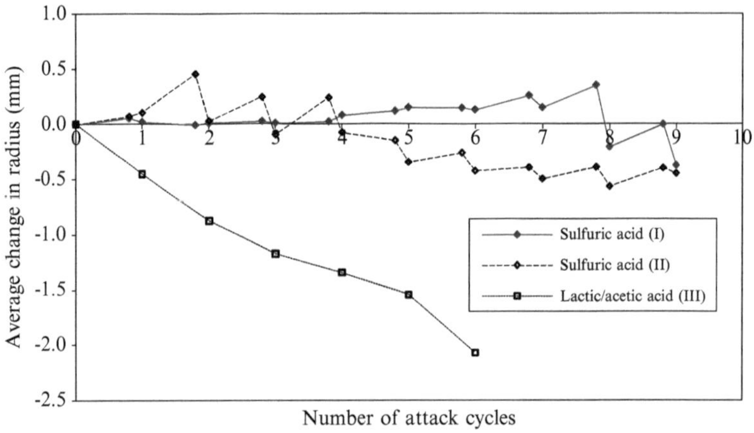

Fig. 11.9 Average change in radius vs. number of attack cycles for a TAP degradation test with a sulfuric or lactic/acetic acid solution, respectively

drying and brushing. In the sulfuric acid test procedure the measurements are performed twice per cycle, before as well as after brushing. In this way it is possible to determine the average change of the radius of the cylinders due to chemical reaction of the concrete with the sulfuric acid solution during the immersion as well as the change of the radius due to mechanical action of brushing the cylinders.

As an illustration of the experimental results obtained with the TAP, Fig. 11.9 shows the average change in radius for concrete specimens subjected to sulfuric or lactic/acetic acid attack. For the sulfuric acid attack, the simulation liquid consisted of 0.5% sulfuric acid in water (pH = 0.8–1.0). The concrete mix proportions of reference specimen I were based on a mix design commonly used for sewer pipes, featuring a high sulfate resistant (HSR) Portland cement (CEM I 42.5 HSR/LA) and a w/c ratio of 0.40. For specimen II, 30 kg/m^3 silica fume was added. The lactic/acetic acid degradation was carried out with a simulation liquid consisting of lactic and acetic acid in water both with concentrations of 30 g/l (pH = 2.0–2.2). The reference concrete specimen III had a mix design commonly used for production of prefabricated concrete slats, including an ordinary Portland cement (CEM I 42.5R) and a w/c ratio of 0.39.

The figure clearly demonstrates the different degradation mechanisms occurring in sulfuric and lactic/acetic acid attack. Submersion of concrete in sulfuric acid causes the formation of expansive gypsum and ettringite. In the experiment discussed here, gypsum plays the major role, since only a small amount of C_3A was present in the HSR cement. Especially for the concrete with silica fume addition (II), the alternating expansion due to gypsum formation during the submersion stage, and the decrease in radius due to material loss by brushing, are pronounced. For the reference concrete, the same effect is noticeable from cycle 7 onwards. Regarding the degradation by the organic acids, there is no expansive reaction, but a process of dissolution and leaching of soluble calcium salts occurs. The decrease of the pH in the pore solution causes the calcium silicate hydrates of the hardened cement paste to become unstable and disintegrate. This results in a very weak external layer which is easily removed by brushing. For the same number of attack cycles (also keeping in mind that the total period of cycling for the sulfuric acid attack is twice as long), the average decrease in radius in the lactic/acetic acid solution is much more pronounced than for the sulfuric acid solution (−2.1 mm for concrete III after 6 cycles, compared to +0.13 mm and –0.42 mm for concrete I and II, respectively). It can be noticed that the pH of the sulfuric acid solution (pH = 0.8–1.0) was significantly lower than the pH of the organic acid solution (pH = 2.0–2.2). However, other authors (Bayoux et al. 1990) have also remarked that the acceptable pH limits for weak acids, such as lactic and acetic acid, are higher than for strong acids, because weak acids must be present in a higher concentration to reach the same pH level.

Figure 11.10 shows the change, caused by the organic acid attack, of a surface profile measured along 200 mm of the circumference of a cylinder from concrete III. This illustrates how, during the first few cycles, the outer mortar layer of about 1 mm thick is eroded. Subsequently, the mortar in between the gravel aggregates starts to disappear. The surface of the samples after 6 cycles is shown in Fig. 11.11.

11.4 General Conclusions

From the test methods discussed, it is clear that the choice of test method has substantial influence on the results. If the test is intended to simulate 'real' conditions being investigated, then careful attention must be paid to the conditions imposed during the test, such as type of acid used, environmental conditions

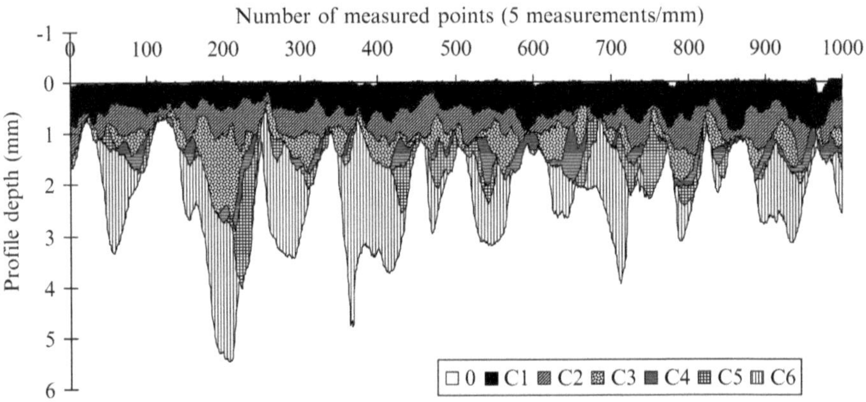

Fig. 11.10 Change in surface profile for reference concrete III submitted to lactic and acetic acid; 0: initial measurement, Ci (with i = 1–6): measurement after attack cycle i

Fig. 11.11 Picture of reference concrete III subjected to 6 cycles of lactic and acetic acid attack

and cycling, mechanical abrasion or removal of corrosion products, methods of measuring specimen dimensions, etc. The issue of removal of adhering layers during testing, e.g. the gypsum layer in sulfuric acid testing, is important, since it affects the results substantially. The method of removal of such a layer is also important: by direct abrasion, or by soft brushing. For example in a sewer, there may be direct abrasion by flowing water at the water line, and removal of corrosion product by continuous dissolution. The use of different acids and their different effects in terms of corrosion products formed and their solubility, saturation of the test solutions, etc. also need careful consideration.

References

Alexander, M.G., Fourie, C.W.: Performance of sewer pipe concrete mixtures with Portland and calcium aluminate cements subject to mineral and biogenic acid attack. Mater. Struct. **44**(1), 313–330 (2010)

Bayoux, J.P., Letourneux, J.P., Marcdargent, S., Verschaeve, M.: Acidic corrosion of high alumina cement. In: R.J. Mangabhai (ed.) Calcium aluminate cements. E. & F.N. Spon, London (1990)

BS1134: British standard method for the assessment of surface texture, British Standards Institution (1972)

De Belie, N., Monteny, J., Taerwe, L.: Apparatus for accelerated degradation testing of concrete specimens. Mater. Struct. **35**(251), 427–433 (2002)

Fourie, C.W.: Acid resistance of sewer pipe concrete. MSc thesis, University of Cape Town (2007)

Fourie, C.W., Alexander, M.G.: A dynamic hydrochloric acid test for evaluating acid resistance of sewer pipe concrete. In: De Belie, N. (ed.) Workshop on Performance of Cement-Based Materials in Aggressive Aqueous Environments – Characterization, Modelling, Test Methods and Engineering Aspects, Ghent (2007)

Fourie, C.W., Alexander, M.G.: Acid resistant concrete sewer pipes. In: Alexande M.G., Bertron A. (eds.) Proceedings of the International RILEM TC 211 PAE Final Conference, pp. 408–418, Toulouse, Bagneux, RILEM Publications S.A.R.L. (2009)

Chapter 12
Bacteriogenic Sulfuric Acid Attack of Cementitious Materials in Sewage Systems

Karen Scrivener and Nele De Belie

Contents

Keywords Bacteria • *Acidothiobacillus* • Sewer • Biogenic • Calcium aluminate cement • Hydrogen sulfide • Sulfuric acid

12.1 Introduction

Well functioning sewage networks are critical to public health. Increasing urbanisation puts rising demands on existing networks and creates a need for new networks, especially in developing countries. Sewage networks represent a very aggressive environment for cementitious materials. At the same time their relative

K. Scrivener (✉)
Laboratory of Construction Materials, EPFL, Lausanne, Switzerland
e-mail: karen.scrivener@epfl.ch

N. De Belie
Magnel Laboratory for Concrete Research, Department of Structural Engineering, Ghent University, Ghent, Belgium
e-mail: nele.debelie@ugent.be

M. Alexander et al. (eds.), *Performance of Cement-Based Materials in Aggressive Aqueous Environments*, RILEM State-of-the-Art Reports 10,
DOI 10.1007/978-94-007-5413-3_12,

inaccessibility poses challenges for maintenance and repair. The main cause of degradation is the corrosion of concrete due to the in-situ production of sulfuric acid by bacteria. In this chapter, we discuss the mechanisms behind this so-called bacteriogenic corrosion and the state of the art for determination of the relative resistance of different concrete types to this form of degradation. The reader is also referred to Monteny et al. (2000) for a more extensive review on the subject.

12.2 Phenomenon of Bacteriogenic Corrosion

The production of sulfuric acid is the final step in a complex chain of processes which comprise bacteriogenic corrosion, illustrated in Fig. 12.1. Sewage contains organic matter. When the flow of sewage is sluggish, and availability of oxygen limited, anaerobic bacteria break down this organic matter and in so doing produce H_2S gas. This gas in turn serves as a food source for another group of aerobic bacteria, which colonise the walls and crown of the sewer above the water line. These bacteria break down the H_2S to produce sulfuric acid, which attacks the concrete. Damage is usually concentrated at the crown and water line of the sewer, Fig. 12.1. The nature of the process means that the amount of acid produced is limited by the extent of activity of the bacteria so the phenomenon is very poorly modelled by tests in which mineral acid is used, especially where the quantity of acid is large in comparison with the amount of cementitious materials (Monteny et al. 2000, 2001; De Belie et al. 2004).

In fact the colonisation of the concrete surfaces is progressive. The bacteria responsible, – Thiobacilli (in the year 2000 reclassified as Acidithiobacilli) – is a whole family of bacteria, the various strains of which thrive at different pH levels, as illustrated in Fig. 12.2. Initially the surface of the concrete is highly alkaline, due to the nature of the hydrates, particularly Portlandite, $Ca(OH)_2$. The first colonisation is by the strains of bacteria capable of flourishing at the highest pH. Through the production of sulfuric acid they lower the pH level of the concrete surface and facilitate colonisation by new strains, adapted for lower pH conditions. The final stage of the process is colonisation by *Acidithiobacilli thioxidans,* the most aggressive. These bacteria thrive best at a pH of 2 or less and are capable of generating enough sulfuric acid to reduce the surface pH down to 1, which is highly aggressive to all cementitious materials.

Even when colonies of *Acidithiobacilli thioxidans* are well established, the process of dissolution of the concrete tends to neutralise the acid created. Of the major oxides present in cementitious materials, CaO; SiO_2; Al_2O_3 and Fe_2O_3 (C, S, A, F for short); S and F are virtually insoluble even at very low pH. C is readily soluble at all acid pH-levels and each mole of calcium which dissolves will neutralise 2 moles of acid (Eq. 12.1). Equivalent equations can be written for C-S-H which will also decalcify.

$$Ca(OH)_2 + 2H^+ \rightarrow Ca^{2+} + 2H_2O \qquad (12.1)$$

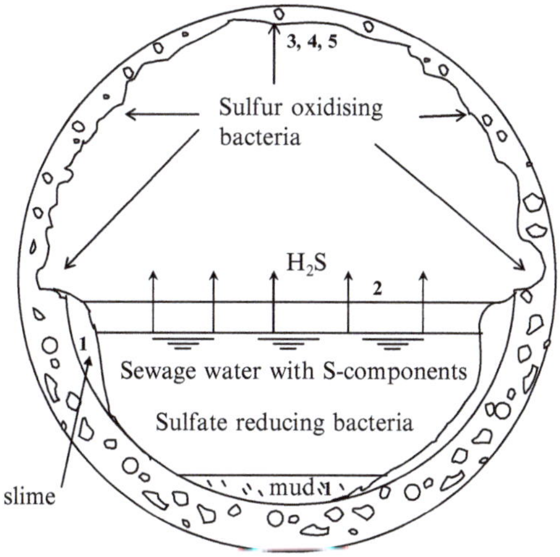

Fig. 12.1 H_2S Bacterial corrosion: (*1*) Reduction of sulfates to sulfide by anaerobic bacteria living in the slime and mud layer; (*2*) Turbulence ejects H_2S dissolved in the effluent into the pipe atmosphere where it is carried by convection to the pipe crown; (*3*) Oxidation of H_2S to sulfur; (*4*) Transformation of sulfur to sulfuric acid by aerobic bacteria; (*5*) Destruction of the concrete

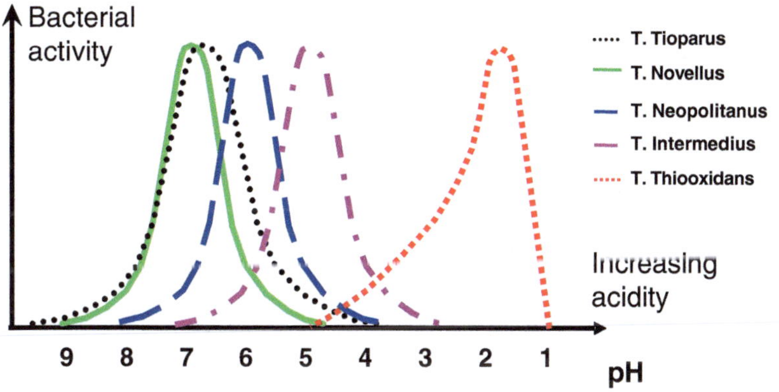

Fig. 12.2 Activity of different bacterial strains as a function of pH

Alumina is virtually insoluble above pH 4, but below this pH dissolves to neutralise 3 moles of acid (Eq. 12.2).

$$Al(OH)_3 + 3H^+ \rightarrow Al^{3+} + 3H_2O \tag{12.2}$$

Although it may be thought advantageous to have insoluble materials, leached siliceous cementitious materials are highly porous and provide no effective barrier to the penetration of acid. In the case of bacteriogenic attack, where the supply of acid is limited, it is preferable to have a material with a high neutralisation capacity, due to the presence of lime and alumina. This concept of neutralisation capacity is even more important in the choice of aggregates, which make up more than 60% of a concrete and here it has been widely shown that use of a "sacrificial" limestone or dolomitic aggregate gives much better performance in the case of bacteriogenic attack (De Belie et al. 2004).

12.3 Field Experience of Different Concrete Types

Given the complexity of bacteriogenic attack, it is important to consider evidence from the field, before discussing the appropriateness of different tests. For durability problems, simulation tests are never completely satisfactory. It is difficult to reproduce all natural conditions and interactions under artificial conditions in a laboratory. In fact, the only way to study the real process of biogenic sulfuric acid corrosion, while taking into account all factors and interactions, are in situ exposure tests and observations. To investigate the resistance of different concrete mixes, some researchers have performed in situ tests. The Virginia test sewer in South Africa is an excellent source of data in this respect (Alexander and Fourie 2010). The Virginia test sewer is a length of real sewer, with a bypass to allow the test section to be examined, along which lengths of different concrete pipes were laid in the 1980s. The final inspection was made after 12 years, at which time the test had to be stopped due to the dangerous state of several lengths of pipe.

Three concrete types were tested:

1. Portland cement with siliceous aggregate
2. Portland cement with dolomitic (calcareous) aggregate
3. Calcium aluminate cement with siliceous aggregate

After 12 years the Portland/siliceous section was completely corroded through the 60 mm thickness of the pipe in many places, Fig. 12.3. The Portland/calcareous section fared marginally better, but was still heavily corroded beyond the depth of the reinforcement. The CAC/siliceous section, however was only slightly attacked to a maximum depth of 12 mm at the water line. The superior performance of the CAC concrete, cannot be explained only on the basis of neutralisation capacity (neutralisation capacity is only higher below pH 4) and indicates that other mechanisms, such as biological activity, are important in bacteriogenic corrosion.

In other cases, mortar or concrete samples were placed in sewer pipes in which biogenic sulfuric acid corrosion occurred. The samples were monitored for a few years. The large amount of parameters involved in the corrosion process and the great diversity in the sampled sites and concrete mixes used make it difficult to compare the results of the different studies. Mori et al. (1992) used a heavily corroded sewer pipe to expose mortar specimens to biogenic sulfuric acid. The mortar specimens were made with Portland cement and a water/cement ratio of 0.65. The test samples measured $40 \times 40 \times 160$ mm. The atmospheric concentration of hydrogen sulfide in the pipe ranged from 5 to 400 ppmv and the temperature ranged from 10 to 30°C. The mortar specimens were placed in a manhole for 8 months. They measured a corrosion rate of 5.7 mm/year.

From 1988 till 1991, in situ experiments were performed in Rotterdam (the Netherlands) (Pluym-Berkhout et al. 1989; van Mechelen and Polder 1991). Samples of four different concrete mixes were exposed to biogenic sulfuric acid corrosion in ten different locations. Three of the four mixes were made in the laboratory and had a water/cement ratio of 0.32 and a cement content of 400 kg/m^3.

Fig. 12.3 State of concrete pipe section in test sewer after 12 years. The Portland/siliceous section (*left*) is heavily corroded, in parts the earth behind the pipe is visible; the CAC/siliceous section (*right*) is relatively unaffected; only a black part around the water line shows up to 10 mm of erosion (Photographs courtesy of A. Goyns)

The only difference between the three mixes was the type of cement: blast furnace cement, Portland cement, or Portland fly ash cement. The test samples measured 140 × 45 × 45 mm and were placed in a basket, which was put in a manhole just above the dry weather sewage level. After 3.5, 23, and 35 months, the test specimens were examined. The degradation of the test specimens was followed by measuring the porosity, the flexural strength, the density, the mass loss, the neutralisation depth with phenolphthaleine, and by visual inspection. The sum of the neutralisation depth and mass loss (in mm) was called the degradation of the sample. The different exposure locations were characterised by measuring the pH, the temperature, the oxygen content, and the sulfide concentration of the sewage. After 3.5 months, a small increase in mass was measured for some samples. This was probably due to the uptake of water. After 23 months, the samples with Portland cement showed the lowest degradation rate of 0.8 mm/year (measured as the sum of the loss in material and the neutralization depth). The samples with blast furnace cement showed an average degradation rate of 1.0 mm/year. The largest degradation rate of 1.3 mm/year was found for samples with Portland fly ash cement. After 35 months of exposure, the samples with blast furnace cement showed less mass loss than the other samples: 0.4 mm/year. When the neutralisation depth was also taken into account, there was no difference between all samples made in the laboratory (1.0 mm/year). The samples of the concrete pipe made in a factory generally showed a slightly better resistance (0.8 mm/year) than the samples made in the laboratory.

12.4 Laboratory Testing

12.4.1 Inorganic Acid Tests

As already indicated, tests carried out in mineral acids do not well represent the performance of cementitious materials subject to bacteriogenic corrosion. For example the concretes in the Virginia sewer test were tested in sulfuric acid prior to exposure (Alexander and Fourie 2010). This test indicated that the performance of CAC/siliceous and OPC/siliceous were comparable and that of OPC/dolomite superior. These differences did not correspond to the differences seen in the field. In general testing in mineral acids has two main problems – frequency of renewal of solution, due to neutralisation by the dissolution of the cementitious material, and how to deal with softened material or deposits of the sample surface – should these be brushed off, or left (which will make tracking of performance by methods such as weight loss difficult).

12.4.2 Test Developed at Hamburg University

To the authors' knowledge the most representative test of bacteriogenic corrosion developed to date is that used for many years at the University of Hamburg (Sand et al. 1987). Unfortunately since the retirement of Professor Bock it is thought that the method is no longer in use.

This test had the major advantage of letting colonisation of bacteria occur on the surface of concrete samples, in conditions as near to reality as possible. The only accelerating factors were the maintenance of an optimal air flow and temperature for the bacteria of 30°C and the provision of H_2S gas and of nutrients in the form of an aerosol spray. In this test, concrete blocks of dimensions 60 × 11 × 7 cm were stored in a chamber (chest freezer compartment) with the base standing in 10 cm of water. The upper part of the test blocks was pre-sawn in small cubes (18 × 18 × 20 mm) to facilitate the sampling.

At set times from 3 months to 1 year samples were removed and tested for surface pH, mass loss and appearance. The increase in temperature and optimal supply of H_2S and nutrients were estimated to accelerate corrosion rate by at least 16 times, but even so the test duration was 1 year.

This test well captures the difference in performance of cement types seen in the Virginia Sewer study (Fig. 12.3). Moreover, as seen in Fig. 12.4, the development of the surface pH is dramatically different. The surface pH of Portland cement based samples reduces to about 1 over the course of 6 months, after which mass loss increases rapidly, so that the samples are completely destroyed after 350 days. The CAC based sample on the other hand show a similar (or slightly more rapid) decrease in pH at first, but this levels off at around pH 3–4 after about 100 days and the weight loss of these samples is considerably lower. This pH level coincides with that at which $Al(OH)_3$ increases in solubility and it suggests that the entry of aluminium ions into solution inhibits the activity of the bacteria

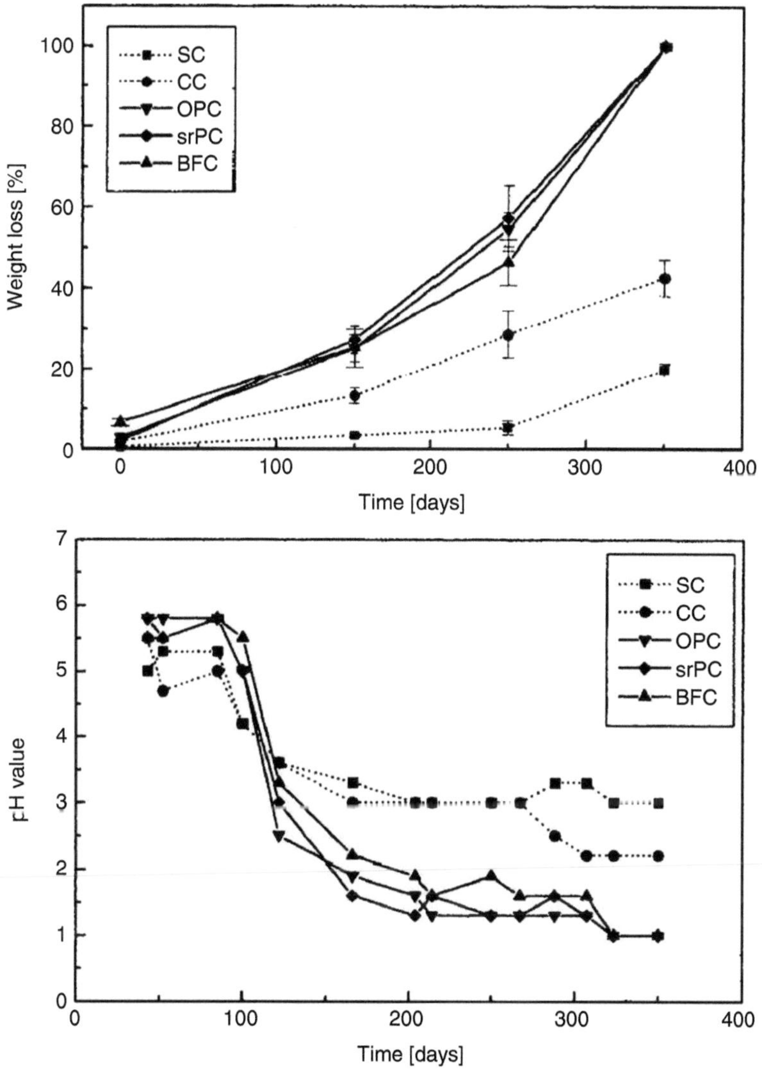

Fig. 12.4 Test results from Hamburg Chamber (Ehrich et al. 1999). SC and CC, *dotted lines*, are calcium aluminate cement mortars. The samples with *solid lines* are all Portland based systems, OPC = plain Portland; srPC = sulfate resisting Portland cement and BFC = Portland/Blast Furnace slag blend

12.4.3 Other Tests Using Bacteria

As mentioned above, the Hamburg test still requires a year to produce useful results. In order to shorten the testing period various researchers have tried to develop faster test methods.

A simulation system has been built by the Research, Development and Consulting Department of *Heidelberger Zement*, in which timespans of 3 ± 5 months are sufficient to investigate the resistance of several concrete types against biogenic sulfuric acid (Hormann et al. 1997; Schmidt et al. 1997). The test system is separated in two parts: a growth and a reaction part. In the growth part, a 25 L fermentor, monoculture of T. thiooxidans is cultivated at optimal conditions. The reaction part consists of a glass bio-reactor where the test specimens are stored. A warm and humid environment was sustained in the reactor. In contrast with the other test system, the test specimens were, in this case, mortar briquettes with dimensions of 10 × 10 × 60 mm. These specimens were flooded periodically for 5 min by an aliquot of the fermentor content followed by a 1 h interruption. The experiment was followed up by measuring the weight loss of the test specimens and by determining the cell density on the surface of the specimens. The difference between high alumina cement and ordinary Portland cement and the influence of the type of aggregates on the resistance to corrosion was investigated. All aggregates were tested using high alumina cement mortar. The specimens with high alumina cement (3–4% loss of weight) showed better resistance against severe sulfuric acid attack than ordinary Portland cement specimens (18–31% loss of weight). These values were obtained after 5 months of testing. After 1 month, it was already possible to see a difference between the two mortars. In comparison with the type of cement, the influence of the aggregate that was used (high alumina cement-clinker, quartz and broken traprock) was very small.

Another investigation, which used a simulation chamber, was carried out by Mori et al. (1992). They investigated the effects of nutrients on the corrosion of concrete. In comparison with other researchers, they worked with a very high concentration of H_2S (400 ppmv) in the chamber. This was in accordance with measurements in heavily corroded sewer pipes. The test specimens were mortar samples of size 4 × 4 × 16 cm. They were placed in sewage, a special culture medium (a solution containing all the nutrients and minerals needed by the bacteria but without thiosulfate), or distilled water. The experiment lasted for 6 months and during the first 2 months, the test specimens were inoculated with T. thiooxidans every 2 weeks. The corrosion rate was determined by measuring the reduction in cross-section of the specimens. Using a scanning electron microscope, the corroded samples were investigated. The number of T. thiooxidans was determined by plate counting of the bacteria. Mori et al. (1992) found a corrosion rate of 6.1 mm/year when the specimens were submerged in the sewage and 3.8 mm/year for the specimens placed in the basal medium. The most severe corrosion occurred approximately 20 mm above the liquid level. This confirmed their statement that the greatest corrosion occurs close to the liquid level because this location provides a constant supply of moisture and nutrients.

These researchers also simulated the corrosion process in a second way. They used a demonstration plant in which the conditions to cause biogenic sulfuric acid corrosion were created. It consisted of a pipe with a diameter of 15 cm and a length of 20 m. The temperature in the system ranged from 12 to 30°C and the H_2S concentration ranged from 25 to 300 ppmv. They used the same test specimens as those in the simulation chamber. After 6 months, the mortars started to corrode just above the sewage level. Corrosion rates of on average 3.8 mm/year were measured. Corrosion reached its maximum level just above the sewage level.

As part of a wide-ranging project to investigate deterioration of concrete caused by metabolites of aerobic microorganisms, Tazawa et al. (1994, 1996) performed some experiments with sulfide-producing and sulfur-oxidizing bacteria. These bacteria were isolated from an underground structure in which biological deterioration occurred. The test mortar specimens were completely immersed in culture bottles with 300 ml of ordinary liquid medium. After the bottles were sterilised, bacteria were inoculated. Two types of hydrogen sulfide-producing bacteria were used (*Xanthomonas* sp., the other type of bacteria could not be identified) and one type of sulfur-oxidizing bacteria (*T. intermedius*). The *Acidithiobacillus* species were inoculated 9 days after the inoculation of hydrogen sulfide-producing bacteria. The experiments were monitored by measuring the pH, the calcium concentration, the changes in dissolved hydrogen sulfide concentration, and changes in sulfate concentration of the culture medium. Growth curves of the micro-organisms were measured by light absorbance of the medium. They also investigated the calcium content of the surface area and the internal area of the mortar specimens. The authors first thought that the deterioration of the concrete was caused by sulfuric acid. However, they found little gypsum on the surface of the specimens and little change of the concentration of sulfate ions in the culture medium. Analysis of the compositions and the concentrations of the metabolites of the micro-organisms in the culture medium showed that considerable amounts of acetic acid (85.6 ppm after 3 h), propionic acid (55.0 ppm after 3 h), and carbonic acid (343 ppm after 1 day) were generated. The authors concluded that the deterioration of concrete was mainly caused by carbonic acid and organic acids (acetic acid, propionic acid) produced by the bacteria. High levels of bicarbonate and carbonic acid excreted by respiration of the bacteria decomposes cement hydrates, producing calcium carbonate, silica gel and alumina gel. The calcium carbonate (solubility in water: 6.86 mg/l) is easily soluble in the presence of organic acids, produced by the bacteria.

In the test procedure developed by the team of De Belie and Verstraete (De Belie et al. 2004), concrete cylinders of 80 mm diameter and 15 mm height, are glued on plastic plates of 120 × 90 mm. These samples are incubated in a 20 fold diluted aqueous solution of a biological sulfur suspension. The dilution solution contains tap water and an additional N- and P-source (100 mg L^{-1} $(NH_4)_2SO_4$ and 10 mg L^{-1} K_2HPO_4). Biologically produced sulfur is the end-product of the microbiological sulfide oxidation, a process carried out by mixed cultures of Acidithiobacillus-like bacteria. It consists of complex aggregates containing elemental sulfur, biomass and biopolymers and has an initial pH of 7. For the control samples, an inactivated suspension is used, consisting of the basic suspension, inactivated by dosing 1,000 mg L^{-1} of the biocide glutaraldehyde. One cycle of 17 days, performed at 28°C, consists of the following steps:

- Step 1: Uptake of H_2S by the concrete. The air-dry concrete samples were placed in H_2S incubation chambers of 10 L for 3 days. The gas concentration was generated by 100 ml of a 4% Na_2S solution and 100 ml of a 1.5 N HCl solution. The initial gas concentration was ca. 250 ppmv.

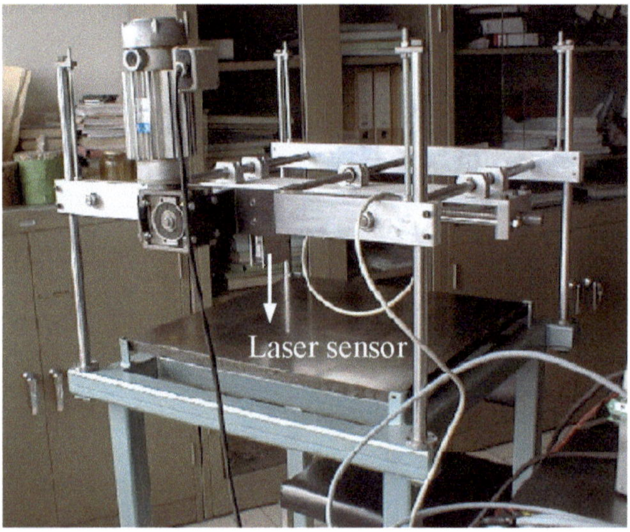

Fig. 12.5 Automated laser measurement system (ALM)

- Step 2: Incubation of concrete in the solution; sulfuric acid production by the *Thiobacilli* and concrete degradation. Each block was hung in a separate glass recipient and submerged in 600 ml of the culture medium. The recipients were covered and placed for 10 days on a rotary shaker (90 rpm). A control treatment for each concrete composition consisted of a concrete block hanging in an inactivated suspension.
- Step 3: Rinsing with Milli-Q water. On rainy days, high flow rates of water can occur in sewer pipes and corrosion products and debris can be washed away, which provides a new surface for the corrosive attack. Placing the concrete blocks in separate glass recipients, containing Milli-Q water on the rotary shakers for 2 days simulated this.
- Step 4: Drying. After periods of high loading, dry periods follow in which H_2S can penetrate into the surface layers of the concrete and new reaction products can be formed. This was simulated by drying the concrete samples at 28°C for 2 days.

During step 2, the pH of the suspension is determined daily. The sulfate concentration in the suspension is determined according to Standard Methods. The total Ca-concentration of the suspension during steps 2 and 3 is determined by a destructive method according to Standard Methods, following detection with a flame atomic spectrometer.

After every cycle, the thickness of the concrete blocks is measured. A measurement table with an accurate laser sensor (ALM: "Automated laser measurement system") was developed in the Magnel Laboratory for Concrete Research (Fig. 12.5). The laser can be moved in the horizontal X-Y plane by a motor and measured the vertical (Z) distance to the objects on the measurement table. For each

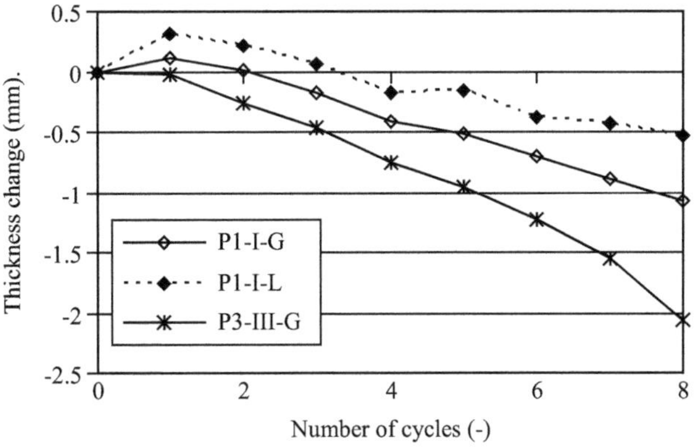

Fig. 12.6 Thickness change of concrete samples during the BSA test

concrete sample, surface profiles are measured in the X-direction for six different Y-positions. Per mm along the concrete surface 3.3 measurements are recorded. After every attack cycle, the plastic plates with concrete samples are put on the same position on the measurement table, using a template. The height of the samples is calculated relative to the plastic plate. The measurements with the laser can also be used to calculate the change in surface roughness of the concrete samples.

As an illustration of the experimental results obtained with the BSA procedure, results from three types of concrete specimens taken some mm below the inner surface of sewer pipes or inspection pits are shown (Fig. 12.6). More results can be found in De Belie et al. (2004). The corrosion process was simulated in two times four cycles. Values presented are averages over six measured profiles of three test specimens per concrete type.

Before the exposure and after each series of four attack cycles, the specimens were dried at 60°C until they reached a constant weight. Concrete types P1-I-G and P1-1-L were produced using the method of immediate form removal (w/c = 0.36–0.40). Both types contained Portland cement with a cement content = 350 kg/m^3. P1-I-G contained inert (gravel) aggregates, while P1-I-L contained limestone aggregates. P3-III-G contained blastfurnace slag cement and inert aggregates and had hardened in the formwork (w/c = 0.43–0.49, cement content = 350 kg/m^3). Compressive strength at a concrete age of 28 days (NBN B15-220 1990), water absorption (NBN B15-215 1989) on three cylinders of 80 mm diameter and 70 mm height, and CaO content of the different concrete types (NBN B15-250 1989) are shown in Table 12.1.

During the second step of each cycle the sulfate concentration increased from about 0 to 2–4 g SO_4^{2-} per litre solution, proving the production of sulfuric acid by the micro-organisms. Simultaneously, the pH of the solution decreased for all concrete types rapidly from about 8.0 to 3.0 in 3 days time, and continued to decrease to pH 1.0 on day 10. In a control treatment, in which concrete blocks were suspended in the solution inactivated by means of glutaraldehyde, no change

Table 12.1 Characteristic parameters of the different concrete specimens, compressive strength and water absorption at 28 days; all cement types used were high sulfate resistant and low alkali HSR/LA

Code	Production method[a]	Cement type	Aggregate type[b]	Compressive strength (N/mm^2)	Water absorption (%)	CaO content (%)
P1-I-G	P1	CEM I 42.5	G	85.0	3.6	10.7
P1-I-L	P1	CEM I 42.5	L	84.4	3.6	35.6
P3-III-G	P3	CEM III/B 42.5	G	56.5	6.2	7.8

[a]*P1* immediate form removal, *P3* hardening in the formwork
[b]*G* gravel or porphyry, *L* limestone

in sulfate concentration could be determined, while the pH increased slowly to about 9.0 due to the high alkalinity of the concrete specimens. The average calcium loss during step 2 and 3 of each cycle indicated, as expected, that more calcium was released by the samples with limestone aggregates, compared to samples with inert aggregates.

When a chemical test with sulfuric acid and microbiological test method were compared, it could be noticed that the microbiological tests were more aggressive to concrete: the chemical tests resulted in a maximum degradation depth of 0.4 mm after ten cycles of 5 days in acid solution, while the microbiological tests resulted in a maximum degradation depth of 0.8 mm after four cycles of 10 days in the biological suspension. This difference, however, is due to the different experimental procedure and not to the origin of the acid. Concrete degradation by chemical and biogenic sulfuric acid using exactly the same sample size and test procedure were compared by (Vincke 2001). This was realised by performing a chemical test, mimicking the microbiological test described higher. During the second step of every cycle of the microbiological test procedure, pH and sulfate concentrations were measured and based on these profiles, sulfuric acid was dosed in the chemical test procedure in order to obtain the same values. These data showed that for Portland cement samples, thickness and weight losses were more than twice as high for the chemically treated concrete samples in comparison with the microbiologically treated samples. The continuous shaking of the samples during this procedure, resulting in immediate removal of deteriorated material, may be one of the important causes of the faster degradation in this procedure compared to the TAP procedure (which is described in detail in Sect. 11.3).

12.4.4 Differences Between Biogenic Sulfuric Acid and Chemical Sulfuric Acid Attack

Although most researchers, except for Tazawa et al. (1994, 1996), agree that attack in sewers is mainly caused by sulfuric acid produced by S-oxidising bacteria, large differences are found between purely chemical tests with H_2SO_4 and tests involving microbially produced H_2SO_4. The bacteria play a distinctive role. Even more important is the relationship between the bacteria and the substratum (Alexander and Fourie 2010). The latter influences the activity of the bacteria, which, in turn, affects the amount of sulfuric acid formed. In all instances, gypsum and ettringite are the main corrosive products. Yet, the ratio of the latter products could be one of the differences between the two types of degradation. A soft and pulpy layer with an increased porosity is formed on the eroded surface. In the case of chemical corrosion by sulfuric acid, the formation of this layer constitutes an extra barrier for further attack. In the case of biologically produced sulfuric acid, the soft layer creates excellent conditions for the growth of the bacteria. Due to increased porosity, the bacteria penetrate into the layer and the formation of more sulfuric acid can occur near the "un-attacked concrete". Due to the gypsum layer, the humidity of the environment stays high and it protects the bacteria against dry conditions. Brushing the specimens may have opposite effect on the two types of attack. For chemical degradation, it creates new "un-attacked" concrete and removes the extra barrier. On the contrary, brushing the specimens disturbs the microbiological balance and retards corrosion. Once the gypsum layer is too large, it may be that the oxygen content at the site of the "un-attacked" concrete is too low for the growth of the aerobic bacteria. In this case, brushing should increase the degradation rate (Monteny et al. 2000).

12.5 Conclusions

Although much research has been performed on the issue of biogenic sulfuric acid corrosion of concrete, more research should be performed focusing on the relationship between the substratum, in this case, concrete, and the bacteria responsible for the production of the sulfuric acid. The role of gypsum as a growth matrix and provider of moisture and potentially other nutrients to the bacteria deserves an in-depth study. In previous studies, parameters such as weight loss, reduction in compressive strength, and change in dynamic modulus were often measured. Yet, change in the dimensions of the specimens, penetration rate, and depth of the acid and/or the bacteria are also important parameters and need to be taken into account.

References

Alexander, M.G., Fourie, C.W.: Performance of sewer pipe concrete mixtures with Portland and calcium aluminate cements subject to mineral and biogenic acid attack. Mater. Struct. **44**(1), 313–330 (2010)

De Belie, N., Monteny, J., Beeldens, A., Vincke, E., Van Gemert, D., Verstraete, W.: Experimental research and modelling of the effect of chemical and biogenic sulfuric acid on different types of commercially produced concrete sewer pipes. Cem. Concr. Res. **34**(12), 2223–2236 (2004)

Ehrich, S., Helard, L., Letourneux, R., Willocq, J., Bock, E.: Biogenic and chemical sulfuric acid corrosion of mortars. Mater. Civ. Eng. **11**(4), 340–344 (1999)

Hormann, K., Hofmann, F., Schmidt, M.: Stability of concrete against biogenic sulfuric acid corrosion, a new method for determination. In: Justnes, H. (ed.) Proceedings of the 10th International Congress on the Chemistry of Cement, Gothenburg, 4vi038 (1997)

Monteny, J., Vincke, E., Beeldens, A., De Belie, N., Taerwe, L., Van Gemert, D., Verstraete, W.: Chemical, microbiological and in situ test methods for biogenic sulfuric acid corrosion of concrete. Cem. Concr. Res. **30**(4), 623–634 (2000)

Monteny, J., De Belie, N., Vincke, E., Verstraete, W., Taerwe, L.: Chemical and microbiological tests to simulate sulfuric acid corrosion of polymer-modified concrete. Cem. Concr. Res. **31**(9), 1359–1365 (2001)

Mori, T., Nonaka, T., Tazaki, K., Koga, M., Hikosaka, Y., Noda, S.: Interactions of nutrients, moisture and pH on microbial corrosion of concrete sewer pipes. Water Res. **26**(1), 29–37 (1992)

NBN B15-215.: Proeven op beton – Wateropslorping door onderdompeling (in Dutch). NBN, Brussels, Belgium (1989)

NBN B15-220.: Proeven op beton: bepaling van de druksterkte (in Dutch). NBN, Brussels, Belgium (1990)

NBN B15-250.: Chemisch onderzoek van verharde mortel en beton (in Dutch). NBN, Brussels, Belgium (1989)

Pluym-Berkhout, J.M., Mijnsbergen, J., Polder, R.B.: Rioleringen (II), biogene zwavelzuuraantasting. Cement **9**, 16–20 (1989)

Sand, W., Bock, E., White, D.C.: Biotest system for rapid evaluation of concrete resistance to sulphur oxidising bacteria. Mater. Perform. **26**(3), 14–17 (1987)

Schmidt, M., Hormann, K., Hofmann, F.J., Wagner, E.: Beton mit erhöhten Widerstand gegen Säure und Biogene Schwefelsäurekorrosion (Concrete with greater resistance to acid and to corrosion by biogenous sulfuric acid). Betonw. Fert. Tech. **4**, 64–70 (1997)

Tazawa, E.I., Morinaga, T., Kawai, K.: Deterioration of concrete derived from metabolites of microorganisms. In: Malhotra, V.M. (ed.) Proceedings of 3rd CANMET/ACI International Conference on Durability of Concrete, SP 145–59, pp. 1087–1097, Nice (1994)

Tazawa, E.I., Morinaga, T., Kawai, K.: Il degrado del calcestruzzo delle opere fognarie causato dai metaboliti dei microorganismi aerobici e le misure di prevenzione (The deterioration of concrete in sewerage works caused by metabolites of aerobic microorganisms, and preventive measures). L'industria italiana del cemento **11**, 792–805 (1996)

van Mechelen, A.C.A., Polder, R.B.: Rioleringen (XI), biogene zwavelzuuraantasting, verslag van een vervolgonderzoek in Rotterdam. Cement **2**, 26–30 (1991)

Vincke, E.: Biogenic sulfuric acid corrosion of concrete: microbial interaction, simulation en prevention. Ph.D. thesis, Faculty of Agricultural and Applied Biological Sciences, Ghent University, 208 p (2001)

Chapter 13
Test Methods for Chloride Transport in Concrete

Qiang Yuan and Manu Santhanam

Contents

Keywords Chloride • Diffusion • Fick's law • Migration • Salt ponding • Conductivity

13.1 Introduction

Chlorides *per se* are not generally thought of as being damaging to concrete. However, chlorides can alter the products of cement hydration, and also cause other forms of concrete deterioration. For these reasons, this chapter is included here.

Chlorides from seawater and groundwater can attack the hydrated cement paste phases in concrete, primarily leading to the lowering of the alkalinity of concrete.

Q. Yuan (✉)
School of Civil Engineering, Central South University, Changsha, China
e-mail: dhxyyuan@yahoo.com.cn

M. Santhanam
Department of Civil Engineering, Indian Institute of Technology, Chennai, India
e-mail: manus@iitm.ac.in

M. Alexander et al. (eds.), *Performance of Cement-Based Materials in Aggressive Aqueous Environments*, RILEM State-of-the-Art Reports 10,
DOI 10.1007/978-94-007-5413-3_13, © RILEM 2013

The C_3A in the cement can bind external chlorides to produce chloroaluminate compounds, such as Friedel's salt, which do not cause any expansion (Verbeck 1975). This can help reduce the level of free chlorides. Chlorides at the surface of embedded reinforcing steel can result in the depassivation of the steel, since the passivating iron oxide layer is readily attacked by chlorides to form soluble iron chloride. Pitting corrosion is also associated with the action of chlorides (Bentur et al. 1997).

There are numerous mechanisms that govern chloride ingress into concrete (Poulsen and Mejlbro 2006): (1) hydrostatic advection or permeation, (2) diffusion, (3) electrical migration, (4) capillary suction and (5) thermal migration. Permeability is the process where external agents are transported into concrete with hydraulic pressure (pressure differential) as driving force. The coefficient of permeability describes the permeation characteristics through porous materials under a pressure head. Both gases and water penetrate into concrete due to a pressure difference. For example, structures such as dams are subjected to pressure on one side, causing water to penetrate under pressure. Permeation is governed by Darcy's law for laminar flow.

Diffusion is the process by which external agents are transported into the concrete due to concentration differences. Here mass transfer occurs due to random movement of free molecules or ions in pore solution from higher concentration regions to lower concentration regions. Rate of mass transfer is proportional to the concentration gradient. Rate of diffusion is faster in gases than in liquids. Diffusion is measured using flux which is the rate of transfer per unit area. The diffusion coefficient is ideally a material property, which describes the transferability of the given ionic species. Depending on the location of the structure (exposed to tides, totally submerged in water, exposed to atmosphere) either ionic diffusion, water vapour diffusion or gas diffusion can occur. The diffusion equation in steady state can be expressed by Fick's first law. The transport processes in concrete are time dependent, since the microstructure of concrete continually evolves. For this reason, Fick's first law (Eq. 13.1), which requires steady state diffusion, is not really applicable. Most diffusion processes in concrete, including chloride diffusion, can be modelled using Fick's second law of diffusion, which states that the change in chloride content per unit time is equal to the change of flux per unit length (Eq. 13.2).

$$J = -D\frac{\partial c}{\partial x} \tag{13.1}$$

$$\frac{\partial c}{\partial t} = \frac{\partial J}{\partial x} = -\frac{\partial}{\partial x}\left(D\frac{\partial c}{\partial x}\right) = -D\frac{\partial^2 c}{\partial x^2} \tag{13.2}$$

Where:

J = Flux

D = diffusion coefficient

c = chloride concentration

x = depth

t = time

Migration is the process whereby ion transport is driven by the difference in electrical potential. In an electric field, positive ions will move preferentially to the negative electrode and negative ions to the positive one. Migration generates a difference in concentration in a homogenous solution or may provoke a species flux in the direction of concentration gradients. Ions always move from zones of high electrical potential to zones of lower electrical potential. This phenomenon of migration is mainly important in the case of chloride transport. When concrete is exposed to a stray current, chloride migration occurs. For example, the negative chloride ions from soil in contact with the concrete will migrate towards the positive anode zones of the reinforcement and increase the corrosion rate and cause pitting corrosion. The rapid chloride permeability test (RCPT), described in ASTM C1202 (2008), is one test where this mechanism is employed to find the resistance of concrete to chloride penetration under an applied potential difference.

When a fluid is in contact with a solid, there is a free interfacial energy created by the difference between the forces that attract the molecules towards the interior of each phase and those that attract them to the contact surface. As the difference increases, fluids such as pore water are retained in the porous medium above the elevation of the water table. The interfacial tension or suction force in a capillary tube causes water to rise and form a meniscus. This transport mechanism is called capillary suction, and is typically limited to a shallow cover region, and will not, by itself, bring chlorides to the level of the reinforcing steel unless the concrete is of extremely poor quality and the reinforcing steel is shallow. However, it does bring chlorides to some depth quickly in the concrete and reduces the distance that they must diffuse to reach the rebar.

It is well known that ions or molecules in a hot environment move faster than those in a cold environment. If a saturated concrete sample has initial uniform chloride concentration, when one portion is heated, chloride ions in the hotter region will move towards colder regions. The most obvious situation for this process to occur is when a concrete structure, which has been contaminated with deicing salt, heats up in sunlight. Salt-saturated water in the surface pores of the concrete will migrate rapidly into the cooler parts of the structure under the temperature gradient.

13.2 Measurement of Chloride Transport

13.2.1 Theoretical Basis

Chlorides are available from various sources, such as salty aggregates, water, ground water, admixtures containing chlorides, seawater, de-icing salts etc. All the mechanisms, namely diffusion, migration, permeation, capillary suction and thermal migration, result in penetration of chlorides into the concrete. When the concrete is fully saturated, the transport of chlorides is commonly assumed to be governed by diffusion.

During the past two decades, many methods have been proposed and developed to measure chloride transport in concrete. However, there is no one strict way to categorize the existing methods. Streicher and Alexander (1995), Stanish et al. (2000) and Shi et al. (2007) have given reviews on the existing methods. Recently, Yuan (2009) presented a review on the existing methods.

Actually, there are two purposes for the measurement of chloride transport in concrete:

- Ranking the chloride resistance of concrete.
- Obtaining input parameters for the prediction of the service life of the concrete structure subjected to chloride environments.

Some test methods can only fulfil the first purpose, such as ASTM C 1202, in which the charge passed through a concrete specimen within 6 h is measured. This cannot be directly used as an input parameter for the prediction of the service life of the concrete structure. This type of test is more suitable to be a quality control tool. In contrast, the diffusion coefficient of concrete can be measured by other test methods. Some test methods can fulfil both purposes, and are therefore more desirable.

It is well known that chloride ions travel in concrete at a very slow rate; rapid results, however, are always desired in engineering applications. Various techniques are used to accelerate chloride transport in concrete to shorten the test duration and obtain rapid results. Accordingly, different theoretical bases are applied to assess the chloride resistance of concrete. Yuan (2009) proposed the classification of the existing methods on a theoretical basis, as shown in Table 13.1. Using this classification, the test methods are grouped into six categories: Fick's first law; Fick's second law; Nernst-Planck equation; Nernst-Einstein equation; formation factor; other.

13.2.2 Summary of Common Test Methods

Chloride diffusion is characterized by the chloride profile, i.e. a graph of the chloride content of the concrete vs. the depth from the concrete surface. The chloride profile is obtained by various tests, which are standardized as ASTM C 1556-03 in the US, and NT Build 443 in Europe respectively. In ASTM C 1556-03, specimens are subjected to pre-conditioning at 50°C for 7 days. After removal from the oven, all the surfaces are coated with epoxy, except that one is left uncovered. 3% NaCl solution is ponded on the specimens for a specified period of 3 months and due to diffusion the chloride ions are transported into the concrete (in order to avoid wick action, the bottom surface is coated with epoxy paint). After completion of the exposure period, powder samples are taken from various depths and free chloride

Table 13.1 Summary of test methods for chloride transport in concrete (Yuan 2009)

Theoretical base	Test method	Measurement	Test duration	Reference
Fick's first law	Steady state diffusion test	Chloride flux	Several months	Page (1981)
Fick's second law	NT build 443	Chloride profile	>35 days	NORDTEST NT BUILD 443 (1997)
	(Bulk diffusion)			ASTM C 1556-03
	ASTM C 1556-03			
Nernst Planck equation	NT build 355	Chloride flux	Several weeks	NORDTEST NT BUILD 355 (1995)
	Truc's method	Chloride flux	Several days	Truc et al. (2000)
	NT build 492 or Tang's method	Penetration depth	24–72 h	NORDTEST NT BUILD 492 (1995), Tang and Nilsson (1992), Tang (1996)
	Breakthrough time method	Breakthrough time	Several weeks	Halamickova et al. (1995)
	Multi-regime method	Conductivity at the anolyte	Several days	Castellote et al. (2001a, b) UNE 83987 (2009)
	Samson's method	Current	120 h	Samson et al. (2003)
	Friedmann's method	Current	Several weeks	Friedmann et al. (2004)
Nernst Einstein equation	Lu's method	Resistivity	Several minutes	Lu (1998)
	Andrade's method	Resistivity	Several minutes	Andrade et al. (2000)
Formation factor	Formation factor Method	Resistivity	Several minutes	Streicher (1995)
Other	ASTM C1202 or AASHTO T227	Charge passed	6 h	AASHTO T 259-80 (1980)
	Conductivity test method	Conductivity	Several minutes	South African Durability Index Test Proc. Manual
	AASHTO T 259 90 day ponding test	Chloride profile	90 days	AASHTO T 259-80 (1980)
	Water pressure method	Penetration depth	Several weeks	Freeze and Cherry (1979), Stanish et al. (2000)
	AC impedance method	impedance	Several minutes	Shi et al. (1999)
	Integral Method	Electrical potential	Several days	Castellote et al. (2002)

contents are obtained using a water soluble chloride determination. The chloride profile is used for finding the diffusion coefficient, using Fick's second law. The differences between ASTM C 1556-03 and NT Build 443 lie in:

- NT Build 443 requires a completely saturated specimen to ensure pure diffusion during the test. However, ASTM C 1556-03 conditions the specimen at 50°C for

7 days prior to immersion. Thus, chloride will travel through diffusion and capillary suction.
- 165 g/l NaCl is specified in NT Build 443, and 3% in ASTM C 1556-03.
- Total chloride is used in NT Build 443, while water-soluble chloride is used in ASTM C 1556-03.

Various techniques, which include DC, AC and high pressure, are used to accelerate the process. Among the accelerated techniques, use of DC is most common. As can be seen from Table 13.1, Nernst-Planck equation is the most often used theory in the calculation of migration coefficient. Only a few representative test methods are discussed in detail here.

Chloride transport under an applied electric field is measured using many test methods, two of which are the rapid chloride permeability test and the conductivity test. Although these two test methods do not give information on the diffusion coefficient, they have the advantage of short test duration, simple manipulation and good precision. They are quite suitable for engineering application to rank the permeability of concrete. Streicher and Alexander (1995) developed a chloride conductivity test at the University of Cape Town, in which virtually all ionic flux occurs by conductivity due to a 10 V potential difference. This test measures the conductive ionic flux through a concrete disc under a potential difference, and is related to the chloride diffusion properties of the concrete. In the test, chloride ions move through all pores of sufficient size, without favouring the larger pores as with gaseous permeation. The chloride conductivity test therefore provides an indication of overall diffusivity, being sensitive to changes in pore structure and concrete chemistry, which might appear to be insignificant when using the permeation process. The lower the index, the lower is the diffusivity and the better the potential durability of the concrete. The test is very sensitive to binder type with blended binders such as fly ash or slag showing superior properties. The apparatus consists of a two-cell conduction rig, each cell containing a 5 M NaCl solution so that there is no concentration gradient across the sample and chloride migration is due to conduction from the applied potential difference.

Mackechnie (1996) showed that 28-day chloride conductivity indexes reduced with increasing concrete grade, but were more affected by the extent of curing and the type of binder. When properly cured, the addition of fly ash (FA) or ground granulated blast furnace slag (GGBS) had the effect of refining the pore structure and changing its chemistry, and the chloride conductivity test has been found to be extremely sensitive to such changes.

The Rapid Chloride permeability Test (RCPT) method is standardized under:

- AASHTO T277, "Standard Method of Test for Rapid Determination of the Chloride Permeability of Concrete", and
- ASTM C1202, "Standard Test Method for Electrical Indication of Concrete's Ability to Resist Chloride Ion Penetration"

The RCPT is an electrical test in which the result is a direct function of the resistance of the test specimen (Julio-Betancourt and Hooton 2004). In principle, the use of electrical properties to measure the ionic transport properties of concrete is well grounded (McCarter et al. 2003) and is affected by two aspects of the concrete: (1) the connectivity of the capillary pore system, and (2) the electrolytic capacity (ionic concentration) of the pore solution. Of primary interest with respect to assessing the durability of concrete is the capillary pore system. RCPT results correlate fairly well with chloride penetration by diffusion (McGrath and Hooton 1999) in which the pore system has the primary influence on results. Two notable exceptions where the ionic concentration of the pore solution greatly affects RCPT results occur with silica fume (very low) (Shi et al. 1998) and calcium nitrite (very high) (ASTM C 1202). Julio-Betancourt and Hooton (2004) showed that increase in temperature causes increase in rate of charge passing. This test is further complicated by the fact that a number of mechanisms, including migration, diffusion and permeation are at play.

Tang developed a method in 1992 for accelerating chloride ingress into a concrete specimen, which was standardized in 1999 by NordTest as NT Build 492. The test method recommends varied applied voltages and testing durations based on measured initial currents through tested specimens so as to avoid significant heating during the test and to obtain a reasonable chloride penetration depth. At the end of testing, the specimens are axially split into two and 0.1 mol/l silver nitrate solution is sprayed on the freshly split section to determine the chloride ion penetration depth. The chloride migration coefficient can be calculated using the chloride penetration depth.

Most of the above-mentioned tests occur under non-steady-state conditions; that is to say chloride concentration at any location in the specimen varies with time during the test. By contrast, NT Build 355 is an electrically accelerated test method for the measurement of the steady-state migration coefficient of concrete. In this test, a concrete specimen is placed between two cells containing chloride solution and alkaline solution respectively. Electrical field is applied across the specimen, and the transport of chloride is thus accelerated. Monitoring the change in chloride concentration with time in the downstream cell with alkaline solution, the slope of the line of chloride concentration against time can be obtained. The chloride steady-state migration coefficient can be calculated using the slope.

The Multi-Regime Method (MRM) developed by Castellote et al. (2001a, b) allows the determination of both diffusion coefficients, the steady state D_s and non-steady state, D_{ns}, and therefore the binding ability of the matrix. The test is based on measuring the amount of chlorides arriving in the downstream cell (anolyte) by means of measuring the conductivity of that solution instead of analysing chlorides in it. The concentration of chlorides is calculated through an empirical correlation between chloride concentration and conductivity for the specific conditions of the test. The steady-state coefficient is calculated from the flux of chlorides through the specimen, obtained from the measurement of the conductivity of the anolyte in the anodic compartment. The calculation of the non steady-state diffusion coefficient is made from the time taken by the chloride ions to establish a constant flux, that is to say, from the so-called time-lag. The Integral method, (previously called Threshold Migration Induced Method (TMIM)) [Castellote et al. 2002], allows to determine the threshold value of corrosion of steel and simultaneously the non steady state diffusion coefficient. In this method, a specimen with a rebar is used. Chloride ions are driven to the rebar by migration due to an external voltage and during the test, the potential of the rebar is monitored by means of a reference electrode placed on the surface of the specimen as close as possible to the rebar, and the current intensity passing through the specimen is registered. The electrical field has to be switched off when changes in polarization of the steel are detected, which indicates that depassivation is going to take place, followed by waiting until depassivation occurs in natural conditions.

13.3 Results from Studies Using Different Methods

In the framework of the RILEM Technical Committee TC 178-TMC: "Testing and Modelling Chloride penetration in Concrete", extensive work on chloride penetration was done, e.g. a Round-Robin on methods for determining chloride transport parameters in concrete. Twenty-seven different laboratories around the world participated, using 13 different methods, in triplicate specimens, for 4 different mixes cast with different binders. Three groups of methods were used: Natural diffusion methods, Migration methods and Resistivity methods. Statistical treatment of the data was done according to ISO 5725-2:1994 for the determination of the accuracy (trueness and precision) of measurement methods and results (Part 2: Basic method for the determination of the repeatability and reproducibility of a standard measurement method).

In order to make an evaluation of these methods, four indicators were identified and within each of them, several sub-indicators were assigned. According to this system of classification, the methods were classified following each indicator (trueness, precision, relevance and convenience), and also globally, by assigning different factors of importance (F.I.) to the different indicators. The results were published in Castellote and Andrade (2006).

The scope of this section of the chapter, however, is to present a summary of results from different types of tests conducted at two laboratories – IIT Madras (India) and Magnel Laboratory (Belgium) – which throws some light on the comparative effectiveness of the various methods.

13.3.1 Study Conducted at IIT Madras

In a recent research study at IIT Madras, various durability parameters (relating to water, gas, and chloride ingress of concrete) were measured for three different strength grades of concrete – M20, M45, and M70 – where the numbers represent the characteristic compressive strength of the concrete in MPa. All concretes were

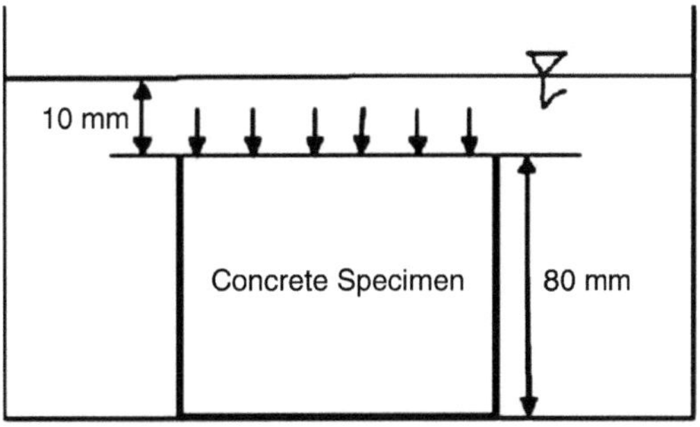

Fig. 13.1 Ponding of specimen in salt solution

prepared using 53 grade OPC (equivalent to CEM I 52.5). The sections below explain the procedures and results for the various tests pertaining to chlorides.

13.3.1.1 Salt Ponding Test

This test can be performed according to the procedure prescribed in ASTM C 1556-03 (2008). Concrete cylinders of size 100 mm diameter and 200 mm height were cast and stored in a moist room for 7 days. After 7 days of curing, test specimens of size 100 mm diameter and 80 mm thickness were obtained by slicing the cylinder specimens and pre-conditioning them at 50°C for 7 days. After removal from the oven, circumference of the specimen was coated with epoxy and allowed to dry. Specimens were placed in the container having 3% NaCl solution and a level of 10 mm solution above the specimen was maintained, as shown in Fig. 13.1.

The solution in the container was replaced at 1 month intervals. After exposure periods of 28, 56 and 90 days, the specimens were removed from the container and allowed to dry. Powder samples were collected from the various depths of the specimen using a lathe machine. The depths from which the samples were taken depended on the grade of the concrete (w/c ratio) as per ASTM C 1556 (2008) and are given in Table 13.2. Chloride content in the concrete powder sample was obtained by titration with $AgNO_3$ solution, using potassium chromate as indicator.

For finding the diffusion coefficient and initial chloride content, a graph (as shown in Fig. 13.2) was plotted between Chloride content (%) and depth (mm). A chloride profile was obtained and the values were fitted to an error function given by the Eq. (13.3).

$$C(x,t) = C_s - (C_s - C_i).\mathrm{erf}(x/\sqrt{(4D_a t)}) \qquad (13.3)$$

Where

$C(x,t)$ = chloride concentration, measured at depth x and exposure time t, %

C_s = projected chloride concentration at the interface between the exposure liquid and test specimen that is determined by the regression analysis, mass %

C_i = initial chloride concentration of the cementitious mixture prior to submersion in the exposure solution, mass %

Table 13.2 Recommended depth Intervals (mm) for powder grinding

Design strength of concrete (MPa)		
20	45	70
0–1	0–1	0–1
1–3	1–3	1–2
3–6	3–5	2–3
6–10	5–7	3–4
10–15	7–10	4–6
15–20	10–13	6–8
20–25	13–16	8–10
25–30	16–20	10–12

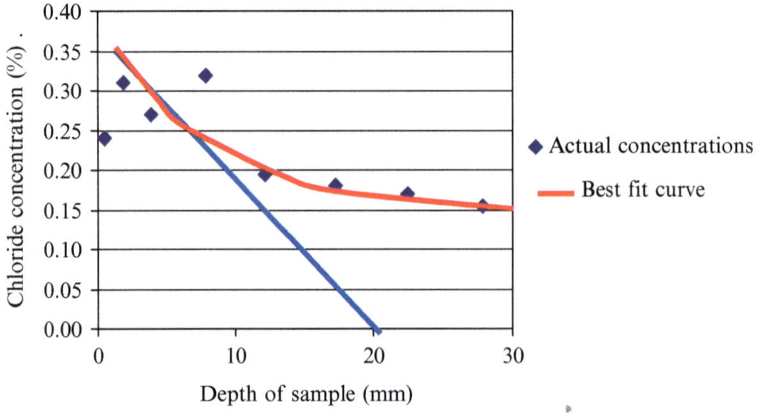

Fig. 13.2 Plot of chloride content against depth

x = depth below the exposed surface (to the middle of a layer), (m)
D_a = apparent chloride diffusion coefficient, m^2/s
t = the exposure time, s
erf = the error function

Figure 13.2 shows a typical plot of concentration of chlorides (%) with depth (mm). A curve which fits best through the points was drawn as shown in the figure. Tangent from the point (0, C_0) was drawn and its intercept with the x-axis was marked as 'a'. The diffusion coefficient was found using Eq. (13.4).

$$\text{Diffusion Coefficient} = a^2/(\pi.\ t) \tag{13.4}$$

Where
t = time, s
a = intercept on x-axis

The results of the salt ponding test of M20, M45 and M70 concretes at 28, 56 and 90 days are presented in Table 13.3 and Fig. 13.3. Table 13.3 presents the diffusion

Table 13.3 Diffusion coefficient and initial surface concentration as per Fick's second law

	Diffusion coefficient ($\times 10^{-12}$) m^2/s			Initial surface concentration (%)		
Grade of concrete	28 days	56 days	90 days	28 days	56 days	90 days
M20	38.1	36.3	33.2	0.35	0.50	0.52
M45	14.5	12.9	11.8	0.19	0.36	0.72
M70	4.6	3.7	3.3	0.40	0.60	0.65

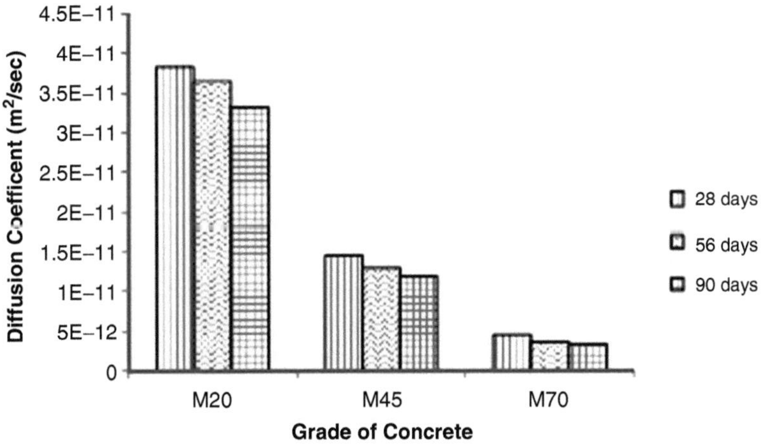

Fig. 13.3 Effect of grade of concrete on diffusion coefficient with respect to age

coefficient values as well as initial surface concentration, as calculated using Fick's second law equation. It is observed from Fig. 13.3 that the diffusion coefficient decreases as the grade of concrete increases. This can be attributed to the difference in pore structure of the concretes. As the grade of concrete increases, it is more resistant to salt penetration.

It is observed from Fig. 13.3, for M20 grade of concrete, that there is a decrease in diffusion coefficient value by 5% from 28 to 56 days and by 9% from 56 to 90 days. For M45 grade of concrete, there is a decrease in diffusion coefficient value by 11% from 28 to 56 days and by 9% from 56 to 90 days. For M70 grade of concrete decrease in diffusion coefficient value is 20% from 28 to 56 days and 11% from 56 to 90 days. Thus, the decrease in diffusion coefficient is more for high grade concretes than the lower grade concrete, for this particular type of concrete.

13.3.1.2 Rapid Chloride Permeability Test (RCPT)

This test was performed according to the procedure prescribed in ASTM C1202-07 (2008) on specimens of size 100 mm diameter and 50 mm thickness, which were obtained by slicing 100 × 200 mm cylinder specimens. On the day of testing, the curved surface of the specimens was coated with epoxy paint, avoiding epoxy on

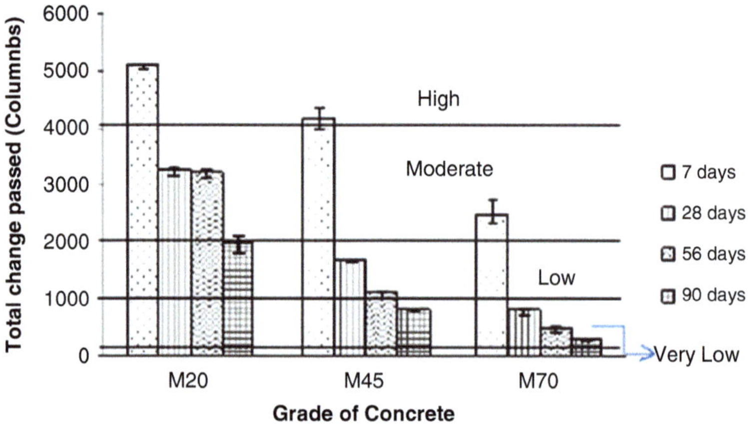

Fig. 13.4 Effect of grade of concrete on chloride permeability

the flat test surfaces and covering any pinholes. The epoxied specimens were allowed to cure for 18–24 h in a desiccator. After drying of the specimens, they were placed in a water bath which was stored inside a desiccator connected to a vacuum pump. The desiccator was evacuated to between −75 and −80 kPa and this vacuum was maintained for 3 h ± 15 min. After 18 h soaking, the specimen was removed from the water. Then it was placed in the test device between the two cells for the conduct of the test.

The RCPT results are presented graphically in Fig. 13.4. The permeability classes as per ASTM C1202 are also marked on the figure. The results clearly indicate the effect of the grade and age of concrete on the chloride permeability. At 28 days, the M20, M45, and M70 concretes were in the Moderate, Low, and Very Low categories of permeability as per ASTM C1202. However, at 90 days, even the M20 concrete was in the Low category, while the higher grade concretes were both in the Very Low category.

13.3.1.3 Chloride Conductivity Test

This test was performed according to the procedure of Alexander et al. (1999). The apparatus was modified to suit the available cells (which are normally used for RCPT) as shown in Fig. 13.5. Specimens of size 100 mm diameter and 50 mm thickness were obtained by slicing of cylinder specimens. The specimens (discs) were kept in an oven at a temperature of 50°C for a time period of 7 days. After removal of specimens from the oven, the vertical curved surface of the specimens was coated using epoxy paint, avoiding epoxy on the flat test surfaces and any pinholes. The specimens were allowed to dry for 18–24 h in a desiccator. The specimens were then placed in a bath containing 5 M NaCl solution which was stored inside a desiccator connected to vacuum pump. The desiccator was

Fig. 13.5 Modified cells for chloride conductivity test

evacuated to between −75 and −80 kPa and this vacuum was maintained for 3 h ± 15 min. After 18 h soaking, the specimens were removed from the solution and were placed between the two cells which were both filled with 5 M NaCl solution. DC power supply was adjusted to maintain the voltage applied across the specimen to approximately 10 V. Current was recorded after 15 min. For each test specimen, chloride conductivity was calculated using Eq. (13.5).

$$\sigma = (\mathrm{i.t})/(\mathrm{V.A}) \qquad (13.5)$$

Where
σ = conductivity of the specimen (mS/cm)
i = electric current (mA)
V = voltage difference (V)
t = average thickness of specimen (cm)
A = cross-sectional area of the specimen (cm^2)

Results of chloride conductivity for the different grades of concrete are given in Fig. 13.6. The difference in conductivity between the M20 and M45 concretes is greater compared to that between M45 and M70 concretes. While the reduction in conductivity with age is significant for the lower grade concretes, there is still a large difference between the M20, M45 and M70 concretes at 90 days. This is contrary to the qualitative result provided by the ASTM C1202 test, which suggests that even the M20 concrete has a 'Low' permeability at 90 days. The conductivity test appears to give a better quantitative characterization of the differences between the concretes.

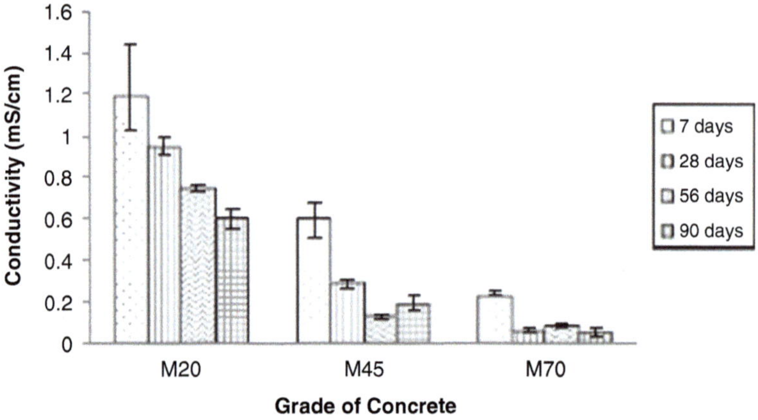

Fig. 13.6 Effect of grade of concrete on conductivity

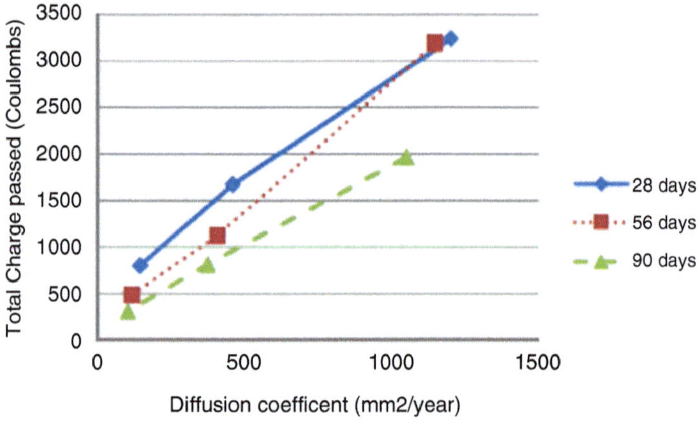

Fig. 13.7 Interrelation between diffusion coefficient in salt ponding test and charge passed in RCPT

13.3.1.4 Correlations Between Results from Different Test Methods

Salt Ponding Test and RCPT

Correlations between the diffusion coefficient from the salt ponding test and RCPT for M20, M45 and M70 concretes at 28, 56 and 90 are shown in Fig. 13.7. Regression coefficients for 28, 56, 90 days are 0.997, 0.995 and 0.999 (not shown on graph). There is good correlation between the results of the salt ponding test (diffusion coefficient) and Rapid chloride permeability test. From Fig. 13.7, it is observed that the change in total charge passed significantly changes with age of concrete for M45 and M70 grades of concrete. But there is not much significant change in total charge passed for M20 from 28 to 56 days.

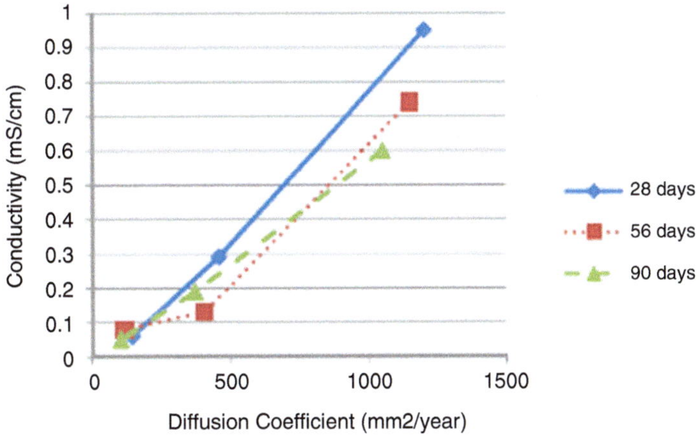

Fig. 13.8 Interrelation between Salt ponding test and Conductivity test

Salt Ponding Test and Chloride Conductivity

The correlation between the diffusion coefficient from Salt ponding test and Chloride Conductivity Test for M20, M45 and M70 at 28, 56 and 90 days is shown in Fig. 13.8. The regression coefficient values for 28, 56, 90 days are 0.998, 0.957 and 0.999 (not shown on graph) i.e., there is a good correlation between the results of the salt ponding test (diffusion coefficient) and Chloride Conductivity. From Fig. 13.8, it is observed that for higher grades of concrete (M45 and M70), the change in conductivity values with age is small.

Chloride Conductivity and RCPT

From the previous sections, it is observed that both RCPT and Conductivity tests are well correlated with the diffusion coefficient from salt ponding test. Here an attempt has been made to find the relationship between RCPT and conductivity test, so that RCPT can be replaced with the conductivity test. The correlations between conductivity test and RCPT at 7, 28, 56 and 90 days age for M20, M45 and M70 are shown in Fig. 13.9.

Figure 13.9 shows that as the age of the concrete increases the curves move towards the left, which indicates that there is decrease in total charge passed and conductivity values with age. The regression coefficient values for 7, 28, 56, 90 days are 0.910, 0.974, 0.987 and 0.997 (not shown on graph) i.e., there is a good correlation between the results of the total charge passed and Chloride Conductivity. The slope of the best fit line does not change significantly with age, indicating consistent behaviour. Thus, it can be concluded that the conductivity test, which is less prone to errors arising from the multiple mechanisms of transport

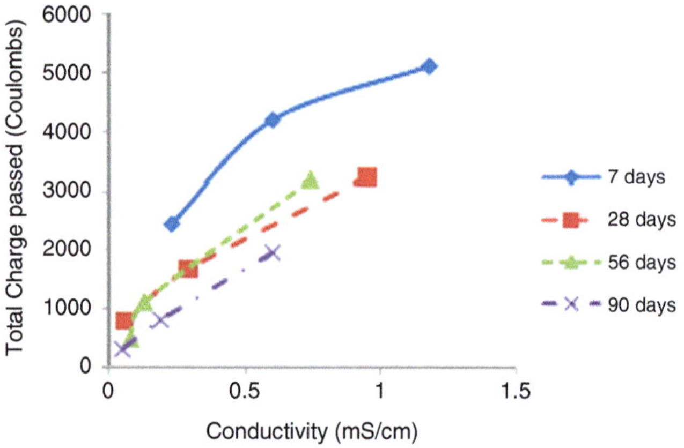

Fig. 13.9 Interrelation between total charge passed in RCPT and conductivity test

involved in RCPT (such as migration, diffusion, and permeation), can be used as a replacement for RCPT.

13.3.2 Study Conducted at Magnel Lab for Concrete Research

A study on the evaluation of test methods for chloride transport in concrete was conducted at the Magnel Lab at Ghent University. Concretes containing various supplementary cementing materials, with water-to-binder ratios of 0.35, 0.48 and 0.6 were tested by different test methods.

13.3.2.1 Bulk Diffusion Test

The bulk diffusion tests were conducted according to NT Build 443. The technique of vacuum saturation was applied to ensure the saturation of concrete. Before vacuum saturation, test specimens were dried at room temperature to a surface-dry condition and then all faces of the test specimens except the one to be exposed were given an approximate 1 mm thick coating of epoxy. The surface closer to the cast-surface was the surface exposed to the chloride. When the coating materials hardened, the test specimens were vacuum saturated with lime solution, and then immersed in chloride solution. The containers were covered by film to prevent evaporation, and the containers were shaken once every week. The ratio between the exposed area (3×78.5) cm^2 and the volume of exposure liquid (11.78 l) was 20.

After the immersion period of 6 weeks, the specimens were removed from the solutions. The chloride profile was measured immediately after the exposure by successively grinding off layers (0.5–1 mm) parallel to the exposed surface. Not

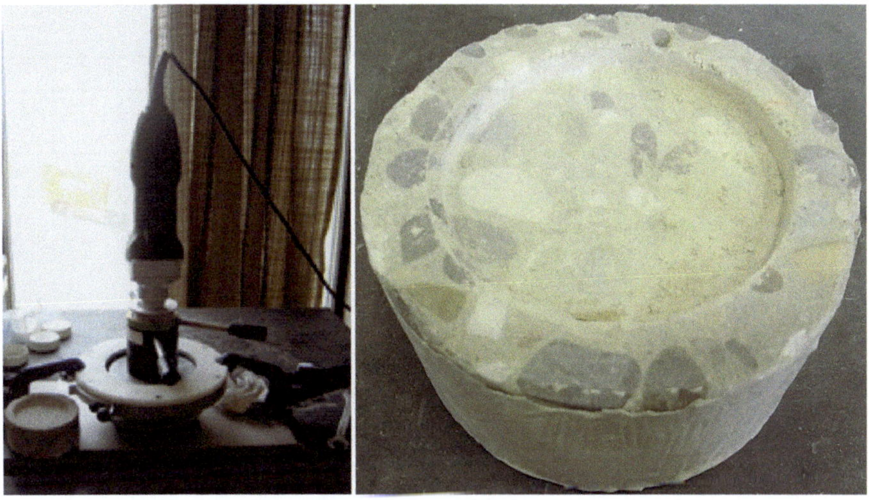

Fig. 13.10 Profile grinding machine and specimen after the process of grinding

every layer was collected for analysis. A good penetration profile can be obtained with a minimum of approximately five or six points in the penetrated zone. The grinding was performed by Profile Grinder 1100, as shown in Fig. 13.10. The grinding area is 73 mm, and Fig. 13.10 shows a ground specimen.

Acid-soluble chloride content was determined, and was regarded as total chloride. At the same time, water-soluble chloride was also determined. In this study, the chloride concentration determined by pore expression method is regarded as free chloride. Chloride concentrations determined by water extraction method and pore expression method are different, and their relationship was established in literature (Yuan 2009). In this study, the water-soluble chloride was converted to free chloride (determined by pore expression method) by a linear relationship (Yuan 2009).

The typical total, water-soluble and free chloride profiles are shown in Fig. 13.11. As expected, total, water-soluble and free chlorides showed the same trend, and decreased gradually along the penetration direction.

Chloride diffusion coefficients are obtained by fitting both free chloride and total chloride profiles to Fick's second law – see Table 13.4. As can be seen, chloride diffusion coefficient decreases with decreased w/b ratio. Supplementary cementing materials behave differently in decreasing chloride diffusion coefficient. Silica fume is the most effective supplementary cementing material, followed by slag. Fly ash is effective only in the case of low w/b ratio.

A very good linear relationship with a slope of 0.76 between diffusion coefficients is obtained from fitting total and free chloride profiles, as shown in Fig. 13.12.

Table 13.4 Diffusion coefficients obtained from fitting total and free chloride profiles of concrete ($\times 10^{-12}$ m^2/s)

Mix	B6	B48	B35	FA6	FA48	FA35	SL6	SL48	SL35	SF6	SF48	SF35
Total	13.24	5.97	6.35	15.3	9.34	2.55	4.81	6.5	2.46	5.62	5.12	2.42
Free	10.52	4.95	5.07	9.49	6.54	2.25	4.68	5.77	2.32	4.91	3.84	2.33

Note: *B* represents Portland cement concrete; *FA, SL* and *SF* represent fly ash, slag and silica fume concrete respectively (FA/C = 0.25; S/C = 0.66; SF/C = 0.05); 6, 48 and 35 represents w/b ratios of 0.60, 0.48 and 0.35

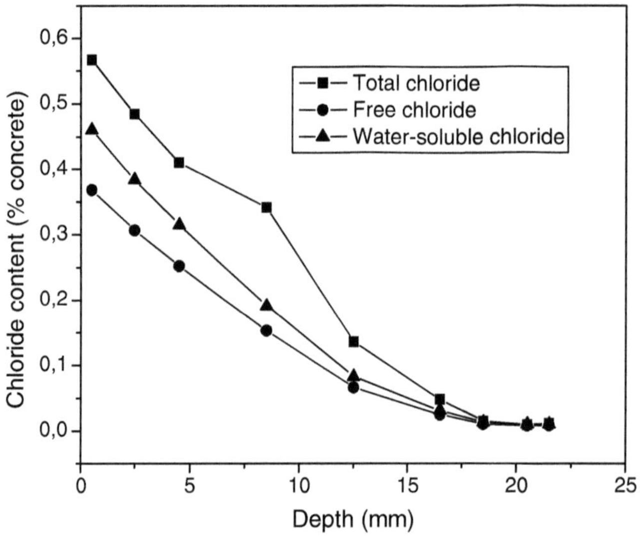

Fig. 13.11 Typical total, free and water-soluble chloride profiles (B6, 20°C, 165 g/l)

13.3.2.2 Non-Steady-State Migration Test

NT Build 492 non-steady-state migration test was carried out. Specimens were first vacuum saturated with lime solution, and then an electrical field was applied across the specimen. Testing setup is shown in Fig. 13.13. After migration test, the specimens were axially split into two pieces and 0.1 mol/l silver nitrate solution was sprayed on the freshly split section to determine the chloride ion penetration depth, as shown in Fig. 13.14. Based on the chloride penetration depths, the migration coefficients are tabulated in Table 13.5.

Initial current is also an indicator of chloride resistance of concrete (Feldman et al. 1999), and has been proposed to replace the 6 h charge passed specified in ASTM C 1202. During the test, the initial current was measured under 30 V. A reasonably good linear relationship can be found between migration coefficient and initial current, as shown in Fig. 13.15.

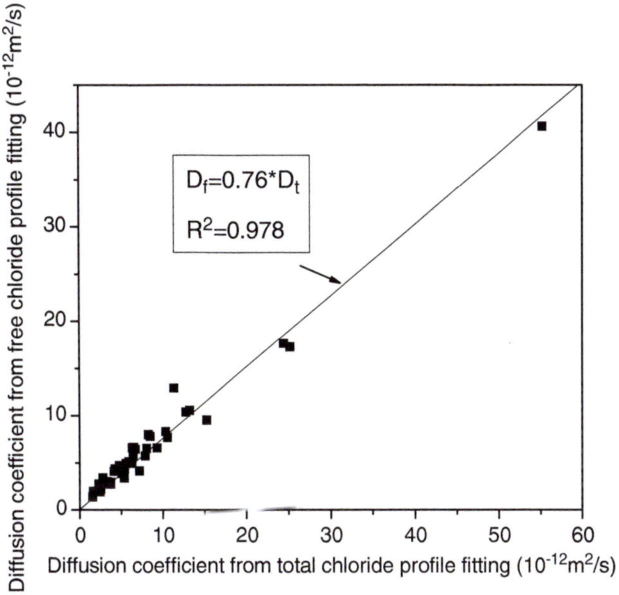

Fig. 13.12 Relationship between diffusion coefficients obtained from fitting total and free chloride profiles

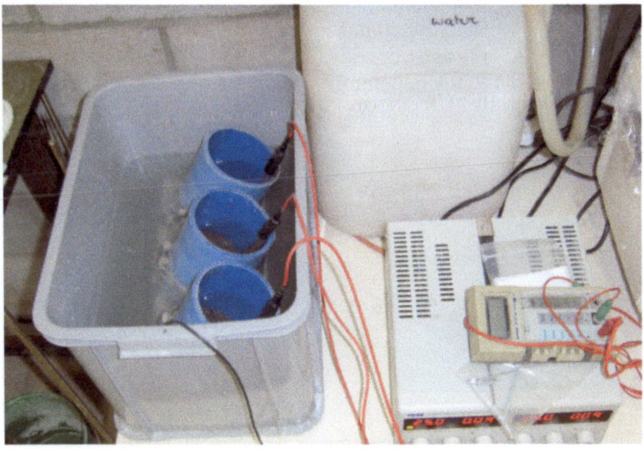

Fig. 13.13 Testing setup of non-steady-state migration test

13.3.2.3 Steady-State Migration Test

Steady-state migration tests were carried out according to NT Build 355. The testing setup is shown in Fig. 13.16.

Fig. 13.14 Spraying $AgNO_3$ solution onto tested specimens and the colour appearance after spraying

Table 13.5 Migration coefficient obtained from NT Build 492 ($\times 10^{-12}$ m^2/s)

Mix	B6	B48	B35	FA6	FA48	FA35	SL6	SL48	SL35	SF6	SF48	SF35
D	18.15	9.55	5.65	20.64	8.99	3.39	6.92	8.29	1.62	5.21	4.63	0.91

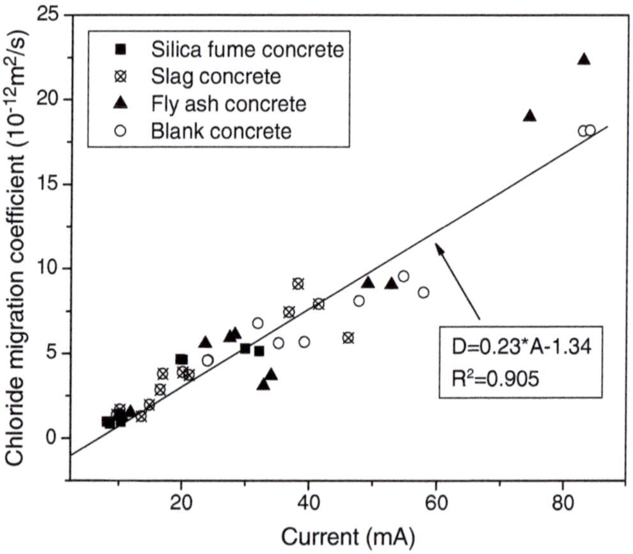

Fig. 13.15 Relationship between initial current and D_{nssm}

The chloride concentrations both in upstream and downstream were monitored, as shown in Fig. 13.17. The slope obtained from upstream is greater than that from downstream. This is due to the effect of chloride binding. The migration coefficients obtained from steady-state migration tests are given in Table 13.6. Silica fume and slag are efficient in reducing the chloride migration coefficient.

Fig. 13.16 Testing setup of steady-state migration test

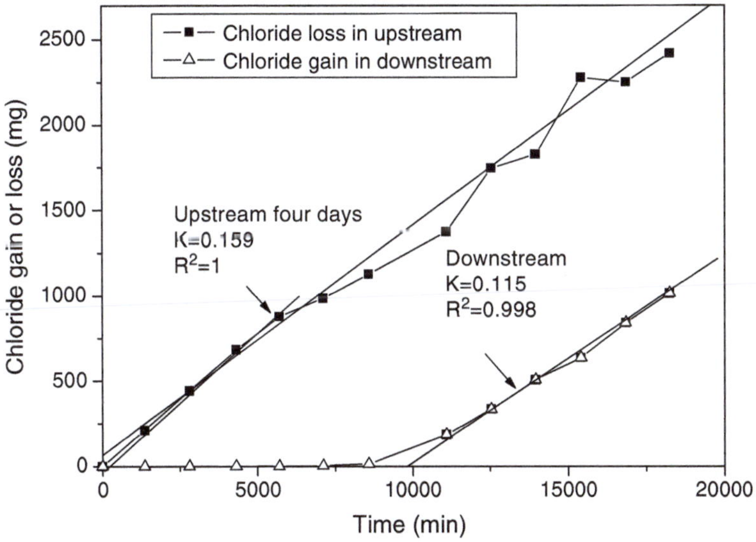

Fig. 13.17 Evolution of upstream and downstream chloride concentrations with time

Table 13.6 Steady-state migration coefficients and initial current

Mix	Steady state migration ($\times 10^{-12}$ m^2/s)	Initial current (mA)
B48	0.67	39.4
FA48	0.76	31.6
SL48	0.46	22.9
SF48	0.29	11.2

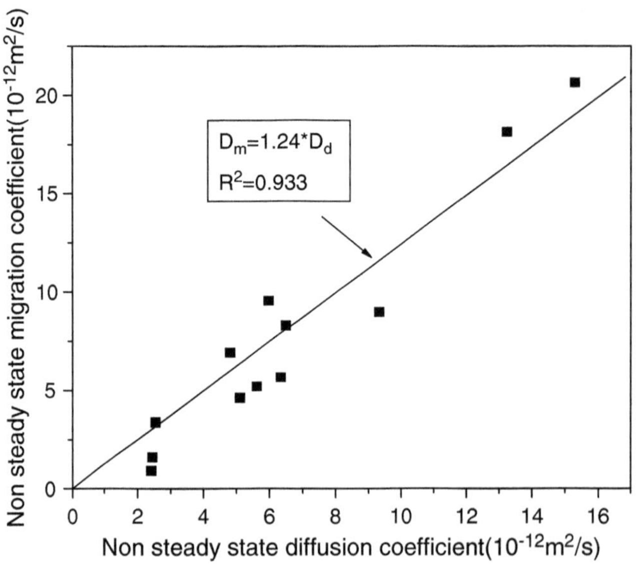

Fig. 13.18 Relationship between migration coefficient obtained from NT Build 492 and diffusion coefficient obtained from NT Build 443

13.3.2.4 Relation Between Results from Different Tests

Non-Steady-State Migration Test and Bulk Diffusion Test

The results from NT Build 492 were generally higher than the results from diffusion test, and they are well correlated linearly, as shown in Fig. 13.18. The migration test seems more sensitive to differences between mixes with low w/b ratio and containing different binders than diffusion test. For example, the migration coefficients for the mixes FA35, SL35 and SF35 were 3.39, 1.62 and 0.91×10^{-12} m^2/s respectively, and the difference was quite remarkable; on the other hand, the diffusion coefficients for the mixes FA35, SL35 and SF35 were 2.55, 2.46 and 2.42×10^{-12} m^2/s respectively, indicating very little difference.

If NT Build 443 is taken as reference method, the conclusion that NT BUILD 492 gives a wrong ranking of chloride resistance can be easily reached. To illustrate this, the chloride profile of the three concretes – FA35, SL35, and SF35 – are compared in Fig. 13.19. It can be seen that the three concretes have very similar chloride profile. This means the three concretes do have a similar chloride resisting ability, as concluded by NT Build 443. However, as per NT Build 492, the chloride resistance increased from the fly ash mix to the slag mix to the silica fume mix. The initial current may help to explain this. The initial current (or resistivity of concrete) with voltage preset at 30 V was 33 mA for FA35, 13 mA for mix SL35 and only

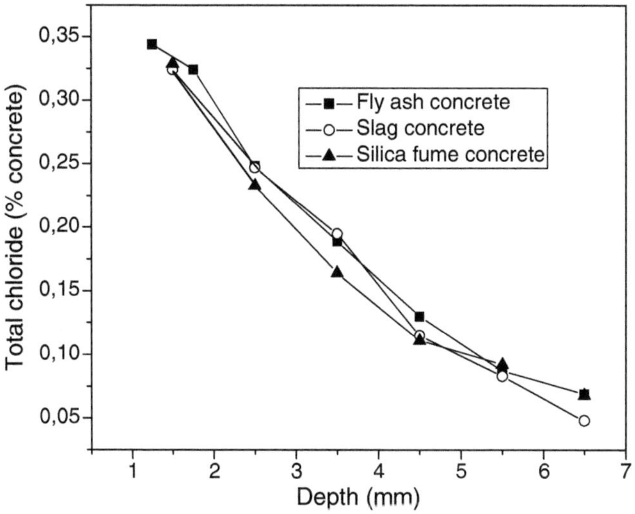

Fig. 13.19 Chloride profiles of concretes with w/b ratio of 0.35

Table 13.7 Steady state migration and non steady state migration/diffusion coefficients ($\times 10^{-12}$ m^2/s)

Mix	NT BUILD 355	NT BUILD 492
B48	0.5	10.66
FA48	0.41	7.95
SL48	0.21	6.08
SF48	0.22	4.30

9 mA for mix SF35. Thus, the initial current seems to be related to the migration coefficient. Initial current depends on chemistry of pore solution and pore structure of concrete, while permeability mainly depends on pore structure. Silica fume reduces ionic concentration of the pore solution greatly. For a given voltage, even when silica fume, slag and fly ash concretes have the same pore structure, the initial current of silica fume will be the lowest. Under the lowest current, the least chloride will be migrated into silica fume concrete by electrical potential. This explains why NT Build 492 gives a different ranking of chloride resistance. However, this problem was not observed at high w/b ratios.

Steady-State Migration Test and Non-Steady-State Migration Test

Compared to the non-steady-state migration coefficients, steady-state migration coefficients are one order of magnitude lower, as shown in Table 13.7.

13.4 Conclusions

A number of methodologies are used to determine the transport of chlorides into concrete, such as salt ponding test, rapid chloride permeability test, chloride conductivity test, steady-state migration test and non-steady-state migration test, and these methodologies cover a number of mechanisms of chloride transport. These have to be used in conjunction with microanalytical measurements of types and amount of phases inside the cement paste (under chloride attack), such as depletion of CH, and formation of chloroaluminates (such as Friedel's salt). Measurement of pH could also be a useful addition to this list of test methods.

References

AASHTO T 259-80: Standard method of test for resistance of concrete to chloride ion penetration. American Association of State Highway and Transportation Officials, Washington, DC (1980)

AASHTO T 277-86: Rapid determination of the chloride permeability of concrete. American Association of States Highway and Transportation Officials, standard specifications – Part II tests, Washington, DC (1990)

Alexander, M.G., Ballim, Y., Mackechnie, J.R.: Concrete durability index testing manual, Research Monograph 4, University of Cape Town (1999)

Andrade, C., Alonso, C., Arteaga, A., Tanner, P.: Methodology based on the electrical resistivity for the calculation of reinforcement service life. In: Nalhotra, V.M. (ed.) Proceedings of the 5th CANMET/ACI International Conference on Durability of Concrete, pp. 899–915. Barcelona (2000)

ASTM C 1202-07: Standard test method for electrical indication of concrete's ability to resist chloride ion penetration. ASTM annual books of standards, ASTM International, Philadelphia (2008)

ASTM C 1556-03: Standard test method for determining the apparent chloride diffusion coefficient of cementitious mixtures by bulk diffusion. ASTM annual books of standards, ASTM International, Philadelphia (2008)

Bentur, A., Diamond, S., Berke, N.S.: Corrosion of Steel in Concrete. E&FN Spon, London (1997)

Castellote, M., Andrade, C.: Round-robin test on methods for determining chloride transport parameters in concrete. Mater. Struct. **39**(10), 955–990 (2006)

Castellote, M., Andrade, C., Alonso, C.: Measurement of the steady and non-steady-state chloride diffusion coefficients in a migration test by means of monitoring the conductivity in the anolyte chamber. Comparison with natural diffusion tests. Cem. Concr. Res. **31**(10), 1411–1420 (2001a)

Castellote, M., Andrade, C., Alonso, C.: Modelling of the processes during steady-state migration tests: quantification if transference numbers. Mater. Struct. **32**(3), 180–186 (2001b)

Castellote, M., Andrade, C., Alonso, C.: Accelerated simultaneous determination of the chloride depassivation threshold and of the non-stationary diffusion coefficient values. Corros. Sci. **44** (11), 2409–2424 (2002)

Feldman, R.F., Luiz, R., Prudencio Jr., U., Chan, G.: Rapid chloride permeability test on blended cement and other concretes: correlations between charge, initial current and conductivity. Constr. Build. Mater. **13**(3), 149–154 (1999)

Freeze, R.A., Cherry, J.A.: Groundwater. Prentice-Hall, New Jersey (1979)

Friedmann, H., Amiri, O., Aït-Mokhtar, A., Dumargue, P.: A direct method for determining chloride diffusion coefficient by using migration test. Cem. Concr. Res. **34**(11), 1967–1973 (2004)

Halamickova, P., Detwiler, R.J., Bentz, D.P., Garboczi, E.J.: Water permeability and chloride ion diffusion in portland cement mortars: relationship to sand content and critical pore diameter. Cem. Concr. Res. **25**(4), 790–802 (1995)

Julio-Betancourt, G.A., Hooton, R.D.: Study of the Joule effect on rapid chloride permeability values and evaluation of related electrical properties of concretes. Cem. Concr. Res. **34**(6), 1007–1015 (2004)

Lu, X.: Rapid determination of the chloride diffusivity in concrete. In: Gjørv, O.E., Sakai, K., Banthia, N. (eds.) Proceedings of the Second International Conference on Concrete under Severe Conditions, Tromsø, pp. 1963–1969, (1998)

Mackechnie, J.R.: Predictions of reinforced concrete durability in the marine environment. Ph.D. thesis, University of Cape Town (1996)

McCarter, W.J., Chrisp, T.M., Starrs, G., Blewett, J.: Characterization and monitoring of cement-based systems using intrinsic electrical property measurements. Cem. Concr. Res. **33**(2), 197–206 (2003)

McGrath, P., Hooton, R.D.: Re-evaluation of the AASHTO T259 90-day salt ponding test. Cem. Concr. Res. **29**(8), 1239–1248 (1999)

NORDTEST NT BUILD 355: Chloride diffusion coefficient from migration cell experiments, Finland (1995)

NORDTEST NT BUILD 443: Accelerated chloride penetration, Finland (1997)

NORDTEST NT BUILD 492: Chloride migration coefficient from non-steady-state migration experiments, Finland (1995)

Page, C.L., Short, N.R., El-Tarras, A.: Diffusion of chloride ions in hardened cement pastes. Cem. Concr. Res. **11**(3), 395–406 (1981)

Poulsen, E., Mejlbro, L.: Diffusion of Chloride in Concrete. Taylor & Francis, London (2006)

Samson, E., Marchand, J., Snyder, K.A.: Calculation of ionic diffusion coefficients on the basis of migration test results. Mater. Struct. **36**(257), 156–165 (2003)

Shi, C., Stegemann, J.A., Caldwell, R.: Effect of supplementary cementing materials on the rapid chloride permeability test (AASHTO T 277 and ASTM C1202) results. ACI Mater. J. **95**(4), 389–394 (1998)

Shi, M., Chen, Z., Sun, J.: Determination of chloride diffusivity in concrete by AC impedance spectroscopy. Cem. Concr. Res. **29**(7), 1111–1115 (1999)

Shi, C., Yuan, Q., Deng, D., Zheng, K.: Test methods for the transport of chloride ion in concrete. J. Chin. Ceram. Soc. **35**(4), 522–530 (2007) (in Chinese)

Stanish, K.D., Hooton, R.D., Thomas, M.D.A.: Testing the chloride penetration resistance of concrete: a literature review. FHWA Contract DTFH61-97-R-00022, University of Toronto, Toronto, 31 p. (2000)

Streicher, P.E., Alexander, M.G.: A chloride conduction test for concrete. Cem. Concr. Res. **25**(6), 1284–1294 (1995)

Tang, L.: Chloride transport in concrete – measurement and prediction. Ph.D. thesis, Department of Building Materials, Chalmers University of Technology, Göteborg (1996)

Tang, L., Nilsson, L.: Rapid determination of the chloride diffusivity in concrete by applying an electrical field. ACI Mater. J. **89**(1), 49–53 (1992)

Truc, O., Ollivier, J.P., Carcassès, M.: A new way for determining the chloride diffusion coefficient in concrete from steady state migration test. Cem. Concr. Res. **30**(2), 217–226 (2000)

UNE 83987: Durabilidad del hormigón, Métodos de ensayo, Determinación de los coeficientes de difusión de los iones cloruro en el hormigón endurecido, Método multirrégimen. Concrete durability, test methods, measurement of chloride diffusion coefficient in hardened concrete, multiregime method (2009)

Verbeck, G.J.: Mechanisms of corrosion of steel in concrete. In: Corrosion of Metals in Concrete, vol. SP-49, pp. 21–38. ACI Publication, Detroit (1975)

Yuan, Q.: Fundamental studies on test methods for the transport of chloride ions in cementitious materials, Ph.D. thesis, Department of Structural Engineering, Ghent University; School of civil engineering and architecture, Central South University (2009)

Chapter 14
Test Methods for Magnesium Attack

Manu Santhanam

Contents

Keywords Magnesium sulfate • Magnesium chloride • Seawater • Brucite • Ettringite • Friedel's salt • pH • Expansion

14.1 Introduction

In chemical attack research, the role of the cation is often not addressed adequately, particularly when the cation is magnesium, Mg^{2+}. Conventional tests that determine the influence of aggressive chemicals such as sulfates or chlorides do not address the deterioration brought about by the presence of Mg.

Analysis of concrete damage from different Mg bearing solutions can help improve the understanding of Mg attack. A comparative study of attack of Portland cement mortars by magnesium sulfate and magnesium chloride solutions was conducted by Moukwa (1990). Pore structure analysis by Mercury Intrusion Porosimetry (MIP) showed that the attack by magnesium sulfate actually led to a densification of the pore structure at early ages, and a corresponding increase in strength. On the other hand, attack by magnesium chloride caused an increase in the porosity of the mortar, as well as in the threshold diameter (in MIP, the threshold

M. Santhanam (✉)
Department of Civil Engineering, IIT Madras, Chennai, India
e-mail: manus@iitm.ac.in

M. Alexander et al. (eds.), *Performance of Cement-Based Materials in Aggressive Aqueous Environments*, RILEM State-of-the-Art Reports 10,
DOI 10.1007/978-94-007-5413-3_14,

diameter refers to a critical pore size, beyond which there is a sudden increase in the intrusion – similar to a 'bottleneck').

The primary difference in the pattern of attack by magnesium sulfate and magnesium chloride is in the rates of deterioration (Santhanam 2001). The attack by $MgSO_4$ starts slowly, owing to the limited penetration of the sulfate ions into the mortar because of the formation of a surface layer of brucite. The reaction at this stage is diffusion controlled. The sulfate ions diffusing through the brucite layer react with the hydrated cement products to produce gypsum and ettringite. The reduction in pH due to the consumption of calcium hydroxide (CH) can ultimately lead to the decalcification of the C-S-H (Bonen and Cohen 1992). The formation of M-S-H occurs when the C-S-H is directly attacked by the magnesium sulfate solution. The brucite layer tends to break down leading to an increased penetration at later ages. This is reflected in the increased rate of attack as seen from the sudden increase in mass, and a sudden drop in the compressive strength seen at later ages with $MgSO_4$ attack. The presence of M-S-H is primarily in those areas within the specimen where the brucite layer is cracked. This could imply that the damaging effects of the attack can be delayed considerably if a thick brucite layer is formed. But on the other hand, the finding that M-S-H occurs in areas where the brucite layer is cracked, could also indicate that brucite is used to make M-S-H.

Literature presents no real evidence of the time for removal/abrasion of brucite layer, or a change in the rate of attack after removal of the layer. Generally, the softening of the cement paste under this layer, because of the formation of gypsum, and later on M-S-H, would lead to a gradual weakening of the layer, causing it to break or separate, allowing further ingress of the solution. Typically, in seawater, thickness of the brucite layer is around 30 μm (Skalny et al. 1999).

$MgCl_2$ attack occurs very rapidly at first, but slows down ultimately. The high quantity of brucite formed in $MgCl_2$ attack, along with the rapid consumption of CH indicates that the primary effect of the Mg ion is to drive the reaction towards brucite formation. The rate of attack, from mass change and compressive strength measurements, slows at later ages, possibly because of a decreased formation of brucite. $MgCl_2$ attack depletes the CH very fast to form brucite, and triggers the decalcification of C-S-H, rendering the paste porous. Due to the absence of sulfate ions, the only secondary product that forms as a result of decalcification is M-S-H. Because no other secondary products are deposited, the structure becomes porous, causing a rapid drop in strength at early ages. The chlorides enter into reaction with C_3A to produce chloroaluminate compounds such as Friedel's salt. Chlorides also become incorporated within the C-S-H. The $CaCl_2$ that is formed as a result of the reaction between $MgCl_2$ and CH is highly soluble and does not precipitate.

It may be concluded from literature that the presence of the Mg ion is responsible for the following:

1. Formation of an insoluble surface layer of brucite on the surface of the mortar specimens that temporarily prevents further ingress of the attacking solution. However, the formation of brucite leads to a rapid consumption of calcium hydroxide. When no secondary products (such as gypsum) are precipitated, the

consumption of calcium hydroxide can render the structure porous leading to a rapid drop in strength.

2. Direct or indirect attack of C-S-H and its subsequent conversion to M-S-H (reported to be non-cementitious) causes the ultimate degradation. In the case of a direct attack, such as in the case of magnesium sulfate solution, the C-S-H is transformed progressively into silica hydrate and M-S-H through reactions with the attacking solution. On the other hand, indirect attack pertains to the situation where decalcification of C-S-H occurs as a result of the prevalent low pH levels (created by the consumption of calcium hydroxide). Subsequent conversion of decalcified C-S-H to M-S-H causes the ultimate failure.

14.2 Test Parameters Used in Magnesium Attack Studies

From the previous section, the role of Mg in chemical attack is primarily assessed through $MgSO_4$ and sometimes, $MgCl_2$ solutions. Several parameters are varied in these studies, and there is no test method that provides a clear picture of Mg attack. In fact, the ASTM C1012 (2010) test method allows the user to choose the type of solution, and bases the analyses on measurement of length change alone. This does not truly bring out the damage caused by the Mg ion attack.

Table 14.1 (at the end of this chapter) provides a summary of some important papers that deal with sulfate (and combined sulfate and chloride) attack, which explain the effect of Mg. The parameters chosen for these investigations can be summarized as follows:

1. Choice of the type of mix – paste/mortar/concrete
2. Choice of the binder – OPC/OPC with extenders
3. Type and size of specimens – cylinders, cubes, prisms etc.
4. Type and concentration of attacking solution
5. Test methods:
   (i) Physical and mechanical properties: length change, mass change, compressive strength, elastic modulus, visual inspection, ultrasonic pulse velocity
   (ii) Mineralogy – X-ray diffraction
   (iii) Thermal analysis – differential scanning calorimetry (DSC), differential thermal analysis (DTA) and thermogravimetry (TGA)
   (iv) Microstructural alterations – scanning electron microscopy (SEM).

In addition to these primary test methods, mercury intrusion porosimetry is used to detect changes in pore structure, and X-ray microtomography to understand the nature and extent of cracking has also been attempted.

In terms of identifying the potential of concretes to withstand Mg ion attack, the test methods need to address the dual issues of alteration in microstructure and the manifestation of damage in the physical sense, i.e. loss in strength, or expansion. It would be suitable to devise methodologies where concrete (or mortar) specimens

Table 14.1 Summary of some papers dealing with magnesium attack

Author	Mix details	Specimens	Attacking solution	Tests performed	Remarks
Bonen (1993)	Mortar, 0.4:1:3 SF 10% 5% C_3A cement	5 × 15 × 80 prisms	2.1% and 4.2% $MgSO_4$ 7 days initial curing	Backscatter SEM – (i) Observation of deposits, (ii) variation of C, S, A, SO_3 across the specimen by EDAX	(i) Mortars more damaged than paste, (ii) Greater decalcification with SF
Nehdi and Hayek (2005)	Mortar, 0.30, 0.45, and 0.6 w/cm Binder to sand ratio 1:2.75 8% SF, 25% PFA and 25; % BFS	50 × 100 mm cylinders for strength 25 × 25 × 285 mm bars for length 100 × 200 mm cylinders for salt hydration test (diametrically cut into two half cylinders and immersed partially)	10% $MgSO_4$ after attainment of 20 MPa Constant and cycling RH used for salt hydration test	Apart from strength and expansion, XRD and SEM of efflorescence in the salt hydration test was performed	No efflorescence (salt crystallization) in $MgSO_4$
Tumidajski and Chan (1996)	Concrete, 0.40 w/c, replacements by slag and fly ash	25 × 75 × 300 mm flat beams for Young's modulus (ASTM C78-84) Chloride ingress measured on 75 × 75 × 300 mm beams	Two brine solutions-Brine 1 with 0.11% Mg and Brine 2 with 1.46% Mg – both with similar Cl (~16.8%) and SO_4 (~0.16%)	Engineering properties; Cl profile; MIP	Influence of Mg less for good concretes
Skaropoulou et al. (2009)	Mortar, 0.5 w/c, replacement of cement by 15 and 30% limestone, using calcareous and siliceous sand	40 × 40 × 53 mm prisms	1.8% $MgSO_4$ at 5 and 25°C	Visual inspection, strength, UPV, mass, XRD, TGA, and SEM	No visible damage at 25°C Thaumasite damage at low temperature

Hekal et al. (2002)	Paste, 0.3 w/b, with replacements of cement by SF, GGBS, limestone, and combinations of the three	1 in. cubes (28 day curing)	10% $MgSO_4$ solution (a) room temp, (b) 60 C, (c) 60 C with dry-wet cycles	Compressive strength and mass change	More damage at 60°C, and even more for dry-wet cycles as a result of crystallization of $MgSO_4.\,nH_2O$
Bonen and Cohen (1992)	Paste, 0.4 w/b, 10% replacement by SF	5 × 15 × 80 mm prisms	2.1 and 4.2% $MgSO_4$ after 7 days control curing	SEM, line traverse, element mapping by EDX	OPC – brucite gypsum double layer; SF paste– no internal gypsum layers
Gollop and Taylor (1992)	Paste, 0.3 w/b	25 mm cubes	7 days control curing, followed by 0.25 mol/l $MgSO_4$	SEM and EDX	Similar observations as above; deposition of $CaCO_3$ as aragonite favoured by Mg^{2+}
Naik et al. (2006)	Paste, 0.485 w/c	12 mm dia. and 40 mm height cylinders for analytical study; 12.5 mm cubes for compressive strength; 25 × 25 × 285 mm prisms for length (mortar, s/c = 2.75)	33,800 ppm SO_3 (control curing details not provided)	X-ray microtomography and energy dispersive XRD	EDXRD gives chemical profile along the specimen depth, while microtomography shows internal cracking
Ferraris et al. (2005)	Mortar	Comparison of expansion rates between 10 × 10 × 100 mm prisms and 25 × 25 × 285 mm prisms			Smaller size prisms (10 × 10 × 100 mm) increase expansion rate, giving the test results in much less time than the standard 25 × 25 × 285 mm specimens
	Mortar				

(continued)

Table 14.1 (continued)

Author	Mix details	Specimens	Attacking solution	Tests performed	Remarks
Ferraris et al. (1997)		Cylinders with constant length of 152 mm and varying diameters (25, 50, 75 mm) exposed to solution from the sides only			Ultimate expansion values similar but obtained in different times Concluded that the expansion from external sulfate attack is mostly governed by ionic diffusion and could hence be accelerated using smaller specimens

subjected to attack by different Mg bearing solutions are periodically monitored for the physical properties (length, mass, and compressive strength) and microstructure (by scanning electron microscopy – using adequate quantification). The results should then be analysed for relationships between physical properties and microstructure. Some examples are given below:

1. Change in mass vs. quantity of brucite formed (by DSC/DTA)
2. Change in mass, or expansion, vs. thickness of brucite layer (by BSEM)
3. Change in mass, or expansion, vs. amount of calcium hydroxide consumed (by DSC/TGA)
4. Loss in strength vs. depth of concrete from the surface where decalcification has occurred (EDAX of C-S-H to quantify the C/S ratios) or where M-S-H formation has occurred.

14.3 Prediction of Field Behaviour from Laboratory Studies

Literature shows that using laboratory investigations to predict field behaviour could have several problems. Reliable models, in addition to appropriate laboratory test techniques, are necessary in order to predict service life and to select suitable materials for structures exposed to sulfate-rich environments. It is not always possible to create field-like conditions in the laboratory. The primary difficulties are related to the scale of testing, as well as the need for results within a short time period. However, it is often possible to design adequate accelerated tests, which can mimic, to a certain degree, the behaviour of concrete under field conditions, and give some guidance on mix designs. According to ASTM E 632-82 (Standard Practice for Developing Accelerated Tests to Aid Prediction of Service Life of Building Components and Materials) (2008), accelerated aging tests are ones in which the degradation of building materials is intentionally accelerated over that expected in service. One basic assumption in this methodology is that the mechanisms of action in the laboratory and the field are identical. In the case of sulfate attack, however, the mechanism of concrete deterioration depends on numerous factors, including the effects of the service environment, such as concentration of sulfate in the solution, and temperature and humidity variations. Thus, in order to predict concrete performance in situations involving sulfate attack, it is essential to incorporate these effects into any model that is developed for prediction purposes.

The current test method for sulfate attack, ASTM C-1012 (2008), suggests the use of a 50-g/l Na_2SO_4 solution (5 mass%, or 33,800 ppm SO_4^{2-}). It also states that other sulfate solutions, such as $MgSO_4$, and other concentrations, may be used to simulate the desired exposure conditions. It should be noted that a concentration of 5% leads to a rapid attack, which is desired in the laboratory. Thus, most studies tend to use sulfate solutions that are of nearly the same concentration. The exposure levels commonly found in field conditions are an order of magnitude lower in concentration (in the range of 0.5–1%). The effect of varying concentrations

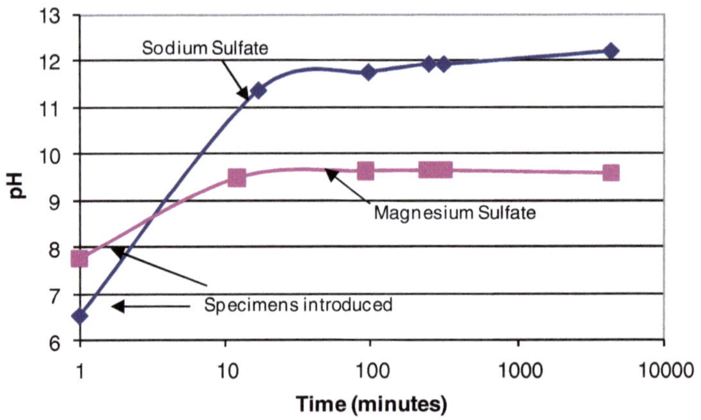

Fig. 14.1 pH of attack solutions as a function of time of immersion (for both solutions, equivalent $SO_3 = 25{,}000$ ppm) (Santhanam 2001)

becomes critical especially in field structures where the same concrete column may be attacked in different regions by different concentrations of sulfates due to the development of drying and wetting fronts. Sulfate concentration gradients may also exist through the depth of slabs on grade.

Varying results are produced depending on whether the attacking solution is stagnant or continuously moving. The pH levels in case of flowing water are expected to be lower, and would tend to drive the mechanism towards gypsum formation. Investigations by Mehta and Haynes (1975) and Brown (1981) suggest that continuous immersion of test specimens, as suggested by the ASTM standards, is not a valid representation of the field situation, since the pH of the attacking solution rapidly changes from neutral (~7) to basic (~12) due to a leaching of the alkalis from concrete into the surrounding sulfate solution. This is shown in Fig. 14.1, where the pH of the attacking solution is plotted as a function of time for both sodium and magnesium sulfate solutions. Furthermore, the sulfate concentration in the solution decreases during continuous immersion (Brown 1981). This could lead to a discrepancy in using lab experiments to predict field behaviour. In general, lab specimens subjected to continuous immersion are able to withstand the attack longer than the corresponding field exposure specimens. This is because the field specimens are subjected to atmospheric effects such as wetting and drying, in addition to the attack by an almost constantly maintained concentration (and pH) of the sulfate solution. In other words, attack in field conditions occurs under an almost constant flow rate of the sulfate carrying groundwater.

Figure 14.1 indicates that the pH of the surrounding solution stabilizes between 9 and 10 for magnesium sulfate attack, possibly because of the formation of the insoluble brucite layer. As stated in Chap. 3 of Part I of this Report, in seawater, a layer of aragonite will be formed in addition to the brucite, which could lead to a temporary protection against further ingress of sulfates.

Another aspect of the attack in a real structure is that there will be considerable physical abrasion (and related loss) of the cover concrete because of the continuous flow of the sulfate bearing groundwater. Gypsum and ettringite formation, and subsequently, direct attack on the C-S-H would render the surface soft and weak, causing it to erode under the action of the flowing water.

14.4 Conclusion

Literature clearly indicates that Mg ion attack is highly detrimental to the long term performance of concrete. The attack can also be roughly divided into two phases: (1) when the underlying concrete is possibly protected by the brucite layer formed on the surface – this prevents further entry of the solution, and (2) final stage when the C-S-H is gradually converted to M-S-H.

In terms of identifying the potential of concretes to withstand Mg ion attack, the test methods need to address the dual issues of alteration in microstructure and the manifestation of damage in the physical sense, i.e. loss in strength, or expansion. It would be suitable to devise methodologies where concrete (or mortar) specimens subjected to attack by different Mg bearing solutions are periodically monitored for the physical properties (length, mass, and compressive strength) and microstructure (by scanning electron microscopy – using adequate quantification). The results should then be analysed for relationships between physical properties and microstructure.

References

ASTM C1012-04: Standard Test Method for Length Change of Hydraulic-Cement Mortars Exposed to a Sulfate Solution. ASTM Annual Book of Standards, Philadelphia (2010)

ASTM Standard Designation C 1012: Standard Test Method for Length Change of Hydraulic-Cement Mortars Exposed to a Sulphate Solution. American Society of Testing and Materials, Philadelphia (2008)

ASTM Standard Designation E 632: Standard Practice for Developing Accelerated Tests to Aid Prediction of the Service Life of Building Components and Materials. American Society of Testing and Materials, Philadelphia (2008)

Bonen, D.: A microstructural study of the effect produced by magnesium sulfate on plain and silica fume-bearing Portland cement mortars. Cem. Concr. Res. **23**(3), 541–553 (1993)

Bonen, D., Cohen, M.D.: Magnesium sulphate attack on Portland cement paste – I. Microstructural analysis. Cem. Concr. Res. **22**(1), 169–180 (1992)

Brown, P.W.: An evaluation of the sulfate resistance of cements in a controlled environment. Cem. Concr. Res. **11**(5–6), 719–727 (1981)

Ferraris, C.F., Clifton, J.R., Stutzman, P.E., Garboczi, E.J.: Mechanisms of degradation of Portland cement-based systems by sulfate attack. In: Scrivener, K.L., Young, J.F. (eds.) Mechanisms of Chemical Degradation of Cement-Based Systems. E & FN Spon, London (1997)

Ferraris, C.F., Stutzman, P.E., Peltz, M., Winpigler, J.: Developing a more rapid test to assess sulfate resistance of hydraulic cements. Res. Natl. Inst. Stand. Technol. **110**(5), 529–540 (2005)

Gollop, R.S., Taylor, H.F.W.: Microstructural and microanalytical studies of sulphate attack: I. Ordinary Portland cement paste. Cem. Concr. Res. **22**(6), 1027–1038 (1992)

Hekal, E.E., Kishar, E., Mostafa, H.: Magnesium sulphate attack on hardened blended cement pastes under different circumstances. Cem. Concr. Res. **32**(9), 1421–1427 (2002)

Mehta, P.K., Haynes, H.H.: Durability of concrete in seawater. J. Struct. Div. **101**(8), 1679–1686 (1975)

Moukwa, M.: Characteristics of the attack of cement paste by $MgSO_4$ and $MgCl_2$ from the pore structure measurements. Cem. Concr. Res. **20**(1), 148–158 (1990)

Naik, N.N., Jupe, A.C., Stock, S.R., Wilkinson, A.P., Lee, P.L., Kurtis, K.E.: Multi-mode X-ray study of sodium and magnesium sulfate attack on Portland cement paste. Advances in C-Ray Analysis Vol. 49, JCPDS-International Center for Diffraction Data, pp 63–72 (2006)

Nehdi, M., Hayek, M.: Behavior of blended cement mortars exposed to sulfate solutions cycling in relative humidity. Cem. Concr. Res. **35**(4), 731–742 (2005)

Santhanam, M.: Studies on sulphate attack: mechanisms, test methods, and modeling. Ph.D. dissertation, Purdue University, W. Lafayette (2001)

Skalny, J., Marchand, J., Odler, I.: Sulfate Attack on Concrete. E&FN Spon, London (1999)

Skaropoulou, A., Tsivilis, S., Kakali, G., Sharp, J.H., Swamy, R.N.: Long term behaviour of Portland limestone cement mortars exposed to magnesium sulfate attack. Cem. Concr. Compos. **31**(9), 628–636 (2009)

Tumidajski, P.J., Chan, G.W.: Durability of high performance concrete in magnesium brine. Cem. Concr. Res. **26**(4), 557–565 (1996)

Chapter 15
Methods for Testing Cementitious Materials Exposed to Organic Acids

Alexandra Bertron

Contents

Keywords Agro-industrial effluents • Silo • Weak acids • pH • Concentration • Solution renewal • Kinetics • Degradation mechanism testing

15.1 Introduction

Concrete in agricultural and agro-industrial settings suffers severe degradation, notably linked with attack by waste waters (De Belie et al. 1997a, 2000a, b; O'Donnell et al. 1995a, b). Waste waters such as liquid manure, silage juices,

A. Bertron (✉)
LMDC (Laboratoire Matériaux et Durabilité des Constructions), Université de Toulouse; UPS, INSA, 135, avenue de Rangueil, F-31 077 Toulouse Cedex 04, France
e-mail: alexandra.bertron@insa-toulouse.fr

M. Alexander et al. (eds.), *Performance of Cement-Based Materials in Aggressive Aqueous Environments*, RILEM State-of-the-Art Reports 10,
DOI 10.1007/978-94-007-5413-3_15,

whey, molasses and distillery residues contain organic acids that attack the concrete. Concrete is the most widely used material for the construction of structures intended for the production, collection, storage and treatment of such effluents as it complies with several requirements: it is economical, watertight, ensures good thermal inertia and respects the health standards imposed in agro-food industries.

There is a need to understand the mechanisms of degradation by the various organic acids and to design materials that perform well in these environments. However, the modes of exposure to the real media and the types of acids vary greatly. Moreover, no specific standard exists for assessing the performance of concrete exposed to organic acids. Consequently a wide range of tests are described in the literature, from original designs implemented to mimic the real situations to more theoretical ones inspired, for example, by standardized tests such as ASTM D 3987 (Standard Test Method for Shake Extraction of Solid Waste with Water). The durability parameters measured (degraded depths, mass variations, surface roughness, mechanical strengths, etc.) are also varied.

This chapter first lists the main types of concrete structures in agricultural and agro-food industries and the consequent modes of exposure to organic acids. It then details the specificities of organic acids that influence test methods. The various types of tests involving organic acids in the literature are also presented and the influence of important experimental parameters is discussed. Finally, the various types of tests performed to assess degradation mechanisms and performance of materials are reviewed (Chap. 6 in Part I of this Report should also be referred to for further details).

15.2 Context and Aims

15.2.1 Composition of Aggressive Media

Agricultural and agro-food industry waste waters contain a wide range of organic acids. Liquid manure contains acetic acid and other volatile fatty acids (less than five carbon atom chains) such as propionic, butyric or valeric acids. Silage juices contain lactic, acetic and propionic acids. Whey (effluent from the making of dairy products) includes lactic, citric, acetic and propionic acids. In molasses, wastes produced by the sugar industry contain lactic, malic, acetic, oxalic, aconitic and succinic acids. Distillery residues such as vinasses contain tartaric, succinic or acetic acids.

These acids have very varied chemical characteristics in terms of mono- or poly-acidity, dissociation constants and chemical and physical properties of salts, and show very varied alteration mechanisms towards concrete.

Concentrations of acid in the effluent vary considerably and can reach several tens of grams per litre. The pH is between 3.5 and 4.5 for silage juice or acid whey, and between 6 and 8 for liquid manure or soft whey.

The compositions of the effluents and the impact of the various acids on the cementitious matrix are detailed in Chap. 6 of Part I.

15.2.2 *Inventory of Structures Exposed to Organic Acids and Exposure Mode*

There are numerous types of concrete structures in agricultural and agro-food installations. They are in contact with organic-acid-bearing aqueous media under various exposure situations, corresponding to the multiple steps of production, collection, storage and treatment of the effluents. The main types of structures and the resulting exposure modes are listed below:

- Production areas, such as animal houses, milking rooms, and plants, combine: (i) intermittent run-off of waste waters, which can lead to an acceleration of the chemical attack on concrete because of the wetting-drying cycles, and (ii) moderate mechanical actions on surfaces by animal or machine traffic and/or high pressure cleaning resulting in erosion, mainly of the concrete skin (De Belie et al. 2000a, b).
- Waste water collection systems, where dynamic conditions of exposure develop. The intrinsic severity of this exposure mode is linked with the constant renewal of the aggressive medium and also with the carriage of solid particles that can cause erosion of concrete surfaces. Another specificity is linked to the presence of microorganisms, and notably sulfate reducing and sulfur oxidising bacteria, which are involved in the H_2S cycle and lead to sulfuric acid attack of cementitious materials. This is developed in Chap. 12 of this Part of the Report (Part III).
- Storage silos provide static conditions with no or only slight movement of the liquid medium, which is, consequently, heterogeneous. This aggressive medium is periodically, totally or partially, renewed. In any case, the liquid volume/concrete area is so large that the aggressive conditions toward concrete are maintained even if the composition of the medium changes with time, notably because of bacterial activity. Hydrostatic conditions enable local chemical equilibria to become established in the vicinity of the concrete, which can be favourable to the material. On the other hand, biofouling of the concrete surface is made easier. The formation of biofilms on solid substrates leads to higher concentrations of organic acids (and other metabolites) in direct contact with the concrete surface as the microorganisms' metabolites are retained by the biofilm substances. Moreover, depending on the depth in the silo, the composition of the effluent, notably in acids, can vary, because of sedimentation but also because of changes in the metabolisms of microorganisms depending on the amount of oxygen or the type of nutrients available locally.
- The zones of the structures close to the effluent's surface can be subjected to wetting and drying cycles with amplitudes and frequencies that vary depending on several factors (precipitation or windy/dry period, filling with fresh

materials/removal from silos, etc.). This situation can favour the precipitation of acid salts that may, or may not, worsen the degradation of the cementitious matrix.

- Treatment plants provide a variety of situations. The common scheme is a containment basin in which successive treatment operations are performed (decantation, filtration, aeration and biological degradation, etc.). The treatment processes are mainly based on bacterial degradation in free culture in batch operation (activated sludge). The physical conditions are close to what can occur in storage silos but with more intense bacterial activity and certainly less heterogeneity of the liquid medium because of aeration, which ensures a certain level of mixing.

15.2.3 Various Objectives of the Studies

Experimental studies exposing cementitious specimens to organic acids have a variety of objectives:

- To qualify degradation mechanisms by characterizing chemical, physical and mineralogical modifications of cementitious materials,
- To quantify aggressiveness of the acids, generally through comparison to a reference (another acid or aggressive substance),
- To quantify the performance of materials and the degradation kinetics.

Depending on the aim of the test, the procedure can vary. When analysing the degradation mechanisms through the investigation of chemical, mineralogical and physical modifications of the cementitious materials, it is preferable that the degraded zone of the specimens is preserved (Pavlík 1994b; Bertron et al. 2004, 2005a, 2009a). When quantifying degradation kinetics, for example, the degraded part of the specimen may be removed regularly to accelerate the kinetics or to simulate mechanical actions on specimens (De Belie and Monteny 1998; De Belie et al. 2002).

15.2.4 Standards

No specific standard exists for the testing of cementitious specimens in organic acids. Standardized tests intended for strong acids (nitric acid) may be used but the specificity of organic acids should be considered.

In particular, attention should be paid to the similarity in degradation mechanisms between the selected organic acid(s) and the strong acid. This is particularly important in the case of tests considering calcium concentration in an aggressive solution. The anion(s) of organic acid with slightly soluble or insoluble calcium salts may combine with calcium ion released by the cement paste (Bertron et al. 2009a). So the calcium concentration of the aggressive solution is no longer a reliable indicator of the degradation. Moreover, relationships between concentration and pH are different for strong and weak acids, involving different aggressiveness and kinetics of degradation for the same concentration of acids or the same pH. Section 15.3 below gives details of these specificities.

15.3 Specificity of Organic Acids Influencing Testing Methods

Organic acids are weak acids. This means that they are partially dissociated in water. Their dissociation reactions are characterized by one or more dissociation constants depending on their mono- or poly-acidity. The typical dissociation reactions are given in Sect. 6.3, Chap. 6 of Part I of this Report. The initial pH of a solution made of a given organic acid is linked with the concentration of acid c through relation (15.2).

$$pH = \frac{1}{2}(pKa - \log(c)) \tag{15.1}$$

Compared to strong acids, which totally dissociate in solution, sensibly higher concentrations of organic acids have to be used to reach a given pH. As a reminder, the relationship between pH and concentration of hydronium ions H_3O^+ and so of acid (as it is completely dissociated), is given in (15.2).

$$pH = -\log(c) = -\log([H^+]) \tag{15.2}$$

As an example, making an acid solution with a pH of 4 requires $1.00.10^{-4}$ M of nitric acid (HNO_3) but $5.75.10^{-4}$ M of acetic acid (CH_3COOH – pKa = 4.76). All other things being equal (notably the solubility and physical properties of the organo-metallic salts of the two acids), the weak acid solution can thus be more aggressive to the cementitious matrix than the strong acid, as was shown by Shi and Stegemann (2000) with pH 3 nitric and acetic solutions (the salts of the two acids are soluble in water). It should not be forgotten that, for a weak acid, the concentration of acid is a more significant parameter of the aggressiveness than pH.

Weak acids are characterized by their buffer zone. For pH within [pKa − 1, pKa + 1], the addition of large quantities of base increases the pH of the solution only slightly (Fig. 15.1).

Consequently, when cementitious specimens are exposed to weak acids, large amounts of hydroxyl ions can be released by the specimen with only a slight increase in the pH of the solution. This is useful in practical terms when the experiments are performed by monitoring the pH over time. For example, for an experiment with a pH 4-aggressive solution of 0.3 M acetic acid (addition of 0.05 M NaOH to bring the initial pH up to 4), one can choose to use the pH interval [4.0; 4.5], which matches XA3 exposure class in standard EN: 206-1. In this interval,

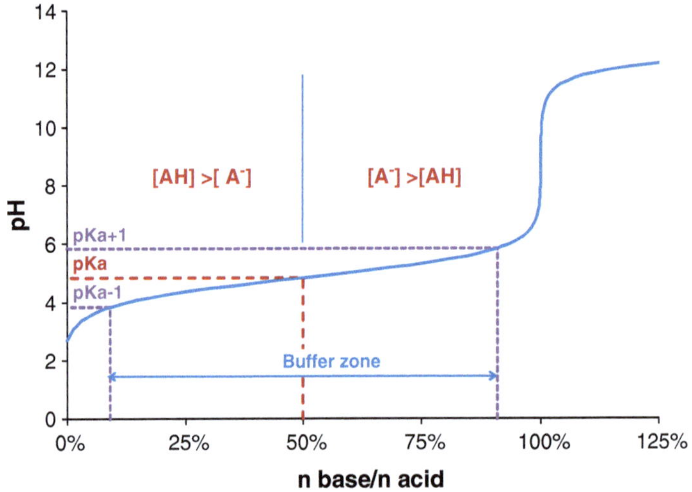

Fig. 15.1 Titration curve of acetic acid (0.3 M) with sodium hydroxide (0.1 M)

18% of the acid will have been consumed when the solution is replaced at pH 4.5 (experiment performed by Bertron et al. 2005a). One can also choose to monitor the concentration of acid by periodic titration of the solution. Pavlik (1994a, b, 1996) chose to replace his solutions each time the concentration of acid decreased by 5%. The first procedure is simpler in terms of carrying out the experiment because, first, measuring a pH is far easier and quicker than titrating a solution and, second, the aggressive solution had to be changed every 3 or 4 days during the first weeks in Bertron's experiment but every 4 h in Pavlik's. On the other hand, degradation kinetics were necessarily higher in Pavlik's methodology, which is an appreciable advantage in accelerated tests.

The chemical specificity of organic acids also influences the way the aggressive properties of the acid solutions are maintained. Different methods can be used to keep the pH/concentration of the solution at the target values: renewal of the solution, addition of concentrated acid, or a combination of the two methods.

The addition of concentrated acid allows the solid/liquid volume ratio – or the exchange surface/liquid volume ratio – to be kept constant by small changes in the volume of the solution (specimens must be taken out of the solution during the operation). This leads to less experimental handling than the renewal of the solution. However, chemical species, notably calcium ions, released by the paste accumulate in the solution and the conditions of the attack can vary. For example, slightly soluble calcium salt can precipitate if suitable concentrations and pH are reached in the solution. Degradation mechanisms and kinetics can then be affected.

Besides, the use of one method or the other has different consequences in terms of quantities of acid used and thus degradation kinetics. Larreur-Cayol (2012) compared the two methods through tests in which cement paste specimens were immersed in acetic acid solution (0.28 M) with a pH of 4 (addition of sodium

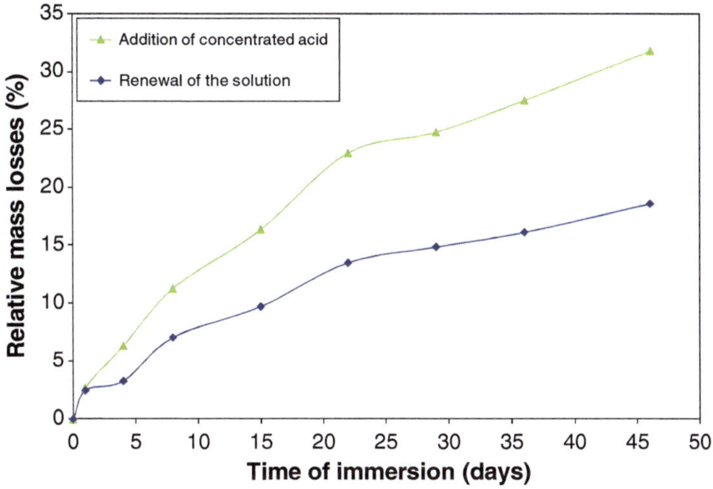

Fig. 15.2 Degradation kinetics of CEM I pastes immersed in acetic acid solutions with pH maintained either by addition of concentrated acid or by renewal of the solution (From Larreur-Cayol et al. 2011)

hydroxide 0.05 M) with a solid/liquid volume ratio of 1/17, and a solid area/liquid volume of 0.102 cm^{-1}. The solution renewals, or additions of acetic acid, were performed at the same time for both solutions (with specimens removed from the solutions). It was observed that both the quantity of acid consumed and the degradation kinetics (mass losses – Fig. 15.2, degraded depths), were noticeably higher for the solution to which concentrated acid was added.

The quantity of acid consumed during the whole experiment was 2.8-fold higher in the case of addition of concentrated acid than in that of solution renewal (6.2 vs. 2.2 mol per litre of aggressive solution). The high quantities of concentrated acid that must be added in the aggressive solution to bring the pH down to the target value may be linked with Oswald law for organic acids (the more concentrated the acid, the less dissociated it is). In contrast, when the solution is simply renewed, only the initial concentration of acid (0.28 M) is added. Consequently, relative mass losses were at least twice as high in the solution to which concentrated acid was added than in the solution that was renewed periodically. Shi and Stegemann (2000) also noted that much larger quantities of acid were consumed to maintain the pH of acetic acid solution (pH 3) than for nitric acid (pH 3). This result by Shi et al. shows that the use of concentrated acid should be avoided for tests aiming to compare attack by organic and strong acids to prevent this side effect; the difference in aggressiveness between the acids could be overestimated.

As some organic acids are volatile (all acids in which the carbon chains have less than 5 carbon atoms: acetic, propionic, butyric, etc.), it is preferable to use closed containers to limit evaporation during tests. Also, some organic acids being mal-odorous (butyric, iso-butyric, valeric, iso-valeric), it is sometimes appropriate, for comfort in the laboratory, to use other acids (with equivalent effects on the matrix), in replacement of the odorous ones, Bertron et al. (2005b) have, for example, shown that butyric, iso-butyric and valeric acids are equivalent to acetic acid in terms of aggressiveness and degradation mechanisms, in accordance with their similar pKa and solubility of calcium salts. More details on the degradation mechanisms by various organic acids are given in Chap. 6 of Part I of this Report. For any other cases, preliminary tests should be performed before using one acid instead of another. Moreover, it should be noted that if all the organic acids are corrosive and/or irritant substances (contact with eyes and skin and ingesting should be avoided), some of them are furthermore toxic (oxalic acid for example).

15.4 Different Types of Tests and Influencing Parameters

15.4.1 General Considerations

15.4.1.1 Overview of the Various Types of Tests Mentioned in the Literature and the Range of Real Conditions

Section 15.2.2 listed the main types of concrete structures encountered in agricultural and agro-food industries and the related modes of exposure to organic acids. These real situations can combine chemical, physical and/or mechanical actions. Ideally, the architecture of tests, and particularly the mode of specimen exposure to organic acids, should correspond to the real situation to be simulated. However, researchers may wish to depart somewhat from the real conditions so as to place themselves in more ideal – and thus controlled – conditions to help with the analysis of degradation mechanisms, for example, or to accelerate kinetics.

The main categories of tests found in the literature, classified according to the type of attack on concrete and the corresponding real situations, are as follows:

- Purely chemical attacks are performed through immersion tests in which the specimens are constantly immersed. The experiment can be carried out:
  - with the organic acid solution kept static or mixed and the solution's characteristics periodically maintained either through renewal of the solution or addition of concentrated acid to keep the pH constant (Pavlik 1994a, b, 1996; Bertron et al. 2005a, 2007, 2009b; Bertron and Duchesne 2007; Larreur-Cayol et al. 2011; Larreur-Cayol 2012). These types of tests can be carried out to simulate conditions in storage silos or some treatment plants, for example.
  - with the solution continuously renewed to simulate a collecting system, for example. This dynamic exposure mode is not much described in the literature as far as organic acids alone are concerned (O'Donnel 1995a, b).
  - in the form of batch extraction tests, where a portion of a granular cementitious sample is mixed with the aggressive solution fluid and agitated until a steady-state condition is achieved (no renewal of the solution). This type of

test does not simulate a real situation. For organic acids, it was notably used by Bertron et al. (2005b) and by Oueslati et al. (2009) in order to identify the release of specific chemical elements from the cementitious matrix or to compare the aggressiveness of various acids.

- Combined chemical and physical attack, occurring on concrete floors or in tidal zones of storage silos, can be simulated using, for example, alternate wetting and drying cycles in the same way as the test designed by Weydert et al. (2001) or using semi-immersion tests (Bertron 2004).
- Combinations of chemical, physical and mechanical attacks, such as those taking place in animal houses or production areas, have been implemented by De Belie et al. (2002) in the "TAP test" (test equipment for accelerated degradation) and by Gorninski et al. (2007a, b).

15.4.1.2 Inventory from the Literature of Various Organic Acids Tested

A large range of organic acids have been tested; literature mainly reports tests using organic acids with soluble salts, which have no influence on the degradation mechanisms and kinetics: acetic and other volatile fatty acids (propionic, butyric, valeric acids), lactic and formic acids (Pavlik 1994a, b; Bertron et al. 2005a, b; Israel et al. 1997; De Belie et al. 1996; De Belie and Monteny 1998). In contrast, acids with slightly soluble or insoluble salts (citric, oxalic, succinic, malic or tartaric acids) have been investigated far less (Medgyesi 1969; Bertron et al. 2009a; Larreur-Cayol et al. 2011; Larreur-Cayol 2012). Degradation mechanisms, notably the influence of various chemical properties influencing their aggressiveness, is thus not well understood (see Chap. 6, Part I).

15.4.1.3 Nature, Shape and Size of Specimens

Nature of Specimens: Cement Paste, Mortar or Concrete

This problem, of the choice of specimen, is common to many types of attacks by aggressive aqueous environments. It may be preferable to analyse degradation mechanisms on cement pastes (Bertron et al. 2005a, 2007, 2009b; Pavlík 1994b; Israel et al. 1997) as the presence of aggregates complicates the task of performing chemical analyses (extraction of aggregates may be necessary before analysis, for example) or the exploitation of mineralogical analyses (cementitious matrix X-ray diffraction peaks are of far lower intensity than those of aggregates in most cases). For experiments intended to qualify the performance of materials, the use of smaller specimens made of mortar allows smaller quantities of aggressive solution to be used (for given solid to liquid ratio) and thus facilitates the experimental work.

However, the issue arises whether studies on pastes are properly representative. First, the presence of reactive aggregates may modify the mechanisms and kinetics of degradation. Secondly, the presence of aggregates in the matrix results in ITZ (interfacial transition zones) around aggregates. In ordinary concretes or mortar, ITZ are more porous and richer in calcium hydroxide and ettringite (Ollivier et al. 1995) than the surrounding paste and should logically be altered more rapidly. However, in high performance materials, and notably those containing pozzolanic additions, ITZ are no longer detectable.

It seems that no specific study has been published that assesses the influence of the ITZ on the degradation mechanisms and kinetics induced by organic acids. However, it is worth mentioning the study by Bourdette et al. (1995), who investigated the influence of ITZ on the alteration of CEM I mortars immersed in aqueous solution, the pH of which was maintained at pH 8.5 by the addition of nitric acid. The mortars, made with quartz aggregates (inert to acid attack), water/cement 0.38 and sand/cement 2, showed chemical and mineralogical zonations identical to those identified by Adenot and Buil (1992) on cement pastes altered in the same conditions. ITZ were not altered more rapidly than the rest of the paste. Moreover, the texture and structure of ITZ in the altered part of the specimens were the same as in the surrounding paste, which was not the case in the initial mortar. Carde and François (1997) modelled the leaching of mortars and pastes exposed to NH_4NO_3 and highlighted an ITZ effect on the alteration of mechanical properties (residual strengths of degraded zones of mortars were lower than those of pastes). This effect was not observed for mortars and pastes including silica fume.

It is reasonable to expect the same kind of results for organic acids with soluble salts (acetic, propionic, butyric, formic acids) as their degradation mechanisms on cementitious materials are similar to that of nitric acid or NH_4NO_3 (See Chap. 4, 5, 6, part I).

Shape of Specimen

All sizes of specimen can be used provided they ensure a significant zone of homogenous attack where degradation measurement can be performed (kinetics, mechanisms). In cases of prismatic specimens, measurements in the edge zones should be avoided.

As the attack by organic acids often results in progressive dissolution of the outer part of the specimens and rarely or never in swelling or expansion (Chap. 6 of Part I), most test methods do not consider expansion measurement (except De Belie et al. with the 'TAP' test fitted with sensors set up to monitor radial changes over time – see Sect. 11.3, Chap. 11, Part III). This constraint, which can influence the chosen shape of specimens (in the case of sulfate attack for example, see Sect. 10.5.2, Chap. 10), seems pointless in the case of organic acids.

For mechanical testing, in order to avoid heterogeneity of the specimen in the axial direction, the loading zones of specimens should be protected during exposure to the aggressive solution. If the loading zones are altered, they should be abraded before testing. For compressive strength tests, cylindrical specimens should be preferred to prisms (4 × 4 × 16 cm, for example).

15.4.1.4 Composition of the Aggressive Solution

The problem of the acceleration of kinetics in test methods – necessity and advantages, drawbacks and risks – was addressed by N. De Belie in the introductory section of this Part (Chap. 8, Part III) of the Report. Regarding the specific case of organic acid attack, several methods can be used:

- Modify the chemical characteristics of the aggressive solution: decrease pH, increase concentration(s) of acid(s) or use a more aggressive acid. In this case, it should be carefully checked that the changes have no or little impact on the degradation mechanisms. For example, in the framework of the analysis of alteration caused by liquid manure on concrete, Bertron et al. (2005a) wanted to validate the use of solutions with a pH of 4 to replace solutions with pH 6, the lowest pH occurring in real manure. The two solutions were made using the same concentrations of acids but different concentrations of sodium hydroxide to adjust the pH to either 4 or 6. It was found that the degradation mechanisms were practically identical, the minor differences being linked with the stability of residual anhydrous grains C_2S, C_4AF and slag. The change of pH from 4 to 6 enabled ninefold higher degradation kinetics to be obtained. Likewise, if the acceleration consists of replacing an acid by a more aggressive one, it should be checked that the degradation mechanisms, notably in terms of intensity of decalcification or precipitation of acid salt, are identical. On the occasion of a large experimental campaign on the deterioration kinetics of cement paste specimens immersed in acetic acid, Pavlik (1994a) calculated, from experimental data, a general formula (15.3) giving the degraded depth *d* according to the time of immersion t and the concentration of acid *c*:

$$\begin{gathered} d = 1.232c^{0.525}t^{0.52} \\ \text{with d (mm), c (mol/l) and t (days)} \end{gathered} \tag{15.3}$$

  This relation is an example of the influence of an increase in acetic acid concentration on the degradation kinetics, but this relation cannot be extended to another acid.
- Use a physical effect, by imposing very frequent drying/wetting cycles to accelerate the penetration of the aggressive species (De Belie et al. 2002; Weydert et al. 2001). However, this can lead to the formation of salt that can increase the deterioration through additional physical strains on the cementitious matrix, notably if the concentrations of acids are higher than in the real medium.
- Use a mechanical effect (abrasion). Acceleration can be obtained by regularly removing altered parts of the degraded specimens, which could otherwise slow down the kinetics of further degradation. Removal is done by automated or manual action, often by brushing (De Belie and Monteny 1998; De Belie et al. 2002; Gorninski et al. 2007a, b).

For tests aiming to qualify degradation mechanisms, the design of the test should allow the altered part of the specimen to be preserved for subsequent chemical, mineralogical, and physical analyses.

15.4.1.5 Solid-to-Liquid Ratio

The solid-to-liquid ratio is a dimensionless, and thus easy-to-use variable. However, the ratio of the specific surface of the solid specimen to the aggressive solution volume may be more relevant, as it provides information about the exchange area between solid and solution and thus reveals leaching kinetics, for example.

15.4.1.6 Mixing of the Aggressive Solution/Frequency of Renewal of the Solution

This parameter also influences the results. In static conditions, a gradient of concentration is naturally created in the vicinity of the specimen: the concentrations of leached cementitious species decrease and the concentration of acid increases when the distance to the specimen increases. This gradient is favourable to a chemical equilibrium between the liquid and solid phases and exchange reactions slow down. Mixing of the solution destabilizes this equilibrium and can increase alteration kinetics. Larreur-Cayol et al. (2012) investigated this influence using immersion tests of cement paste specimens in acetic acid solution (0.3 M) at pH 4 (addition of 0.05 M NaOH). Tests comparing the kinetics between mixed (with a magnetic stirrer) and static aggressive solutions were performed. It appeared that the mixing increased the degradation kinetics slightly. At the end of a 68-day test, mass losses were $26.07 \pm 0.04\%$ with mixing but $21.63 \pm 0.04\%$ in static conditions (20% higher mass losses with mixing).

The aggressive characteristics of the solution can be maintained or not (to achieve a steady-state equilibrium). As mentioned in Sect. 15.3, the method for keeping the aggressive characteristics of the solution constant, either by renewal of the solution or by addition of concentrated acid, has noticeable consequences on the aggressiveness of the attack and consequently, on the kinetics. In all cases, the pH or concentration of acids should be monitored over time in the aggressive solution to obtain an indicator of the aggressiveness of attack.

The frequency of renewal of the aggressive solution also influences the kinetics. As an example, Bertron et al. performed immersion tests of CEM I cement paste specimens with different conditions of solution renewal. In a first test, the specimens were put in a column which was fed with constantly renewed aggressive solution (downstream of the column, the aggressive solution was removed into a bin). In a second test involving the same column geometry, the aggressive solution was made to circulate in a closed circuit but the solution was renewed daily. The flow rate in the two columns was set at 40 ml/h, which imposed quasi-static hydraulic conditions. The two experiments were performed with the same solid/liquid volume

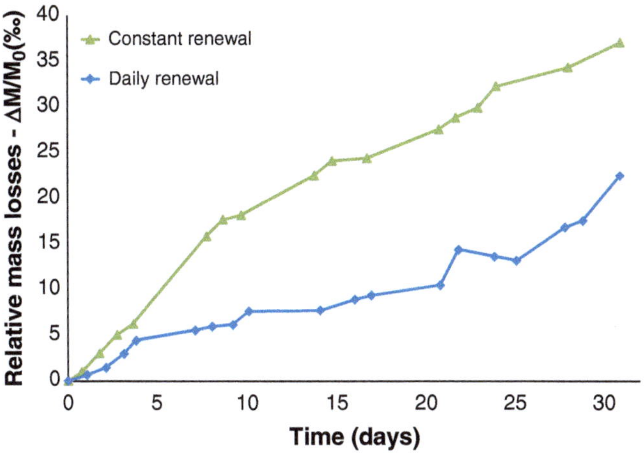

Fig. 15.3 Influence of the renewal frequency of the aggressive solution on the degradation kinetics (CEM I pastes immersed in pH 6 acetic acid solution) – (original data)

ratios of 0.117 and the same solid area/liquid volume of 0.18 cm^{-1}. The aggressive solution with a pH of 6 was made of acetic acid (0.120 M) with added NaOH (0.113 M). The average pH of the aggressive solution before the daily renewal was 10.4. Specimen mass losses were measured daily. The results are shown in Fig. 15.3. The mass losses were much more rapid with constantly renewed solution: the mass loss rate was, on average, twice as high in the constantly renewed solution as in the solution renewed daily.

15.4.2 Description of the Tests

This section describes the various tests encountered in the literature. They are classified according to the type of attack:

- purely chemical attack during immersion tests on monoliths, or dynamic tests involving extraction
- tests combining chemical and physical attack (alternate wetting and drying cycles, to accelerate kinetics, since kinetics of transport by capillary suction is higher than that by diffusion)
- tests combining chemical, physical and mechanical attacks.

For each test, this section gives details on the objective and the architecture of the tests, the type of materials, the aggressive solution, and the way the aggressive conditions are maintained (renewal/partial renewal/addition of concentrated acid).

15.4.2.1 Purely Chemical Attack

In this category, many test methods are inspired by leaching tests performed using soft water attack. The classification of the various types of leaching tests and the definitions are detailed in Chap. 9 of this Part (Part III). The literature on organic acid attack comprises the following types of tests:

- immersion tests involving monolithic specimens (flow-around leaching tests) and that are performed:
  - under dynamic conditions (intermittent renewal of the solution to maintain a driving force for leaching)
  - as extraction tests (no renewal of the solution: at the end of the test, the experiment reaches steady-state conditions)
- leaching tests that use granular specimens and that bring a given amount of solution into contact with a given amount of material, with no renewal of the solution, to obtain steady-state conditions at the end of the test (extraction tests).

Immersion Tests of Monoliths

Objectives

Immersion tests are the most widespread and the most simple exposure procedure reported in the literature. The attack normally occurs in saturated conditions. It is the most polyvalent test in the sense that it allows for all kinds of objectives:

- investigation of degradation mechanisms (Pavlík 1994b; Israel et al. 1997; Bertron et al. 2005a, 2007a, b, 2009b; Larreur-Cayol et al. 2011; Larreur-Cayol 2012),
- analysis of chemical leaching kinetics (Oueslati et al. 2009; Oueslati 2011)
- assessment of degradation kinetics (Bertron et al. 2005a; Pavlík 1994a; Shi 2003; Shi and Stegemann 2000; Medgyesi 1969),
- performance comparisons for various binders or concrete/mortar designs (Mehta 1985; Pavlik and Uncik 1997; Pavlík 1996; Roy et al. 2001; Shi and Stegemann 2000; Shi 2003; Bertron et al. 2006, 2007; Oueslati et al. 2009; Oueslati 2011),
- comparison between degradation kinetics in organic acids or mechanisms with those in strong acids (Pavlík 1994a, 1996; Mehta 1985; Israel et al. 1997).

Tests with Periodical Renewal of the Solution or Addition of Concentrated Acid (Dynamic Flow-Around Tests)

Size and Shape of Specimens: Most authors consider small cylindrical specimens (diameter $\phi = 25$–90 mm, height = 30–90 mm) and either a radial attack (Bajza 1992; Mehta 1985; Bertron et al. 2005a, 2007) or an axial one (Pavlík 1994a, b, 1996; Pavlik and Uncik 1997; Shi and Stegemann 2000; Shi 2003). Pavlik used

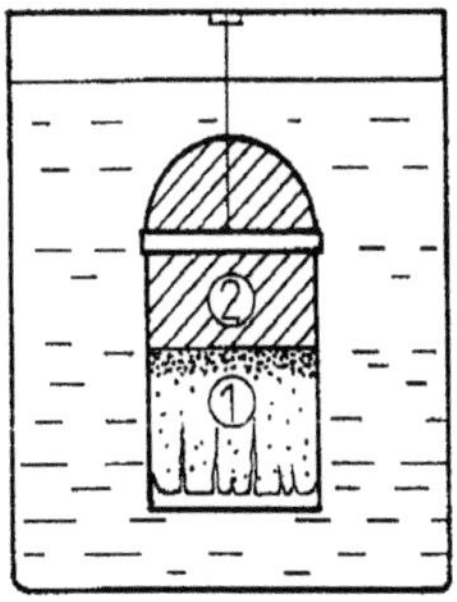

Fig. 15.4 Schematic arrangement of Pavlik's test (Pavlik 1994a, b). Hardened paste specimen cast in glass tubes open on one side and immersed in the acid solution

cementitious specimens ($\phi = 4.5$ cm, h = 7.5 cm) that were cast in glass tubes open on one plane side so that axial attack could take place. The glass tubes containing the cement pastes were suspended in the aggressive solutions, plane side downwards. The advantages of this test arrangement, according to the authors, are notably to allow constant measurement (i) of the corrosion depth through the glass surface (the degraded depth measurement is then no longer a destructive test), (ii) of the shrinkage of the degraded layer, and also a mass loss measurement referred to a constant reaction area (Fig. 15.4).

Composition of Aggressive Solution (Concentration of Acids/pH): Aggressive solutions are made with one organic acid or a mix of several depending on the objective of the test. Bertron et al. (2005a, 2007, 2009b) and Larreur-Cayol et al. (2011), Larreur-Cayol (2012) used aggressive solutions to simulate the acidic fraction of agricultural or agro-food effluents (liquid manure, silage juices, molasses, etc.) using acids found in these types of effluents. A large range of acids was tested: acetic, propionic, butyric, iso-butyric, valeric, lactic, oxalic, citric, tartaric, malic, and succinic acids. Depending on the case, each organic acid was taken separately, for example to compare relative aggressiveness of the various acids or to identify the degradation mechanisms, or several acids were mixed in the proportions occurring in real media. In all cases, the same concentration of acid was taken (0.28 mol/l, matching the highest concentration of acids found in liquid manure) for comparison purposes. Then, to reach the target pH (4 or 6), sodium hydroxide was added to the solutions except for oxalic acid that forms salts with Na^+.

Pavlik (1994a, b, 1996; Pavlik and Uncik 1997) investigated mechanisms and kinetics of degradation of cement pastes and of silica fume- blended mortars immersed in acetic acid (0.025, 0.05, 0.1, 0.2, 0.5 M, pH between 3.73 and 2.54) and nitric acid (0.025, 0.05, 0.1 M, pH between 1.61 and 1.10). Shi et al. (2000, 2003) used acetic acid solution with a pH of 3 or 5 and nitric acid of pH 3.

Bajza (1989, 1992) considered formic (0.5, 3 and 5%) and acetic acid solutions (3 and 5%) in order to compare the performance of various binders exposed to several aggressive aqueous media. Roy et al. (2001) investigated the durability of mortars prepared with various additions (silica fume, metakaolin and fly ash) and exposed to solutions of acetic (5%) and lactic acids (1%) but also to hydrochloric and sulfuric acids, ammonium nitrate and sodium and magnesium sulfates.

Israel et al. (1997) investigated the comparative durability and degradation mechanisms of pore-reduced cement pastes exposed to sulfuric (0.20 M) pH 0.9, hydrochloric (0.27 M) pH 0.5, and acetic (0.05 M) pH 3 acids.

S/L Ratio, Mixing, Methods for Maintaining Aggressive Conditions: The influence of mixing and of the frequency of the aggressive solution renewal on the degradation kinetics in the specific case of organic acids is discussed in Sects. 15.2 and 15.4.1.6.

Bertron et al. (2005a) performed tests with a solid/liquid volume ratio of 1/17 and an exchange area/liquid volume ratio (A/V) of 0.102 cm^{-1}. Solutions were renewed each time the pH reached 4.5, i.e. every 2 days at the beginning of the experiment and every 4 or 5 days at the end.

In the experiment by Pavlik (1994a, b), with $S/L = 1/10$ and $A/V = 0.013\,cm^{-1}$, the aggressive solution was replaced each time the concentration of acid decreased by about 5% which resulted in huge experimental requirements (replacement being required every 2 or 3 h in the first few days!). Shi et al. (2000, 2003) used the same S/L and A/V as Pavlik but the pHs of the solutions were maintained by the addition of concentrated acid when the concentration of acid decreased by 5%. The author mentions the difficulty of reaching pH 3 because of the buffering action of acetic acid: large quantities of acids had to be added in comparison to nitric acid solutions.

In Mehta's experiment (1985) and in Israel's one (1997), the solutions were renewed every week, with no monitoring of the pH over time. Bajza et al. (1992) replaced the aggressive solutions every 28 days during the first 6 months and, during the remaining experiment, every 84 days.

Whatever the frequency of solution renewal, it is recommended to regularly monitor the pH or the concentration of acid over time to record the intensity of the aggression as the experiment progresses.

Depending on the experiment, the solution is constantly mixed (Mehta 1985) or not (Bertron et al. 2005a, 2007, 2009b) depending on the decision of the authors either to have homogeneous conditions or to reproduce a static aggressive medium, as occurs in silos for example.

Tests with No Renewal of the Solution

Tests by Oueslati and Duchesne (2009): Oueslati and Duchesne (2009), Oueslati (2011) performed immersion tests of blended cement paste specimens in order to assess the relative performance of various binders (OPC partially substituted by silica fume, fly ash, slag or metakaolin). Cylindrical paste specimens (diameter = 100 mm; height = 150 mm) made with a W/B mass ratio of 0.27 were immersed for 3 months in 0.25 M acetic acid solution at a pH of 4 with a solid-to-liquid ratio of 0.9. In this test, pH was carefully monitored over time as was the concentration of the main chemical species released by the cement paste (Ca, Si, Al, Mg, Al, Fe) until the steady-state condition was reached in the various aggressive solutions. For each chemical element, the authors exploited the results by defining a leaching rate (LR) defined as the element mass in solution divided by the initial element mass in the specimen. The aim was notably to analyse the comparative leaching behaviour

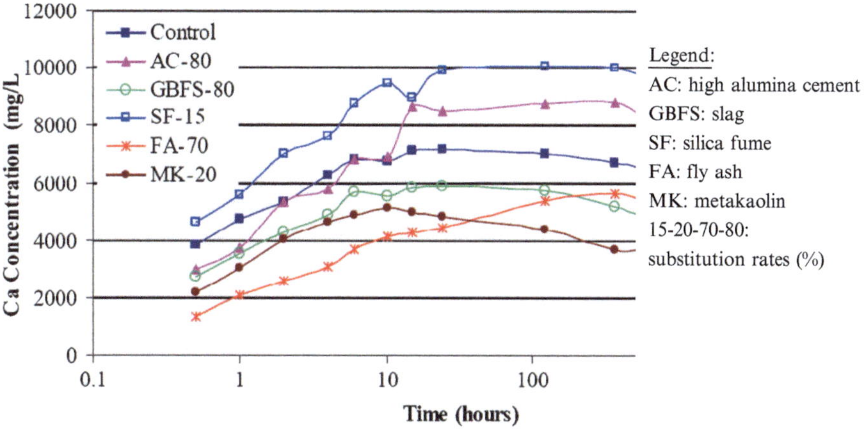

Fig. 15.5 Monitoring over time of the concentration of calcium in the aggressive solution for various types of binders in the experiment by Oueslati and Duchesne (2009). *AC* high alumina cement, GBFS slag, *SF* silica fume, *FA* fly ash, MK metakaolin, 15-20-70-80 substitution rates (%)

of the various cementitious chemical species (as exemplified for calcium in Fig. 15.5).

Tests Inspired by Köch-Steinegger Method (1970): Some tests using organic acids are performed according to the Köch-Steinegger method, initially designed for sulfate attack (Medgyesi 1969; Macias et al. 1999). Macias et al. (1999), for example, immersed small cement paste specimens (1 × 1 × 6 cm) (i) in an acid buffer solution made of 0.87 M CH_3COOH/CH_3COONa with pH 4.5, and (ii) in water for comparison. In the Köch-Steinegger test, the solution is not renewed and the flexural strength of specimens is measured after 56 days of immersion. Macias clearly criticizes this method by demonstrating that flexural strength measurements taken alone are not sufficient to evaluate the degree of cement paste degradation in an acid medium. The authors explain that two phenomena with opposing consequences occur and make the criterion unreliable: a densification of the cement paste in the specimen core by continuing hydration and a degradation of the outer surface. They finally recommend taking the remaining calcium content and the growth of the corroded layer into account to evaluate the degree of corrosion.

Flow-Test – Constantly Renewed Aggressive Solution

An original experimental facility was implemented by O'Donnell et al. (1995a, b) in the context of an on-farm trial. The facility aimed to simulate the exposure conditions of horizontal concrete silos exposed to the flow of silage juices. This test method was also used by Richardson et al. (1999). The authors wanted to reproduce the constantly renewed conditions of flowing acid occurring at the surface of concrete in well-drained silos.

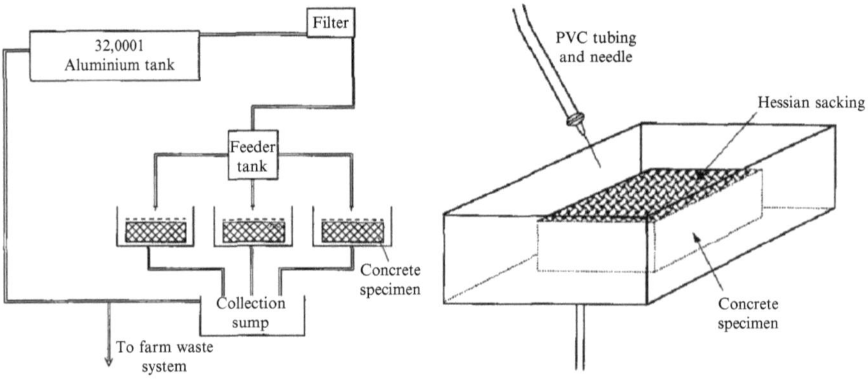

Fig. 15.6 Experimental facility implemented by O'Donnell et al. (1995a, b)

Aggressive solutions were real undiluted silage juices produced from an unwilted grass silage. These juices were regularly analysed: the pHs were between 3.8 and 4.4, with concentrations of a lactic, acetic and volatile fatty acids between 12 and 24 g/l, 3.5 and 5.5 g/l, and 3.5 and 6.3 g/l respectively.

The effluent flow in the experiment was fixed to simulate real conditions in silo facilities (143 l/m^2d). Silage juices were stored in an aluminium tank from which they were extracted and passed through a 40-μm-fabric filter (Fig. 15.6). Then, the filtered effluent flowed by gravity through piping to receptacles containing concrete specimens. The ends of the pipes were equipped with needles fitted with gauges allowing flow rate to be controlled to the target value. The concrete specimens were covered with hessian sacking in order to favour homogenous distribution of the flowing effluent over the surface for more homogenous surface attack. The effluents were then collected and either released into the farm waste system or recycled in the tank for further attack. The same effluent could be used for several 28-day periods.

Concrete specimens (28 × 28 × 10 cm) were exposed to nine 28-day-periods of effluent flow simulating 1 year of corrosion. Between each period, a 2-week stop was performed to proceed with various measurements.

At the end of each 28 day-flow period, specimens were washed with water and brushed with a nylon brush to remove loose matter on the surfaces prior to characterization tests. However, in this case brushing was not performed to accelerate the kinetics.

Extraction Test – Tests with Granular Material

Batch leaching tests were also performed by Bertron et al. (2005b). The tests were quite similar to leaching tests described in ASTM D 3987 (Standard Test Method for Shake Extraction of Solid Waste with Water). Bertron et al. performed dissolution tests on cement paste specimens in organic acids. The purposes of these tests, as stated by the authors, were:

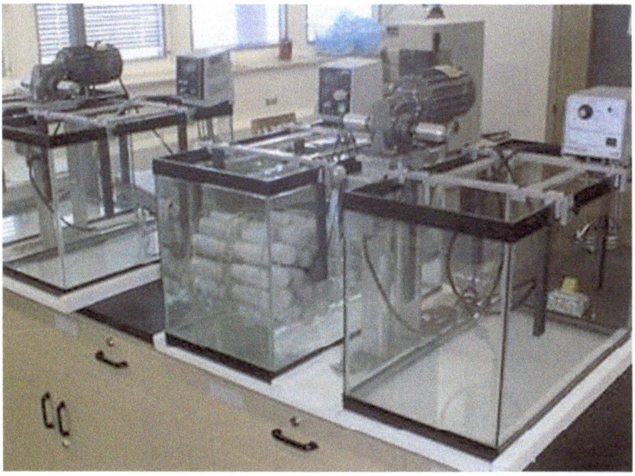

Fig. 15.7 Experimental device used by Bertron et al. (2005b) to perform extraction tests

- to analyse the mechanisms of cementitious matrix degradation by organic acids by considering the dissolution behaviour of each chemical element in cement. The aim was to identify the parameters of cement composition influencing the durability of cementitious materials in agricultural and agro-food environments,
- to compare the relative aggressiveness of each acid tested (acetic, propionic, butyric, iso-butyric, lactic) toward cement paste, and improve the knowledge of the chemical parameters of the acids (pKa, solubility of salts, complexation phenomenon) influencing their aggressiveness. The authors notably aimed to determine whether acetic acid was representative of the other acids tested and whether or not the use of one acid to make synthetic aggressive solutions in laboratory conditions was a valid approach,
- to investigate the behaviour of several binders (CEM I, CEM III, CEM I with 10% silica fume, CEM I + 10% fly ash) exposed to organic acid solution (acetic acid).

A series of batch leaching experiments were performed by immersing samples of cementitious material (roughly crushed cement paste specimens with granular size >2.36 mm) in aggressive solutions made of organic acids in 60-ml high-density polyethylene containers. The containers were mounted horizontally on a Plexiglas carousel ensuring constant stirring (Fig. 15.7). The carousel was suspended in a bath maintained at 20 °C and was rotated at 15 rpm for 13 weeks. Stirring caused friction between grains, thus breaking down the altered zone, a difference from static immersion tests (as presented in Sect. 15.4.2.1.1). The experiments were performed with a solid-to-liquid ratio of 1/20.

The aggressive solutions were made with several acids found in agricultural and agro-industrial effluents, each acid being taken separately. The pH was not maintained during the experiment (initial pH of the solutions = 4). The pHs of

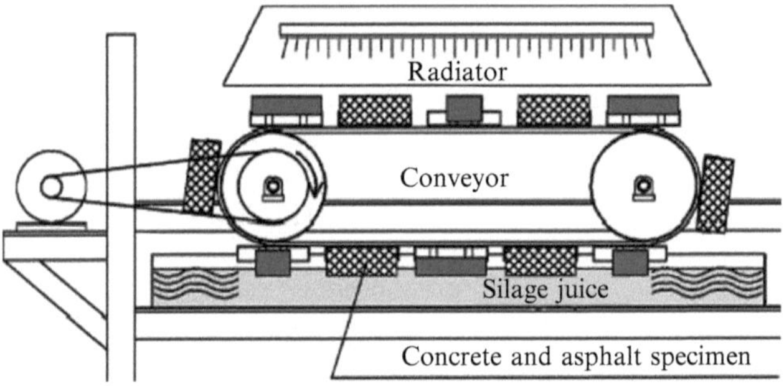

Fig. 15.8 Automated test device by Weydert et al. (2001)

the solutions and the concentrations of Ca, Si, Al, Fe, Mg in the solutions were monitored over time (first sampling: after ½ hour, overall duration of the experiment and last sampling: 13 weeks). The authors exploited the results by defining a dissolution rate (DR) as the ratio of the mass of an element in solution to the initial element mass in the specimen. This ratio is equivalent to the leaching rate defined by Oueslati et al. (2009). This allowed the stability of each chemical element to be analysed. At each sampling period, the solutions were filtered and, if a precipitate had formed in sufficient quantity, it was analysed by XRD and observed by SEM.

15.4.2.2 Combination of Chemical and Physical Attack

The authors mentioned below combine chemical and physical attack either to accelerate the kinetics (alternate wetting/drying) or to reproduce real conditions (alternate wetting and drying on structures, tidal zones, etc.).

Automated Test Device by Weydert et al. (2001)

Weydert et al. (2001) developed a test to investigate the comparative durability of materials (concrete, asphaltic concrete and mastic asphalt) exposed to agricultural effluents (maize silage in particular). The effluents were analysed for concentrations of organic acids (lactic, acetic, propionic, and butyric acids). To accelerate degradation kinetics, the authors implemented automated wetting and drying cycles by using a conveyor (Fig. 15.8). Specimens were alternately immersed in silage juice at pH 3.8 and then dried by heat lamps (rated at 1 kW/m^2). The conveyor speed was one rotation per hour, the immersion period lasting about 20 min per rotation. The behaviour of 20 $\times$ 20 $\times$ 8-cm prisms was compared.

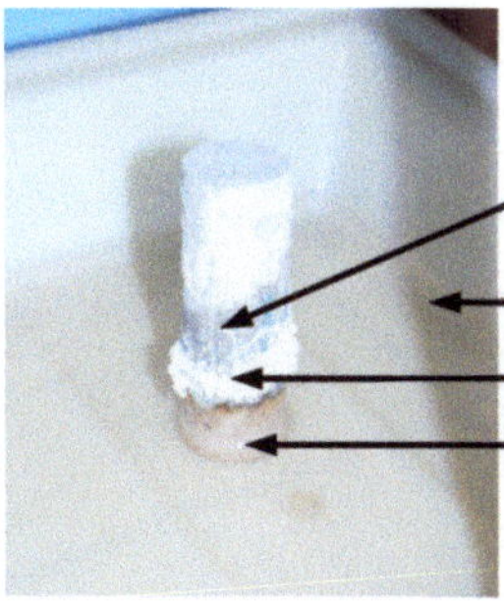

Fig. 15.9 Semi-immersion test (acetic acid) by Bertron (2004)

Semi-immersion Test

Semi-immersion tests were used by Bertron (2004) to analyse the impact of the precipitation of calcium salts that could occur in silage juice or liquid manure storage silos in the tidal zone of structures. Most acids in these effluents have salts that are soluble in water, which means that the salts do not precipitate in fully-saturated conditions. However, in the tidal zones, the formation of these salts is made possible by the increase of the concentration of the various species in the liquid phase, linked with the evaporation of the liquid phase. The authors wanted to assess the possible supplementary degradation induced by precipitation on the matrix. The principle of the test is very simple: cementitious specimens are partially immersed in acid solutions made of each acid taken separately. Some acid solution naturally migrates through the porosity of the specimen by capillarity from the immersed part to the emerged parts (Fig. 15.9).

The precipitation of salts (mainly calcium salts in the case of ordinary cements) occurs after only a few days. The specimens are then observed with a scanning electron microscope to identify degradation symptoms linked with the precipitation of the salts. Bertron performed the tests with acetic, propionic, butyric and lactic acids, all of which have salts soluble in water.

15.4.2.3 Combination of Chemical, Physical and Mechanical Attack

TAP Test by De Belie et al. (2002)

De Belie et al. (2002) implemented an automated test named 'Testing Apparatus for Accelerated Degradation Tests', also known as 'TAP', combining chemical attack by organic acids, alternate wetting and drying cycles and mechanical abrasion. The aim was to simulate the attacks on structural concrete in agricultural and agro-industrial environments such as concrete floors of animal houses or silage silos. Alternate wetting and drying cycles simulate real conditions in the tidal zones of storage silos or on floors which are alternately wet by feed effluents mixed with cleaning products and dried. Mechanical abrasion simulates animal or machine

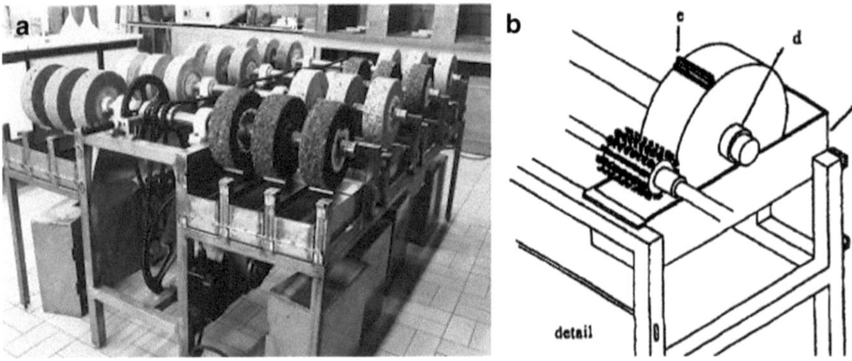

Fig. 15.10 'Testing Apparatus for Accelerated Degradation Tests' or 'TAP' (De Belie et al. 2002). (**a**) Overview of the device, (**b**) view of a specimen cylinder and the brushing system

traffic, mechanical or high-pressure cleaning etc. The aim of the test was to evaluate the performance of various concrete or mortar designs in organic acid environments.

The automated design of the test is intended to avoid laborious handling and user-induced errors. The apparatus consists of stainless steel horizontal rollers on which concrete or mortar cylinders ($\phi = 270$ mm, h = 70 mm) are mounted (Fig. 15.10). The rollers are driven by an electric motor that can ensure different speeds of rotation. The cylinders are partially immersed in vessels filled with the aggressive solution (e.g. a mixture of lactic and acetic acids). The vessels can be positioned at different heights so that wetting and drying cycles can be applied to the specimens. The various speeds of rotation enable a more or less continuous wet environment to be maintained.

The various concrete designs are tested in separate vessels in order to avoid mutual interferences between compositions. The authors notably mention the case of highly buffering materials which can protect other specimens. The disadvantage of this method, used by many authors (Bertron et al. 2007; Oueslati et al. 2009), is that the various compositions are not exposed to the same aggressive conditions during the tests: low buffering capacity matrixes suffer more severe pH or concentrations of acids throughout test, as the aggressive solutions are renewed on the same schedule. Abrasion of the outer part of the specimens, altered by the chemical and physical actions described above, is performed by retractable rotary nylon brushes. The pressure exerted on the specimens through the brushes can also be varied according to the severity of the mechanical actions to be simulated. The majority of the particles from the specimens are collected in a vessel and weighed.

The test procedure developed by De Belie et al. consists of a chosen duration (between a few days and several weeks) of wetting and drying cycles with the aggressive solution followed by a 2-h drying of the specimen (containers with acids in the lowest position) and then brushing. Measurements can be performed once after the wetting and drying cycles and once after specimens have been brushed. This allows the changes in radius of the specimens due to chemical and physical attack to be distinguished from those due to mechanical abrasion. The aggressive solution is renewed at the beginning of each wetting and drying cycle. The pH of the different solutions can be recorded during the test.

The comparative degradation kinetics of concrete or mortar specimens are assessed through laser measurements that quantify (i) surface roughness (an important consideration for animal welfare when they move around on concrete floors), (ii) progression of the concrete surface inward from the initial state (giving the evolution of the thickness of the removed part, respectively the expansion in case sulfuric acid or sulfate attack is considered). Three laser sensors are used. The data acquisition rate is selected so that five measurements are recorded per millimetre along the concrete surface. For every cylinder, several parallel profiles can be measured, equally distributed over the cylinder height.

An example of monitoring the average change of radius of concrete specimens exposed to the TAP degradation test with a lactic/acetic acid solution is given in Fig. 8.2 in Chap. 8, Part III.

De Belie et al. (1996, 1997a, b, De Belie and Monteny 1998) used aggressive solutions made from a mixture of acetic (30 g/l – 0.5 M) and lactic (30 g/l – 0.22 M) acids. The initial mix with a pH of 2.0–2.2 (simulation liquid SL1) had various sodium hydroxide concentrations added to reach the target pH of 3.8 (for SL2), 4.5 (for SL3), and 5.5 (for SL4). The authors explained that pH 3.8 was the lowest pH encountered in the simulated media, pH 4.5 and 5.5 matched the pH limits between highly aggressive (XA3), moderately aggressive (XA2) and low aggressive (XA1) environment classes according to the Dutch standard NEN 5996 (now replaced by EN 206-1). The corresponding additions of sodium hydroxide were 8.22 g/l (0.21 M), 18.06 g/l (0.45 M), and 29.97 g/l (0.75 M) respectively for SL2, SL3 and SL4.

A large range of binders (CEM I, low-C_3A-CEM I, CEM II, CEM III, CEM V, fly ash or silica fume as cement replacements) and concrete designs (polymer cement concretes) was tested by De Belie. Jiang (2002) used the same experimental device and the same composition of aggressive environments to analyse the durability of blast furnace slag concretes.

Other Tests

Gorninski et al. (2007a, b) also performed combined chemical/physical attack and mechanical abrasion on cementitious specimens, but manually. The study investigated the effect of different concentrations of fly ash on the durability of polymer concrete made with orthophthalic and isophthalic polyester. The test method consisted of 14-day exposure cycles. During each cycle, 4 × 4 × 16 cm mortar specimens were fully immersed in various aggressive solutions made of acetic, citric, formic or lactic acid for 7 days and then they were dried for 7 days in the laboratory environment. The specimens were rinsed with a water jet to remove corrosion products. For each exposure cycle, a fresh solution was used. The specimens were weighed before and after each drying period. Five exposure cycles were performed. The solid/liquid volume ratio was 1:4. The authors monitored flexural strength and weight losses over time.

15.5 Characterization of Degradation Mechanisms and Kinetics

15.5.1 Qualification of Degradation Mechanisms

15.5.1.1 Macroscopic Observations

Before undertaking more sophisticated tests, visual inspection of specimens provides much information on the degradation mechanisms: presence/absence of acid salt, changes in dimensions (erosion/swelling of specimen), change in colour of the specimens and coloured zonation of cross-sections.

15.5.1.2 Chemical Changes

Analyses of the solid: chemical changes of altered specimens can be investigated by classical wet chemical analysis. In this case, a representative volume of specimen is sampled in each zone identified visually and analysed to obtain the average composition of the volume (Pavlík 1994b; Bajza 1992; Bertron et al. 2009b). Chemical changes can also be investigated by spot analysis of the solid, performed by SEM coupled with EDS or with an electron microprobe (EPMA) (Pavlík 1994b; Bertron et al. 2005b, 2007, 2009a, b). These techniques consist of bombarding a solid specimen with a high-energy (typically a few kV: 5–30 kV) focused electron beam, which ionizes the first micrometres of the specimen and induces the emission of characteristic X-rays. With EPMA, these emissions are processed to obtain quantitative analyses using standards for the comparison of the concentration of a given element in the specimen to the known concentration of the same element in the standard. With EDS, semi-quantitative analyses are obtained. The techniques are particularly interesting for the analysis of the chemical alteration mechanisms of cementitious matrices by aggressive environments since they can be used to obtain chemical profiles from the surface in contact with the aggressive medium to the core of the specimen. EPMA was notably used by Bertron et al. (2005a, 2007, 2009a, b), Israel et al. (1997) and Oueslati (2011) for the investigation of cement paste specimens exposed to various organic acids. Note that, the analyses being obtained as mass percentages, the real evolution of each element is not visible when the attack causes partial or total leaching of one or more elements. The loss of an element (calcium, for example) leads to an increase in the proportion of a more stable element in the attack (silicon, for example). Bertron et al. (2009b) proposed a method for processing electron microprobe data in order to display the absolute variation of the matrix element concentrations from the sound core to the surface in contact with the aggressive solution. The aim was to improve the analysis and the understanding of the mechanisms of alteration by these media. The processing is based on the use of the titanium content of the specimen since this element is known to be stable in acidic media.

Analyses of the aggressive solution: the leaching behaviour of the cementitious specimen can also be obtained from the chemical analysis of the aggressive solution by monitoring the various chemical elements Ca, Si, Al, Fe, Mg, Na, K, S... (Bertron et al. 2005b; Oueslati et al. 2009; Shi and Stegemann 2000).

15.5.1.3 Mineralogical and Microstructural Changes

To investigate the mechanisms of degradation by the various organic acids, X-Ray diffraction (XRD) analyses can be performed to follow (i) the evolution of crystallized species of the cementitious matrix (hydrated and residual anhydrous phases) and (ii) the possible formation of acid salts during the attack (Bertron et al. 2004, 2005a, 2007, 2009a; Israel et al. 1997).

Pavlik (1994b), Bajza (1992) and Medgyesi (1969) used differential thermal analysis (DTA) and thermal gravimetry (TG) to monitor the dissolution of calcium hydroxide and/or amorphous phases in degraded specimens.

Many authors have performed scanning electron microscope (SEM) observations, mainly to identify the various dissolution/precipitation fronts in altered specimens from changes in density of the matrix, observable in back-scattered electron mode (Israel et al. 1997; Bertron et al. 2005b, 2007, 2009a).

Nuclear Magnetic Resonance: ^{29}Si and ^{27}Al has been used by some authors to investigate microstructural changes in cementitious matrices exposed to organic acids (Bertron 2004; Oueslati 2011). This technique provides information on the material structure at the atomic scale and allows low-crystallized or quasi-amorphous solids to be examined. Bertron (2004) used CEM III paste to analyse the structure of the altered part of specimens and the evolution of residual slag grains. Oueslati (2011) used pastes made with CEM I partially substituted by slag or metakaolin and investigated the evolution of slag and metakaolin residual anhydrous grains during attack by pH 4 acetic acid solutions.

15.5.1.4 Physical Changes

The evolution of physical characteristics and notably the porosity of the cementitious matrix has been assessed by mercury intrusion porosimetry (MIP) (Bajza 1992; Bertron 2004; Beddoe and Dorner 2005; Oueslati et al. 2009, 2011) and by water intrusion porosimetry (Bajza 1992, Bertron et al. 2007). However, care is needed when exploiting mercury porosimetry results if the technique is used on the altered specimens as the pressures exerted on the specimens could modify the microstructure.

The weak point of water and mercury intrusion porosimetry techniques for the investigation of altered specimens is that they give average values over the selected sample. When porosity gradients occur over small distances, sampling becomes impossible. A new technique for the characterization of porosity was tested by Sardini et al. (2007): the ^{3}H-MMA (methyl methacrylate doped with tritium) impregnation method. This was initially developed by Hellmuth et al. (1993) for crystalline rocks. The main characteristic of this method is its aptitude to identify local microstructure variation of the matrix from the scale of about 10 μm up to the scale of decimetres, thus allowing the problem of altered cementitious specimen micro-sampling to be resolved. The ^{3}H-MMA method is based on the impregnation of a volume of material by the liquid monomer ^{3}H-MMA (the small size, the very low dynamic viscosity and the high wettability of the MMA molecule enables high impregnation rates). The tagged resin is then polymerized by exposure to a gamma source. After the specimen has been heated (to avoid luminescence of irradiated materials), autoradiography is performed on it, which allows the spatial distribution of the tagging agent to be viewed. The autoradiography film is scanned and grey levels are converted into porosity using a calibration function (Hellmuth et al. 1993; Sammartino et al. 2002; Prêt et al. 2004). The average porosity and the 2D-spatial distribution of porosity are thus obtained. The quantification is checked by comparing the calculated average porosity to another saturation porosimetry method. This method has been applied to cement paste specimens altered by organic acid solution at a pH of 4 (Sardini et al. 2007). The variation of porosity obtained from the ^{3}H-MMA method was globally well correlated to the chemical and mineralogical modification characterized by EPMA and XRD. Moreover, the method highlighted a porosity gradient in the so-called "sound zone", in which chemical and mineralogical analyses were very similar to those of control specimens.

15.5.2 *Qualification of Degradation Kinetics or Performance of Materials*

15.5.2.1 General Considerations

Several parameters can be measured to evaluate the alteration kinetics and the durability of binders exposed to organic acids: altered depth, loss of mechanical strength, relative mass loss, dissolution of cement paste chemical elements, etc. The choice of criteria should be made carefully since different rankings for durability may be obtained depending on the criteria elected (Bertron et al. 2006; De Belie et al. 1996). Moreover, the rankings should also take the mechanical strength of the unaltered material into account.

Considering reinforced concrete structures exposed to organic acids, the first criterion is that of altered depth in relation to time of immersion, which directly influences the time for the degradation front (the front of cement paste major hydrates dissolution) to reach the reinforcement. Since the residual resistance of the altered zone is very low for concrete exposed to organic acids such as acetic, propionic, or lactic acids, the loss of mechanical strength is directly linked with altered depth.

The criterion of relative mass loss is a secondary indicator and becomes significant when related to the altered depth if the altered zone was preserved prior to the weighing. It then reveals the leaching intensity of the altered zone. For a specimen immersed in volatile fatty acids or lactic acid, and for a given altered depth, the mass loss depends mainly on the calcium content of the binder – when siliceous aggregates are used – since the alteration of the matrix is characterized by an almost total decalcification and the preservation of silicon, iron and aluminium oxide. It is important to bear this in mind when using mass losses as the only indicator to evaluate the performance of a binder.

15.5.2.2 Tests to Assess the Performance of Materials

Relative Mass Variations

Many authors use relative mass losses to monitor degradation of cementitious specimens over time as it is an easy, quick and non-destructive measurement. The specimen can be kept intact prior to the weighing (Bertron et al. 2006; Oueslati 2011) or the loose parts of altered zones can be removed by washing and brushing (Mehta 1985; De Belie et al. 1996, 1997a, b; De Belie and Monteny 1998; O'Donnel et al. 1995a, b). It is also possible to make both measurements: weighing before and after brushing in order to obtain mass loss by pure chemical attack and mass loss linked with abrasion (De Belie et al. 1996; Gorninski 2007a, b; Teoreanu et al. 1999, 2000)

Altered/Degraded Depths/Thicknesses

Phenolphthalein has been used as an indicator to distinguish the altered zone from the sound zone (Bertron 2004; Bertron et al. 2005b, 2007; Larreur-Cayol et al. 2011; Larreur-Cayol 2012; Oueslati 2011). The change in colour occurs when the pH is higher than about 9.5. The pH of the unaltered cementitious matrix being 12.5, some alteration can occur in the so-called "sound zone". Bertron et al. analyzed that the sound zone as defined by the phenolphthalein indicator was in fact composed of two zones: the unaltered core, zone 1 on Fig. 15.11 (with the same chemical composition as the control specimen) and a lighter grey transition zone, zone 2. The chemical and mineralogical changes in zone 2 were moderate: zone 2 was slightly decalcified because of calcium hydroxide dissolution and enriched with sulfate precipitating in the form of ettringite. The change in colour of phenolphthalein occurred at the limit between zone 2 and zone 3 and was correlated to an abrupt drop in calcium content linked with the dissolution of C–S–H. Zone 3, with a

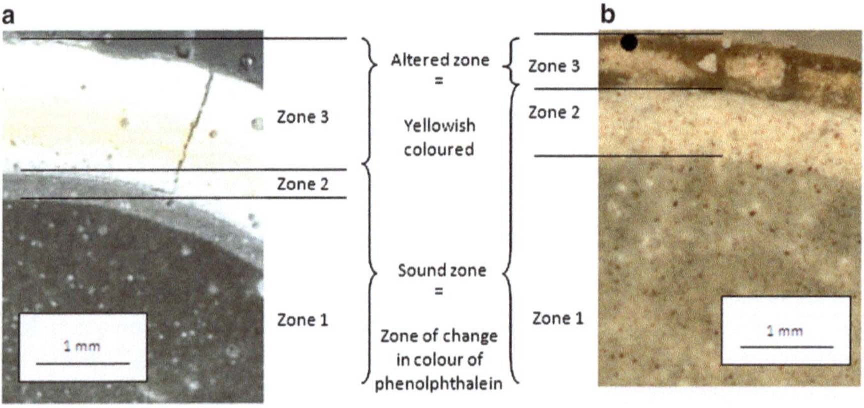

Fig. 15.11 Cross-sections of altered hardened CEM I pastes after 4 weeks of immersion in acetic acid solutions (0.28 M) with: (**a**) pH 4 and (**b**) pH 6 (Bertron et al. 2005a)

yellowish colour, was very porous and totally decalcified. The influence of zone 2 on the overall properties of the altered specimen was low as the chemical and mineralogical changes were small compared to those of zone 3. However, the influence of zone 2 increased as the aggressiveness of the attacking solution decreased. For experiments performed at pH 4 with a solution of acetic acid (0.28 M and addition of 0.05 M NaOH), the thickness of zone 2, after 4 weeks of immersion, was 200 μm for a 2.1-mm-thick zone 3. At pH 6 (0.28 M acetic acid solution with addition of 0.24 M NaOH), the thickness of zone 2 was increased threefold (600 μm) relative to pH 4, for a 600 μm-thick zone 3.

It should be noted that the measurement of the altered depth as performed by Bertron is valid when the zone is not dissolved during the attack. When it is dissolved, the difference in dimensions between the initial specimen and the altered one should be used for thickness measurements (De Belie 1996; Larreur-Cayol et al. 2011, Pavlík 1994a).

The distinction between the corroded and the unaltered zone was made visually by Pavlik by considering the change in colour of the cementitious matrix (from grey to yellowish). The corroded zone, as defined in Fig. 15.12, matches zone 3 of Bertron's zonation (Fig. 15.11). The methodology used by Pavlik (1994a) (casting cement pastes in glass cylinders open on one plane side and providing one-dimensional attack) enables possible dissolution of the outer part of the specimen to be measured reliably thanks to the reference provided by the lower extremity of the glass tube (L_s on Fig. 15.12). Zone 2, a lighter grey transition zone, is observed inside the uncorroded zone and is called the "core layer".

When the degraded zone of the specimen is abraded after the exposure to the aggressive solution, the measurement of the corroded depth is obtained by subtraction between the initial and final dimensions of the specimens. For example, De Belie et al. (2002) measured the change in radius of cylindrical specimens

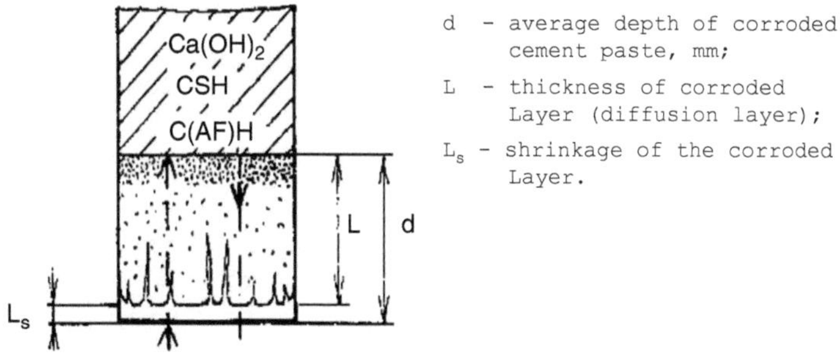

Fig. 15.12 Schematic observation of corroded layer of cement pastes exposed to acetic acids as defined by Pavlik (1994a)

exposed to acetic and lactic acid solutions. The measurements were performed using laser sensors. The device is described by De Belie et al. (2002).

In O'Donnell et al.'s tests (1995a, b) with prismatic concrete blocks, the thickness of the specimen after the treatment was measured at nine points located using a removable template fitted with a grid, ensuring consistent location of the measurement points. The same method was used by Richardson et al. (1994). The direct depth was recorded at 16 points on the top surface of each panel using dial gauges mounted on magnetic bases, giving a reported accuracy of ±0.01 mm

Surface Roughness

In the context of agricultural structures, the surface roughness of concrete floors is an important parameter as it has consequences for animal welfare. Moreover, in the context of microorganism-bearing media, surface defects are favourable to the deposit of microorganisms. This parameter is therefore measured by some authors, notably in comparative studies of concrete designs.

De Belie et al. (1998, 2002) and Jiang (2002) made laser-sensor measurements of the surface roughness of concrete specimens exposed to the TAP test. The surface roughness is expressed through the R_a-value (arithmetical average value of the departure of the profile above and below the centre line throughout the prescribed sample length L) as defined by ISO 25178.

O'Donnell et al. (1995a, b) and De Belie (1995, 1997) used or adapted the sand-patch test as described by the Road Construction Centre (Belgium). The advantage of the method is that it is easy to implement. It consists of spreading a known volume of sand in an approximate circle on the concrete surface and measuring the area of the patch (the greater the roughness, the smaller the area).

Mechanical Testing

The compressive and tensile strengths of specimens exposed to organic acid attack have been monitored by several authors (Bajza 1989, 1992; Teoreanu et al. 2000; Bertron et al. 2006; Gorninsky et al. 2007a, b)

To ensure consistency in the results, the design of test procedures (notably the specimen shape) should take the limitations of mechanical testing into account. The chemical attack should be performed perpendicularly to the loading direction. This means that specimens intended for splitting or bending tensile tests must be designed differently from those intended for compressive or direct tensile tests. Whatever the type of test, the loading zones should be protected from the attack or the attacked portion should be removed by specimen profiling prior to mechanical testing.

Others: Chemical Measurements (pH of Aggressive Solution, Chemical Analyses of the Aggressive Solution)

Comparative leaching behaviour of the cementitious specimens may also be obtained from the chemical analysis of the aggressive solution by monitoring the relevant chemical elements: Ca, Si, Al, Fe, Mg, Na, K, S (De Belie et al. 1997b; Bertron et al. 2005b; Oueslati et al. 2009; Shi 2003). When comparing performances of materials on the basis of the release of chemical elements, the initial content inside the specimens must be taken into account.

The quantity of acid consumed to maintain the target pH and/or pH evolution of the aggressive solution may also be measured (Shi and Stegemann 2000; Shi 2003; De Belie et al. 2002), which reveals the buffering capacity of the material.

15.6 Conclusion

Concrete structures in agricultural and agro-industrial environments are exposed to organic acids in a wide variety of exposure conditions: static or dynamic exposure with periodically or constantly renewed aggressive media, combination of wetting and drying cycles, and/or abrasion by animal or machine traffic, etc.. The large range of test methods found in the literature covers these varied situations quite well.

Organic acids are weak acids and this specificity influences test methods. In particular, attention should be paid to the way the aggressive conditions are maintained (addition of concentrated acid/renewal of the solution) during tests as the partial dissociation of organic acids and their buffer zones greatly influences the quantity of acids consumed and thus the degradation kinetics.

Among the various tests carried out to assess degradation mechanisms and kinetics, it should be remembered that the performance of cementitious materials should be ranked primarily using degraded depths, an indicator revealing the time for the degradation front to reach the reinforcement, rather than mass variations, which, in the case of dissolution degradation mechanisms, are more related to initial calcium content in the material. Moreover, when performing mechanical testing, particular care should be given to the design of the specimens intended for the tests so as to ensure consistency of results.

Finally, although their degradation mechanisms and kinetics are specific, some acids are either little studied in the literature or not at all. These are mainly acids having calcium salts that are slightly soluble to insoluble in water, such as malic, tartaric, and succinic acids, notably found in molasses or distillery residues.

References

Adenot, F., Buil, M.: Modelling of the corrosion of the cement paste by deionized water. Cem. Concr. Res. **22**(2–3), 489–496 (1992)

Bajza, A.: Corrosion of hardened cement paste by NH_4NO_3 and acetic and formic acids. In: Roberts L.R., Skalny J.P. (eds.) Pore Structure and Permeability of Cementitious Materials, Materials Research Society Symposia Proceedings, vol. 137, pp. 325–334 (1989)

Bajza, A.: Corrosion of hardened cement pastes by formic acid solutions. In: Proceedings of the 9th International Congress on the Chemistry of Cement, vol. 5, pp. 402–408, New Delhi (1992)

Beddoe, R.E., Dorner, H.W.: Modelling acid attack on concrete: part I. The essential mechanisms. Cem. Concr. Res. **35**(12), 2333–2339 (2005)

Bertron, A.: Durabilité des matériaux cimentaires soumis aux acides organiques, cas particulier des effluents d'élevage. Ph.D. thesis, INSA, Toulouse, 256 p (2004)

Bertron, A., Duchesne, J.: Attack of cementitious materials by organic acids in agricultural and agrofood effluents – a review. In: De Belie N. (ed.) Proceedings of the RILEM TC 211-PAE Workshop, pp. 11–24, Ghent (2007)

Bertron, A., Escadeillas, G., Duchesne, J.: Cement pastes alteration by liquid manure organic acids: chemical and mineralogical characterization. Cem. Concr. Res. **34**(10), 1823–1835 (2004)

Bertron, A., Duchesne, J., Escadeillas, G.: Accelerated tests of hardened cement pastes alteration by organic acids: analysis of the pH effect. Cem. Concr. Res. **35**(1), 155–166 (2005a)

Bertron, A., Duchesne, J., Escadeillas, G.: Attack of cement pastes exposed to organic acids in manure. Cem. Concr. Compos. **27**(9–10), 898–909 (2005b)

Bertron, A., Escadeillas, G., Duchesne, J.: Durability of various binders exposed to organic acids in liquid manure. In: Malhotra V.M. (ed.) Proceeding of the 7th CANMET/ACI International Conference on Durability of Concrete, Montreal (2006)

Bertron, A., Escadeillas, G., Duchesne, J.: Degradation of cement pastes by organic acids. Mater. Struct. **40**(3), 341–354 (2007)

Bertron, A., Larreur-Cayol, S., Le, T., Escadeillas, G.: Degradation of cementitious materials by some organic acids found in agroindustrial effluents. In: Alexander MG, Bertron A (ed.) Proceedings of the RILEM TC 211-PAE Final Conference, vol. 1, pp. 96–107, Toulouse (2009a)

Bertron, A., Escadeillas, G., De Parseval, P., Duchesne, J.: Processing of electron microprobe data from the analysis of altered cementitious materials. Cem. Concr. Res. **39**(10), 929–935 (2009b)

Bourdette, B., Ringot, E., Ollivier, J.P.: Modelling of the transition zone porosity. Cem. Concr. Res. **25**(4), 741–751 (1995)

Carde, C., François, R.: Effect of ITZ leaching on durability of cement-based materials. Cem. Concr. Res. **27**(7), 971–978 (1997)

De Belie, N.: Measurement of surface erosion of concrete slats. J. Agric. Eng. Res. **61**, 63–66 (1995)

De Belie, N.: On-farm trial to determine the durability of different concrete slats for fattening pigs. J. Agric. Eng. Res. **68**(4), 311–316 (1997)

De Belie, N., Monteny, J.: Resistance of concrete containing styrol acrylic acid ester latex to acids occurring on floors for livestock housing. Cem Concr. Res. **28**(11), 1621–1628 (1998)

De Belie, N., Verselder, H.J., De Blaere, B., Van Nieuwenburg, D., Verschoore, R.: Influence of the cement type on the resistance of concrete to feed acids. Cem. Concr. Res. **26**(11), 1717–1725 (1996)

De Belie, N., Debruyckere, M., Van Nieuwenburg, D., De Blaere, B.: Attack of concrete floors in pig houses by feed acids: influence of fly ash addition and cement-bound surface layers. J. Agric. Eng. Res. **68**(2), 101–108 (1997a)

De Belie, N., De Coster, V., Van Nieuwenburg, D.: Use of fly ash or silica fume to increase the resistance of concrete to feed acids. Mag. Concr. Res. **49**(181), 337–344 (1997b)

De Belie, N., Richardson, M., Braam, C.R., Svennerstedt, B., Lenehan, J.J., Sonck, B.: Durability of building materials and components in the agricultural environment: part I, the agricultural environment and timber structures. J. Agric. Eng. Res. **75**(3), 225–241 (2000a)

De Belie, N., Lenehan, J.J., Braam, C.R., Svennerstedt, B., Richardson, M., Sonck, B.: Durability of building materials and components in the agricultural environment, part III: concrete structures. J. Agric. Eng. Res. **76**(1), 3–16 (2000b)

De Belie, N., Monteny, J., Taerwe, L.: Apparatus for accelerated degradation testing of concrete specimens. Mater. Struct. **35**(7), 427–433 (2002)

Gorninski, J.P., Dal Molin, D.C., Kazmierczak, C.S.: Comparative assessment of isophtalic and orthophtalic polyester polymer concrete: different costs, similar mechanical properties and durability. Construct. Build Mater. **21**(3), 546–555 (2007a)

Gorninski, J.P., Dal Molin, D.C., Kazmierczak, C.S.: Strength degradation of polymer concrete in acidic environments. Cem. Concr. Compos. **29**(8), 637–645 (2007b)

Hellmuth, K.H., Siitari-Kauppi, M., Lindberg, A.: Study of porosity and migration pathways in crystalline rocks by impregnation with 14C-polymethylmethacrylate. J. Contam. Hydrol. **13** (1–4), 403–418 (1993)

Israel, D., Macphee, D.E., Lachowski, E.E.: Acid attack on pore-reduced cements. J. Mater. Sci. **32**(15), 4109–4116 (1997)

Jiang, S.: Use of blast furnace slag cement for the durability of precast concrete in agricultural applications. In: De Belie N. (ed.) Proceedings of the IVth International Symposium: Concrete for a Sustainable Agriculture, pp. 139–147, Ghent (2002)

Steinegger, H.: Prüfung und beurteilung der sulfatwiderstandsfähigkeit van zementen. Zement-Kalk-Gips **23**(2), 67–71 (1970)

Larreur-Cayol, S.: Altération des bétons en milieux agricoles et agro- industriels. Ph.D. thesis, Université Paul Sabatier, Toulouse (2012)

Larreur-Cayol, S., Bertron, A., Escadeillas, G.: Degradation of cement-based materials by various organic acids in agro-industrial waste-waters. Cem. Concr. Res. **41**(8), 882–892 (2011)

Macias, A., Goñi, S., Madrid, J.: Limitations of Köch-Steinegger test to evaluate the durability of cement pastes in acid medium. Ceme. Concr. Res. **29**(12), 2005–2009 (1999)

Mehta, P.K.: Studies on chemical resistance of low water/cement ratio concretes. Cem. Concr. Res. **15**(6), 969–978 (1985)

Medgyesi, I.: Comparaison de l'effet corrosif de certaines acides organiques. In: Valenta O. (ed.) Final Report of the International Symposium on Durability of Concretes, pp. C-431–445, Prague (1969)

O'Donnell, C., Dodd, V.A., O'Kiely, P., Richardson, M.: A study of the effects of silage effluent on concrete: part 1, significance of concrete characteristics. J. Agric. Eng. Res. **60**(2), 83–92 (1995a)

O'Donnell, C., Dodd, V.A., O'Kiely, P., Richardson, M.: A study of the effects of silage effluent on concrete: part 2, significance of environmental factors. J. Agric. Eng. Res. **60**(2), 93–97 (1995b)

Ollivier, J.P., Maso, J.C., Bourdette, B.: Interfacial transition zone in concrete. Adv. Cem. Based Mater. **2**(1), 30–38 (1995)

Oueslati. O.: Durabilité des matériaux cimentaires soumis aux acides organiques – Résistance chimique, mécanique et de corrosion, Ph.D. thesis, Université Laval, Québec (2011)

Oueslati, O., Duchesne, J.: Acetic attack of blended cement paste. In: Alexander M.G., Bertron A. (eds.) Proceedings of the RILEM TC 211-PAE Final Conference, vol. 1, pp. 108–115, Toulouse (2009)

Pavlík, V.: Corrosion of hardened cement paste by acetic and nitric acids part I: calculation of corrosion depth. Cem. Concr. Res. **24**(3), 551–562 (1994a)

Pavlík, V.: Corrosion of hardened cement paste by acetic and nitric acids part II: formation and chemical composition of the corrosion products layer. Cem. Concr. Res. **24**(8), 1495–1508 (1994b)

Pavlík, V.: Corrosion of hardened cement paste by acetic and nitric acids part III: influence of water/cement ratio. Cem. Concr. Res. **26**(3), 475–490 (1996)

Pavlik, V., Uncik, S.: The rate of corrosion of hardened cement pastes and mortars with additive of silica fume in acids. Cem. Concr. Res. **27**(11), 1731–1745 (1997)

Prêt, D., Sardini, P., Beaufort, D., Zellagui, R., Sammartino, S.: Porosity distribution in a clay gouge by image processing of 14C-PolyMethylMethAcrylate (14C-PMMA) autoradiographs: case study of the fault of St. Julien (basin of Lodève, France). Appl. Clay Sci. **27**(1–2), 107–118 (2004)

Richardson, M., Dodd, V.A., Lenehan, J.J., Conaty, S., O'Kiely, P.: The influence of cement content and water/cement ratio on the durability of Portland cement concretes exposed to silage effluent. J. Agric. Eng. Res. **72**(2), 137–143 (1999)

Roy, D.M., Arjunan, P., Silsbee, M.R.: Effect of silica fume, metakaolin, and low-calcium fly ash on chemical resistance of concrete. Cem. Concr. Res. **31**(12), 1809–1813 (2001)

Sammartino, S., Siitari-Kauppi, M., Meunier, A., Sardini, P., Bouchet, A., Tevissen, E.: An imaging method for the porosity: adjustment of the PMMA method – example of a calcareous shale characterization. J. Sediment. Res. **72**(6), 937–943 (2002)

Sardini, P., Bertron, A., De La Asunción-Parreira, C., Siitari-Kauppi, M., Prêt, D., Escadeillas, G., Sammartino, S., Amiri, O.: Distribution spatiale de la porosité des matériaux cimentaires. Une approche méthodologique pour leur caractérisation. Eur. J. Environ. Civ. Eng. **11**(6), 739–749 (2007)

Shi, C., Stegemann, J.A.: Acid corrosion resistance of different cementing materials. Cem. Concr. Res. **30**(5), 803–808 (2000)

Shi, C.: Corrosion resistance of alkali-activated slag cement. Adv. Cem. Res. **15**(2), 77–81 (2003)

Teoreanu, I., Andreescu, E., Ciochina, M.: Acid corrosion of Portland cement-based materials and Portland cement with admixtures. (I) Kinetics and structural-compositional approaches of the corrosion process. Chem. Bull. Politechnica Univ. Timisoara. **44**, 1–9 (1999)

Teoreanu, I., Andreescu, E., Ciochina, M.: Acid corrosion of Portland cement-based materials and Portland cement with admixtures. (II) effects of acid corrosion on structural and mechanical stability of hardened mortars (cement rock). Materiale de Constructii **30**(3), 225–230 (2000)

Weydert, M., Van Caenegem, L., Dehon, P.: Resistance of concrete, asphaltic concrete and mastic asphalt to silage juice. J. Agric. Eng. Res. **79**(2), 151–157 (2001)

Part IV
Cementitious Materials Performance in Aggressive Aqueous Environments – Engineering Perspectives

Authors of chapter in Part IV

Chapter 16:
A. Chatterjee
A. Goyns

Reviewers for Part IV

M.G. Alexander

Chapter 16
Cementitious Materials Performance in Aggressive Aqueous Environments – Engineering Perspectives

Anjan Chatterjee and Alaster Goyns

Contents

A. Chatterjee (✉)
Conmat Technologies Private Limited, Kolkata, India
e-mail: anjan.k.chatterjee@gmail.com

A. Goyns
PIPES CC, Pretoria, South Africa
e-mail: mcegg@pix.co.za

M. Alexander et al. (eds.), *Performance of Cement-Based Materials in Aggressive Aqueous Environments*, RILEM State-of-the-Art Reports 10,
DOI 10.1007/978-94-007-5413-3_16, © RILEM 2013

Keywords Deterioration • Concrete construction • Durability design • Environmental exposure • Probabilistic approach

16.1 Introduction

Concrete forms an indispensable element of structures throughout the world. It is essential that the production of concrete or cement-based materials meets predetermined expectations, which may be set out in the form of specifications or codes, the key considerations for which can be summarized as follows:

- the intended design life of the structure or its constituent elements
- the environmental exposure conditions
- the essential requirements of serviceability
- the fixation of measurable criteria to define serviceability failure
- the planned maintenance requirements
- the risk levels and safety factors.

Practically, there are degrees of arbitrariness in applying the above considerations as the engineering community is still awaiting the following developments:

- reliable tests for concrete durability properties for performance prediction
- mathematical models of deterioration that can be applied in practice

- a probabilistic approach to durability design with values of the acceptable probability of failure, and an understanding of their meaning in practice.

There are sporadic publications on the needed above developments, yet they have not gained any significant foothold in practice perhaps due to lack of harmonization of results on the one hand and durability guidance for concrete being mostly qualitative in most codes of practices and product standards, on the other.

This chapter is aimed at understanding the magnitude of the problem of deterioration of concrete structures in aggressive aqueous environments, in field conditions. The chapter also highlights limitations of the prescriptive status of durability design, differences in the various codes of practice particularly in defining the exposure classes, and varying needs of different constructional domains. The objective is to determine, based on available scientific inputs, an approach to more reliable and realistic design and maintenance of concrete structures for performance. Rebar corrosion is excluded from this review as it has elsewhere been extensively reported in the literature. In general, unless otherwise noted, concrete in this chapter can be assumed to be based on Portland cement as the primary binder.

16.2 Deterioration of Concrete Structures in Selected Aggressive Aqueous Environments

16.2.1 Constructional Domains and Primary Aqueous Agencies of Deterioration

Structures that may be exposed to aggressive aqueous environments are numerous and varied. For practical purposes they may be grouped into the following domains:

(a) marine structures including ports and harbours, etc.
(b) hydroelectric structures including dams, diversion tunnels, headrace tunnels, power houses, etc
(c) sewers and sewage treatment facilities
(d) industrial and chemical plants
(e) water supply conduits and storage facilities
(f) other liquid-containing structures
(g) bridges, culverts, roads and motorways
(h) residential low- and high-rise buildings
(i) farm buildings including agro-industrial structures.

The interrelation of the aggressive aqueous agents with these constructional domains is shown in Table 16.1. Present design practice is to impart overriding importance to loads and stability rather than to materials degradation. However, exposure conditions and materials selection have received greater attention for marine structures, sewerage systems, chemical plants, etc.

In recent decades, the technological advancement of underwater concreting has resulted in tremie concreting for bridge foundations, deep tremie placements for offshore platforms, in-the-wet construction of navigation structures, etc. Here, the interaction of the materials with the aqueous environments is complex and needs special attention, more so as the technology is still evolving.

16.2.2 Trends of Concrete Deterioration by Water and Aqueous Chemicals

Various types of water-mediated chemical reactions that lead to concrete deterioration are known phenomenologically. They can be depicted as in Fig. 16.1.

From the published information and data (Beaudoin et al. 2001; Mehta and Gerwick 1982; Detwiler and Taylor 2005; Mehta 1986; Mindess and Young 1981; Bertron and Duchesne 2007), it is evident that the deterioration trends of concrete in aqueous environments are governed by the following parameters:

(a) Concrete is not acid-resisting by itself. It can be attacked at pH less than 7.0 but the attack is severe at pH below 6.5 and very severe below 4.5.
(b) Although each unit decrease in pH represents a ten-fold increase in acidity, the degree of dissociation of the acid is more significant and not directly related. For example, concrete in contact with oxalic acid of pH 3 reacts to form calcium salts that are insoluble and durability is not impaired.
(c) The aggressiveness of aqueous solutions is not solely dependent on pH or concentration of dissolved ions, but also on the availability of the solution for reactions with concrete. Flowing and percolating water will cause more rapid deterioration than still water.
(d) Crystalline hydration products in hydrated pastes, such as calcium hydroxide, are the primary sources of potential expansion when immersed in chloride and sulfate solutions. The process of expansion appears to be influenced by both the solubility and the nature of reaction products.

    The order of dissolution of the hydrated phases is as follows:

$$\mathrm{CH > AFm > AFt > C\text{–}S\text{–}H}$$

    The C–S–H phase suffers gradual decalcification after removal or alteration of CH, due to the inability to maintain the Ca^{2+} ion concentration. Since CH is more soluble in cold water, leaching is more rapid at low temperatures.
(e) The rate of sulfate attack increases with increase in the strength of solution. The concentration of sulfates up to 1,000 mg/L is considered moderately severe and up to 2,000 mg/L as severe, especially if magnesium sulfate is present.

Table 16.1 Interrelation of aggressive aqueous environments with constructional domains

Nature of Aggressive Solutions		Constructional domains								
Type	Constitutional aspect of importance	Marine	Hydel	Sewers	Indus. & chem. plants	Water supply conduits	Liquid containing strucs	Roads	Residential buildings	Farm buildings
Purewater	H_2O					√	√		√	
Soft water	H_2CO_3		√			√	√		√	
Weak acids	Low H_3O^+ concentration						√			√
Strong acids	High H_3O^+ concentration & $HCl/H_2SO_4/HNO_3$			√	√		√			
Saline water	Presence of K^+, Na^+, SO_4^{2-}		√		√		√	√		√
Sea Water	Presence of Cl^-, Na^+, SO_4^{-2}, Mg^2+, Ca^2, K+, HCO_3, Br	√								

Note:
Column headings correspond to the domains listed above
√ = significant interaction

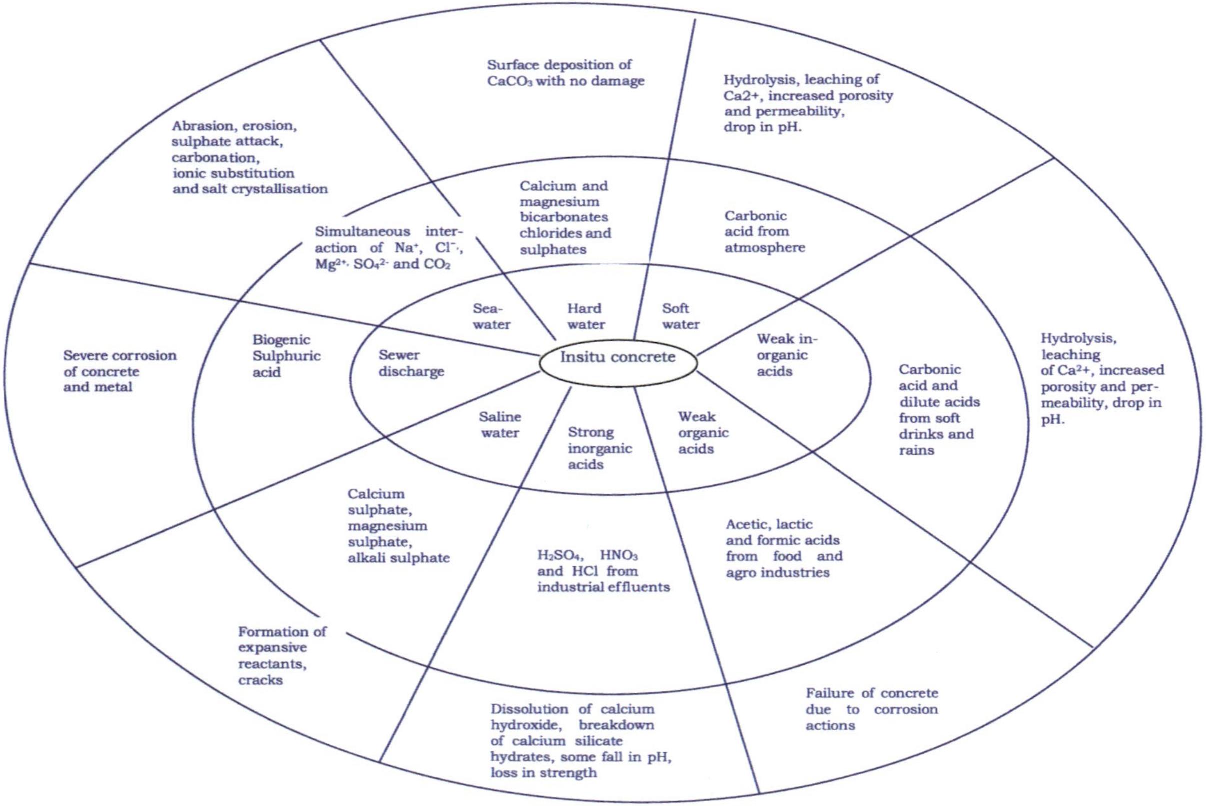

Fig. 16.1 Various water-mediated deterioration trends in concrete

A saturated solution of magnesium sulfate can cause serious damage to concrete with high w/c ratio, e.g. above 0.6. It is generally observed that if the concrete is made with low w/c ratio, the concrete can withstand the action of magnesium sulfate for 2–3 years.

(f) The organic acids found in the effluents of food-processing and agro-industrial plants, such as liquid manure, silage juice, whey, white waters, molasses, etc. induce severe attack on concrete structures. These effluents contain carboxylic acids such as acetic and propionic acids (volatile fatty acids), lactic acid, citric acid and oxalic acids which are mostly accompanied by micro-organisms. The aggressiveness of these weak acids depends on the solubility of the resultant calcium salts, as already mentioned above.

16.2.3 Concrete for Marine Structures

The forces acting upon wharves, jetties, dolphins, piers, docks, sea-walls and similar marine and estuarine structures include those due to the wind and waves, blows and pulls from vessels, loads from cranes, railways, roads, storage of goods and other live loads imposed on decks, and the pressure of earth retained behind the structure. In addition to the complexity of these physical and mechanical forces, the effect of sea water on concrete deserves special attention.

Sea water generally contains 3.5% of salt by mass. The ionic concentrations of Na^+ and Cl^- are the highest, typically 11,000 and 20,000 mg/L respectively. It also contains Mg^{2+} and SO_4^{2-}, typically 1,400 and 2,700 mg/L respectively. The pH of sea water in a number of measurements ranged from 7.5 to 8.4. Sea water also contains some carbon dioxide.

Various processes of transport and deposition of salt in the marine environment are shown in Fig. 16.2 (CS 109 1996). As a result concrete undergoes several concurrent reactions when subjected to sea water. It has been observed that the deterioration of concrete in sea water is often not characterized by sulfate expansion but takes more the form of erosion or loss of constituents from the parent mass. It is generally accepted that the presence of chlorides in sea water retards the expansive reaction of the sulfates. It is further observed that both calcium hydroxide and calcium sulfate are considerably more soluble in sea water, resulting in increased leaching action.

Some other observations in this context are the following:

(a) Relatively slender concrete members exposed to sea water show more leaching than expansion, while massive concrete members such as dock walls etc. show expansion.
(b) The most severe attack of sea water on concrete occurs just above the level of high tide. The parts below the low-tide water level are the least affected, while the intermediate zone shows moderate degradation.

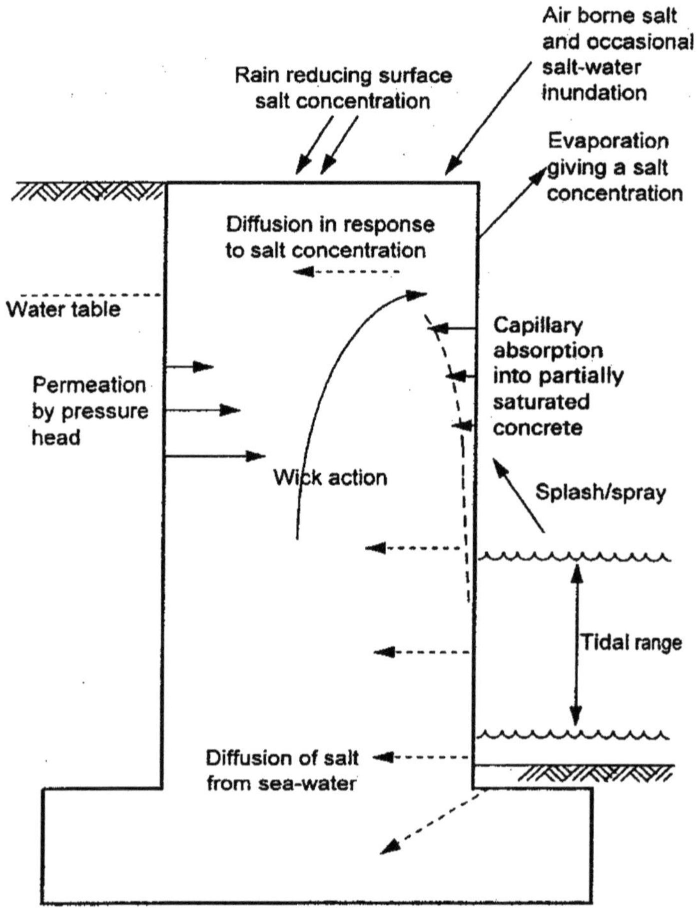

Fig. 16.2 Transport process and movement of salts in a sea wall (CS 109 1996)

(c) Sea water that penetrates into reinforced concrete ultimately leads to reinforcement corrosion.
(d) The velocity of wave action containing sand and silt may cause severe abrasion of concrete.

The chemistry of sea water and concrete interactions are still not fully understood in terms of mechanisms and kinetics. However, certain phenomenological studies have been carried out to ascertain the performance of concrete in field conditions. One extensive study was reported in Malhotra and Bremner (1996). A series of more than 250 concrete prisms, 350 × 305 × 91.5 mm, were cast over a period of 16 years and installed at Treat Island, Maine, within a few km of the Canadian border. The prisms of the first phase of the investigation were installed at the site in 1978 with the remaining specimens being installed at almost yearly intervals, the last phase of investigation being 1994. The test prisms were installed at mid-tide level on a rack and were exposed to repeated cycles of wetting and drying, and to about 100 cycles of freezing and thawing per year. Thus, the exposure conditions were very severe. The test specimens were evaluated annually by visual inspection and by applying an arbitrary damage rating scale. Some of the principal conclusions were:

- Use of non-air-entrained concrete was found unsuitable for the given site conditions
- Concretes made with ASTM Types I, II and V cements (i.e. general purpose, moderate sulfate resistance, and high sulfate resistance respectively), performed equally well under the exposure conditions
- Concrete prisms containing 65% blast furnace slag as replacement of Portland cement but with W/B ratio of 0.40 showed satisfactory performance.
- Limiting the use of silica fume to 10% was found necessary
- Both normal-weight and semi-lightweight concretes containing fly ash, slag or silica fume or a combination of these materials were observed to be in good condition, when W/B ratio was less than 0.50 and Portland cement content was kept at 200–250 kg/m^3.

However, in a recent study (Khatri and Sirivivatnanon 2004) on the use of a statistical method to estimate the characteristic service life of normal Portland cement concretes of different grades (20, 32, 40 and 50 MPa) in marine environments with a confidence level of 90%, the variation in service life due to variation in concrete quality and cover depth was quantified. It was found that service life is more sensitive to cover depth than diffusion coefficient and more sensitive to surface chloride concentration (0.7%) than critical chloride level (0.2%).

16.2.3.1 Bridge Exposed to Seawater

The Confederation Bridge was constructed across the Northumberland Strait between New Brunswick and Prince Edward Island in Canada between 1993 and 1997. During winter the strait regularly freezes and the concrete used in the bridge was required to resist the effects of sea water, salt scaling, freezing/thawing and abrasion from floating ice for a life of 100 years.

In order to withstand this aggressive marine environment a concrete was designed with 480 kg/m^3 total cementitious content, of which 10% was a class F fly ash, and 6.75% was silica fume. The water-cementitious materials ratio was less than 0.30 (Dunaszegi 1998). The average 90-days strength of the concrete was 80 MPa against the structural requirement of 60 MPa, with an air content of 5–8%. Due care was taken for curing up to 5 days followed by the application of a curing compound. It appears that the above prescription of concrete has served well in the aggressive marine environment, although the specification achieved was perhaps quite conservative.

The marine environment here had all the ingredients for the following deterioration mechanisms:

- Corrosion
- Sulfate attack
- Magnesium ion substitution
- Abrasion and erosion
- Frost damage
- Carbonation
- Salt crystallization

When several deterioration mechanisms occur simultaneously, the net effect is often greater than the sum of the individual factors. The study of the aggressiveness of the marine environment, therefore, is highly complex and the material design turns out to be equally conservative, as in this specific case.

16.2.4 Concrete for Hydroelectric Projects

Hydroelectric structures consist of dams (concrete or rockfill), diversion tunnels, headrace tunnels, powerhouse structures, etc. Large quantities of concrete are used in these structures. The grade of concrete for dams is generally M15[1] or lower. For reinforced concrete structures in powerhouse and tunnel linings, concrete of M20 and above is generally used. The spillways of concrete dams are of higher grade (M60–M80) in order to resist impact and abrasion. Dam structures and their appurtenances must be safeguarded against distress due to chemical reactions or physical attack (see Fig. 16.3a, b).

Fracture mechanics, rock mechanics and finite elements techniques are used in studying the progressive foundation failure of concrete dams involving: initiation of cracks on the upstream face of the dam; crack propagation inwardly due to a sustained large hydraulic load and subsequent breakthrough; frictional sliding of monoliths, and; ultimate failure. Thus, overriding reliance is laid on physical parameters with little consideration of chemical causes. Since hydroelectric structures are continuously in contact with an aqueous environment, the process of leaching of concrete has to be considered as an important distressing factor. The attack of sulfate and chloride bearing aqueous solutions is widely known. The occurrences of deleterious alkali-aggregate reactions are reported from many countries (Shayan and Grimstad 2006; Maiti et al. 2005). However, the likely changes in the chemical reactions under high hydraulic load are not reported at all.

[1] In the Indian Code of Practice the concrete grades are designated with letter ‘M’, while in Eurocode the letter used is ‘C’ and the value is given in MPa.

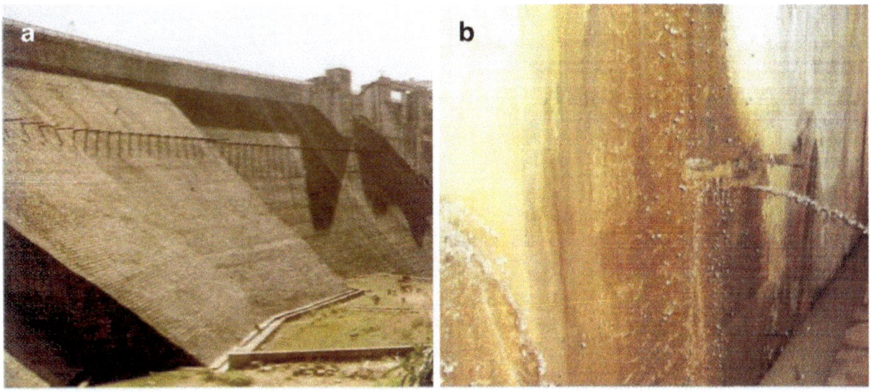

Fig. 16.3 (**a**) A panoramic view of Muran Dam in India, (**b**) Water leakages in the drainage gallery of Muran Dam

16.2.5 Underwater Concrete

In recent decades, engineering advances have opened up a variety of new applications for underwater concrete as a structural material. A review has been provided in Yao and Gerwick (2004). Tremie concreting for bridge foundation retrofitting, deep tremie placements for offshore platforms, in-the-wet construction of navigation structures, and various underwater repairs are some of the examples of the still evolving underwater concrete technology.

In-place concrete quality is closely related to the way the concrete flows under water. When concrete is highly flowable and cohesive, it flows in a bulged flow pattern, which develops concrete with a relatively flat and smooth top surface and good in-place quality. On the other hand, less flowable concrete flows in a layered pattern. Apparently the layered flow pattern exposes more concrete surfaces to water and is usually associated with a steeply sloped and rugged top surface with large quantities of laitance (Fig. 16.4).

In a 19 km long below-grade railway trench project in Los Angeles, CA, more than 9,000 reinforced tremie concrete drilled shafts were constructed as a part of a secant pile/retaining wall system. After excavation of the railway tunnel, a large number of the exposed drilled shafts were found to be defective (Fig. 16.5). The vertical and horizontal grooves in the concrete cover demonstrated that loss of tremie concrete flowability prevented it from flowing around and fully encasing the reinforcing cages during placement. This kind of concrete defect has obvious impacts on the structural durability.

Since concrete placed under water is inherently susceptible to cement washout, as well as laitance, segregation, cold joints and water entrapment, it must possess some unique workability and durability characteristics. The workability requirements include easy flow, cohesiveness against washout and segregation, and self-consolidation characteristics (as no mechanical vibration can be applied

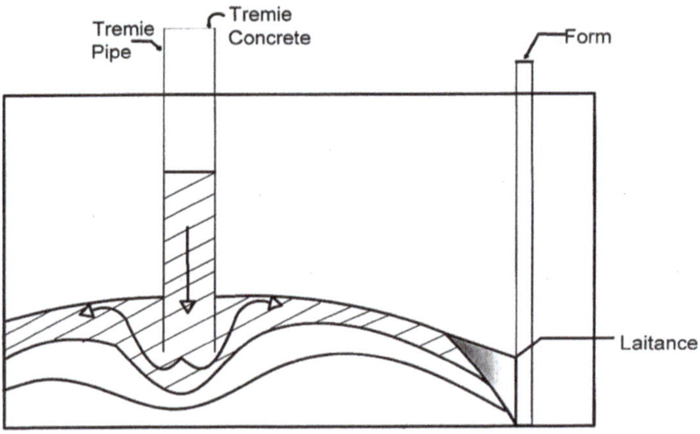

Fig. 16.4 Less flowable underwater concrete exhibiting a layered flow pattern

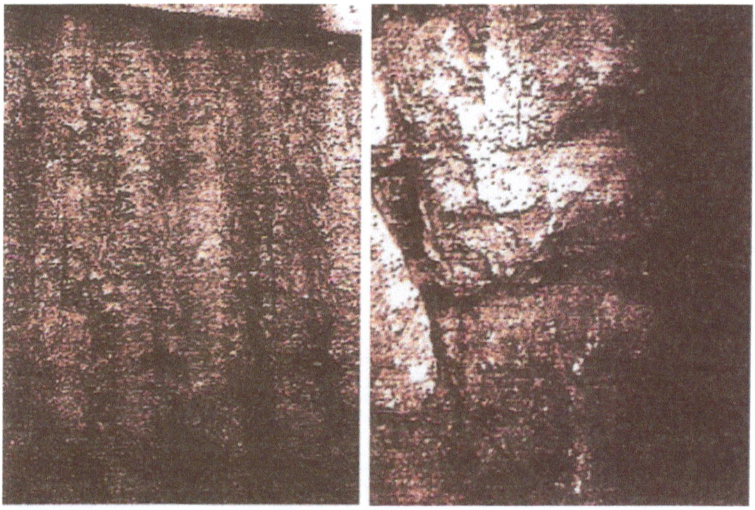

Fig. 16.5 Defects in drilled shafts due to lack of concrete flowability

to consolidate under water). As far as the durability requirements are concerned, resistance against water entrapment and penetration of water-borne aggressive ions is obviously required. In addition, for massive underwater concrete placements, the thermal stress is a special concern because of the high cement content (360–500 kg/m^3) and the consequent high heat of hydration in a usually cooler environment with constant temperature. The water-to-fines ratio strongly affects the workability of underwater concrete. Fines include cement and other mineral admixtures as well as inert fillers such as limestone powder. A water-to-fines ratio in the range of 0.85–1.0 by volume often provides a good balance between workability and cohesion requirements. It may also be relevant to mention that the recent trends of underwater concreting show increasing use of ternary blends of mineral admixtures in concrete, which include fly ash, slag and silica fume. For the flowability requirements the simultaneous use of chemical admixtures such as high-range water reducers as well as anti-washout agents of saccharide or cellulose polymers has become standard practice.

The critical issues of designing underwater concrete are summarized in Table 16.2. The mix proportions for underwater concrete must be determined with sound understanding of the application techniques and environmental demands.

16.2.6 Concrete for Sewers and Sewage Treatment Facilities

It is well known that corrosion of concrete, metal and electrical equipment in urban drainage systems can be caused by the generation of hydrogen sulfide (H_2S). Particularly susceptible locations are points of turbulence following long-retention times e.g., back-drop manholes, wet wells of pumping stations, and outlets of wastewater rising mains. Problems are particularly serious in hot and arid climates. In addition, hydrogen sulfide can cause odour nuisance when escaping into the atmosphere, danger to sewer workers, and acute toxicity to aquatic organisms. Wastewater naturally contains sulfur as inorganic sulfate or organic sulfur compounds. The sulfate is usually derived from the mineral content of the municipal water supply or from saline groundwater infiltration. Organic sulfur compounds are present in excreta and household detergents, and in high concentrations in certain industrial effluents such as from the leather, brewing and paper industries.

Bacterial activity in slow flowing or stagnant water quickly depletes any dissolved oxygen that is present and septicity develops. Under these anaerobic conditions complex organic substances are reduced to form volatile fatty acids resulting in a drop in pH. Desulphovibrio bacteria in pipe biofilms and sediment reduce organic sulfur compounds and sulfates to sulfides:

$$SO_4^{2-} + C, H, O, N, P, S, \rightarrow S^{2-} + H_2O + CO_2 \quad (16.1)$$

H_2S results from a reaction with hydrogen ions in the water and hence is pH dependent, with more H_2S being formed when more severe acidic conditions occur in the effluent.

$$S^{2-} + 2H^+ \rightarrow H_2S \quad (16.2)$$

In pipes flowing under gravity, H_2S escaping into the atmosphere from the solution in the wastewater tends to rise and accumulate in condensation water on

Table 16.2 Key underwater concrete mixture parameters and recommended ranges (Yao and Gerwick 2004)

Parameter	Recommended ranges for general usages	Comments
Water-cementitious material ratio	0.35–0.45 (by mass)	Critical to concrete strength and durability
Fines content	360–500 kg/m^3	Fines include cement, pozzolans, slag cement, and proper inert fillers such as limestone powder
Water-to-fines ratio	0.85–1.0 (by volume)	Critical to concrete workability
Sand-to-total aggregate ratio	45–50%	Moderately high sand content improves cohesiveness of concrete
Nominal maximum size aggregates	10–19 mm for structural concrete; 19–25 mm for mass concrete	Preferably less than 1/5 the clear spacing between steel bars
Silica fume	0–6%	A mineral-based anti-washout admixture. At 6% or less, it improves workability; above 6%, it reduces workability
Limestone powder	0–20% of cement content	Supplement to cement to improve the workability
Anti-washout admixture (AWA) e.g., Saccharide or cellulose polymers	1. Not essential for tremie seal/drilled shafts 2. Moderate dose for reinforced concrete with long flow distance 3. High dose for self-levelling, thin layer placements over large areas or placements in flowing water	In general, flowable and self-compacting underwater concrete can be made without AWA. But AWA significantly enhances anti-washout performance of concrete to meet certain special requirements AWA should be used with HRWRA and preferably added to concrete after HRWRA has been fully mixed in concrete
High range water-reducing admixture (HRWRA)	Adjust dosage to produce a proper slump flow in the range 350–680 mm, depending on the application requirements	HRWRA is generally used together with set-retarder or low-range water reducer to ensure the workability retention with required time period

the exposed surface of the pipe. There, it is oxidized by thiobacillus bacteria to form sulfuric acid (H_2SO_4):

$$H_2S + 2O_2 \rightarrow H_2SO_4 \tag{16.3}$$

Absorption of H_2S readily occurs on the moist surfaces above the sewage level, with higher concentrations giving rise to higher rates of absorption. Oxidation of

H_2S to sulfuric acid occurs in two ways: naturally (auto-oxidation) by a spontaneous reaction of hydrogen sulfide with oxygen in an aqueous solution to form sulfuric acid (Eq. 16.3), or biologically due to aerobic micro-organisms, namely thiobacillus bacteria, that thrive in the moisture layer and oxidise the sulfide (Eq. 16.4) to generate energy to assimilate carbon for synthesis of cell materials (Alexander et al. 2008).

$$2H_2S + O_2 \rightarrow 2S + 2H_2O + 3O_2 \rightarrow 4H^+ + 2SO_4^{2-} \qquad (16.4)$$

The aerobic respiratory process utilises large amounts of sulfide that results in the production of a substantially higher amount of H_2SO_4 than that produced via auto-oxidation. A mixed flora of micro-organisms develops in the moisture layer on the concrete walls above the sewage level. An alkaline environment (pH 10.5–12.5) from the dissolution of calcium hydroxide in the cement paste initially stifles the growth of the thiobacillus species. However, acidic conditions that favour growth of the acid-generating thiobacillus species gradually develop in the moisture layer due to the absorption of carbon dioxide and hydrogen sulfide. Increasingly acid-tolerant species are stimulated as the pH reduces, leading ultimately to the growth of thio-oxidans in environments of pH 1.0 and below (Alexander et al. 2008).

Sulfuric acid can cause serious damage to pipe materials. The acid first reacts with calcium hydroxide in concrete to form gypsum, followed by reaction of gypsum with tricalcium aluminate to form ettringite.

$$Ca(OH)_2 + SO_4^{2-} = CaSO_4 \cdot 2H_2O + 2OH^- \qquad (16.5)$$

$$C_3A + 3CaSO_4(Aq.) = C_3A \cdot 3CaSO_4 \cdot 32H_2O \qquad (16.6)$$

Both the products involving volume expansion result in cracking. If all CH is consumed, decalcification of C–S–H can take place, aggravating the destruction of the paste structure. However, in flowing sewers, acid corrosion products can be removed from the concrete walls by the flowing sewage, thus continually exposing fresh concrete to direct acid attack. The attack is most severe on the pipe walls just above the low flow level, due to the combined effect of corrosion and water erosion (Fig. 16.6).

Although the above mechanisms are widely accepted to explain the degradation of concrete in sewers, certain investigations have shown that resistance of certain concrete to inorganic and biogenic sulfuric acid may be different (Sand et al. 1992; Schmidt et al. 1997). In particular, calcium aluminate cement concretes have a greater resistance to biogenic sulfuric acid attack. However, further confirmation is required for engineering applications.

16.2.6.1 Sulfide Build-Up and the Rate of Concrete Deterioration

A formula published by Davey in 1950 was later modified by Pomeroy and has been in use for estimating the sulfide build-up in gravity sewers less than 600 mm

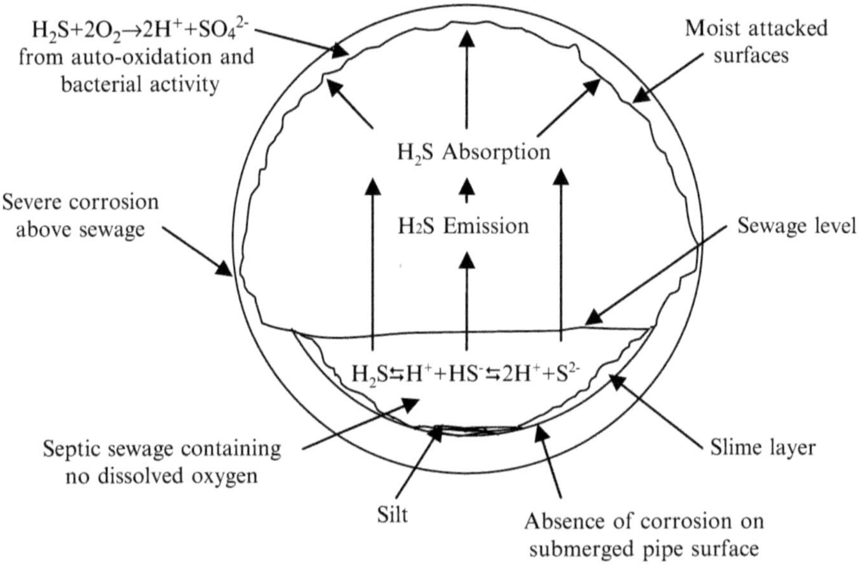

Fig. 16.6 Pattern of degradation in sewer pipes (Alexander et al. 2008)

Table 16.3 Probability of sulfide development

Z	Prevalence of sulfide
<5,000	Rarely present
~7,500	Low concentrations likely
~10,500	May cause odour and corrosion problems
~15,500	Frequent problems with odour and significant corrosion problems

diameter that are flowing partly full (Pomeroy and Parkhurst 1977). Known as the Z Formula it is written

$$Z = [EBOD / S_o^{1/2} Q^{1/3}] \times [P/B] \tag{16.7}$$

Where,

EBOD = Effective BOD = BOD $\times$ $1.07^{(T-20)}$ (mg/L)
T = Wastewater temperature (° C)
S_o = Sewer gradient (m/100 m)
Q = Flow rate (L/s)
P = Wetted perimeter (m)
B = Flow width (m)

The values of Z and their interpretation are given in Table 16.3.

As indicated elsewhere in this chapter, the sulfur cycle in sewers results in the formation of H_2S in the sewer atmosphere. The chemical or biological oxidation of this sulfide gas results in generation of sulfuric acid on the walls of the sewer, and consequent corrosion. Further research in this direction led to the development of quantitative methods for predicting the rate of sulfide generation and the resultant rate of concrete corrosion, which later became known as the "Life Factor Method" (McLaren 1984).

In the Pomeroy model, applicable to concrete sewer pipes, the rate of concrete deterioration is calculated by the equation,

$$C_r = [11.5k\ \Phi_{sw}]/Alk \tag{16.8}$$

Where,

- C_r = corrosion rate (mm/year)
- k = coefficient of efficiency for acid reaction considering estimated fraction of acid remaining on sewer wall
- Φsw = H_2S release [g/(m^2/h)] to the sewer wall
- Alk = alkalinity of the pipe material (expressed as $CaCO_3$ equivalent).

The alkalinity of the pipe material is based on the $CaCO_3$ equivalent and is a measure of its neutralization capacity. It can be calculated from the experimentally determined content of soluble CaO of concrete or its constituents, cement and aggregates, by the following equation:

$$Alk = \frac{\left(c \cdot CaO_{cement} + a \cdot CaO_{aggr}\right)}{d}(100/56) \tag{16.9}$$

Where,

c = Cement content (kg cement/m^3 concrete);
CaO_{cement} = Soluble CaO in the cement (kg CaO/kg cement);
a = aggregate content (kg/m^3 concrete);
CaO_{aggr} = content of soluble CaO in the aggregate (kg CaO/kg aggregates)
d = concrete density (kg/m^3).

The designer needs to know how long the sewer will last, and consequently the properties of the sewer pipe of alkalinity, and the additional cover to steel to account for the corrosion loss z, are combined to give the Life Factor Equation:

$$Alk \cdot z = 11.5k\ \Phi swL \tag{16.10}$$

Where z is the additional cover to the steel (=C_rL) and L is the required design life in years.

The use of 'Alk.z' allows various materials with different 'Alk' values and the appropriate amount of cover 'z' to be compared to give the most cost effective option.

16.2.6.2 Materials Testing and Selection for Sewer Designs

Research performed on the resistance of concrete has been reviewed in Monteny et al. (2000). A detailed study on a new procedure to examine chemical and microbiological corrosion tests is presented in De Belie et al. (2004), Saucier and Lamberet 2009. The findings were:

(a) The aggregate type had the largest effect on degradation. Limestone aggregate concrete was found to be more resistant than siliceous aggregate concrete under experimental conditions, due to the creation of a buffering environment that protected the cement paste. Calcareous aggregate concrete was found to be more effective in dealing with corrosion than siliceous aggregate concrete, because both aggregate and binder were corroded. This meant that the concrete was corroded away fairly uniformly and not selectively, resulting in a slower rate of material deterioration. With this concrete under experimental conditions, the deterioration appeared to be further retarded due to buffering that protected the cement paste.
(b) Concrete pipes produced by a centrifugal process showed a higher resistance than those hardened in the formwork. This was due to the lower w/c ratio in such products.
(c) While the influence of the cement type was observed to be limited, high sulfate resisting Portland cement or Portland slag cement displayed some advantage. But in microbiological tests, SRC behaved better than PSC.
(d) In the Pomeroy model, the addition of water absorption as a measure of porosity as an additional factor over alkalinity showed better correlation.
(e) Calcium aluminate cements have a long history of effective use in protecting sewer networks from biogenic corrosion. Since the 1940s this type of cement has been utilized for concrete pipes, as lining on OPC pipes, as inside lining of ductile iron pipes, as well as for repair of in-service sewers and associated wastewater installations. Examples are available from Australia, South Africa, Malaysia, Egypt, USA, etc. It has been reported and subsequently proven by the test method developed at Hamburg University that with calcium aluminate concrete or mortar the pH value stabilizes around 3–4, rather than going down to 1.

A very comprehensive and practical approach to developing durable sewer lining materials has been reported in a series of publications (Goyns 2006; Alexander and Goyns 2008; Goyns et al. 2008; Fourie et al. 2009). According to the authors, there are three options for preventing or minimizing corrosion in concrete sewers:

- Reducing the acid formation by adjusting the hydraulic design of the sewer.
- Modifying the concrete mix and/or providing additional cover to reinforcement (a sacrificial layer).
- Protecting concrete by using an inert lining or coating under very severe conditions.

Concrete made of calcium aluminate cement and dolomite aggregate for a corrosion control lining has been used based on information gained from the testing of various cementitious materials in very aggressive real life situations. This measure has been used in South Africa since 1997 and has proved to be both technically sound and cost effective. The lining is about four times as effective as the Portland cement dolomite concrete and at least ten times as effective as Portland cement quartzite concrete, in resisting acid attack.

Data obtained from the performance of concrete pipes made with calcium aluminate cement and other cementitious materials in the aforesaid experimental sewer in South Africa has been applied to the design of new, and the rehabilitation of existing concrete sewers. The Life Factor Method, so far applied to Portland cement concretes only, can now be extended with a "Material Factor" to predict the corrosion rates of concrete sewers made using other cements such as calcium aluminate cements. Further, the concept of a host pipe with a corrosion control layer made of CAC and dolomite aggregates has been developed more comprehensively (Goyns et al. 2008). Towards further improvements, CAC concrete with CAC aggregates (tradename 'Alag') demonstrated higher durability than the CAC + dolomitic concrete (Saucier and Lamberet 2009; Fourie and Alexander 2009).

16.2.7 Concrete Exposed to Sulfate Attack

The term "sulfate attack" has classically been associated with the failure of structures due to disruptive expansion of concrete in contact with sulfate bearing soils and groundwaters. There are also two other modes of concrete failure due to sulfate attack – thaumasite sulfate attack caused by external sources, and delayed ettringite formation caused internally.

16.2.7.1 Classical Form of Sulfate Attack

In the classical form of sulfate attack the source of the sulfate ion is typically sodium sulfate, calcium sulfate, magnesium sulfate or potassium sulfate. Ammonium sulfate is occasionally encountered. These sulfates may occur naturally in soils, clays, or groundwaters. Industrial processes may also produce sulfates. Ettringite and gypsum are the principal compounds formed. The following factors are known to influence the intensity of attack (Richardson 2002):

- sulfate concentration
- solubility of sulfates
- groundwater mobility
- concrete permeability
- wetting and drying cycles
- evaporation
- degree of carbonation prior to exposure.

16.2.7.2 Thaumasite Occurrence

Thaumasite sulfate attack results from a reaction between sulfates, calcium silicates in cement and calcium carbonate from aggregates or groundwater percolation. Thaumasite is a complex sulfate-bearing mineral with the composition $CaSiO_3 \cdot CaCO_3 \cdot CaSO_4 \cdot 15H_2O$. It forms more readily in the presence of magnesium sulfate than sodium sulfate but is less likely to form in calcium sulfate solutions. Samples of deteriorated concrete often show white haloes of thaumasite around the affected aggregates. Most of the thaumasite cases in UK motorway bridges involved the presence of pyrite (Hobbs and Taylor 2000). The primary factors that must simultaneously be present for thaumasite sulfate attack are as follows (Hartshorn et al. 1999; DoE 1999):

- sulfates and/or sulfides in the ground
- mobile groundwater
- presence of carbonate
- low temperature ($<15^\circ$C).

16.2.7.3 Delayed Ettringite Formation

Ettringite is formed during normal hydration in plastic concrete except at temperatures above 70°C. Higher temperatures can be encountered during the curing of large pours or during steam curing of precast units (Shimada et al. 2005).

In such circumstances, the sulfate, instead of forming ettringite, is absorbed by the calcium silicate hydrate, forms small amounts of monosulfate and some may go into the pore fluid solution. The alumina content of the calcium silicate hydrates increases. The phases are however unstable at ambient temperatures, and ultimately give rise to ettringite crystals in the hardened concrete later, when the concrete cools in service. Expansion of the cement paste may result followed by associated cracking. The cracks are often uniform in width and may become filled with ettringite. The main influencing parameters for delayed ettringite formation are:

- temperature during hydration
- size and shape of pour
- cement content
- cement chemistry and fineness.

16.2.7.4 Some Illustrations of Sulfate Attack

In an investigation (Limaye et al. 1992), RC beams were cast and cured under laboratory conditions. The details of the concrete are given in Table 16.4.

A solution of sodium sulfate was prepared (350 g/L) and the RC beams in batches were subjected to alternate wetting and drying cycles in this solution for

Table 16.4 Details of concrete and reinforcement characteristics (Limaye et al. 1992)

Materials	Parameters	Values
Concrete	Grade	M20
	Workability (CF)	0.92
	A/C ratio	4.30
	CA: FA ratio	60: 40
	w/c ratio	0.50
	Cube strength, MPa	24.7, 23.1, 23.5, 27.0
Reinforcement	Type	8 mm deformed bar
	Yield stress at 0.2% strain	400 MPa

CA coarse aggregate, *FA* Fine aggregate

2 and 3.5 months. Each cycle consisted of immersing the specimen in solution for 8 h and exposure to dry climate for 16 h. The depth of immersion was maintained at 20 mm. It was observed that the material loss after 60 cycles was 770 g, and after 105 cycles 3,283 g. It was evident that the deterioration progressed at a faster rate only after 60 cycles. The deteriorated beams also exhibited lower load-bearing capacity and stiffness as compared to control specimens, when tested under two-point loading.

The structural integrity of these chemically deteriorated beams was restored by repairing one set of beams with epoxy resin and another with a polymer-based latex system. The repair process and materials used for rehabilitation proved to be effective with an increase of 13–15% in the ultimate load bearing capacity over the original values.

In another instance of sulfate attack of an industrial channel (Detwiler and Taylor 2005) it was observed that the drainage system transporting some of the sulfate-bearing groundwater draining from a mine dump had open rectangular channels (300 × 300 mm) made with 25 MPa concrete. Four years after the construction the channels showed distress; the concrete surface was soft and spongy and could be removed easily with a small hammer. The damage was apparent to a depth of about 10 mm, while the concrete below the damaged zone appeared sound. Analyses of the water showed that it contained approximately 3,000 mg/L sulfate and was acidic. The channels were repaired and lined with a polymeric material to prevent further damage.

The combined acid and sulfate attack on concrete due to the oxidation of iron disulfide in soil has been under in-situ long-term investigation (Siebert and Breitenbucher 2009). After about 18 months storage in iron disulfide soils, the concrete specimens, made with different binders, showed different levels of attack as visually observed. Concretes with high sulfate resistant cements with low C_3A content (viz., CEM 1-HS and CEM III/A) exhibited better durability than CEM-III cement. The concrete with Portland Limestone Cement suffered from intense formation of thaumasite. In soils with sulfate concentration of more than 20,000 mg/L the specimens were completely disintegrated after 18 months of storage.

In spite of considerable advances in recent years, the problem and remedy of sulfate attack are not fully understood. A meaningful review of this complexity is presented in Santhanam et al. (2001), with the following key issues:

(a) Since different sulfate bearing solutions produce different effects in concrete, any research should avoid generalization of results from studying just one type of sulfate solution for all conditions.
(b) The current test methods and standards for sulfate attack need to be re-examined. Issues such as the changing pH of the solution, drying and wetting as well as the effects of temperature and concentration of the sulfate solution need to be addressed.
(c) Estimating the remaining service life of structures exposed to sulfate attack is important in order to develop repair and maintenance schedules for such structures.
(d) As a first step towards modelling the sulfate attack phenomenon, a set of reliable experimental parameters needs to be determined, since the sulfate attack damage does not manifest itself in the same may all the time.
(e) The role of attack products such as gypsum and thaumasite needs to be quantified.
(f) The mitigation approach for magnesium sulfate attack requires a detailed study.

It should also be borne in mind that internal expansion of concrete could occur even as a result of dehydration and rehydration of salts such as sodium sulfate or magnesium sulfate, where the SO_4 ion does not play any active part. For example, the conversion of thenardite (Na_2SO_4) to mirabilite ($Na_2SO_4{\cdot}10H_2O$) is accompanied by volume expansion which can cause cracking in concrete. Although a reaction between cement and sulfate ions is not responsible for the damage in this case, there is still a chemical effect (Santhanam 2007) occurring. Hence it is important to distinguish the causes of damage involving sulfate phases for proper repair or design.

Another recent experimental finding in the context of sulfate environment may be of some practical significance. The experimental data suggest that the presence of carbonate in sulfate environments lead to carbonation of the cement paste which reduces the expansion initially. It appears that no gypsum is formed in presence of carbonates and instead calcite is seen. Ettringite, however, seems to form both in the presence and absence of carbonates (Kunther et al. 2009). Further investigations or confirmations of such findings may help in dealing with complex environments containing both carbonates and sulfates.

Other chapters in this Report (specifically in Part I) deal with the mechanistic aspects of sulfate attack of concrete, and should be consulted.

16.2.8 Concrete Exposed to Organic Acids

Field studies on the deterioration of cementitious matrices by the organic acids encountered in food processing, animal breeding and agricultural constructions

Table 16.5 Percentage of farms showing concrete degradation with age (De Belie 1997a)

Floor age (year)	Slat degradation (%)	Solid floor degradation (%)	Slat replacement (%)
2	15	–	–
5	40	16	1
15	87	59	16
25	–	–	49

have not been reported very extensively (although additional data are now available in Part I of this Report). Although some valuable information and data are available on the nature of attack by such acids as acetic, propionic, butyric, lactic, etc., there is little published information on the alteration mechanisms by citric, oxalic, malic, succinic and tartaric acids. It has been reported (Bertron and Duchesne 2007) that the degradation at pH between 2.5 and 6.0 in the presence of some of these organic acids results in a well-marked zonation of the cement paste. The degraded zone, which is very porous with low mechanical strength, is totally decalcified with the formation of ferro-alumino-siliceous gel. When using normal Portland Cement in such environments, one may have to consider mainly the reactions of calcium hydroxide with such weak acids as lactic and acetic acids. It is possible that smaller quantities of volatile fatty acids as well as aggressive ions like NH_4^-, Mg^{2+}, Cl^- and SO_4^{2-} are also present in the environment. Also, there are mechanical impacts of animals as well as of high-pressure cleaning, which may lead to exposure of aggregates and deterioration of the surface.

In a study reported in De Belie et al. (1996), quantitative data regarding the concentration of acids and other constituents in different sections of a piggery are presented. Lactic and acetic acids were present in rather high concentrations up to 22 and 15 mg/g of fresh samples, respectively. The lactic acid originated from the acidified meal suspension, while the acetic acid originated from the manure. Other aggressive ions were also substantial. Obviously in such an environment special considerations had to be given to concrete mix design and quality.

In a survey of farm buildings in East and West Flanders, the progress of concrete degradation with age was reported (Table 16.5) (De Belie 1997a).

From the above it is evident that the rate of deterioration of solid floors was slower than that of slats. In the slats the nature of deterioration was observed to be the crumbling of edges and consequent enlargement of gap width in the attacked floor. Further studies using a full scale on-farm trial with concrete floor slats made of seven different concrete compositions has been reported (De Belie 1997b). The compositions included a reference ordinary Portland Cement Concrete, four concretes with different cements, and two concretes with fly ash/silica fume addition. Of the slats made with reference concrete, some were treated with an epoxy resin, fluorosilicate coating, siloxane and cement-based ground-tuff coating. After 9 months, concrete deterioration was observed in front of the wet feeders. Concrete treated with epoxy resin and cement-tuff coating showed relatively better performance than the reference or blended cement concretes.

Overall, the presence of aggressive components in acidifying meal/water mixtures and manures on the floor of animal rearing facilities can result in concrete corrosion. Lactic and acetic acids appear to be the most important agents in this process and require further studies.

16.2.9 Degradation of Concrete in Ammonium Nitrate Environments

Ammonium nitrate environments in the fertilizer industry cause significant damage to reinforced concrete structures, with deterioration of both the concrete as well as the reinforcement. Damage occurs in the form of cracking, spalling, delamination, thinning of cross-sections, failure of beams, etc.

The results of laboratory studies performed on a large number of concrete samples extracted from the corrosion-affected elements showed a wide range of concentration of NO_3^- (0.70–9.02%) and NH_4^+ (0.15–2.15%); the pH of aqueous suspension of damaged concrete was seen to be in the range of 5.5–8.5. The possibility of appearance of an expansive phase $3CaO{\cdot}Al_2O_3{\cdot}Ca(NO_3)_2{\cdot}10H_2O$ has been reported (Pepenar 2009). Decalcification, expansion, dealkalization, etc. lead to cracking, strength loss and often structural destruction.

16.3 Approaches for Durability Design of Concrete Structures in Aggressive Aqueous Environments

The fundamental aim of engineering design of concrete structures is to achieve an acceptable probability that the structure will perform satisfactorily during its intended service life. The service life should include a maintenance programme as part of the design, although this is seldom done at present. The method of design should rely on "limit state" of safety and serviceability before failure. The limit states of serviceability are often defined by deflection, cracking, and certain requirements of durability. The achievement of durability in a concrete structure depends on the following measures:

- the appropriate selection of the environment exposure class etc.
- suitable design and detailing appropriate to exposure class
- an appropriate specification for both the concrete and the execution of concreting work
- the supply of fresh concrete according to the specification
- correct workmanship.

Failure to adopt the above measures may result in lower levels of durability.

16.3.1 Specifying Durability of Concrete

Durability is the ability of concrete to resist weathering action, chemical attack and abrasion while maintaining its desired engineering properties. Different concretes require different degrees of durability depending on their environmental exposure conditions and the properties desired.

Often the strength parameter is mistakenly used for durability. Although some of the measures required for durability may result in high strength, the two should not be taken as equivalent. It is important to distinguish between strength and durability and to specify them as directly as possible. In reality, structural engineers use numerical values for mechanical properties of concrete, but do not have the requisite information on how concrete interacts with its environment. Hence, in various codes, emphasis has been on defining the environmental exposure conditions and specifying the materials accordingly, that is a prescriptive approach.

Three main approaches may be conceived for the specification of durable concrete:

(a) general prescriptive approach
(b) deterioration-specific prescriptive approach
(c) performance-based durability-design and performance testing approach.

The method currently widely adopted is the first one. The second approach is a key feature of the introduction to practice of European Standard EN 206-1. It is projected to become the most acceptable methodology in the foreseeable future, despite its shortcomings. The third approach is expected to provide an alternative method which is potentially more reliable but is likely to be used in a minority of cases until research progresses further in the areas of deterioration modelling and developing performance-based specifications (Richardson 2002).

The third approach is radically different from the first two. Durability design involves consideration of each relevant deterioration mechanism and the expected service life of the structure in a quantitative way. Appropriate material parameters may then be determined based on an acceptable probability of failure.

However, for all the above approaches, there is a common link, i.e. the classification of the severity of the environmental exposure of concrete, which is mostly qualitative and widely different from country to country. The challenge therefore is to provide more quantitative measures of environmental aggressivity, and to harmonise these for different regions as far as possible.

16.3.2 Illustrations of Some Environmental Exposure Classifications

The definition of exposure environments is not easy due to both macro-and micro-climatic effects. Notwithstanding such phenomena, practising engineers have generally followed empirical severity classification of exposure conditions for materials prescription, a few examples of which are furnished below.

16.3.2.1 Environmental Exposure Conditions in the Indian Concrete Code

A comprehensive classification is provided in the Indian Code of Practice for Plain and Reinforced Concrete (IS 456 2000), reproduced in Table 16.6. The free water-cement ratio and the cement content are fixed appropriately to resist the exposure conditions. For Indian conditions, the values of the above parameters are given in Table 16.7, while Table 16.8 provides the recommendations for the type of cement, maximum free water-cement ratio and minimum cement content, which are required at different sulfate concentrations.

16.3.2.2 Environmental Exposure Conditions in the EN Code

In EN-206 (2002) the environmental exposure conditions have been quite elaborately classified. The major classes are based on the conditions such as carbonation, non-marine chloride attack, freeze-thaw effects and chemical attack. Each of the classes has been further sub-divided on the basis of severity of the attack as detailed in Table 16.9. In the EN classification it has been emphasized that when concrete containing reinforcement or other embedded metal is exposed to air and moisture, the moisture condition relates to that in the concrete cover adjacent to reinforcement or other embedded metal; in many cases, conditions in the concrete cover can be taken as reflecting that in the surrounding environment. In these cases classification of the surrounding environment may be adequate. This may not be the case if there is a barrier between the concrete and the environment, or where wetting cycles, for example, by short rainfall incidents, are very brief.

The classification for concrete exposed to chemical attack is for natural soils or groundwaters. Sea water classification depends on the geographical location. The EN standard specifically mentions that a special study may be needed to establish the relevant exposure conditions where there are situations outside the specified limits, aggressive chemicals other than specified, chemically polluted ground or water, or high water velocity in combination with the chemical specified. The EN standard distinguishes weak, medium and strong levels of environmental severity corresponding to pH values of 6.5–5.5, 5.5–4.5 and 4.5–4.0 respectively.

16.3.2.3 Eurocode 2 Design Guide

The environmental classification and durability design requirements for concrete are specified in Eurocode 2, with details in Tables 16.10 and 16.11. There are substantial differences in Tables 16.9 and 16.10. The Eurocode now has excluded

Table 16.6 Environmental exposure conditions in the Indian Concrete Code (IS 456 2000)

Sl. no.	Environmental exposure conditions	
I	Mild	Concrete surface protected against weather or aggressive conditions, except those situated in coastal area
II	Moderate	Concrete surfaces sheltered from severe rain or freezing whilst wet
		Concrete exposed to condensation and rain
		Concrete continuously under water
		Concrete in contact with or buried under non-aggressive soil/ground water
		Concrete surfaces sheltered from saturated salt air in coastal area
III	Severe	Concrete surfaces exposed to severe rain, alternate wetting and drying or occasional freezing whilst wet or severe condensation
		Concrete completely immersed in sea water
		Concrete exposed to coastal environment
IV	Very Severe	Concrete surfaces exposed to sea water spray, corrosive fumes or severe freezing conditions whilst wet
		Concrete in contact with or buried under aggressive sub-soil/ground water
V	Extreme	Surface of members in tidal zone
		Members in direct contact with liquid/solid aggressive chemicals

Table 16.7 Minimum cement content, maximum water-cement ratio and minimum grade of concrete for different exposures with normal weight aggregates of 20 mm nominal maximum size, Indian Concrete Code (IS 456 2000)

		Plain concrete			Reinforced concrete			
Sl. no.	Exposure	Min cem. content (kg/m^3)	Max free (w/c)	Min conc. grade	Min cem. content (kg/m^3)	Max free (w/c)	Min Conc. Grade	Min nom cover (mm)
I	Mild	220	0.60	–	300	0.55	M20	20
II	Moderate	240	0.60	M15	300	0.50	M25	30
III	Severe	250	0.50	M20	320	0.45	M30	45
IV	Very severe	260	0.45	M20	340	0.45	M35	50
V	Extreme	280	0.40	M25	360	0.40	M40	75

Notes

1. Cement content prescribed in Table 16.7 is irrespective of the grades of cement and it is inclusive of additions of cement replacement materials. The additions such as fly ash or ground granulated blast furnace slag may be taken into account in the concrete composition with respect to the cement content and water-cement ratio if the suitability is established and as long as the maximum amounts taken into account do not exceed the limit of pozzolana and slag specified in IS 1489 (Part I) and IS 455 respectively

2. Minimum grade for plain concrete under mild exposure condition is not specified

3. For a longitudinal bar in a column, nominal cover shall not be less than 40 mm, nor less than the diameter of such bar

4. For reinforcement bar up to 12 mm dia. for mild exposure, the nominal cover may be reduced by 5 mm

5. For severe and very severe exposure conditions, reduction of 5 mm may be made, where concrete grade is M35 and above. The actual concrete cover should not deviate from the required nominal cover by +10 mm

Table 16.8 Requirements for concrete exposed to sulfate attack, according to the sulfate concentration Indian Concrete Code (IS 456 2000)

		Concentration of sulfates, as SO_3			Dense, fully compacted conc., 20 mm nom max size aggregates (IS 383)		
		In soil					
Sl. no.	Class	Total SO_3 (%)	SO_3 in 2:1 water: Soil extr. (g/L)	In ground water (g/L)	Type of Cement	Min cem. content (kg/m^3)	Max free w/c
I	1	Traces (<0.2)	Less than 1.0	Less than 0.3	Ordinary Portland cement or Portland slag, cement or Portland pozzolana cement	280	0.55
II	2	0.2–0.5	1.0–1.9	0.3–1.2	Ordinary PC, or Portland- slag or – pozzolana cement	330	0.50
					Supersulfated cement or Sulfate res. Portland cement	310	0.50
III	3	0.5–1.0	1.9–3.1	1.2–2.5	Supersulfated cement or Sulfate res. Portland cement	330	0.50
					Portland pozzolana cement or Portland slag cement	350	0.45
IV	4	1.0–2.0	3.1–5.0	2.5–5.0	Supersulfated or sulfate resisting Portland cement	370	0.45
V	5	More than 2.0	More than 5.0	More than 5.0	Sulfate resisting Portland cement or supersulfated cement with protective coatings	400	0.40

Notes

(a) Cement content irrespective of cement grade. (b) Supersulfated cement generally restricted where prevailing temperature is above 40°˜C. (c) Supersulfated cement gives acceptable life if concrete is dense, with w/c of 0.4 or less, in mineral acids, down to pH 3.5. (d) Cement contents in column 7 are the minimum recommended. For SO_3 contents near upper limit of any class, cement contents above these minima are advised. (e) For severe conditions (e.g. thin sections under hydrostatic pressure on one side only, sections partly immersed), give consideration to a further reduction of w/c. (f) Portland slag cement conforming IS 455 with slag content >50% exhibits better sulfate resisting properties. (g) Where chloride is encountered with sulfate in soils or groundwater, Portland cement with C_3A content from 5 to 8% is desirable in concrete, instead of sulfate resisting cement. Alternatively, Portland slag cement conforming to IS 455 with >50% slag or a blend of Ordinary Portland cement and slag, provided sufficient information is available on performance of such blended cements in these conditions

Table 16.9 Exposure classification in EN 206 (2002)

Class designation	Description of the environment	Informative examples where exposure classes may occur
No risk of corrosion or attack		
X0	For concrete without reinforcement or embedded metal. All exposures except where there is freeze-thaw, abrasion or chemical attack	Concrete inside buildings with very low humidity
	For concrete with reinforcement or embedded metal. Very dry	
Corrosion induced by carbonation		
XC1	Dry or permanently wet	Concrete inside buildings with very low humidity
		Concrete permanently submerged in water
XC2	Wet, rarely dry	Concrete surfaces subject to long-term water contact. Many foundations
XC3	Moderate humidity	Concrete inside buildings with moderate or high humidity
		External concrete sheltered from rain
XC4	Cyclic wet and dry	Concrete surfaces subject to water contact, not within exposures of class XC2
Corrosion induced by chlorides other than sea water		
XD1	Moderate humidity	Concrete surface exposed to air-borne chlorides
XD2	Wet, rarely dry	Swimming pools. Concrete exposed to industrial waters containing chlorides
XD3	Cyclic wet and dry	Parts of bridges exposed to spray containing chlorides
Corrosion induced by chlorides in sea water		
XS1	Exposed to air-borne salt but not in direct contact with sea water	Structures near or on the coast
XS2	Permanently submerged	Parts of marine structures
XS3	Tidal, splash and spray zones	Parts of marine structures
Freeze-thaw attack		
XF1	Moderate water saturation, without de-icing agents	Vertical concrete surface exposed to rain and freezing
XF2	Moderate water saturation, with de-icing agents	Vertical concrete surface of road structures exposed to freezing and air-borne de-icing agents
XF3	High water saturation with de-icing agents	Horizontal concrete surfaces exposed to rain and freezing
XF4	High water saturation with de-icing agents or sea water	Road and bridge decks exposed to deicing agents
		Concrete surface exposed to direct spray containing de-icing agents and freezing. Splash zones of marine structures exposed to freezing.
Chemical attack		
XA1	Slightly aggressive chemical environment	–

(continued)

Table 16.9 (continued)

Class designation	Description of the environment	Informative examples where exposure classes may occur
XA2	Moderately aggressive chemical environment	–
XA3	Highly aggressive chemical environment	–

Table 16.10 Exposure classes related to environmental conditions in Eurocode 2 (2002)

Exposure class	Examples of environmental conditions	
1. Dry environment	Interior of building for normal habitation or offices	
2. Humid environment	(a)	Interior of buildings where humidity is high (e.g. laundries)
	Without frost	Components in non aggressive soil or water
	(b)	Exterior components exposed to frost
	With frost	Components in non aggressive soil and/or water and exposed to frost. Interior components when exposed to frost
3. Humid environment with frost and deicing salts	Interior and exterior components exposed to frost and deicing agents	
4. Sea water	(a)	Components completely or partially submerged in sea water or in the splash zones
	Without frost	Components in saturated salt air (coastal area)
	(b)	Components completely or partially submerged in sea water or in the splash zones and exposed to frost.
	With frost	Components in saturated salt air and exposed to frost
The following classes may occur alone or in combination with the above classes		
5. Aggressive chemical environment	(a)	Slightly aggressive chemical environment (gas, liquid or solid)
		Aggressive industrial atmosphere
	(b)	Moderately aggressive chemical environment (gas, liquid or solid)
	(c)	Highly aggressive chemical environment (gas, liquid or solid)

the sixth category of BS 8110 which referred to conditions of abrasive action of seawater and solids, flowing acid water, machinery or vehicles.

16.3.2.4 Exposure Conditions Stipulated by the American Concrete Institute

ACI 318 (2008) dealing with the building code requirements for structural concrete has been revised in 2008 to include the durability stipulations. The standard differentiates the following four classes of exposure:

Table 16.11 Durability design requirements for the exposure classes of Eurocode 2 (2002)

Exposure class	Nominal cover (mm)				
1	20	20	20	20	20
2a	–	35	35	30	30
2b	–	–	35	30	30
3	–	–	40	35	35
4a	–	–	40	35	35
4b	–	–	40	35	35
5a	–	–	35	30	30
5b	–	–	–	30	30
5c	–	–	–	–	45
Max. free w/c ratio	0.65	0.60	0.55	0.50	0.45
Min. cement content (kg/m^3)	260	280	300	300	300
Lowest concrete grade	C25/30	C30/37	C35/45	C40/50	C45/55

Notes
1. Strength class $C_{25/30}$, for example, refers to cylinder/cube strengths of 25 and 30 N/mm^2
2. Table 16.11 particularly provides the recommendations pertaining to the nominal concrete cover to all reinforcement, minimum cement contents and maximum water/cement ratios, which are also indirectly specified by the grades of concrete

- Class F: freezing and thawing
- Class S: sulfates
- Class C: corrosion
- Class P: requiring low permeability

The subclasses and some of the quantified parameters of the above classes are given in Table 16.12. For each subclass, the maximum water-cementitious materials ratio, minimum compressive strengths and other additional requirements are specified. It is expected that while designing reinforced concrete structures, the designer needs to select relevant exposure conditions for each component of the structure and select the one that requires the strongest durability parameters.

In prescribing the materials, (Detwiler and Taylor 2005) indicate that Type I cement with an appropriate addition of ground granulated blast furnace slag can achieve the required sulfate resistance in lieu of Type II or V cements, if the water-cementitious material ratios are maintained as in Table 16.12. For protection against seawater sulfate attack and steel corrosion, cements with C_3A contents between 4 and 10% have provided satisfactory service. (By way of comparison, in the Indian code of practice, the acceptable range of C_3A is recommended as 5–8% for similar applications).

16.3.2.5 Exposure Classes in the Canadian Standard

CSA A23.1, the concrete specifications of the Canadian Standards Association (rev. 2009), gives the following classes for environmental exposure of concrete:

- Class C: exposed to chlorides
- Class F: exposed to freezing and thawing without chlorides

Table 16.12 Exposure classes specified in ACI 318 (2008)

Exposure class	Subclasses	Max w/cm	Min f_c, MPa	Additional requirements
F (Freezing and thawing)	FO: Negligible; concrete not exposed to freezing and thawing	–	–	–
	F1: Moderate; occasional exposure, no deicing salt	0.45	31	Low entrained air
	F2: Severe; in continuous contact with moisture	0.45	31	Higher entrained air
	F3: Very severe; in continuous exposure to moisture and to deicing salts	0.45	31	Higher entrained air and limits exercised on cementitious materials
S (Sulfate)	SO: Negligible; SO_4 [0.10%(soil)] SO_4 <150 ppm (water)	–	–	–
	S1: Moderate; $0.10 \leq SO_4 \leq 0.20\%$(Soil) $150 \leq SO_4 \leq 1{,}500$ ppm (water)	0.50	31	Cement types: ASTM II, IP (MS), IS(<70)(MS); Maximum expansion (ASTM C 1012): 0.10% at 180 days
	S2: Severe; $0.20 \leq SO_4 \leq 2.00\%$ (soil) $1{,}500 \leq SO_4 \leq 10{,}000$ ppm (water)		31	Cement type: ASTM V; HS; no calcium chloride admixture. Maximum expansion (ASTM C 1012): 0.05% at 180 days or 0.10% at 365 days
	S3: Very severe; $SO_4 > 2.00\%$ (Soil) $SO_4 > 10{,}000$ ppm (Water)	0.45	31	Cement type: ASTM V + pozzolan or slag; no calcium chloride admixtures. Maximum expansion (ASTM C 1012): 0.10% at 545 days
C (Corrosion)	CO: Negligible; concrete dry and protected in service.	–	–	Chloride ion limit (water soluble chloride by % wt of cement): 1.00 in RCC 0.06 in prestressed concrete
	C1: Moderate; Concrete exposed to moisture but not to external source of chloride in service	–	–	Chloride ion limit (water soluble chloride by % wt of cement): 0.30 in RCC concrete and 0.06 in Prestressed concrete
	C2: Severe; Concrete exposed to moisture and an external source of chlorides	0.40	34.5	Chloride ion limit (water soluble chloride by % wt of cement): 0.15 in RCC and 0.06 in Prestressed concrete. Cover to be adequate.

(continued)

Table 16.12 (continued)

Exposure class	Subclasses	Max w/cm	Min f_c, MPa	Additional requirements
P (Permeability)	PO: Concrete where low permeability to water is not required	–	–	–
	P1: Concrete required to have low permeability to water	0.30	27.6	–

- Class N: exposed neither to chlorides nor to freezing and thawing
- Class A: exposed to severe manure and/or silage gases and liquids
- Class S: exposed to sulfates.

There are five subclasses in Class C, designated as C-XL, C-1, C-2, C-3 and C-4; two subclasses in 'F'; four subclasses in 'A'; and three subclasses in 'S'. Class N refers only to concrete not exposed to chlorides or freezing and thawing. The prescriptive requirements of different classes and subclasses are briefly summarized in Table 16.13. The Canadian Standard allows for a 56-day rather than the usual 28-day compressive strength, different curing regimes, and limits of chloride ion permeability (ASTM C1202), where appropriate. W/CM ratio should not be exceeded for a given class of exposure, regardless of exceeding the strength requirements. Although LH cements are for low heat, they are allowed for moderate sulfate resistance based on C_3A content. Guidance is also provided for type 'N' concrete for industrial trowelled floors exposed to wear, with minimum binder content of 265 kg/m^3. For different curing regimes, refer to the standard. All classes of concrete exposed to sulfates should comply with the minimum requirements of class 'S' in Table 16.13. For more details on severity of sulfates, cementing materials, and binder performance requirements, refer to the standard.

16.3.2.6 Exposure Classes in the Australian Standard

AS-3600 (2009) deals with environmental exposure based on geographical location. The country is divided into tropical, arid and temperate zones, with superstructures further differentiated on the basis of their distance from the coastline: "coastal" up to 1.0 km from the coast, "near coastal" from 1.0 to 50.0 km, and "inland" beyond 50 km. There are five main classes and 17 subclasses in the standard, some details of which are given in Table 16.14. Classes A1, A2, B1, B2 and C represent increasing severity of exposure, while class "U" indicates that its severity is not fully known and needs to be assessed before specifying the concrete.

The standard also specifies: characteristic comprehensive strength for different exposure conditions, minimum initial period of continuous curing, average compressive strength at the end of curing, cover thickness in relation to characteristic strength, use of standard or rigid formwork and normal or intense compaction, and minimum characteristic compressive strengths of different kinds of wear surfaces.

Table 16.13 Exposure classes and concrete design parameters extracted from CSA A 23.1 (2009)

Exposure class	Description or illustration	Max. W/CM	Min. f_c (MPa) and test age (days)	Air content (for 20 mm MSA) (%)	Curing schedule	Cement restrictions, if any	Chloride ion penetration value as per ASTM 1202 (Coulomb)
C-XL	Reinforced concrete exposed to chlorides or other severe environments with or without freeze-thaw conditions, requiring the highest durability performance	0.40	50 at 56 days	4–7 or 5–8 if exposed to freezing	Extended	–	<1,000 coulomb in 56 days
C-1	Reinforced concrete exposed to chlorides with or without freeze-thaw conditions. E.g. bridge decks, parking decks, marine structures, salt water pools, etc.	0.40	35 at 28 days	4–7 or 5–8 if exposed to freezing	Additional	–	<1,500 coulombs in 56 days
A-1	Reinforced concrete exposed to severe manure/silage gases, industrial effluents, H_2S etc.						
C-2	Plain concrete exposed to chlorides and freeze-thaw. E.g. garage floors, porches, pavements, gutters, etc.	0.45	32 at 28 days	5–8	Additional	–	–
A-2	Reinf. conc. exp. to moderate to severe manure, silage gases, & liquids with/without freeze thaw. E.g. reinf. walls in exterior manure tanks silos, feed bunkers etc.						
C-3	Continuously submerged concrete exposed to chlorides but not to freeze-thaw. E.g. underwater portions of marine structures.	0.50	30 at 28 days	4–7	Basic	–	–

A-3	Reinf. Conc. exposed to moderate or severe manure, silage gases with/ without freeze- thaw. Conc. continuously submerged in municipal or indus. effluents						
C-4	Non-structurally reinforced concrete exposed to chlorides but not to freeze-thaw. E.g. underground parking slabs on grade	0.55	25 at 28 days	4–7	Basic	–	–
A-4	Non-structurally reinforced concrete exposed to moderate manure and/or silage gases and liquids without freeze-thaw. E.g. interior slabs on grade.	0.55	25 at 28 days	4–7	Basic	–	–
F-1	Concrete exposed to freeze-thaw in saturated conditions but not to chlorides. E.g. pool decks, patios, fresh water pools, etc.	0.50	30 at 28 days	5–8	Additional	–	–
F-2	Concrete in an unsaturated condition exposed to freeze-thaw but not to chlorides. E.g. extension walls and columns.	0.55	25 at 28 days	4–7	Basic	–	–
S-1	Concrete subjected to very severe sulfate exposure	0.40	35 at 56 days	4–5	Additional	HS or HSb	–
S-2	Concrete subjected to severe sulfate exposure	0.45	32 at 56 days	4–7	Basic	HS or HSb	–
S-3	Concrete subjected to moderate sulfate exposure	0.50	30 at 56 days	4–7	Basic	MS or MS6	
N	Concrete not exposed to chlorides and freeze- thaw. E.g. : footing and interior slabs, etc.	As required for structural design		Not specified	Basic	–	–

Table 16.14 Exposure classes specified in AS 3600 (2009)

		Severity classification				
		Reinforced/prestressed concrete				Plain conc.
Sl. no.	Exposure environment	Class	Min. fc, MPa	Curing days	Strength on curing, MPa	Class
1.	In contact with ground					
	(a) Members protected with damp-proof membrane	A-1	>20	3	$\nless 15$	A-1
	(b) Residential footings in non-aggressive soils	A-1	>20	3	$\nless 15$	A-1
	(c) Other members in non-aggressive soils	A-2	$\nless 25$	3	$\nless 15$	A-1
	(d) Members in aggressive soils	U	–	–	–	U
2.	In interior environment					
	(a) Fully enclosed, except during construction	A-1	>20	3	$\nless 15$	A-1
	(b) In industrial buildings the member being subject to repeated wetting and drying	B-1	$\nless 32$	7	$\nless 20$	A-1
3.	In above-ground exterior environments in areas					
	(a) Inland (>50 km from coastline), environment being					
	(i) non-industrial and arid climate zone	A-1	>20	3	$\nless 15$	A-1
	(ii) non-industrial and temperate climate zone	A-2	$\nless 25$	3	$\nless 15$	A-1
	(iii) non-industrial and tropical climate zone	B-1	$\nless 32$	7	$\nless 20$	A-1
	(iv) industrial and any climatic zone	B-1	$\nless 32$	7	$\nless 20$	A-1
	(b) Near-coastal (1–50 km from the coastline and any climatic zone)	B-1	$\nless 32$	7	$\nless 20$	A-1
	(c) Coastal (up to 1 km from coastline but excluding tidal and splash zone and any climatic zone)	B-2	$\nless 40$	7	$\nless 25$	A-1

(continued)

Table 16.14 (continued)

		Severity classification				
		Reinforced/prestressed concrete				Plain conc.
Sl. no.	Exposure environment	Class	Min. fc, MPa	Curing days	Strength on curing, MPa	Class
4.	In water					
	(a) fresh water	B-1	≮32	7	≮20	A-1
	(b) sea water					
	(i) permanently submerged	B-2	≮40	7	≮25	U
			≮50	7	≮32	U
	(ii) in tidal and splash zones	C	Specifically def. requirements	As specified	As indicated	U
	(iii) soft or running water	U				
5.	In other environments					
	Any exposure environment not otherwise described in items 1–4	U	Specifically def. requirements	As specified	As indicated	U

16.3.2.7 Environmental Classification of Bridge Sites in the State of Florida in the US

Because of the nearness to the sea the environmental classification for both the superstructure and substructure of the bridges has been drawn up with some rigour and using multiple parameters. The details are furnished in Fig. 16.7a, b (Structural Design Guidelines 2005).

All superstructures that are within line-of-sight and also within 2,500 feet (ca. 750 m) of the Atlantic Ocean or the Gulf of Mexico are subjected to increased chloride intrusion rates. The intrusion rate decreases with distance from open waters or when obstacles such as rising terrain, foliage or buildings alter the wind patterns. In order to optimize the materials selection process, the designer and/or the District Materials Engineer have the option of obtaining representative cores to determine chloride intrusion rates for any superstructure within about 750 m of any major body of water containing more than 11,200 ppm chlorides. The chloride intrusion rates in excess of 0.0016 lbs/cy/year (ca 9.6 g/m^3/year) observed in the representative samples taken and tested classify the environment as "extremely aggressive".

16.3.2.8 Environmental Exposure Conditions in the Gulf Region

The environmental severity of the Arabian Gulf Region demands consideration for a special design code to make concrete more durable (Haque et al. 2005, 2007). A climatic divide was identified in the Arabian Peninsula, viz. hot-and-dry, and hot-and–humid zones, which were further classified as shown in Table 16.15. The Gulf Region is not influenced by a single environmental pattern, being extreme and humid near the coast and less humid in the inland areas. This accounts for the accelerated progress of chlorides, sulfates and carbonation in structures near the coast while the deterioration mechanism inland is different.

16.3.3 Multi-parametric Quantified Environmental Severity Classifications

16.3.3.1 Multi-ionic Severity Classification in Dutch Standard

In one of the earlier versions of the Dutch Standard (NEN 5996 1988) the aggressive environment was subdivided into four different categories, based on both pH of the aqueous solution and other ionic concentrations as shown in Table 16.16 (NEN 5996 – 1988).

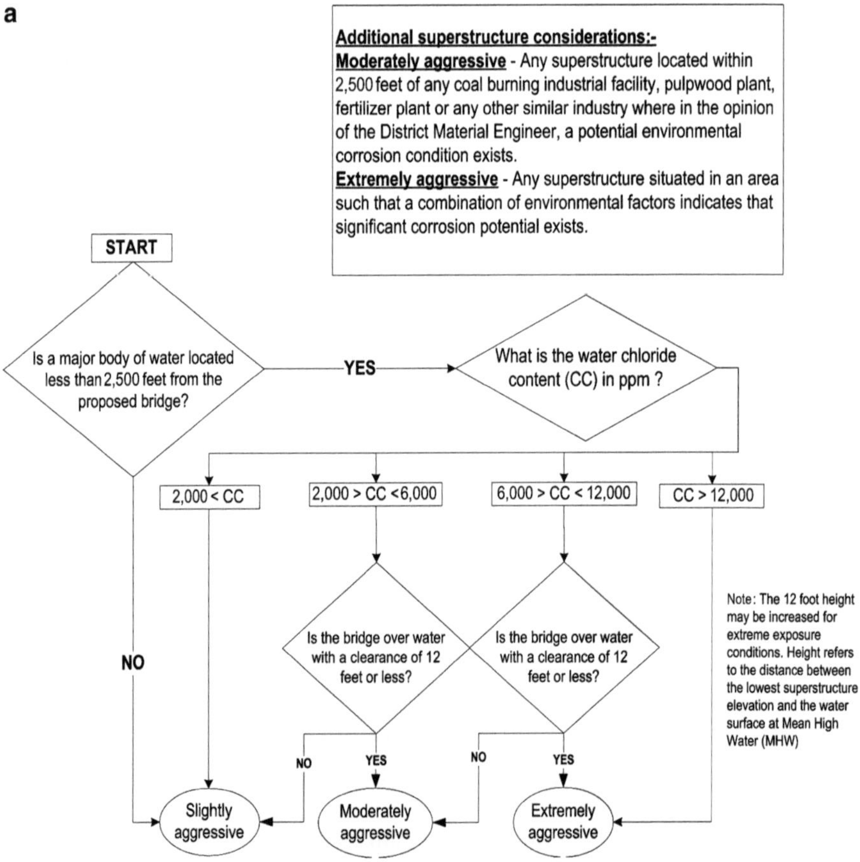

Fig. 16.7 (**a**) Environmental classification for bridge superstructures – State of Florida (Structural Design Guidelines 2005), (**b**) Environmental classification for bridge substructure – State of Florida (Structural Design Guidelines 2005)

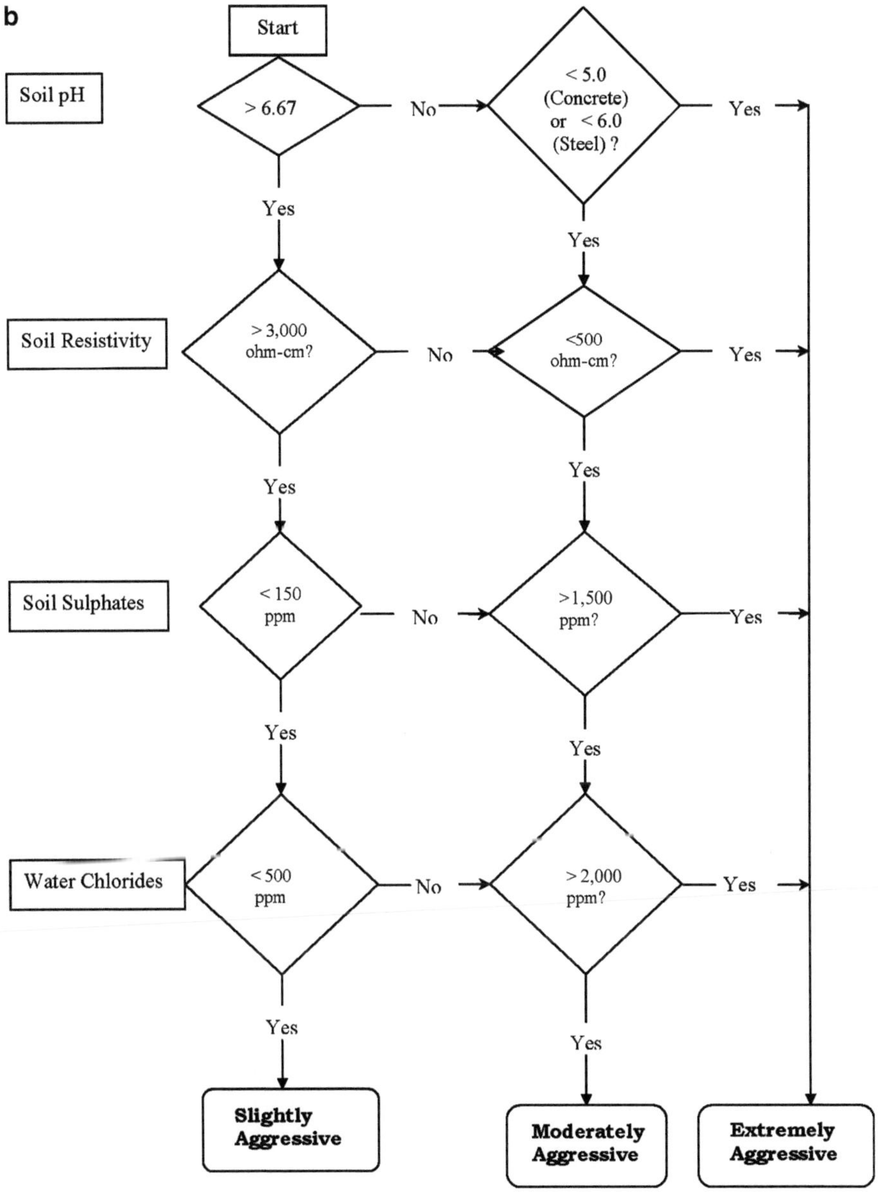

Fig. 16.7 (continued)

16.3.3.2 Quantitative Guidelines in EN 206

Similar quantitative guidelines are also available to interpret some of the exposure classes of EN 206 in a quantitative manner as shown in Tables 16.17 and 16.18 (CS 109 1996).

Table 16.15 Exposure classification based on distance from the sea (Haque et al. 2005, 2007)

Exposure	Distance from sea	Subdivision	Description of attack
Gulf Marine (GM) Zone	0–100 m within the shore	GM1 Spray	Active corrosion due to aerosols and salts
		GM2 Splash/tidal	Acute chloride-induced corrosion due to sea waves and current abrasion
		GM3 Submerged	Minimum corrosion risk
			Chloride and sulfate decomposition
			Biological attack
Gulf Coastal (GC) Zone	100 m from the shore up to 10 km		Dampness on structures attracting salts and fungal growth
			Chloride build-up from salt spray, soils and ground water
			Carbonation due to high R. H. (55–75%)
			Sulfate-rich coastal soils induce sulfate attack
Gulf Inland (GI) Zone	10–15 km	GI_A Within capillary rise zone (i.e. 3 m or less above water table)	Attack due to sulfates and chlorides present in soil and groundwater from either natural or industrial sources
		GI_B Above capillary rise zone	Deterioration due to salt-weathering/carbonation and/or dry winds carrying aggressive salts
Gulf Low-Risk (GL) Zone	50 km and above		Occurrence of contamination or attack is low

Table 16.16 Sub-classification of the aggressive environment (NEN 5996 1988)

		Ion concentration (mg/L)		
Severity classes	pH	NH^{4+}	Mg^{2+}	SO_4^{2-}
5a	>6.5	15–30	100–500	200–600
5b	6.5–5.5	30–60	300–500	600–2,000
5c	5.5–4.5	60–100	1,500–3,000	3,000–6,000
5d	4.5–4.0	>100	>3,000	>6,000

Table 16.17 Proposed values of relative humidities to differentiate exposure classes under category XC in EN-206

Exposure class and description	Range of ave. R.H. (%)	UK examples
XC1: Dry low risk of corrosion	≤ 65	Indoor, air-conditioned
XC2: Wet, rarely dry	91–100	Some buried foundations
XC3: Moderate humidity	66–74	Some residential buildings
XC4: Cyclic wet and dry	75–90	Outdoor

Table 16.18 Limiting values of the aggressiveness of chemical species between 5 and 25°C and moderate water velocity in defining chemical exposure severity in EN-206

Aggressive ions	Exposure XA1	Exposure XA2	Exposure XA3
SO_4 in water, mg/L	≥200 and ≤600	>600 and ≤3,000	>3,000 and ≤6,000
Total SO_4 in soil, mg/kg	≥2,000 and ≤3,000	>3,000 and ≤12,000	>12,000 and ≤24,000
pH	≤6.5 and ≥5.5	<5.5 and ≥4.5	<4.5 and ≥4.0
CO_2, mg/L (aggressive in water)	≥15 and ≤40	>40 and <100	>100
NH_4^+, mg/L in water.	≥15 and ≤30	>30 and ≤60	>60 and ≤100
Mg^{2+}, mg/L in water	≥300 and ≤1,000	>1,000 and ≤3,000	>3,000

Table 16.19 Definition of the aggressive classes for H_2S gas in moist environment (>75% in presence of oxygen) (Saucier and Lamberet 2009)

		Severity classes in NF EN 206-1		
Aggressive agent	Test method	XA 1	XA 2	XA 3
SO_2 in air (mg/m^3)[a]	NF EN 14791	≥0.15 ≤0.5	>0.5 <10	>10 ≤200
H_2S in air (mg/m^3)[b]	ISO 19739	<0.1	≥0.1 ≤10	>10 ≤200

Notes
[a]If the aggressiveness of solutions, soils or gas exceeds the concentration of XA3, it is necessary to plan for an external protection (coating, lining) or internal protection (impregnation)
[b]The values are based on uniform concentration. If the concentrations are non-uniform, then the averages for XA2 and XA3 should lie in the ranges 0.1–7.0 and 7–25 mg/m^3 respectively

16.3.3.3 Inclusion of Biogenic Corrosion Exposure in Codes

Biogenic sewer corrosion has been designated indirectly as chemical attack, but the French Standard pr FD P-18-011 (Definition and Classification of Chemically Aggressive Environments – Recommendations for Concrete Mix Design) now includes H_2S exposure classes (Table 16.19) (Saucier and Lamberet 2009).

While dealing with the above exposure classifications it is difficult to ignore that in actual field conditions concrete may be exposed to a combination of the classes. The concrete must be able to resist all the identified exposure conditions. Where the effects of the environmental actions interact with one another, this aspect also needs to be reflected in the performance requirements of the concrete. This is the case with sewers and other structures containing or conveying water, where the inner and outer faces may be exposed to very different exposure conditions.

16.4 Design of Concrete to Resist Aggressive Aqueous Environments

The approaches to specifying and designing concrete to withstand a given aggressive environment involve the consideration of the exposure severity, the penetrability of concrete and the choice of cement type. A typical example drawn from the

Indian Code of Practice is given in Tables 16.7 and 16.8. While the prescriptions given in Table 16.7 are all-encompassing, those in Table 16.8 are specific to sulfate attack. An essential aspect of producing durable concrete in general, and sulfate-resisting concrete in particular, is the achievement of low permeability concrete. With increased aggressiveness of the environment, concretes with w/c ratios of 0.40 or less are particularly beneficial. Considerable attention must also be paid to the selection of cement types which include low C_3A sulphate resisting cement, Portland slag cement with high slag content, and Portland Pozzolana cement with high fly ash content.

16.4.1 Criticality of Concrete Cover

Failure to achieve the specified cover is probably the greatest single factor influencing the premature corrosion of reinforcement and all other forms of reinforced concrete degradation. As the protective capacity of a given concrete is broadly related to the square of the cover, performance can be highly sensitive to deficiency in cover. Measures to control cover may secure more positive benefits than the pursuit of any other controlling parameter. The required cover will depend upon the requirements in relation to aggregate size, bond, surface crack width, fire resistance or durability.

Any standard design is based on the nominal cover, which is shown on the drawings and is assumed to be the mean cover. Durability design should be based on characteristic minimum cover which will depend upon the exposure conditions and the durability properties of the concrete selected for the purpose. The accuracy of placement and the choice of the allowance depend primarily on how good the site practices are. The Eurocode recommends that for the in-situ building work an allowance of 5–10 mm should be considered, while for precast work the allowance should be 0–5 mm.

The Swiss Standard SIA 262:2003 on "Concrete Construction" presents a newer approach for ensuring an appropriate service life of concrete structures, based on specifying and controlling the permeability of the cover concrete on the finished structure (Torrent and Jacobs 2009). The complementary Swiss Standard SIA 262/1 provides a non-destructive test method to determine the air-permeability of concrete constructions on site (Romer 2005/2009). The test provides the coefficient of permeability to air of the cover concrete (kT), based on which a classification has been proposed as given in Table 16.20 (Torrent and Jacobs 2009).

Further work is being carried out to correlate the exposure class, cover thickness, type of steel and mode of compliance (Torrent and Jacobs 2009).

Table 16.20 Classification of the permeability of the cover concrete (Torrent and Jacobs 2009)

Class	kT (10^{-16} m^2)	Permeability level
PK 1	<0.01	Very low
PK 2	0.01–0.10	Low
PK 3	0.10–1.0	Moderate
PK 4	1.0–10.0	High
PK 5	>10.0	Very high

16.4.2 Design of Concrete in Aggressive Ground

Detailed design advice exists for concrete exposed to water soluble sulfates and acidic conditions in BS 5328: 1982 and BRE Special Digest 1: 2001. Some illustrations of these guidelines are given in Tables 16.21 16.22 and 16.23 which are reproduced from McArthur and Spalding 2004.

Firstly the site is classified on the basis of the sulfate level [Design Sulfate (DS) Class] and Aggressive Chemical Level [Aggressive Chemical Environment for Concrete (ACEC) Class] of the site in accordance with Table 16.21.

Secondly, the Structural Performance Level (SPL) of the concrete is specified according to the following criteria:

- Low service life <30 years, unreinforced concrete, temporary structures, non-critical structural details, long service life but with low stress levels.
- Medium service life 30–100 years, not falling in low or high Categories.
- High service life >100 years, vulnerable critical details, structures retaining hazardous materials.

Thirdly, suitable types of cement and aggregate and the type and quality of concrete for use below ground level in the aggressive chemical environment are identified. Concrete quality is specified in terms of a Design Chemical Class (DC) corresponding to each ACEC Class and use of concrete below ground (Table 16.22). Each DC class defines concrete quality in terms of minimum cement content and maximum free w/c ratio for each cement group (Table 16.23) and allows for any aggregate carbonate content by additionally specifying concrete in terms of Aggregate Carbonate Range (ACR). The ACR is expressed as calcium carbonate equivalent as a percentage of the total aggregate mass; for carbonate in both the coarse and fine aggregate fraction. Range A is 10–30 to 100%, Range B is 2–12 to 10–30% and Range C is up to 2–12%. The ACR has been introduced to combat the thaumasite form of sulfate attack. Relaxation or increases to the concrete quality requirements, as incorporated in the DC class, are allowed according to the element thickness and the exposure conditions. In some circumstances, Additional Protective Measures (APMs) are recommended to further protect concrete (Table 16.22 These include APM1 (enhanced concrete quality)); APM2 (use of controlled permeability formwork); APM3 (provide surface protection); APM4 (provide sacrificial layer); APM5 (address drainage of the site).

Table 16.21 Aggressive chemical environment for concrete (ACEC) site classification

	Sulfate and magnesium					Natural soil		Brownfield[c]		
	2:1 water/soil extract		Ground water		Total potential sulfate[b]	Static water	Mobile water	Static water	Mobile water	
DS Class	SO_4 (g/l)	Mg (g/l)	SO_4 (g/l)	Mg (g/l)	SO_4 (%)	pH	pH	pH[d]	pH[d]	ACEC Class[e]
						All pH values		All pH values		AC-1s
DS-1							>5.5		>6.5	AC-1
	<1.2		<0.4		<0.24				5.5–6.5	AC 2z
							≤5.5		4.5–5.5	AC-3z
									<4.5	AC-4z
DS-2						>3.5		>5.5		AC-1s
							>5.5		>6.5	AC-2
	1.2–2.3		0.4–1.4		0.24–0.6	<3.5		<5.5		AC-2s
									5.5–6.5	AC-3z
							≤5.5		4.5–5.5	AC-4z
									<0.5	AC5z
DS-3						>3.5		>5.5		AC-2s
							>5.5		>6.5	AC-3
	2.4–3.7		1.5–3.0		0.7–1.2	≤3.5		≤5.5		AC-3s
							≤5.5		5.5–6.5	AC-4
									<5.5	AC-5
DS-4						>3.5		>5.5		AC-3s
							>5.5		>6.5	AC-4
	3.8–6.7	≤1.2	3.1–6.0	≤1.0	1.3–2.4	≤3.5		≤5.5		AC-4s
							≤5.5		≤6.5	AC-5
DS-4m						>3.5		>5.5		AC-3s
							>5.5		>6.5	AC-4m
	3.8–6.7	>1.2[a]	3.1–6.0	>1.0[a]	1.3–2.4	≤3.5		≤5.5		AC-4ms
							≤5.5		≤6.5	AC-5m

DS-5						>3.5		>5.5		AC-4s
	>6.7	≤1.2	>6.0	≤1.0	>2.4	≤3.5	all pH values	≤5.5	all pH values	AC-5
DS-5m	>6.7	>1.2[a]	>6.0	>1.0[a]	>2.4	>3.5		>5.5		AC-4ms
						≤3.5	all pH values	≤5.5	all pH values	AC-5m

Notes

[a]The limit of water soluble magnesium does not apply to brackish groundwater (chloride content between 12 and 18 g/L), this allows suffix m to be omitted from the relevant ACEC classification

[b]Applies only to sites where concrete will be exposed to sulfate ions (SO_4) which may result from the oxidation of sulfides, such as pyrite, following ground disturbance

[c]'Brownfield' is defined as sites which may contain chemical wastes remaining from previous industrial use or from imported wastes

[d]An additional account is taken of hydrochloric and nitric acids by adjustment to sulfate content

[e]Suffixs indicates that, as water has been classified as static, no Additional Progressive Measures (APMs) are generally necessary. Concrete placed in ACEC Classes which include the suffix x have primarily to resist acid conditions and may be made with cements from any of the Cement Groups (refer to the BS standard), subject to the restrictions concerning Portland limestone cement. Suffix m relates to the higher levels of magnesium in Sulfate Classes 4 and 5

Table 16.22 Concrete quality and number of additional protective measures (APMs) recommended for the general use of *in-situ* and precast concrete

	Design Chemical (DC) Class/Number of Additional Protection Measures (APM's)[a,b]							
	Low structural performance level			Normal structural performance level			High struct. perf. level[c]	
	Section thickness							
ACEC Class	<140 mm	140–450 mm	>450 mm[d]	<140 mm	140–450 mm	>450 mm[d]	140–450 mm	>450 mm[d]
AC-1	DC-2/0	DC-1/0	DC-1/0	DC-2/0	DC-1/0	DC-1/0	DC-1/0	DC-1/0
AC-1s	DC-2/0	DC-1/0	DC-1/0	DC-1/0	DC-1/0	DC-1/0	DC-1/0	DC-1/0
AC-2	DC-3/0	DC-2/0	DC-1/0	DC-3/0	DC-2/0	DC-1/0	DC-2/0	DC-1/0
AC-2s	DC-3/0	DC-2/0	DC-1/0	DC-3/0	DC-2/0	DC-1/0	DC-3/0	DC-1/0
AC-2z	DC-3z/0	DC-2z/0	DC-1/0	DC-3z/0	DC-2z/0	DC-1	DC-2z/0	DC-1/0
AC-3	DC-3/2[e]	DC-3/1[e]	DC-2/1	DC-3/3[e]	DC-3/2[e]	DC-2/2	DC-3/3[e]	DC-2/3
AC-3s	DC-4/0	DC-3/0	DC-2/0	DC-4/0	DC-3/0	DC-2/0	DC-3/0	DC-2/0
AC-3z	DC-4z/0	DC-3z/0	DC-2z/0	DC-4z/0	DC-3z/0	DC-2z/0	DC-3z/0[e]	DC-2z/0[e]
AC-4	DC-4/2	DC-4/1[g]	DC-3/1[e]	DC-4/3[e]	DC-4/2[g]	DC-3/2[g]	DC-4/3	DC-3/3
AC-4s	DC-4/0[e,f]	DC-4/0	DC-3/0	DC-4/0	DC-4/0	DC-3/0	DC-4/0	DC-3/0
AC-4z	DC-4z/l[g]	DC-4z/0	DC-3z/0	DC-4z/1[g]	DC-4z/0l	DC-3z/0	DC-4z/0	DC-3z/0
AC-4m	DC-4m/2[e,f]	DC-4m/1[e]	DC-4m/0	DC-4m/3[e]	DC-4m/2[g]	DC-4m/1[e]	DC-4m/3[e]	DC-4m/2[e]
AC-4ms	DC-4m/0	DC-4m/0	DC-3/0	DC-4m/0	DC-4m/0	DC-3/0	DC-4m/0	DC-3/0
AC-5	DC-4/1[g]	DC-4/1[g]	DC-3/1[g]	DC-4/3[e,b]	DC-4/2[e,b]	DC-3/2[e,b]	DC-4**/1[g]	DC-3**/1[g]
AC-5m	DC-4m/1[g]	DC-4m/1[g]	DC-4m/1[g]	DC-4m/3[e,b]	DC-4m/2[e,b]	DC-4m/1[g]	DC-4m**/1[g]	DC-4m**/1[g]
AC-5z	DC-4z/1[g]	DC-4z/1[g]	DC-3z/1[g]	DC-4z/1[g]	DC-4z/1[g]	DC-3z/1[g]	DC-4z/1[g]	DC-4z/1[g]

Notes

[a]Where carbonation of the concrete prior to exposure can be assured, the recommended DC Class/APM (other than those carrying a z suffix) may be relaxed by 1DC Class or 1 APM, provided that for AC-5 and AC-5m conditions the resulting package includes APM3

[b]Where the hydrostatic head of groundwater is greater than 5 times the section thickness, one APM over and above the number indicated in the Table should be applied (preferably APM5 but, if not practicable, (APM1 or APM3))

[c]For High SPL, a section thickness of <140 mm is generally not recommended in aggressive ground conditions (although specific precast units are an exception)

[d]Use of this column is inappropriate where the surface of the concrete is required to remain unaffected by aggressive ground (e.g. friction piles)

[e]Where DC-3, DC-4 or DC-4m is given, the number of APM may be reduced (provided the reduction does not override the recommendation to use APM3) by either one provided DC-3*, DC-4* or DC-4m* concrete is specified, or by two provided DC-3**, DC-4** or DC-4m** concrete is specified

[f]If APM3 is selected for low SPL concrete, no further APM are necessary

[g]Only APM3 is recommended (not applicable to bored piles)

To include APM3, where practicable, as one of the recommended number of APM

Table 16.23 Concrete quality to resist chemical attack

			Dense fully compacted concrete made with agg. conforming to BS 882 or BS 1047	
Design Chemical (DC) Class	Aggregate carbonate range (ACR)	CEM cement or combination group[a]	Min cement cont. (kg/m^3)	Max free w/c ratio
DC-1	No restriction	1, 2, 3	–	–
DC-2	A[d], B, C	1[b]	340	0.50
	A[d], B, B	2, 3	300	0.55
DC-2z[c]	No restriction	1[b], 2, 3	300	0.55
	A	2z	400	0.40
DC-3	A	2b, 3	380	0.45
	B, C	2, 3	340	0.50
DC-3*f	B	2, 3	380	0.45
DC-3**[l]	C	2, 3	380	0.45
DC-3z[c]	No restriction	1[b] ,2, 3	340	0.50
	A	2s	400	0.35
DC-4	A	2b, 3	400	0.40
	BC	2, 3	380	0.45
DC-4*f	B	2, 3	400	0.40
DC-4**[g]	C	2, 3	400	0.40
DC-4z[c]	No restriction	1[b], 2, 3	380	0.45
DC-4m	A	2b, 3	400	0.40
	B, C	3	380	0.45
DC-4m*[f]	B	3	400	0.40
DC-4m**[g]	C	3	400	0.40

Notes

[a]For cement of combination groups refer to the B.S. standard

[b]Portland Limestone cement may be used only in concrete specified as DC-2, DC-2z, DC-3z, DC-4z where the DS class (Table 16.21) of the site does not exceed DS-1

[c]Classes DC-2z, DC-3z and DC-4z apply where chemical resistance is recommended primarily to resist acid attack

[d]In addition to the requirement for a minimum cement content and maximum free w/c ratio, a minimum concrete grade of C35 is recommended when using Range A aggregate combination

[e]Use of APM (enhanced concrete quality) can be satisfied by using the recommendations of the next numerically higher DC Class (where the starting point is starred DC Class, the step is to a similarly starred higher DC Class (e.g. DC-3 plus a 1 step enhancement in concrete quality leads to a specification of DC-4**))

[f]Using a single-starred DC Class may permit the recommended number of APMs to be reduced by (Table 16.22)

[g]Using a double-starred DC Class may permit the recommended number of APMs to be reduced by 2 (Table 16.22)

16.4.3 Durability Verification of Concrete Structures

In the Japanese construction industry certain guidelines are available to deal with the performance verification for the deterioration of structures on account of the processes involving various water-mediated aggressive agencies such as carbonation, chloride ingress, freeze-thaw action, chemical attack and alkali-aggregate reaction. Guidelines are also available to verify the water-tightness and fire-resistance of structures (JGC 2002). An illustration of the approach for water-tightness verification is given below.

For concrete structures such as storage facilities, underground structures, hydraulic structures, water tanks, sewage facilities, tunnels, etc., water-tightness is of great importance since the structural safety, durability, functionality, maintenance; appearance, etc. are greatly affected by the permeation of water. It is also known that leaching of calcium from concrete over a sustained period of time can impair the required performance of concrete structures.

It is suggested that the verification should not be carried out for the entire structure but on individual parts with some verification index for interrelating the values of individual parts. In the case of tank structures, the reduction rate of stored liquid can be treated as the verification index aimed for the whole structure. For example, the water level in the sewage digestion tank must vary within just 5 mm from the specified water level in 24 h. In another instance, permeation amount within 0.1% of water in water tank during circulation can be disregarded.

It should be borne in mind that the water-tightness of a concrete structure is also related to joints and cracks which should be taken care of through use of sealants.

The verification of water-tightness should be carried out by ensuring that

$$\gamma_i = \frac{Q_d}{Q_{max}} \leq 1.0 \quad (16.11)$$

Where,

γ_i is a constant, representing the importance of the structure. In general, it may be taken as 1.0 but may be increased to 1.1 for important ones.
Q_{max} is the allowable flow rate (m^3/s)
Q_d is the design flow rate (m^3/s) which is obtained from

$$Q_d = \gamma_{pn}(K_d.\ A.) = \frac{h}{L} + Q_{cjd} \quad (16.12)$$

Where

K_d is the design value of permeability coefficient of concrete in the structure (m/s) $= K_k \cdot \gamma_c$
K_k is the characteristic value of the permeability coefficient of concrete (m/s)
A is the total surface area of concrete through which water may penetrate (m^2)
h is the difference in the water levels between the interior and exterior surfaces of the structure (m)
L is the expected length of water permeation. (In general, L may be taken as the design thickness of a member)
Q_{cjd} is the design value of the flow rate from cracks or joints (m^3/s)
γ_m is the safety factor to account for the variation in Q_d. Normally it should be taken as 1.15
γ_c is the concrete material factor. In general it may be taken as 1.0.

As mentioned earlier, the above is just an approach for durability verification of concrete structures. It is known that the durability of concrete structures is affected by various causes which may act independently or simultaneously. However, in many cases it is sufficient to evaluate only the effect of the most prominent cause. It should be realized that suitable methodologies for performance verification involving simultaneous actions of multiple causes has not been developed sufficiently at present.

16.4.4 Concrete Performance Tests for Durability – Present Limitations

To the engineering community, concepts of performance testing suffer from divergent views and lack of clarity. For all practical purposes durability performance tests are understood as

- Either tests that directly assess the resistance to a standardized deterioration process, e.g. freeze-thaw actions, or carbonation process, or sulfate, acid or alkali attacks;
- Or tests that directly assess the material parameters such as 'cover' 'adiabatic temperature' etc. which, in turn, singly or jointly impart resistance to concrete to withstand environmental exposure conditions.

The most important condition is that the performance test(s) must represent actual and measurable performance criteria for the structure.

The first category of laboratory tests often suffer from limitations of extendibility to field conditions because of the standardized testing conditions which differ from field conditions, or due to their accelerated mode of testing which introduce significant changes from the field observations. The carbonation test is one such that is unsuitable for production control.

The second category of testing suffers from the limitations of realizing the laboratory generated recommendations or design values in the field. For example, failure to achieve the specified cover is probably the greatest single factor influencing the durability of concrete structures. The global finding is that the 5 mm allowance provided in quite a few codes of practices between the nominal cover and the minimum cover does not reflect the accuracy of placing reinforcement generally achieved in practice.

In the context of the present subject, some observations may be made on the following two test requirements:

(a) Leaching of cement reaction products;
(b) Sulfate/sea water attack

Although a separate Working Group of the present RILEM TC has been focusing on this aspect and have presented their findings and recommendations on this test requirement (see Part III), certain observations from the engineering perspectives are given below.

Table 16.24 Leaching test results after 56 days of the reference concrete (C_1) and test concrete (C_2)

Concrete specimens	Indicators	Lab 1	Lab 2	Lab 3	Lab 4	Mean	Standard deviation	Variation (%)
C_1	DDpH (mm)	0.5	1.0	1.4	1.3	1.0	0.4	34
	Ca leached (mol/m^2)	3.0	3.5	3.1	2.3	2.9	0.4	14
C_2	DDpH (mm)	0.5	0.5	0.6	0.7	0.6	0.1	15
	Ca leached (mol/m^2)	0.9	1.1	0.9	0.6	0.9	0.2	24

16.4.4.1 Leaching of Cement Reaction Products

In order to deal with the performance of cementitious materials in aggressive aqueous environments, it is highly relevant to evolve a test for leaching of cement reaction products by permeating aggressive (e.g. acidic) waters. Leaching should be a function of the concrete quality as well as of the cement type. But since the concrete permeability is likely to dominate leaching performance, any method of test should preferably be based on testing concrete.

A methodology to evaluate the durability of concrete exposed to acid attack and demineralised water as defined in the XA exposure classes of NF – EN 206-1 has been reported in Jacquemont et al. 2009. The test is based on immersing concrete specimens in an acid solution and keeping the pH constant by adding small amounts of nitric acid. In order to ensure the leachability of the system, the solution is periodically renewed. The test results are given in the form of the depth of damaged concrete (DDpH) corresponding to a pH gradient in the concrete pore solution as well as by the amount of leached calcium ion per unit area of concrete [Ca leached (mol/m^2)]. The leaching test results were subjected to reproducibility assessment through a round-robin test programme (see Table 16.24).

The leaching performance test was found to be fairly representative of the in-situ degradation of a 9-year old water tank exposed to comparable aqueous acid attack. Further work is necessary for adopting this test method as a standard.

16.4.4.2 Test for Sulfate and Seawater Attack

No direct rapid test procedure has yet been devised and fully tested for use on the structure or on concrete test specimens for assessing the durability of aggressive ground or water bearing sulfates. The recommendations for protection of concrete are based on many years of long-term testing as illustrated in this report. Prescriptive requirements for suitable concretes are based on type and content of cement and w/c ratio of the mix.

Comprehensive reviews are available in Santhanam et al. (2001), Bellman et al. (2006), Van Tittelboom and De Belie (2008), Hooton and Brown (2009). Most of the tests are based on mortar prism specimens stored in sodium sulfate or artificial sea water. The test duration mostly extends up to 1 year.

If, on the basis of storage in sodium sulfate, a suitable test method is agreed, a number of issues of concern will have to be addressed:

- the relationship between the resistance to sodium sulfate solution and the resistance to solutions of other sulfates and more particularly, magnesium sulfate.
- the need for the test to cover combined attack of sulfate and acid.
- the need to standardize an approach to distinguish the concrete requirements for sulfate resistance and sea water resistance.

This subject also receives attention elsewhere in this report (see Part III).

16.4.4.3 Tests for Acid Attack

Aggressive chemical environments other than sulfates or sea water generally involve different acids – both organic and inorganic. However, the frequency of attack by such acids is not high and, therefore, the field information and data are rather limited. Hence, to devise and standardize a test procedure that would have a realistic relationship with the field conditions will not be easy.

In the absence of standard tests, research-based specially designed test set-ups in expert laboratories may be utilized to arrive at prescriptive designs of concrete at this stage. Part III of this Report also gives further information on acid testing of concrete.

16.5 Explicit Durability Design – The Way Forward

A method of durability design that specifically considers each relevant deterioration process, the design life and the criteria that define the end of this design life in a quantitative way, has been called the explicit durability design (in CS 109 1996). The aim of explicit durability design is to have quantitative methods of analysis so that all options can be investigated on a rational basis. The overall understanding at present is that this stage has not yet been reached. When fully developed, these methods will apply to all types of exposure, concrete structures and elements. At this stage of evaluation, the objective of an explicit durability design method is to determine a concrete quality, the minimum cover and, where appropriate, the need for an additional protection system that will give a high probability of the structure achieving at least the required service life.

To start with, explicit durability design should be based on conservative assumptions. Monitoring of the structure during its life and analysis of data obtained could be used to predict residual service life. With time and experience, the assumptions used in explicit durability design can be refined and made less conservative.

The traditional design approach incorporating all-encompassing prescription of materials of construction was introduced at a time when the deterioration mechanisms were less well understood. Research during the last three decades or so has identified the dominant mechanisms and the key parameters controlling the rates of deterioration. Further work on the proposal and refinement of mathematical models of deterioration is continuing (Cement and Concrete Research 2000; Samson et al. 2000; Neithalath 2003; Baroghel-Bouny 2007; Jooss and Reinhardt 2002; Beddoe and Dorner 2005), including the contribution of a working group in this RILEM TC (Part III). The models for carbonation and chloride-induced corrosion of reinforcement are the most advanced to date but still require refinement. The other deterioration processes dealt with in this report, involving soft water, acids, salts and sea water, are yet to pass through the following stages of development:

- collation of adequate field data leading to specifications based on satisfactory experience
- evolution of performance-based test methods
- development of predictive models for practical applications.

It may be relevant to mention that the translation of models into design aids may involve incorporation of simplifying assumptions without diluting the integrity and rigour of the approach. It is often not possible to assume the worst values as the output could become unrealistic. Hence, there has to be a safety factor developed appropriately for explicit durability design.

Further, the use of statistical methods is highly appropriate in durability design because the fundamental requirement is to minimize the risk of failure. This includes the concept of risk analysis. Durability design can be achieved in the context of a defined probability of failure. Equally one may refer to a defined level of reliability. The approach is directly comparable with that used in design for structural resistance.

In this context the adoption of the probabilistic approach, which is commonly used in structural design, for determination of durability parameters may be appropriate.

16.5.1 Probabilistic Approach to Durability Design

It is understood today that the only way to minimize the risk of failure on account of durability is to adopt statistical probability methods. The approach is directly comparable with that used in design for structural resistance. However, the mathematical solutions to the problem of design for durability cannot be as straightforward as those adopted for structural design. Allowance for different variables and their distributions in durability design can sometimes lead to complex mathematical solutions. Thus, a number of approaches have been proposed. Although the terminology in the literature varies for similar approaches, they can be broadly described as "Lifetime Safety Factor Method", "Intended Service Period Design" and "Lifetime Design" (Richardson 2002).

The above methods do not have any radical difference and, broadly speaking, there are a few common elements of approach:

(a) introducing the concept of service life of the structure
(b) determining the design life based on the lifetime safety factor and the target service life
(c) applying the relevant degradation model or performance model, using the design service life and selecting the appropriate material properties, section sizes and/or the protective measures.

Based on the above inputs, if one finds great reduction in safety margin, one takes recourse to redesign, using materials of higher performance, larger sections or additional protective measures.

In the "Lifetime Safety Factor method", all the lifetime safety factors must be calibrated. In the "Intended Service Period Design" one might consider the probability of the chloride level at the steel of, say, 0.4% by weight of the binder during the target service life. In the "Lifetime Design" one considers the probability of the service life being less than the target service life for the condition where the chloride level is 0.4% by weight of the binder at the reinforcement.

The translation of the above approaches to practice is still a great challenge and involves the following (Richardson 2002):

- development of accepted models of deterioration and conversion of these models into design tools
- calibration of the models with experience of real structures;
- selection of favoured methods of probabilistic durability design
- definition of applicable limit states;
- determination of acceptable levels of failure probability.

Notwithstanding the abovementioned challenges facing the engineering community in adopting the probabilistic approach to durability design, a specific case study may be highlighted in this context. In South Africa durability indexes have been adopted as engineering measures of the potential resistance of concrete cover to the transport mechanisms of gaseous diffusion, water absorption and chloride diffusion. In a specific study (Muigai et al. 2009), a combination of a service life prediction (SLP) model and accelerated testing results from the chloride conductivity test (CCT) were applied in order to ensure the required durability of concrete elements in marine environment during their intended service life. The SLP model was based on modified Fick's second law of diffusion and included such parameters as the surface chloride concentration and critical chloride concentration derived from both field and laboratory tests, CCT values of the material, as well as the cover depth from field measurements of existing structures. The probability theory was applied to include uncertainties in the model parameters. The methodology was introduced for reliability design of a concrete pier to be cast in-situ in an extreme splash and tidal marine environment using CEM I:GGBS (50:50). The final outcome of this model study showed that for the above RC pier to have a 6.7% probability of corrosion initiation during its 100 year life time, the designer should specify a chloride conductivity value of not greater than 0.9 mS/cm and cover depth of 50 mm. It was observed that accounting for the uncertainty in the model parameters helped to refine the service life prediction.

16.6 Concluding Remarks

In the context of this report, consideration has been given to engineering aspects of the durability of concrete structures in aggressive aqueous environments. While the basic science is making steady progress, engineering aspects such as specification and design for durability are slower in coming. The following general concluding remarks are appropriate.

(a) In broad terms, durability of concrete is its ability to resist weathering action, chemical attack and abrasion while retaining its required engineering properties over its design life. During the service life of a concrete structure, degradation and deterioration may take place but the structure is considered safe for use as long as the performance requirements are satisfied. In this context the exposure assessment, design and specification of materials and the structure, and detailing and execution are important in the achievement of durability.

(b) Degradation and deterioration of concrete occur due to many factors and processes, the most widespread being carbonation and chloride induced rebar corrosion, which is outside the purview of the present report. This leaves consideration of agents such as soft water, acids, saline fluids and sea water, which affect the durability of concrete though simultaneous occurrence of aqueous phase ionic reactions, reactions involving ion exchange between the solution and the surface of the solid phase as well as heterogeneous reactions involving precipitation, dissolution, oxidation or reduction processes. Since all these reactions are primarily mediated through water, the transport of moisture in mortars and concrete under field conditions is immensely relevant and significant. Ultimately, performance of concrete depends on its environmental exposure conditions, transport of waterborne chemicals through its mass, and quality as designed, specified, supplied and used.

(c) The overall exposure environment for a concrete structure during its service life is variously classified in different national and other codes. Traditionally the classifications have been descriptive and qualitative. However, there has been awareness in recent times to provide quantitative guidelines to the provisions of exposure classifications as in EN 206. There is sufficient justification for harmonizing the guidelines, taking into account geo-climatic factors.

(d) Apart from the exposure conditions, the constructional domains and their relationships with aggressive aqueous agents should be given more extensive consideration. In fact, only on account of such focus, has there been a significant development in the materials system design for the durability of sewers. Similarly focus is necessary for other domains such as underwater concreting, marine construction, chemical plants, dams, roads and agri-farm sector.

(e) Specifying durable concrete for a given domain exposed to a given set of exposure conditions is presently carried out by adopting a prescriptive approach. The quantified well-defined exposure classifications under development may lead the engineering community to adopt a deterioration-specific prescriptive approach in the foreseeable future. But the ultimate objective is to move towards "explicit durability design", which is a method of durability design that specifically considers each relevant deterioration process, the design life, and the criteria that define the end of this design life in a quantitative way.

(f) Since the durability design approach allows consideration of the type of structure, material properties, microclimates, required life, quality of site practice and the probability of failure, it demands that the following prior steps are taken in order to develop the approach on a reliable basis:

- Collation of data on long term satisfactory experience with concrete specified for similar works in a particular environment
- Development of performance test methods for specified performance
- Use of predictive models for complex chemical and physical interactions occurring in a deterioration process
- Converting the models into workable design aids.

(g) In the context of this report, while there is progress in the testing and modelling of rebar corrosion processes, there is inadequate focus on tests concerning (a) leaching of cement reaction products, (b) sulfate and sea water attacks and (c) acid attacks. Of these three performance test requirements, the first two need serious and urgent attention for durability design purposes. The third area of acid attack can follow, as it is still in the research arena.

(h) The test procedures and compliance requirements must be correlated with satisfactory performance in practice. This should take into account the differences between standard test conditions and field situations. From this difference will emerge the safety factors to be adopted for durability design.

(i) Probability analysis is already implicit in routine design for structural resistance of load-bearing elements and it should be integrated with design for durability.

(j) Finally, it is important to adopt a quantitative approach for durability verification of concrete structures, following the Japanese practice, at least in respect of carbonation, chloride ingress, freeze-thaw process, chemical attack and alkali-aggregate reaction.

References

ACI 318: Building Code Requirements for Reinforced Concrete. American Concrete Institute, Farmington Hills (2008)

Alexander, M.G., Goyns, A.M.: Durability Considerations for Concrete Pipes. ICCX, Sun City (2008)

Alexander, M.G., Goyns, A., Fourie, C.: Experiences with full-scale experiment sewer made with CAC and other cementitious binders in Virginia, South Africa, Calcium Aluminate Cements. In: Proceedings of the Centenary Conference, pp. 279–292. IHS-BRE Press, Avignon, UK (2008)

AS 3600: Concrete Structures. Standards Australia International, Sydney (2009)

Baroghel-Bouny, V.: Water vapour sorption experiments on hardened cementitious materials, Part I: essential tool for analysis of hygral behaviour and its relation to pore structure. Cem. Concr. Res. **37**, 414–437; Part II: Essential tool for assessment of transport properties and for durability prediction. Cem. Concr. Res. **37**, 438–454 (2007)

Beaudoin, J.J., Catinaud, S., Marchand, J.: Volume stability of calcium hydroxide in aggressive solutions. Cem. Concr. Res. **31**, 149–151 (2001)

Beddoe, R.E., Dorner, H.W.: Modelling acid attack on concrete: Part I. The essential mechanism. Cem. Concr. Res. **35**, 2333–2339 (2005)

Bellmann, F., Moser, B., Stark, J.: Influence of sulphate solution concentration on the formation of gypsum in sulphate resistance test specimen. Cem. Concr. Res. **36**, 358–363 (2006)

Bertron, A., Duchesne, J.: Attack of cementitious materials by organic acids in agricultural and agro food effluents – a review. In: Workshop Proceedings on Performance of Cement Based Materials in Aggressive Aqueous Environment, pp. 15–28. Ghent, Belgium (2007)

Cement and Concrete Research: Special issue on symposium on transport properties and microstructure of cement-based systems, Boston, Massachusetts. Cem. Concr. Res. **30**(12) (2000)

CS 109: Discussion Document on Development in Durability Design & Performance-Based Specification of Concrete. Concrete Society Special Publication, Slough (1996)

CSA A 23.1: Concrete Materials and Methods of Concrete Construction. Canadian Standards Association, Toronto (2009)

De Belie, N.: A survey on concrete floors in pig houses and their degradation. J. Agric. Eng. Res. **66**, 151–156 (1997a)

De Belie, N.: On-farm trial to determine the durability of different concrete slats for fattening pigs. J. Agric. Eng. Res. **68**, 311–316 (1997b)

De Belie, N., De Blaere, B., Verschoore, R.: Compounds Aggressive to Concrete Floors in Pig Houses. pp. 22–26. Landwards, Autumn (1996)

De Belie, N., Monteny, J., Beeldens, A., Vincke, E., Van Gemert, D., Verstraete, W.: Experimental research and prediction of the effect of chemical and biogenic sulphuric acid on different types of commercially produced concrete sewer pipes. Cem. Concr. Res. **34**, 2223–2236 (2004)

Department of the Environment: Transport and the Regions, The Thaumasite Form of Sulphate Attack: Risks, Diagnosis, Remedial Works and Guidance on New Construction. Report of the Expert Group, Rotherham, UK (1999)

Detwiler, R.J., Taylor, P.C.: Specifier's Guide to Durable Concrete, PCA. Engineering Bulletin 221, pp. 1–62. Portland Cement Association, Skokie (2005)

Dunaszegi, L.: High performance concrete in the confederation bridge. Concr. Int. **20**(4), 66–68 (1998). ACI

EN 206: Guide to the Selection of Concrete Quality and Cover to Reinforcement for Normal Concrete Structures, British Cement Association. The Meadows Business Park, Camberley (2002)

Eurocode 2: Design of Concrete Structures. The Concrete Centre, Blackwater (2002)

Fourie, C.W., Alexander, M.G.: Acid resistant concrete sewer pipes. In: Alexander, M.G., Bertron, A. (eds.) Proceedings International RILEM TC 211-PAE Final Conference on Concrete in

Aggressive Aqueous Environments: Performance, Testing and Modelling, pp. 408–418. Toulouse, France (2) (2009)
Goyns, AM.: A model for selecting sewer lining materials, In: RILEM Workshop on Performance of Cement-Based Materials in Aggressive Aqueous Environment, Quebee, Canada (Unpublished) (2006)
Goyns, A.M., Alexander, M.G., Fourie, C.W.: Applying experimental data to concrete sewer design and rehabilitation, Calcium Aluminate Cements. In: Proceedings of the Centenary Conference, pp. 293–308. IHS BRE Press, Avignon, UK (2008)
Haque, M.N., Al Khaiat, H., John, B.: Proposals for a draft code for designing durable concrete structures in the Arabian Gulf. Arabian J. Sci. Eng. **31**(IC), 205–214 (2005)
Haque, M.N., Al Khaiat, H., John, B.: Climatic zones – a prelude to designing durable concrete structures in the Arabian Gulf. Build. Environ. **42**, 2410–2416 (2007)
Hartshorn, S.A., Sharp, J.H., Swamy, R.M.: Thaumasite formation in Portland limestone cement pastes. Cem. Concr. Res. **29**, 1331–1340 (1999)
Hobbs, C., Taylor, M.: Nature of the thaumasite sulphate attack mechanism in field concrete. Cem. Concr. Res. **30**(45), 529–533 (2000)
Hooton, R.D., Brown, P.W.: Development of test methods to address the various mechanisms of sulphate attack. In: Alexander, M.G., Bertron, A. (eds.) Proceedings International RILEM TC 211-PAE Final Conference on Concrete in Aggressive Aqueous Environments: Performance, Testing and Modelling, pp. 280–297. Toulouse, France (2) (2009)
IS 456:2000: Indian Standard, Plain and Reinforced Concrete Code of Practice. Bureau of India Standards, New Delhi (2000)
Jacquemont, F., et al.: (GEF Working Group on "leaching test", 19 co-authors) Leaching performance test. Assessing durability of concrete exposed to chemical attack. In: Alexander, M.G., Bertron, A. (eds.) Proceedings International RILEM TC 211-PAE Final Conference on Concrete in Aggressive Aqueous Environments: Performance, Testing and Modelling, pp. 372–379. Toulouse, France (2) (2009)
Japan Guide for Construction JGC: No. 6, Standard specifications for concrete structures. In: Materials and Construction, Agency for Natural Resources and Energy, Ministry of Economy, Trade and Industry, Tokyo, Japan (2002)
Jooss, M., Reinhardt, H.W.: Permeability and diffusivity of concrete as function of temperature. Cem. Concr. Res. **32**(9), 1497–1504 (2002)
Khatri, R.P., Sirivivatnanon, V.: Characteristic service life for concrete exposed to marine environments. Cem. Concr. Res. **34**, 745–752 (2004)
Kunther, W., Lothenbach, B., Scrivener, K.: Influence of carbonate in sulphate environments. In: Alexander, M.G., Bertron, A. (eds.) Proceedings International RILEM TC 211-PAE Final Conference on Concrete in Aggressive Aqueous Environments: Performance, Testing and Modelling, pp. 498–499. Toulouse, France (2) (2009)
Limaye, R.G., Kamat, M.K., Holta, S.: Strengthening chemically-deteriorated concrete structures. Indian Concr. J. **66**(9), 517–521 (1992)
Maiti, S.C., Agarwal, R.K., Kumar, R.: Concrete for hydroelectric projects: combating alkali-aggregate reaction. Indian Concr. J. **79**, 33–39 (2005)
Malhotra, V.M., Bremner, T.W.: Performance of Concrete at Treat Island, USA, CANMET Investigations, ACI Special Publications, SP – 163, pp. 1–52, Michigan, USA (1996)
McArthur, H., Spalding, D.: Engineering Materials Science: Properties, Uses, Degradation and Remediation. Horwood Publishing, Chichester (2004)
McLaren, F.R.: Design Manual: Sulphide and Corrosion Prediction and Control. American Concrete Pipe Association, Virginia (1984)
Mehta, P.K.: Concrete: Structure, Properties and Materials, pp. 1–450. Prentice-Hall, Englewood Cliffs (1986)
Mehta, P.K., Gerwick Jr., B.C.: Cracking – corrosion interaction in concrete exposed to marine environment. Concr. Int. **4**(10), 45–51 (1982)

Mindess, S., Young, F.J.: Concrete. Prentice-Hall, Englewood Cliffs (1981)
Monteny, J., Vincke, E., Beeldens, A., De Belie, N., Taerwe, L., Verstraete, W., Van Gemert, D.: Chemical, microbiological and in-situ test methods for biogenic sulphuric acid and corrosion of concrete. Cem. Concr. Res. **30**, 623–634 (2000)
Muigai, R., Alexander, M.G., Moyo, P.: Use of chloride conductivity index in the probabilistic modelling for durability design of RC members. Restor. Build. Monum. **15**(4), 267–276 (2009)
Neithalath, N.: Analysis of moisture transport in mortars and concrete using sorption-diffusion approach. ACI Mater. J. **103**, 209–217, Michigan, USA (2003)
NEN 5996: Determination of the Aggressiveness of Watery Solutions, Soils and Gases. The Netherlands Standards Institution, (NNI), Delft (1988)
Pepenar, I.: Degradation of reinforced concrete structures in Ammonium Nitrate Environments. In: Alexander, M.G., Bertron, A. (eds.) Proceedings International RILEM TC 211-PAE Final Conference on Concrete in Aggressive Aqueous Environments: Performance, Testing and Modelling. pp. 442–448 Toulouse, France (2) (2009)
Pomeroy, R.D., Parkhurst, J.D.: The forecasting of sulphuric build-up rates in sewers. Prog. Water Technol. **9**, 621–628 (1977)
Richardson, M.G.: Fundamentals of Durable Concrete. Spon Press, London/New York (2002)
Romer, M.: RILEM Recommendation TC 189-NEC "Comparative test: Part I – comparative test of permeability methods". Mater. Struct. (**38**) 2005, 895–906; (**41**) 2009, 443–447 (2005/2009)
Samson, E., Marchand, J., Beaudoin, J.J.: Modelling the influence of chemical reactions on the mechanisms of chemical reactions on the mechanisms of ionic transport in porous materials – an overview. Cem. Concr. Res. **30**(12), 1895–1902 (2000)
Sand, W., Dumas, T., Marcdargent, S.: Tests for biogenic sulphuric acid corrosion in a simulation chamber confirm the on site performance of calcium aluminate based concrete in sewages. In: International Symposium on Microbiologically Influenced Corrosion (MIC) Testing, 16/17 November, Miami, Florida, USA (1992)
Santhanam, M.: Sulphate attack from an engineering perspective. In: Rilem Workshop Proceedings on Performance of Cement-Based Materials in Aggressive Aqueous Environments, pp. 106–113, Ghent, Belgium (2007)
Santhanam, M., Cohen, M.D., Olek, J.: Sulphate attack research – whither now? Cem. Concr. Res. **31**, 845–851 (2001)
Saucier, F., Lamberet, S.: Calcium aluminate concrete for sewers: Going from qualitative to quantitative evidence of performance. In: Alexander, M.G., Bertron, A. (eds.) Proceedings International RILEM TC 211 – PAE Final Conference on Concrete in Aggressive Aqueous Environments: Performance, Testing and Modelling, pp. 398–407. Toulouse, France (2) (2009)
Schmidt, M., Hormann, K., Hofmann, F.J., Wagner, E.: Concrete with greater resistance to acid and to corrosion by biogenous sulphuric acid. Betonw. Fert. TI Tech. **4**, 64–70 (1997)
Shayan, A., Grimstad, I.: Deterioration of concrete in a hydroelectric concrete dam and its characterization. Cem. Concr. Res. **36**, 371–383 (2006)
Shimada, Y., Johansen, V.C., Miller, F.M., Mason, T.O.: Chemical Path of Ettringite Formation in Heat-Cured Mortar and its Relationship to Expansion: A Literature Review, RD 136, pp. 1–51. Portland Cement Association, Shokie (2005)
Siebert, B., Breitenbucher, R.: Combined acid and sulphate attack on concrete due to oxidation of iron disulfide in construction ground. In: Alexander, M.G., Bertron, A. (eds.) Proceedings International RILEM TC 211-PAE Final Conference on Concrete in Aggressive Aqueous Environments: Performance, Testing and Modelling, pp. 419–426. Toulouse, France (2) (2009)
Structural Design Guidelines: Florida Department of Transportation (District Engineer's Office), Florida, USA (2005)
Torrent, R., Jacobs, F.: Swiss Standard SIA 262: 2003, a step towards performance based specifications for durability. In: Alexander, M.G., Bertron, A. (eds.) Proceedings International RILEM TC 211-PAE Final Conference on Concrete in Aggressive Aqueous Environments: Performance, Testing and Modelling, pp. 532–539. Toulouse, France (2) (2009)

Van Tittelboom, K., De Belie, N.: A Critical Review on Test Methods for Evaluating the Resistance of Concrete Against Sulphate Attack SOA Report of WG3, Rilem TC-PAE, Ghent, Belgium (2008)

Yao, S.X., Gerwick, Jr. B.C.: Underwater concrete. Conc. Int. Part I: 79–83; Part 2: 77–82; Part 3: 60–64 (2004)

ERRATUM

Chapter 13
Test Methods for Chloride Transport in Concrete

Qiang Yuan and Manu Santhanam

Editors

M. Alexander et al. (eds.), *Performance of Cement-Based Materials in Aggressive Aqueous Environments,* RILEM State-of-the-Art-Reports 10,
DOI 10.1007/978-94-007-5413-3_13, © RILEM 2013

DOI 10.1007/978-94-007-5413-3_17

Some of the sentence in part **13.2.2** has been deleted and **13.2.2 Summary of common test methods** read as follows:

Chloride diffusion is characterized by the chloride profile, i.e. a graph of the chloride content of the concrete vs. the depth from the concrete surface. The chloride profile is obtained by various tests, which are standardized as ASTM C 1556-03 in the US, and NT Build 443 in Europe respectively. In both methods, prismatic or cylindrical specimens of at least 75 mm thickness are sealed on all surfaces except the top finished surface, after 28 days of moist curing. The specimens are placed in a container with 165 g/l NaCl solution for 35 days, with the uncoated surface exposed to chloride penetration from the top. After completion of the exposure period, powder samples are taken from various depths and chloride contents are obtained using an acid soluble chloride determination. The chloride profile is used for finding the apparent diffusion coefficient, using Fick's second law.

Various techniques, which include DC,...

Sentences in the part **13.3.1.1 Salt Ponding Test** read as follows:

Concrete cylinders of size 100 mm diameter and 200 mm height were cast and stored in a moist room for 7 days...

The online version of the original chapter can be found at
http://dx.doi.org/10.1007/978-94-007-5413-3_13

Author Index

C

T

Z

MIX
Papier aus verantwortungsvollen Quellen
Paper from responsible sources
FSC® C105338

If you have any concerns about our products,
you can contact us on
ProductSafety@springernature.com

In case Publisher is established outside the EU,
the EU authorized representative is:
Springer Nature Customer Service Center GmbH
Europaplatz 3, 69115 Heidelberg, Germany

Printed by Libri Plureos GmbH
in Hamburg, Germany